Jon Børre Ørbæk
Roland Kallenborn
Ingunn Tombre
Else Nøst Hegseth
Stig Falk-Petersen
Alf Håkon Hoel

Arctic Alpine Ecosystems and People in a Changing Environment

Jon Børre Ørbæk
Roland Kallenborn
Ingunn Tombre
Else Nøst Hegseth
Stig Falk-Petersen
Alf Håkon Hoel
Editors

Arctic Alpine Ecosystems and People in a Changing Environment

with 86 Figures and 10 Tables

Springer

Dr. Jon Børre Ørbæk
Norwegian Polar Institute
Svalbard
P.O.Box 505
9171 Longyearbyen
Norway

Dr. Else N. Hegseth
Norwegian College of Fishery Science
University of Tromsø
Breivika
9037 Tromsø
Norway

Dr. Roland Kallenborn
Norwegian Institute
for Air Research
Polar Environmental Centre
9296 Tromsø
Norway

Dr. Stig Falk-Petersen
Norwegian Polar Institute
Polar Environmental Centre
9296 Tromsø
Norway

Dr. Ingunn Tombre
Norwegian Institute
for Nature Research
Polar Environmental Centre
9296 Tromsø
Norway

Dr. Alf H. Hoel
University of Tromsø
Department of Political Science
Breivika
9037 Tromsø
Norway

Cover photograph: Bjørn Fossli Johansen, Norwegian Polar Institute, 2005

Library of Congress Control Number: 2006935137

ISBN-10 3-540-48512-4 Springer Berlin Heidelberg New York
ISBN-13 978-3-540-48512-4 Springer Berlin Heidelberg New York

Springer is a part of Springer Science+Business Media
springer.com

Cover design: E. Kirchner, Heidelberg
Production: Almas Schimmel
Typesetting: camera-ready by authors
Printing: Krips bv, Meppel
Binding: Stürtz AG, Würzburg

Printed on acid-free paper 30/3141/as 5 4 3 2 1 0

Preface

This book addresses the significant environmental changes experienced by high latitude and high altitude ecosystems at the beginning of the 21st century. Increased temperatures and precipitation, reduction in sea ice and glacier ice, the increased levels of UV-radiation and the long-range transported contaminants in arctic and alpine regions are stress factors that challenge terrestrial and aquatic ecosystems. The large natural variation in the physical parameters of these extreme environments is a key factor in structuring the biodiversity and biotic productivity, and the effect of the new stress factors can be critical for the population structures and the interaction between species. These changes may also have socio-economic effects if the changes affect the bio-production, which form the basis for the marine and terrestrial food chains.

The book is uniquely multidisciplinary and provides examples of various aspects of contemporary environmental change in arctic and alpine regions. The 21 chapters of the book are organised under the fields of •Climate change and ecosystem response, •Long range transport of pollutants and ecological impacts, and •UV radiation and biological effects, each also including aspects of the •Socio-economic effects of environmental change. The introductory chapter presents and explains the internal connection and integration of all chapters. The added value of these reviews and review-like manuscripts from different disciplines hopefully yields new information about the integrated aspects of environmental change.

The chapters are written on the basis of manuscripts presented at the international conference on "Arctic Alpine Ecosystems and People in a Changing Environment", organized in Tromsø, Norway in February 2003. The conference was multidisciplinary in scope, aiming at creating new links and understandings across disciplinary boundaries and among researchers and research infrastructures, inviting the international marine, terrestrial and atmospheric environmental change research communities to meet and exchange recent research and monitoring results. The emphasis was on the European arctic and alpine environments. The conference was organized as a EURO-CONFERENCE supported by the European Commission. It also served as the final conference of the European Network for Arctic-Alpine Multidisciplinary Environmental Research (ENVINET), the final conference of the Nordic Arctic Research Programme (NARP), the last user meeting of the Ny-Ålesund Large Scale Facility, the first conference of the Arctic Seas Consortium and the final workshop of the EU-project UVAC (The influence of UV-radiation and climate conditions on fish stocks).

The following organizations are acknowledged for their financial support of the conference and the preparation of this book:
•EUROCONFERENCE: High Level Scientific Conferences, European Commission IHP-programme, Contract HPCF-CT-2002-00238, •ENVINET: European Network for Arctic-Alpine Multidisciplinary Environmental Research, European Commission IHP-programme, Contract HPRI-CT-1999-40009. •NARP: Nordic Arctic Research Programme, Nordic Council of Ministers, •Ny-Ålesund LSF: Ny-Ålesund Large Scale Facility for Arctic Environmental Research, European Commission IHP-programme, Contract HPRI-CT-1999-00057, •UVAC: The influence of UV-radiation and climate conditions on fish stocks: A case study of the north-east Arctic cod, European Commission Environment-programme, Contract EVK3-CT-1999-00012, •Norwegian Ministry of Environment, •The University of Tromsø, including the Department of Political Science and the Norwegian College of Fishery Science, •Institute of Marine Research, Tromsø and the •Polar Environmental Centre, Tromsø, including the Norwegian Polar Institute, Norwegian Institute for Nature Research and Norwegian Institute of Air Research.

The editors wish to thank all authors and co-authors for their valuable set of complementary and multidisciplinary chapters, which together hopefully will add value to the reflection of the integrated scientific questions and environmental challenges faced by arctic-alpine environments. We would also like to thank the many reviewers that have provided valuable comments and advice to all manuscripts, as well as Mrs. Ingrid Storhaug for her very competent assistance in editing this volume.

Tromsø 2005, on behalf of the editors
Jon Børre Ørbæk, Norwegian Polar Institute

Contents

Contributors

Amundsen, Per-Arne

University of Tromsø, N-9037 Tromsø, Norway, pera@nfh.uit.no

Armbruster, W. Scott [1,2,3]

[1]Department of Biology, Norwegian University of Science and Technology, N-7491 Trondheim, Norway. [2]Institute of Arctic Biology, University of Alaska, Fairbanks, AK 99775 USA. [3]School of Biological Sciences, University of Portsmouth, Portsmouth PO1 2DY, UK, scott.armbruster@bio.ntnu.no

Berg, Torunn

Norwegian Institute for Air Research, P.O. Box 100, NO-2027 Kjeller, Norway, tbe@nilu.no

Bilger, Wolfgang

Botanisches Institut, Christian- Albrechts Universität zu Kiel, Olshausenstr. 40, 24098 Kiel, Germany, wbilger@bot.uni-kiel.de

Bischof, Kay

Department of Marine Botany, University of Bremen, Leobener Str. / NW 2, 28359 Bremen, Germany

Blumthaler, Mario

Innsbruck Medical University, Department of Medical Physics, Muellerstr. 44, A-6020 Innsbruck, Austria, Mario.Blumthaler@i-med.ac.at

Crawford, R.M.M.

Plant Science Laboratory, Sir Harold Mitchell Building, St Andrews University, St Andrews, KY16 9AL, UK. rmmc@st-andrews.ac.uk

den Outer, Peter

Laboratory for Radiation Research, the National Institute for Public Health and the Environment (RIVM), po box 1, 3720 BA Bilthoven, the Netherlands

Edwards, Mary E.[1,2]

[1]School of Geography, University of Southampton, Highfield, Southampton S017 1BJ, UK. [2]Institute of Arctic Biology, University of Alaska, Fairbanks, AK 99775 USA

Falk-Petersen, Stig

Norwegian Polar Institute, Polar Environmental Centre, N-9296 Tromsø, Norway, stig@npolar.no

Furevik, Tore

Geophysical Institute, University of Bergen, Norway, and Bjerknes Centre for Climate Research, Bergen, Norway. tore@gfi.uib.no

Gabrielsen, Geir Wing

Norwegian Polar Institute, Polar Environmental Centre, N-9296 Tromsø, Norway, geir@npolar.no

Hanelt, Dieter

Biocenter Klein Flottbek, University of Hamburg, Ohnhorststr. 18, D-22609 Hamburg, Germany, dieter.hanelt@botanik.uni-hamburg.de

Hanssen-Bauer, Inger

Norwegian Meteorological Institute, P.O.Box 43 Blindern, NO-0313 Oslo i.hanssen-bauer@met.no

Hegseth, Else Nøst

Norwegian College of Fishery Science, University of Tromsø, N-9037 Tromsø, Norway, else.hegseth@nfh.uit.no

Hessen, Dag O.

University of Oslo, Dept of Biology, P.O.Box 1066, Blindern, 0316 Oslo, dag.hessen@bio.uio.no

Hoel, Alf Håkon

Dep. of Political Science, Universtity of Tromsø, N-9037 Tromsø, Norway, hoel@sv.uit.no

Høgda, Kjell Arild

Norut IT, P.O. Box 6434, N-9294 Tromsø, kjell-arild.hogda@itek.norut.no

Ingvaldsen, Randi

Institute of Marine Research, Bergen, Norway, and Bjerknes Centre for Climate Research, Bergen, Norway. randi.ingvaldsen@imr.no

Jeffree, C. E.

Edinburgh University Institute of Cell and Molecular Biology, Daniel Rutherford Building, King's Buildings, Mayfield Road Edinburgh, EH9 3JH, UK

Kallenborn, Roland

Norwegian Institute for Air Research, Polar Environmental Centre, N-9296 Tromsø, Norway, rok@nilu.no

Karlsen, Stein Rune

Norut IT, P.O. Box 6434, N-9294 Tromsø, stein-rune.karlsen@itek.norut.no

Kelfkens, Gert

Laboratory for Radiation Research, the National Institute for Public Health and the Environment (RIVM), po box 1, 3720 BA Bilthoven, the Netherlands

Klemetsen, Anders

University of Tromsø, N-9037 Tromsø, Norway, andersk@nfh.uit.no

MacDonald, Robie W.

Department of Fisheries and Oceans, Institute of Ocean Sciences
PO Box 6000, Sidney BC, V8L 4B2, macdonaldrob@pac.dfo-mpo.gc.ca

Mauritzen, Cecilie

Norwegian Meteorological Institute, Oslo, Norway.
Cecilie.Mauritzen@met.no

Mulvad, Gert [1,2]

[1]Commission for Scientific Research in Greenland. [2]Center of Primary Health Care, Nuuk, Greenland, GM@gh.gl

Nuttall, Mark

Department of Anthropology, University of Alberta, mnuttall@ualberta.ca

Nybakken, Line

Department of Ecology and Natural Resource Management, Norwegian University of Life Sciences, Box 5003, NO-1432 Ås
line.nybakken@umb.no

Olsen, Jørn [1,2]

[1]Commission for Scientific Research in Greenland. [2]The Danish Epidemiology Science Centre, Department of Epidemiology and Social Medicine, University of Aarhus, Aarhus C, Denmark.

Pavlov, Vladimir

Norwegian Polar Institute, N-9296 Tromsø, Norway, pavlov@npolar.no

Pienitz, Reinhard

Centre d'Études Nordiques, Université Laval, Québec, Canada G1K 7P4

Primicerio, Raul

University of Tromsø, N-9037 Tromsø, Norway. raulp@nfh.uit.no

Rae, David A

Department of Biology, Norwegian University of Science and Technology, N-7491 Trondheim, Norway

Rasmussen, Rasmus Ole

NORS – North Atlantic Regional Studies, Roskilde University 21.2, Box 260, DK4000 Roskilde, Denmark . rasmus@ruc.dk

Rautio, Milla

Dépt de Biologie, Pavillon Vachon, local 4073, Université Laval, Sainte-Foy, Québec G1K 7P4, Canada

Rossetti, Giampaolo

University of Parma, I-43100 Parma, Italy, giampaolo.rossetti@ipruniv.cce.unipr.it

Sargent, John R.

University of Stirling, Stirling, FK9 4LA, Scotland, UK.

Slaper, Harry

Laboratory for Radiation Research, the National Institute for Public Health and the Environment (RIVM), po box 1, 3720 BA Bilthoven, the Netherlands, e-mail: harry.slaper@rivm.nl

Sloth Petersen, Henning [1,2]

[1]Commission for Scientific Research in Greenland. [2]Center of Primary Health Care, Nuuk, Greenland

Timofeev, Sergey

Murmansk Marine Biological Institute, Russian Academy of Science, Murmansk 183010, Russia.

Tombre, Ingunn

Norwegian Institute for Nature Research, Polar Environmental Centre, N-9296 Tromsø, Norway, ingunn.tombre@nina.no

Tømmervik, Hans

NINA, Polar Environmental Centre, N-9296 Tromsø, hans.tommervik@nina.no

Vincent, Warwick F,

Dept de Biologie, Laval University, Quebec City, QC G1K 7P4, Canada warwick.vincent@bio.ulaval.ca

Wiencke, Christian

Foundation Alfred Wegener Institute for Polar and Marine Research, Am Handelshafen 12, D-27570 Bremerhaven, Germany

Ørbæk, Jon Børre

Norwegian Polar Institute, Polar Environmental Centre, N-9296 Tromsø, Norway, jonbo@npolar.no

Abbreviations

AARI	Arctic and Antarctic Research Institute
ACIA	Arctic Climate Impact Assessment
Ah	Aryl hydrocarbon receptor
AHDR	Arctic Human Development Report
Alt	Altitude (in meters above sea level - a.s.l.)
AMAP	Arctic Monitoring and Assessment Programme
AMOC	Atlantic meridional overturning circulation
AMOUR	Assessment Model for UV-radiation and Risks
AO	Arctic Oscillation
AOGCM	Atmosphere-Ocean General Circulation Model
a.s.l.	Above sea level
AVHRR	Advanced Very High Resolution Radiometer
AW	Atlantic Water
BCC	Basal Cell Carcinoma
BFR	Brominated flame retardant
BP	Before present
BWF	Biological weighting factor which expresses the relative damage incurred by exposure to UV radiation
CACAR	NCPs Canadian Arctic Contaminants Assessment Report
CAFF	Conservation of Arctic Flora and Fauna
CDOM	Colored dissolved organic matter
CFC	Chlorofluorocarbons
CFC's	Chloroflurocarbons, major ozone depleting substances
CHL	Chlordanes
Chl	Chlorophyll
CMIP	Coupled Model Intercomparison Project
CO_2	Carbon dioxide
CPD	Cyclobutane-pyrimidine dimers
CYP	Cytochrome P450
D_1-protein	Reaction center protein of PS II
3-D	Three-dimensional
DDD	1,1-dichloro-2,2-bis ethane
DDE	1,1-dichloro-2,2-bis ethylene
DDT	1,1,1-trichloro-2,2-bis(p-chlorophenyl)ethane

DEM	Digital elevation model
DF	Digital Filtration
DHA	Docosahexaenoic acid
DKr.	Danish Kroner
DNA	Desoxyribo Nucleic Acid
DOC	Dissolved organic carbon
DOM	Dissolved organic matter
dw	Dry weight
EANET	Acid Deposition Monitoring Network in East Asia
EC_{50}	Effective concentration causing a response in 50 % of the treated organisms
EC AQFD	European Commissions Air Quality Framework Directive
ECHAM	Atmospheric general circulation model developed partly at the European Centre for Medium-range Weather Forecast and partly at the Max-Plank Institute for Meteorology in Hamburg
EGC	East Greenland Current
EIONET	European Environment Information and Observation Network
ENSO	El Niño-Southern Oscillation
ENVINET	European Network for Environmental Research
EPA	Eicosapentaeneoic acid
EROD	7-ethoxyresorufin-*O*-deethylase
ESOP	European Sub-Polar Ocean Programme
FCT method	Flux-Corrected Transport method
FS	Fram Strait
fT3	Free triiodothyronine
fT4	Free tetraiodothyronine
GAW	Global Atmospheric Watch
GEM	Gaseous Elemental Mercury
GIMMS	Global Inventory Modeling and Mapping Studies
GMES	Global Monitoring for Environment and Security
GMES-GATO	GMES-Global Atmospheric Observations
GSI	Global Solar Irradiiance
HadCM	The Hadley Centre coupled climate model
IABP	International Arctic Buoy Programme
HCA	Hydroxycinnamic acid
HBCD	Hexabromocyclododecane
HCB	Hexachlorobenzene
HCH	Hexachlorocyclohexane
HIV	Human Immunodeficiency Virus

IESA	Inside ellipse -species absent
IESP	Inside ellipse -species present
IgG	Immunoglobulin G
IgM	Immunoglobulin M
ind	Number of individuals
IPCC	Intergovernmental Panel on Climate Change
ISMO	Ice Statistical Model
K_{OA}	Octanol-air partition coefficient
KNMI	Royal Netherlands Meteorological Institute, de Bilt, the Netherlands
Lat	Latitude (in degrees North)
LHC	Light-harvesting complex
LHS	Left-hand side
LOAEL	Lowest observed adverse effect level
LOEL	Lowest observed effect level
LPJC	Lund-Potsdam-Jena dynamic global vegetation model
LRTAP-EMEP	Convention on Long-range Transboundary Air Pollution, European Monitoring and Evaluation Programme
lw	Lipid weight
MAA	Mycosporine-like amino acid
MeO	Methoxylated
$MeSO_4$	Methylsulfonyl
mRNA	Messenger ribonucleic acid
MTCLIM	Montane regional climate model
N	North (latitude)
N maps	The area actually occupied by a species as compared with the distribution as represented by a calculated ellipse
NAC	North Atlantic Current
NAMMCO	North Atlantic Marine Mammal Commission
NAO	North Atlantic Oscillation
NARP	Nordic Arctic Research Programme
NASA	National Aeronautics and Space Administration, USA
NCP	Indian and Northern Affair's, Northern Contaminants Program
NDVI	Normalised Difference Vegetation Index
NERI	National Environmental Research Institute (Denmark)
NIP	National Implementation Plans

NMSC	Non-Melanoma Skin Cancer
NOAA	National Oceanic and Atmospheric Administration
NOAEL	No-observed adverse effect levels
NODC	National Oceanographic Data Centre
NOEL	No-observed effect levels
NPI	Norwegian Polar Institute
NwAC	Norwegian Atlantic Current
NwCC	Norwegian Coastal Current
OESA	Outside ellipse -species absent
OESP	Outside ellipse -species present
OH	Hydroxylated
OSPAR/CAMP	Oslo-Paris Commissions Comprehensive Atmospheric Monitoring Programme
OWSM	Ocean weather station
P maps	Maps showing bands of increasing probability of specified winter and summer temperature combinations being suitable for the survival of the species under discussion
PAH	Polynuclear aromatic hydrocarbons
PAR	Photosynthetically active radiation or visible light, 400-700 nm
PBB	Polybrominated biphenyls
PBDE	Polybrominated diphenyl ethers
PCB	Polychlorinated biphenyls
PCDD	Polychlorinated dibenzo-p-dioxins
PCDF	Polychlorinated dibenzofurans
PCN	Polychlorinated napthalene
PFAS	Perfluorinated alkyl substances
PFOS	Perfluorooctane sulfonate
PNA	Pacific North American pattern
POP	Persistent organic pollutants
PS I	Photosystem I
PS II	Photosystem II
psbA	Gene sequence encoding for the D_1-protein
PSC	Polar Stratospheric Clouds
PUFA	Polyunsaturated fatty acid
Q_A	Plastquinon A, primary electron acceptor of PS II
Q_B	Plastquinon B, secondary electron acceptor of PS II
RAF	Radiation Amplification Factor, percent increase in effective UV for a one percent decrease in the atmospheric ozone column
rbcL	Gene sequence encoding for large subunit of

	RubisCO
RHS	Right-hand side
RIVM	The National Institute for Public Health and the Environment, Bilthoven, the Netherlands
ROS	Reactive oxygen species
RubisCO	Ribulose-1,5-bisphosphate carboxylase/oxygenase
SAT	Surface Air Temperature
SCC	Squamous Cell Carcinoma
SCUP-h	Skin Cancer Utrecht Philadelphia, biological action spectrum for skin cancer development in humans (de Gruijl and van der Leun, 1994)
SCUP-m	Skin Cancer Utrecht Philadelphia, murine biological action spectrum for skin cancer development derived from experiments in hairless mice (de Gruijl et al., 1993)
SLP	Sea level pressure
SOD	Superoxide dismutase
SST	Sea surface temperature
SZA	Solar zenith angle
T*	Weighted transparency. An index that expresses the UV-spectral effects of stratospheric ozone depletion and water column optical characteristics on a biologically relevant scale
T3	Triiodothyronine
T4	Thyroxin
$\mathbf{T_a}$	Air temperature
$\mathbf{T_e}$	Operative environmental temperature
TBT	Tri-butyl-tin
TCDD	2,3,7,8-tetrachlorodibenzo-p-dioxin
TEQ	Dioxin equivalency quotient
Tmax	Maximum temperature (in degrees Celsius)
TOC	Total organic carbon
TOMS	Total Ozone Mapping Spectrometer
TS	Temperature and salinity
TT	Travel time
tT3	Total triiodothyronine
tT4	Tetraiodothyronine
UNEP	United Nation's Environment Programme
US-EPA	United States – Environmental Protection Agency
UV	Ultraviolet
UVA	Ultraviolet A, wavelengths in the range: 315-400 nm
UVB	Ultraviolet B, wavelengths in the range: 280-315 nm

UVR	Ultraviolet radiation
VEINS	Variability of Exchanges in the Northern Seas
VEMAP	Vegetation/Ecosystem Modelling and Analysis Project
WMO	World Meteorological Organisation, Geneva, Switzerland
ww	Wet weight
WWII	World War II
XP	Xeroderma Pigmentosum, genetic skin disorder with extreme UV-sensitivity
Zmax	Max depth (in meters)

Arctic-Alpine Ecosystems and People in a Changing Environment - Introduction

1 Integrated aspects of environmental change: Climate change, UV radiation and long range transport of pollutants

J.B. Ørbæk[1], R. Kallenborn[2], I. Tombre[3], E.N. Hegseth[4], S. Falk-Petersen[1] and A.H. Hoel[5]

[1]Norwegian Polar Institute, Polar Environmental Centre, N-9296 Tromsø, Norway, [2]Norwegian Institute for Air Research, Polar Environmental Centre, N-9296 Tromsø, Norway, [3]Norwegian Institute for Nature Research, Polar Environmental Centre, N-9296 Tromsø, Norway, [4]Norwegian College for Fishery Science, University of Tromsø, Breivika, N-9037 Tromsø, Norway, [5]Department of Political Science, University of Tromsø, N-9037 Tromsø, Norway, e-mail: jonbo@npolar.no

1.1 Introduction

Global warming, changes in climate variability, long-range transport of pollutants, and reduced stratospheric ozone, represent increasing challenges to the arctic and alpine ecosystems. Forced by natural and anthropogenic variability, these key environmental factors are amplified in polar (high latitude) and alpine (high altitude) environments. Climate change and ecosystem impact studies involve a number of different and related forcing factors and interaction processes in the atmospheric, terrestrial and marine environments. They represent multiple stress factors that add to harsh environments with large natural variability, and the changes are also connected through natural links and feedback processes.

The integrated physical and biological effects and interactions on marine and terrestrial ecosystems are complex to understand. UV radiation and its ecosystem effects are shaped by a number of physical parameters in the atmospheric, terrestrial and marine environments which contribute to its total biological impact. And so is true also for the long range transported pollutants. Their transport into the physical and biological environments, involve a number of pathways as well as physical, chemical and biochemical transformation processes. Their impact on ecosystems and humans are complex and cannot be treated in isolation. This introductory chapter provides a framework for the integration of the individual chapters of this book.

1.2 Climate change and ecosystem response

The arctic and alpine areas of Northwest Europe are especially sensitive to climate perturbations, due to the strong influence by the North Atlantic oceanic and atmospheric heat advection processes. As discussed by Furevik et al. (Chap. 8), abrupt climate changes with ~10°C temperature variations over just a few decades occurred during the period of the last glacial maximum, with the termination of the Younger Dryas 11 600 years ago (Dansgaard et al., 1989) representing the last major climate perturbation in the region. They argue that observational evidence suggests that such abrupt changes in climate may be explained by sudden switches in the strength or positioning of the warm and saline Atlantic waters (AW) flowing into the Nordic Seas, forced by increased freshwater discharges making the surface waters fresh enough to inhibit the deepwater formation (Clark et al., 2001). It is therefore not unlikely that the current global warming trends, with enhanced melting of glacier ice and a general intensified hydrological cycle, may in a similar way influence the thermohaline circulation in the North Atlantic and contribute to a destabilization of the stable climate experienced since the last glacial period.

Scientific scenarios suggest that the Arctic temperatures increase almost twice as fast as average global warming (ACIA 2004). Increased summer temperatures leads to more effective ablation. While European alpine glaciers are receding quickly (Beniston et al. 2003), the mass balance of Arctic glaciers show a larger regional variation and variable response (Arendt et al., 2002; Lefeauconnier et al., 1999). The extent of summer melt of the Greenland ice sheet has significantly increased during the past 20 years or so (Steffen et al., 2002). Seasonal snow cover is reduced with increasing length of growing seasons (Høgda et al., Chap. 5), and the continued melt may induce large regional shifts in animal and plant distribution (Crawford and Jeffree, Chap. 6). According to Nuttall (Chap. 2), the results of scientific research and the observations from indigenous peoples suggest that the current climate changes are more pronounced in the Arctic than in any other region of the world (ACIA 2004).

Hansen-Bauer (Chap. 3) provides a review of the climate trends in the European Arctic during the 20th century and provides scenarios of future change. Large changes are seen in the climate records, of which the annual mean precipitation is the most pronounced with significant increase in large parts of the European Arctic. A positive warming trend in temperature is also evident but less significant, due to the large natural variability in this region. According to Hansen-Bauer (Chap. 3), the recent warming

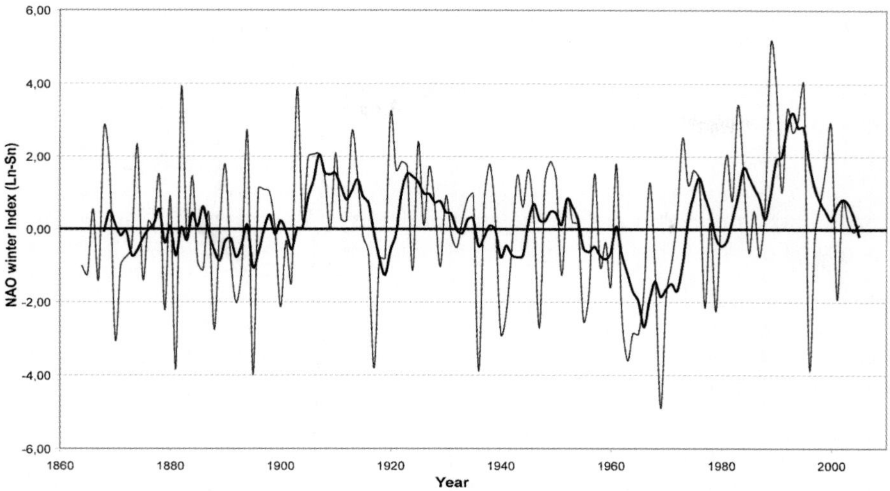

Fig. 1.1. North Atlantic Oscillation (NAO) Winter Index 1864-2003. Data provided by the Climate Analysis Section, NCAR, Boulder, USA, Hurrell (1995).

trend is associated with a positive North Atlantic Oscillation index (NAO, Hurrell 1995) and is evident during the last decades of the 20th century (see Fig. 1.1). This is especially emphasised in central and especially in eastern parts of the European Arctic, probably at least partly triggered by antropogenic forcing of the climate system.

According to Hansen-Bauer (Chap. 3), the Global Climate Models (AOGCMs) show rather different results for the projected changes in atmospheric circulation patterns and do not in general show the observed positive trend in the NAO. However, they project larger warming and precipitation increase in the Arctic than for the global average, with large projected reduction in summer sea-ice during for the 21st century. These projected changes are mainly in conformity with observed changes in temperature and sea-ice concentration during the last decades (Johannessen et al. 2004).

Arctic and alpine lake communities are well suited for studying ecological impacts of climate change, due to their simple structure, sensitivity to variation in ice and snow cover, and the availability of paleolimnological records. According to Primicerio et al. (Chap. 4), the ice phenology is an important driver of a number of ecological parameters for lake biota. Longer ice-free productive periods, induced by climate warming (Magnuson et al. 2000), prolong the seasonal activity of community members with expected increase in production. They claim that the seasonal dynamics of plankton and benthos is likely to change, leading to compositional and structural changes in plankton and benthos as well as the population and community structure of fish.

As for ice and aquatic environments, regional climate change impacts on snow melt and its distribution also significantly affects the length of the growing season. Høgda et al. (Chap. 5) have used the NOAA AVHRR GIMMS NDVI satellite dataset for the last two decades, producing maps for Fennoscandia that define the start and end of the growing season. Their results show that the estimated onset of spring is closely correlation with ground data on the onset of leafing of birch, with high regional differences. In the southern part of Fennoscandia, and on the oceanic west coast of Norway, the earlier spring fits with the pattern for western and central Europe. However, for the mountaineous areas in southern Norway and in the continental parts of northern Fennoscandia, the results indicate a stable or even a slightly delayed trend. Combined with the autumn trends, they find that the growing season is prolonged for the whole area, except the northern continental parts of Fennoscandia.

The tolerance of Arctic and alpine ecosystems to such climate changes can be studied by looking at the geographical limits of plant survival. Crawford et al. (Chap. 6) combine a map-modelling system that is sensitive to changing meteorological data, with comprehensive knowledge of the many interactions between physiology and environment, in their interpretation of plant distribution maps and the relating changing climates to species occurrence. They demonstrate that many species of woody plants of northern distribution have not only northern and southern limits to their distribution, but are restricted also in their east-west dispersal. Their probability models suggest that for some species, the migration patterns are also sensitive to existing temperature *seasonality,* and that the seasonality gradients may present barriers to the migration notwithstanding overall warming.

As plant distribution maps may connect large scale climate variability trends with ecosystem effects, similar variability can also be found on much smaller scales. Topography and the potentially large variability in the physical environmental parameters create a mosaic of microclimatic conditions at landscape scales. According to Armbruster et al. (Chap. 7), the variability in physical conditions, surface inhomogenities and radiation loads are important because all terrestrial biotic response to climate change is mediated by the local microclimate experienced by the organism. They argue that on the scale of meters, the spatial variability in temperature can be of the same order (>2°C) as the estimate of the global warming expected to result from a future doubling of atmospheric CO2 (Houghton et al., 2001), or comparable to moving more than 400 m in elevation or 450 km in latitude. Classical physical climate parameters may therefore not be representative in topographically complex areas and in the high Arctic, where spatial rearrangements of indigenous species with different thermal requirements may in fact be a more prominent biotic response to climate warming than immigration of new species from other regions.

Marine related climate processes significantly force what happens in the atmosphere. Furevik et al. (Chap. 8) focus on the inflow of Atlantic Water (AW) to the Nordic Seas, its pathways and transformation within the Nordic Seas and the Arctic Ocean. These waters are of vital importance for the marine climate, water mass transformation and biomass production in the Nordic Seas. Together with the north-eastward heat flow transport with the numerous North Atlantic cyclones, this heat flow associated with the AW is responsible for the mild and favourable climate of northwest Europe.

The Nordic Seas are also a key area for the conversion of light surface water to dense deep waters, representing the Atlantic meridional overturning circulation (AMOC), or the Atlantic part of the "great conveyor belt" (Broecker, 1991). According to Furevik et al. (Chap. 8), most AOGCMs participating in the third assessment report of the Intergovernmental Panel of Climate Change suggested a 30-40% reduction in the strength of the AMOC during this century (Houghton et al., 2001).

Such dynamical patterns in the inflow of AW to the Arctic are of fundamental importance for Arctic primary production. Falk-Petersen et al. (Chap. 9) postulate that a warmer climate with reduced ice cover will shift zooplankton community structures towards a smaller size spectrum and with lower energy content per individual. This will also lower the potential for seasonal accumulation of lipid stores in their predators such as specialised seabirds. They claim that these effects are based on the very specialized process where carbon fixed photosynthetically in algal blooms is converted into high-energy lipid (oil) reserves by the major Arctic herbivores, a process which varies on all time scales during the Arctic summer, from days to decades and longer due to the variability in sea ice conditions. This lipid-based energy flux, increasing the lipid level from 10-20% of dry mass in phytoplankton to 50 to 70% in herbivorous zooplankton and ice-associated fauna, is probably one of the most fundamental specialisations in Arctic bioproduction (Falk-Petersen et al., 1990). It is therefore also the primary reason for the large stocks of fish and mammals in Polar waters and a key factor in the structuring the biodiversity of Arctic ecosystems.

Such changes in the marine environment and its exploitable resources can have a large impact on social systems. Drawing on recent research from the Arctic Climate Impact Assessment (ACIA 2004) in particular, Nuttall (Chap. 2) provides a brief assessment of climate change impacts on the local livelihoods and traditional resource use practices of arctic indigenous peoples. Rasmussen et al. (Chap. 10) discuss how social changes in Greenland have been heavily influenced by the environmental conditions, and how the social transformations reflect the interaction between different responses to the variations in the natural resource base. Such studies are important for our total understanding of living conditions for people and ecosystems in a changing environment.

1.3 UV radiation and biological effects

Atmospheric UV radiation has potential harmful effects on humans and ecosystems. The depletion of stratospheric ozone observed over the last two decades in the Arctic (Müller et al. 1997; Rex et al. 1997), as well as over parts of Europe (WMO 2003), is of serious concern, since this may lead to an increase in ambient UV radiation. Surface UVB radiation (280-315nm) may induce a wide range of harmful effects on humans (skin cancer, cataracts, suppression of the immune system) and on ecosystems by decreasing biomass and crop yields, growth conditions and algal distribution patterns in unique marine ecosystems ((UNEP 1999, Wiencke et al. 2000). The Vienna Convention (1985) and the Montreal protocol (1987) have successfully led to reduced production and emissions of ozone depleting substances, leading to an expected slow recovery of the ozone layer over the next century (UNEP 2003).

The link between atmospheric ozone and UV radiation is often referred to by Radiation Amplification Factors (Blumthaler et al., 1995), showing that the erythemally weighted UV radiation increases by about 1.1% when ozone decrease by 1%. However, due to its major dependence on other atmospheric and surface related parameters that are also highly variable, trends and future UV levels are difficult to identify. According to Blumthaler (Chap. 11), long term measurements of UVB radiation show a slight increase of a few percent per decade in the 80's and early 90's, most pronounced at high northern latitudes during spring. However, from their 20 years of measurements at a high alpine station, where the variability of UV irradiance under cloudless conditions is dominated by ozone and albedo variations, no significant increase was found.

Surface UV radiation is determined mainly by cloud cover, solar zenith angle, ozone and aerosols. In arctic and alpine environments, altitude and surface albedo are also significant. Blumthaler (Chap. 11) has derived quantitative relations between these parameters and the levels of UV radiation, showing that cloud cover, solar zenith angle, aerosols, altitude, and surface albedo significantly alter the surface UV radiation by the same order of magnitudes as can be the result of ozone variations.

These complex interaction patterns between atmospheric and surface parameters responsible for shaping the surface UV spectrum are important when assessing its ecosystem effects. Plants are expected to be vulnerable to UVB radiation, and according to Nybakken (Chap. 12), a large number of studies has been carried out at all levels of effects, from cellular investigations to large projects concerning the effects on entire ecosystems. Focussing on the possible negative effects of increased UVB radiation and the UV absorbing pigments of plants of both arctic and alpine origin, she contributes to this under-investigated field with only a few previous UVB response studies carried out on plants from arctic and alpine areas, as well as studies conducted in the field or with plants growing in their natural ecosystems (Caldwell et al. 1998). The analysis of screening pigments in a number of arctic and alpine plants agrees well with numerous field and growth chamber studies that show that the increased concentration of phenolic compounds in higher plants is a common response to UVB radiation (Searles et al. 2001).

The arctic and alpine environments also contain a range of waterbodies. In addition to the atmospheric parameters, a number of physiochemical and biological parameters are responsible for shaping the underwater UV spectrum (Ørbæk et al., 2002). Hessen (Chap. 13) focuses on observed and potential effects of UV radiation for the inhabitants of arctic and alpine freshwater ecosystems, as well as the various abiotic challenges that may be superimposed on the UV stress. He claims that the numerous small, shallow and transparent tundra ponds in high arctic localities may support a substantial benthic production, despite a scarcity or absence of benthic macrophytes and fish, with often dense populations of large-sized species of crustaceans. On the other hand, the few deeper (>3m) and larger lakes in the Arctic, which resembles the typically deep and oligotrophic alpine lakes, do not freeze to the bottom and may house populations of fish. In these ecosystems, short wave solar radiation may negatively affect both primary and secondary production. Whereas UV radiation is considered the most harmful part of the spectrum, visible photosynthetic active radiation (PAR) may also cause a suite of cellular damages (Hessen Chap. 13).

Vincent et al. (Chap. 14) show that arctic, antarctic and alpine aquatic ecosystems are particularly vulnerable to climate-induced shifts in underwater UV radiation, and that the controlling effect of snow, ice cover and coloured dissolved organic matter (CDOM) on the biological UV exposure under water, may be larger than those caused by moderate ozone depletion. Although UV radiation may be strongly attenuated in coastal waters by CDOM released from terrestrial ecosystems, the importance of snow and ice cover and the sparse catchment vegetation zones in the arctic typically result in low CDOM concentrations. Based on their paleo-ecological studies of fossil diatoms and UV-screening pigments preserved in lake sediments, a strong landscape influence on the underwater spectral light regime of high latitude and alpine lakes is indicated, in the past and present.

Hanelt et al. (Chap. 15) give a well documented presentation of the recent investigations on biological effects of UV radiation on marine ecosystems in the Arctic. In recent years they have carried out several studies on the distribution, physiology and UV radiation effects on several algal species like seaweeds growing in the Arctic environment (Hanelt et al. 1997a; Bischof et al. 1998). They point out that although the UV radiation is more intense in temperate zones, the polar algae are more sensitive to UV as compared to their temperate relatives. Potential negative effects on primary plant productivity may occur especially in spring, low temperatures and clear water conditions allowing harmful UV wavelengths to penetrate several meters into the marine water column. Hanelt et al. (Chap. 15) also point out that the summer discharge of turbid fresh water into the coastal waters overlays the more dense sea water, causing a stratification in the optical features, salinity and temperature of the water body that strongly attenuate solar radiation in the first meter of the water column. This effect is increased during warm summers with rainfall and intensified runoff from melting snow and ice covers. Organisms in deeper waters are thus more protected against harmful UVB radiation.

Ozone depletion and the induced increased UV radiation levels also have consequences in terms of health risks such as for example skin cancer. In their Assessment Model for UV Radiation and Risks (AMOUR), Slaper et al. (Chap. 16) evaluate the full source-risk chain from production and emission of halocarbons, the resulting stratospheric ozone depletion with changes in ambient effective UV doses, and the corresponding skin cancer risks. Updating his previous analysis of the kind (Slaper et al. 1996), the new model also takes into account the role of climate and ozone interactions in the arctic region on the future risks at mid-latitudes in densely populated areas in Europe. Their analysis predicts that a slow recovery of the ozone layer will occur with a return to 'normal' (1980) levels around 2050, and that skin cancer risks are expected to rise until 2050-2070.

1.4 Ecological impacts of long range pollutants transport

Contaminants are transported from industrialized source areas by ocean currents, sea ice and large scale wind patterns (Fig. 1.2). After entering these transport pathways, the contaminants reach the soils, snow and ice, biota and water of the remote environments in a number of different forms and phases. The soluble and particulate phases as well as the precipitation and scavenging processes involved are highly influenced by changing circulation patterns and other forms of climate change (AMAP 2003). As an example, arctic and alpine regions are especially vulnerable to temperature rise due to potential increased melting of snow and ice, and this phase change to water influence the redistribution and transfer of contaminants from the physical environment to the biota.

MacDonald (Chap. 17) describes thoroughly the processes involved in the long range transport of contaminants as well as the environmental systems that control the further metamorphosis of the pollutants after being deposited and brought into the biological systems. Climate change and variability strongly affect both pathways and the stationary phases (MacDonald et al. 2005), involving complex interaction with temperature, winds, precipitation, runoff patterns, snow and ice, organic carbon cycling, ocean circulation, and human activities.

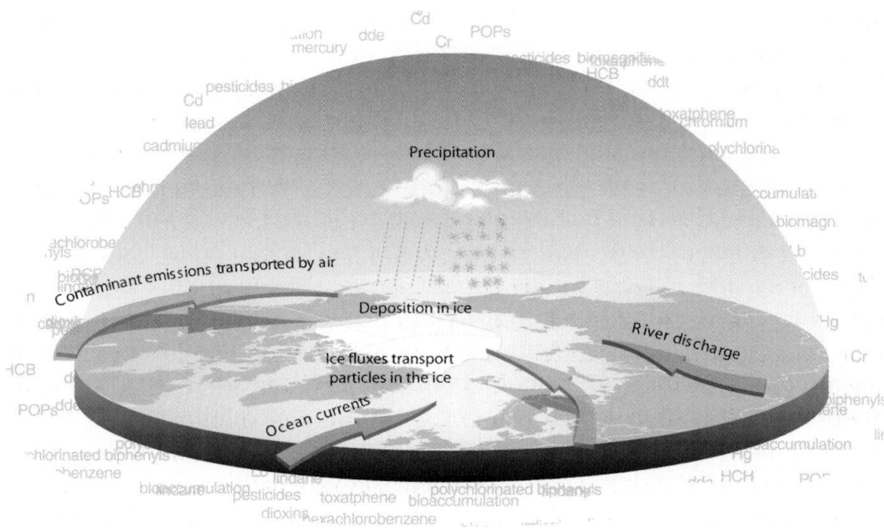

Fig. 1.2. Contamination pathways. UNEP-Grid-Arendal, Vital Arctic Graphics Source: AMAP (2002), ACIA (2004)

Sea ice and water currents in the Polar Ocean is also an important medium for pollutants transport (Korsnes 2002). These pathways undergo significant changes in the Arctic (AMAP 2003). Pavlov et al. (Chap. 18) have made several numerical experiments elucidating the spatial structure of contaminant spreading by water from potential sources in different parts of the coastal zone of the Arctic seas. The experiments have been carried out to estimate the transport of passive tracers by water and ice from potential sources of contaminants in the Arctic Ocean, especially from potential pollutant sources in the vicinity of river-mouths of major rivers flowing into the Arctic Ocean (Pavlov and Pavlov 1999). The numerical experiments, earmarking zones with maximum and minimum contamination, show that some regions, such as for example the northern and western parts of the Laptev Sea and Fram Strait, would be contaminated for all possible source locations in the coastal zone of the Arctic seas.

Kallenborn et al. (Chap. 19) argue that the global pathways of long-range contaminant transport and the principal atmospheric transport of inorganic and organic contaminants, nutrients, aerosols and particulate matter into the polar regions, can only be revealed by the concerted efforts from multinational atmospheric long-term pollution monitoring programs. Their empirically derived monitoring data, involving rapid and effective adaptation for the identification and implementation of priority contaminants, common sampling and analytical protocols etc., is at present the only way to evaluate the accuracy of the predictions calculated by modern models of future contamination scenarios.

As these monitoring programmes give important documentation on pollutant levels in the physical environment, they are a prerequisite to explain the levels of contaminants stored and bio-accumulated in the ecosystems. Gabrielsen et al. (Chap. 20) summarizes recent studies on the levels of heavy metals (HM) and persistent organic pollutants (POPs) in arctic animals, using data on biological effects related to POPs in polar bears *(Ursus maritimus)* and glaucous gulls *(Larus hyperboreus)* from the Svalbard archipelago. According to them, mercury, lead and calsium are the HM of most concern in the arctic environment. The monitoring programmes show that while the global emissions of cadmium and lead have decreased, the emission of mercury is increasing. For POPs, the levels are generally lower in the arctic environment than in more temperate regions. However, high levels of POPs exceeding the critical effect thresholds for effects on behavioural-, biochemical-, physiological- and immunological parameters, as found by laboratory and field studies, have been found in the marine food chain in glaucous gulls from Bjørnøya and in polar bears from Svalbard, Franz Josef Land and Kara Sea (Borgå et al. 2004).

The effective bio-accumulation of POPs in Arctic ecosystems are partly due to their lipid-rich food-chains, and pollutants are therefore also accumulate where Arctic people is at the top of these food chains. According to Gabrielsen et al. (Chap. 20), the levels of cadmium in arctic biota has been stable or is decreasing during the past 5-10 years, while the mercury level is increasing in most marine arctic species. In some arctic areas the levels of mercury and cadmium are high enough to cause health effects in animals and humans. Based on studies of the Greenland Inuit population, Mulvad (Chap. 21) explains that the traditional diet in Greenland to a large extend is based upon marine animals and fish, rich in fat. The partly isolated population with ethnic background provides good conditions for genetic and health impact studies under unique social circumstances, light and extreme cold weather.

1.5 Integrated aspects

Arctic and alpine areas are experiencing significant environmental change related to climate change, pollutant levels, changing pathways, stratospheric ozone depletion and surface UV radiation (AMAP 2003,2004; WMO 2003; ACIA 2004). The environmental changes and effects are in many cases amplified in these areas and closely coupled to the global climate change processes as documented in previous IPCC assessments (Houghton 2001). The Arctic biodiversity and indigenous peoples are vulnerable and constantly under pressure from these changes as well as from the effects of globalization (Nuttall Chap. 2). The decreasing sea-ice in the arctic ocean is one of the most pronounced features of climate change, with a decrease in spring and summer sea ice of the order 10-15% during the last 4 decades (Hougthton et al. 2001). According to Hansen-Bauer (Chap. 3), the variations in sea-ice concentration and air temperature during the last decades are partly accounted for by variations in atmospheric circulation indices such as the Arctic Oscillation (AO). The AO, which describes the general modes of large scale atmospheric circulation over the Northern Hemisphere, has gradually been more positive since the 1970s, with lower than normal surface air pressure anomalies over the Arctic (Thompson et al. 2000).

There is a close relationship between the atmospheric circulation patterns and the different global pathways bringing pollutants into the arctic and alpine environments. According to McDonald (Chap. 17), there is a general agreement that these changing pathways are caused by a combination of natural variability and anthropogenic forcing factors, and that the interdependence between contaminant pathways and climate change patterns, as manifested in anomalies of temperature, winds, precipitation, river flow and ocean circulation, ice and snow cover etc., involves a complex distribution of transport mechanisms, source regions, chemical transformation and magnifying processes.

The bio-accumulation of POPs in the arctic environment is high especially for the marine food chains, due to the fundamental specialisation of the lipid-based energy flux in Arctic bio-production, as discussed by Falk-Petersen et al. (Chap. 9). According to Gabrielsen (Chap. 20), the contaminant levels found in polar bears and glaucous gulls on Svalbard and Bjørnøya exceed critical effect thresholds and affect their health. Sea-ice both directly and indirectly challenge the polar bear population by reducing their habitat along with the forecasted future dramatic reduction of Arctic summer ice during this century (ACIA 2004), as well as by modifying the primary sea-ice related bio-production.

Indigenous peoples of the circumpolar North fundamentally depend on the health of arctic marine and terrestrial ecosystems. As discussed by Nuttall (Chap. 2), changes in climate, weather patterns, migration of animals and human actions all influence their traditional resource use, making e everyday life uncertain and unpredictable. Thus, the integrated effects of pollution, climate change and industrial development have consequences for the ecosystems, food security and human health that seriously may constrain their abilities to achieve sustainable livelihoods.

Atmospheric UV radiation is both physically, politically and biologically closely integrated with the problems of anthropogenic contaminant transport and climate change. The Vienna Convention (1985) with the Montreal protocol (1987) is a successful example of political countermeasures aimed at mitigating anthropogenic environmental change. This framework for reduction of both production and emission of ozone depleting substances are expected to lead to a slow recovery of the ozone layer over the next century (Madronich et al., 1999). However, more evidence of climate related interaction processes at stratospheric altitudes now suggests that ozone depletion is not purely chemically driven. Conditions with stronger and colder than normal polar stratospheric vortices, related to the persistent positive phase of the AO, have lead to an increased abundance of polar stratospheric clouds (PSCs) in the Arctic. And PSCs are effective catalysts in ozone depletion (Shindell et al., 1998). As pointed out by Slaper et al. (Chap. 16), it is therefore assumed that the ozone layer will need longer time to return to the "normal" 1980-levels, even with reduced active chlorine levels in the stratosphere.

This situation will certainly modify future UV levels especially in the Arctic. At the same time a number of atmospheric and surface physical parameters closely modified by climate change processes, significantly influence the surface UV radiation, as described by Blumthaler (Chap. 11). The surface albedo, as determined by the abundance of snow and ice, atmospheric aerosols and the amount and type of clouds represent parameters that are of almost equal importance in shaping the surface UV spectrum, as compared to the influence of ozone. In a similar way, ice cover, sediment particles and coloured dissolved organic matter (CDOM) are the primary controls of UV radiation under water, and these main controlling factors are highly responsive to climate warming or cooling trends. Vincent et al. (Chap. 14) argue that the climate induced shifts in underwater UV radiation can be of much greater magnitude than those caused by moderate ozone depletion. Arctic, antarctic and alpine aquatic ecosystems are particularly vulnerable to such effects because of the importance of snow and ice cover and the typically low CDOM concentrations in these regions. Hessen (Chap. 13) argue that the widespread small, shallow and highly transparent tundra ponds situated on permfrost in the circumpolar Arctic represent distinct ecosystems and unique biota that may be especially vulnerable to changes in UV-radiation, climate warming and changes in the hydrological cycle. Climate change interactions also affect the UV radiation conditions and primary production of seaweeds. According to Hanelt et al. (Chap. 15), organisms in the eulittoral and upper sublittoral zone are affected by UV radiation throughout the polar day. However, during the warm summer seasons with intense runoff, turbid melt water from glaciers and rivers significantly reduce the underwater UV transparency.

1.6 Conclusions

Obviously, the many challenges involving climate and stratospheric change, pollutant transport and social changes are related. A number of other significant dependencies also naturally exist between the various aspects of environmental change parameters and human activities in arctic and alpine areas. The scope here is not to give a complete review of all single processes but to outline the level of interaction between the main contemporary challenges in environmental change, on the basis of the studies presented in the book. Both with respect to the physical processes as well as dealing with their biological impacts and the anthropogenic forcing factors, the changes related to the climate change processes, long range transported pollutants and UV radiation, can not be treated alone or independently but need to be analysed as multiple pressures of the high latitude and high altitude, arctic and alpine environments.

Acknowledgements

This introductory chapter builds on the manuscripts prepared by the many authors contributing to this book. All authors are acknowledged for their high quality contributions and their kind co-operation and patience with the final preparation of this book.

References

ACIA (2004) Impacts of a warming Arctic. Arctic Climate Impact Assessment. Cambridge University Press, Cambridge

AMAP (2003) AMAP Assessment 2002: The influence of Global Change on Arctic Contaminant Pathways to within, and from the Arctic. Arctic Monitoring and Assessment Programme, Oslo, Norway, pp 65

AMAP (2004) AMAP Assessment 2002: Persistent Organic Pollutants in the Arctic. Arctic Monitoring and Assessment Programme, Oslo, Norway pp 310

Arendt, AA, Echelmeyer, KE, Harrison, WD, Lingle, CA and Valentine, VB (2002) Rapid wastage of Alaska glaciers and their contribution to rising sea level. Science 297:383-386

Beniston M, Keller F, Goyette S (2003) Snow pack in the Swiss Alps under changing climatic conditions: an empirical approach for climate impacts studies. Theor Appl Climatol 74: 19-31

Bischof K, Hanelt D, Tüg H, Karsten U, Brouwer PEM, Wiencke C (1998) Acclimation of brown algal photosynthesis to ultraviolet radiation in Arctic coastal waters (Spitsbergen, Norway). Polar Biol 20: 388-395

Blumthaler M, Salzgeber M, Ambach W (1995) Ozone and ultraviolet-B irradiances: experimental determination of the radiation amplification factor. Photochem Photobiol 61:159-162.

Borgå, K, Fisk AT, Hoekstra PF, Muir DCG (2004) Biological and chemical factors of importance in the bioaccumulation and trophic transfer of persistent organochlorine contaminants in Arctic marine food webs. Environ Toxicol Chem, 23:2367-2385.

Broecker WS (1991) The great conveyor belt. Oceanography 4: 79-89

Caldwell MM, Björn LO, Bornman JF, Flint SD, Kulandaivelu G, Teramura AH, Tevini M (1998) Effects of increased solar ultraviolet radiation on terrestrial ecosystems. Photochem Photobiol 46:40-52

Clark P, Marshall S, Clarke G, Hostetler S, Licciardi J and Teller J (2001) Freshwater forcing of abrupt climate change during the last glaciation. Science 293: 283-286

Dansgaard W, White JWC and Johnsen SJ (1989). The abrupt termination of the Younger Dryas climate event. Nature, 339, 532-534.

Falk-Petersen S, Hopkins CCE, Sargent JR (1990) Trophic relationships in the pelagic food web. In: Barnes M, Gibson RN (eds) Trophic relationships in the marine environment. Aberdeen University Press, Aberdeen, pp 315-333

Hanelt D, Wiencke C, Nultsch W (1997a) Influence of UV-radiation on the photosynthesis of Arctic macroalgae in the field. J Photochem Photobiol B: Biol 38: 40-47

Houghton JT, Ding Y, Griggs DJ, Noguer M, van der Linden PJ, Dai X, Maskell K, Johnson CA (eds) (2001) Climate Change 2001: The Scientific Basis. Contribution of Working Group I to the Third Assessment Report of the Intergovernmental Panel on Climate Change (IPCC). Cambridge University Press, Cambridge

Hurrell JW (1995) Decadal trends in the North Atlantic Oscillation: Regional temperatures and precipitation. Science 269: 676-679

Johannessen OM, Bengtsson L, Miles MW, Kuzmina SI, Semenov VA, Alekseev GV, Nagurny AP, Zakharov VF, Bobylev LP, Pettersson LH, Hasselmann K, Cattle HP (2004) Arctic climate change: Observed and modelled temperature and sea-ice variability. Tellus 56 (4): 328-341

Korsnes R, Pavlova O, Godtliebsen F (2002) Assessment of potential transport of pollutants into the Barents Sea via sea ice–an observational approach. Mar Pollut Bull 44(9):861-869

Lefauconnier B, Hagen JO, Ørbæk JB, Melvold K and Isaksson E (1999) Glacier balance trends in the Kongsfjorden area, western Spitsbergen, Svalbard, in relation to the climate. Polar Research 18 (2): 307-313

Macdonald RW, Harner T, Fyfe J (2005) Recent climate change in the Arctic and its impact on contaminant pathways and interpretation of temporal trend data. Science of the Total Environment 342 (1-3): 5-86

Madronich S, Velders GJM, Daniel JS, Murari Lal M, McCullogh A and Slaper H (1999) Chapter 11. In: WMO, Scientific Assessment of Ozone depletion: 1998, Geneva, Switzerland, WMO report 44

Magnuson JJ, Robertson DM, Benson BJ, Wynne RH, Livingstone DM, Arai T, Assel RA, Barry RG, Card V, Kuusisto E, Granin NG, Prowse TD, Stewart KM, Vuglinski VS (2000) Historical trends in lake and river ice cover in the Northern Hemisphere. Science 289: 1743-1746

Müller R, Crutzen PJ, Grooß JU, Brühl C, Russel JM, Gernandt H, Mc Kenna DS, Tuck AF (1997) Severe ozone loss in the Arctic during the winter of 1995-96. Nature 389: 709-712

Ørbæk JB, Svenøe T Hessen DO (2002) Spectral properties and UV-attenuation in Arctic freshwater systems. In: Hessen DO (ed) UV radiation and Arctic Ecosystems, Ecological Series 153, Springer-Verlag, pp 57-72.

Pavlov VK and Pavlov PV (1999) Features of seasonal and interannual variability of sea level and water circulation in the Laptev Sea. In: Land-Ocean systems in the Siberian Arctic. Dynamics and history. Eds Kassens et al., Springer-Verlag, Berlin Heidelberg, pp 3-16

Rex M, Harris NRP, von der Gathen P, Lehmann R, Braathen GO, Reimer E, Beck A, Chipperfield MP, Alfier R, Allaart M, o´Connor F, Dier H, Dorokhov V, Fast H, Gil M, Kyrö E, Litynska Z, Mikkelsen IS, Molyneux MG, Nakane H, Notholt J, Rummukainen M, Viatte P, Wenger J (1997) Prolonged stratospheric ozone loss in the 1995-96 Arctic winter. Nature 389: 835-838

Searles PS, Flint SD, Caldwell MM (2001) A meta-analysis of plant field studies simulating stratospheric ozone depletion. Oecologia 127:1-10

Shindell DT, Rind D, and Lonergan P (1998), Increased polar stratospheric ozone loss and delayed eventual recovery owing to increasing green-house gas concentrations. Nature 392: 589-592

Slaper H, Velders GJM, Daniel JS, Degruijl FR, Vanderleun JC (1996) Estimates of ozone depletion and skin cancer incidence to examine the vienna convention achievements. Nature 384 (6606):256-8

Steffen K, Nghiem SV, Huff R and Neumann G (2004) The melt anomaly of 2002 on the Greenland Ice Sheet from active and passive microwave satellite observations. Geophys. Res. Letters 31 (20):

Thompson DWJ, Wallace JM, Hegerl (2000) Annular modes in the extratropical circulation. Part II: Trends. J Clim 13:1018-1036

UNEP (1999) Environmental effects of ozone depletions: 1998 assessment. United Nations Environmental Program, Photochem. Photobiol. B 46:1-108.

UNEP (2003) Environmental effects of ozone depletion and its interactions with climate change: 2002 assessment. United Nations Environmental Program, ISBN92-807-2312-X

Wiencke C, Gómez I, Pakker H, Flores-Moya A, Altamirano M, Hanelt D, Bischof K, Figueroa FL (2000) Impact of UV radiation on viability, photosynthetic characteristics and DNA of brown algal zoospores: Implications for depth zonation. Mar Ecol Prog Ser 197: 217-229

WMO (2003) Scientific assessment of ozone depletion 2002. Global Ozone Research and Monitoring Project, Geneva, Switzerland: World Meteorological Organization 44

2 An environment at risk: Arctic indigenous peoples, local livelihoods and climate change

Mark Nuttall

Department of Anthropology, University of Alberta,
e-mail: mnuttall@ualberta.ca

2.1 Introduction

Over the last two decades the Arctic has emerged as a region of dramatic environmental change. This vast part of the planet, once seen as pristine and remote, is now represented increasingly as a vulnerable and fragile place, its biodiversity and peoples at risk from climate change, contaminants and globalization. These drivers of far-reaching change have a diffuse distribution around the globe, their origins often difficult to identify and almost impossible to allocate specific moral and legal responsibility to. They are also subject to a contested political debate over whether mitigation or adaptation are the most feasible or possible strategies for protecting Arctic ecosystems and human well-being.

Arctic marine and terrestrial ecosystems provide a variety of ecosystem services which are of fundamental importance to the livelihoods of indigenous peoples (Chapin *et al.*, in press). Yet, the indigenous peoples of the circumpolar North increasingly perceive the Arctic as both an environment *of* risk and an environment *at* risk (Nuttall 1998: 170). It is an environment of risk in that climate variability and local weather events, changes in the movement and behaviour of animals, and human actions all influence traditional resource use activities and make everyday life uncertain and unpredictable. It is also an environment at risk from pollution, global climate change and industrial development. Such threats continue to influence the climate, have an impact on ecosystems, animal habitats and movement, and also have consequences for food security and human health, thus seriously constraining the abilities of indigenous peoples to achieve sustainable livelihoods.

Scientific scenarios suggest that the scale and nature of Arctic climate change in the coming decades may not only be greater than previous changes in the region's history, but within the context of global climate change northern regions will experience a greater degree of change than

countries in the tropics (e.g. ACIA 2005, Weller 2000). Indeed, the results of scientific research and the compelling observations from indigenous peoples are documenting current climate changes that are more pronounced in the Arctic than in any other region of the world (ACIA 2005, Huntington *et al.*2005, Krupnik and Jolly 2002). The significance of these studies goes far beyond the northern reaches of the Earth and enriches our understanding of living on a planet undergoing constant change.
This chapter provides a brief assessment of climate change impacts on the local livelihoods and traditional resource use practices of the Arctic's indigenous peoples. It draws on, and summarizes, recent research on climate change impacts on indigenous livelihoods as well as some of the key work undertaken for the human dimensions chapters of the Arctic Climate Impact Assessment (ACIA 2005).

2.2 Indigenous Peoples and Traditional Livelihoods

The indigenous peoples of the Arctic include the Inupiat, Yup'iit, Alutiit, Aleuts and Athapaskans of Alaska; the Inuit, Inuvialuit, Athapaskans and Dene of northern Canada; the Kalaallit and Inughuit of Greenland; the Saami of Fennoscandia and Russia's Kola peninsula; and the Chukchi, Even, Evenk, Nenets, Nivkhi and Yukaghir of the Russian Far North and Siberia. Arctic peoples have depended for thousands of years on the living resources of land and sea, as hunters, fishers and reindeer herders. Today, many indigenous communities across the Arctic continue to depend largely on harvesting and using living terrestrial, marine and freshwater resources. The most commonly harvested species are marine mammals such as seals, walrus, narwhals, beluga, fin and minke whales, and polar bear and land mammals such as caribou, reindeer and musk-ox; and fish such as salmon, Arctic char, northern pike and other species, such as coregonids (whitefishes). Many of these species are used as food, for clothing and other products, but they also figure prominently in the cash-economy of local households and communities (Caulfield 2000, Dahl 2000, Huntington 1992, Nuttall 1992).
Indigenous peoples maintain a strong connection to the Arctic environment through these activities, which provide the basis for food production, in a way that marks them off from non-indigenous communities. Indeed, one of the defining attributes of being indigenous is that it refers to the quality of a specific people relating their identity to a particular area and to their traditional cultural and economic dependence on local resources, thus distin-

guishing them culturally from other peoples who came to the territory subsequently (Gray 1995).

Historically, cultural adaptations and the ability of the Arctic's indigenous peoples to utilize their local resources have often been associated with or affected by seasonal variation and changing ecological conditions. Climatic variability and weather events often greatly affect the abundance and availability of animals and therefore the abilities and opportunities to harvest and process animals for food, clothing and other uses. Many species are only available seasonally and in localized areas and indigenous cultures have developed the capacity and flexibility to harvest a diversity of animal and plant species. Indigenous cultures have, in many cases, also shown resilience in the face of severe social, cultural, and economic change, particularly in the last 100 years.

The longstanding dependence of present indigenous societies on hunting, herding, fishing, and gathering continues for several critically important reasons. One is the economic and dietary importance of being able to access customary, local foods. Many of these local foods – fish, and meat from marine mammals or caribou and birds, for instance, as well as berries and edible plants – are nutritionally superior to the foodstuffs which are presently imported (and which are often expensive to buy). Another reason is the cultural and social importance of hunting, herding, and gathering animals, fish, and plants, as well as processing, distributing, consuming, and celebrating them (Freeman, 2000).

The Arctic's living resources, therefore, not only provide indigenous peoples with the economic and nutritional bases for survival, they remain important for social identity, cultural survival and spiritual life, and so are as much important cultural resources as they are economic ones. This dependence on animals for food and social, cultural and economic well-being is reflected in community hunting regulations, in herding practices and in patterns of sharing and gift-giving based on family ties and other forms of kinship and close social relatedness (Nuttall 1992). However, Arctic communities have experienced, and are experiencing, stress from a number of different forces that threaten to restrict harvesting activities and sever these relationships. The Arctic regions are tightly tied politically, economically and socially to the national mainstream and are inextricably linked to the global economy. Rapid social, economic and demographic change, resource development, resource management and quota systems, trade barriers and animal-rights campaigns have all had their impacts on hunting, herding, fishing and gathering activities (Caulfield 2000, Nuttall 1998, Wenzel 1991). By and large, hunting, herding, fishing and gathering

Fig. 2.1. Small-scale hunting and fishing continues to be the primary occupation in many Arctic coastal communities. Ilulissat, Greenland, Spring 2005. Photograph by Niels Einarsson.

are also being challenged by environmental changes such as climate variability and change.

As the Arctic Climate Impact Assessment shows (ACIA 2005), scientific projections and scenarios suggest there will be significant changes in the climate of the Arctic, the character of the environment and its resources. Human activities, industrial development, resource use regulations and global economic processes have far-reaching consequences for the environment and therefore magnify the likely impacts on indigenous livelihoods due to variations in weather and climate. Indigenous economies are not self-reliant closed systems and although their involvement in global networks of production and consumption may provide avenues to strengthen and extend the possibilities for Arctic communities, it also introduces greater elements of risk and perhaps makes people and their livelihoods less resilient to coping with and adapting to climate impacts.

2.3 Renewable Resource Use and Climate Change: Risk and Access to Food Resources

Renewable resources and the harvesting of renewable resources by indigenous peoples in the Arctic could be affected by global climate change and increased ultraviolet radiation caused by ozone depletion. Climate change scenarios suggest that climate change will have impacts on marine and terrestrial animal populations, affecting population size, structure, reproduction rates and migration routes (ACIA 2005, IPCC 2001). Arctic residents, particularly indigenous peoples, will feel these climate change impacts first and most intensely.

However, because of the interdependence between Arctic economies and global markets, indigenous peoples are multiply exposed -- to climate change, to changes caused by the global processes affecting markets, technologies and public policies and to local and regional political and economic situations. It is important to contextualise climate change impacts with reference to other changes experienced by Arctic residents. Being able to access traditional food resources and ensuring food security will be one major challenge in an Arctic affected increasingly by climate change and global processes.

The Arctic Climate Impact Assessment (ACIA 2005) shows that the results of scientific research and evidence from indigenous peoples are increasingly documenting climatic changes that are more pronounced in the Arctic than in any other region of the world. Although this indicates that the physical environment, as well as the flora and fauna, has been undergoing noticeable change, the impacts felt throughout the Arctic will be unique and will vary from region to region. Different climatic trends have been observed in different parts of the Arctic – while average temperatures in the North American western Arctic and Siberia have been increasing in the last 30 years (e.g. annual temperatures in the Canadian western Arctic have climbed by 1.5°C and those over the central Arctic have warmed by 0.5°C), temperatures in Canada's Hudson Bay and in Greenland, particularly in the Davis Strait area have decreased (Chapman and Walsh 1993), suggesting that climate change involves regional cooling as well as global warming.

If the scientific predictions and scenarios are realised, climate change could have a potentially devastating impact on the Arctic environment and on the peoples who live there, particularly those indigenous peoples whose livelihoods and cultures are inextricably linked to the Arctic environment and its wildlife. Some scenarios suggest that the most direct changes will be noticeable in the reduction of the extent of sea ice and permafrost, less ice in lakes and rivers, pronounced reductions in seasonal snow, and the disappearance of the existing glacier mass, leading to a corresponding shift in landscape processes (Lange, 2000, Siegert and Dowdeswell 2000, Weller 2000).

Scientific research shows that over the last 100 years there has already been a significant reduction in Arctic ice cover extent and thickness. Since 1979 alone, the extent of sea ice throughout the Arctic has decreased by 0.35%, and record reductions of sea ice coverage were observed in the Beaufort and Chukchi seas in 1998 (Johannessen *et al.*, 1999, Maslanik *et al.*, 1999). Sea ice is highly dependent on the temperature gradient between ocean and atmosphere and on near-surface oceanic heat flow and will react swiftly to changes in atmospheric conditions (Lange 2000). Atmosphere-ocean climate models project a reduction in sea ice of around 60% in the next 50-100 years under a scenario in which CO_2 is doubled. Permafrost will thaw more quickly in spring, but take longer to refreeze in autumn, and permafrost boundaries will gradually move poleward, with most of the ice-rich discontinuous permafrost disappearing by the end of the 21st century.

Climate variability appears to have caused relatively rapid shifts in the organisation of Arctic marine ecosystems. In the Bering Sea ecosystem and the Barents Sea ecosystem climate-driven variability is significant (Sakshaug and Slagstad 1992). There are difficulties, however, in determining what biological changes in marine ecosystems are due to natural environmental fluctuations or human activities. In the eastern Bering Sea upper trophic levels have undergone significant changes in the past 100-150 years, largely due in part to commercial exploitation of mammals, fish and invertebrates (ACIA 2005). Climatic changes may have contributed in part to the changes in animal populations. Higher ocean temperatures and lower salinities, changes in seasonal sea ice extent, rising sea levels and many other (as yet undefined) effects are certain to have significant impacts on marine species, with implications for Arctic coastal communities dependent on hunting and fishing (Weller and Lange 1999).

Most Arctic marine species depend upon the presence of sea ice and many indigenous coastal communities depend on harvesting marine species (marine mammals and fish). The ice-edge is unique among the world's ecosystems in that it moves thousands of kilometres each year, north in spring and south in autumn. Walrus, numerous species of seals and cetaceans such as beluga and narwhals all follow the ice-edge as it moves, taking advantage of the ready access to food and (for the walrus and seals) the availability of ice to haul-out on for sunning, mating and raising pups in late winter and spring (an important time for Inuit hunting communities).

The almost complete elimination of multi-year ice in the Arctic Ocean is also likely to be immensely disruptive to ice-dependent micro-organisms, which will lack a permanent habitat. Preliminary results from research in the Beaufort Sea suggest that ice algae and other micro-organisms may have already been significantly affected by warming over the last 20 years. Research indicates that most of the larger marine algae have died out, and have been replaced with a much less productive community of microorganisms more usually associated with freshwater ecosystems.

It is anticipated that marine mammals such as walrus, seals and whales are likely to undergo shifts in range and abundance, while the migration routes of caribou will be disrupted (Nuttall *et al.* 2005). Such changes could impact upon the hunting, trapping and fishing economies of many small, remote Arctic settlements. Although warming may increase biological production in some wildlife species, the distribution of many species crucial to the livelihoods and well-being of indigenous peoples could change. Important wetlands may disappear, or drainage patterns and tundra landscapes will be altered significantly, which could affect ducks and other waterfowl. Changes in terrestrial vegetation will have consequences for reindeer herding and subsistence lifestyles.

Terrestrial wildlife such as caribou and reindeer, two major species important for indigenous communities throughout the Arctic, would be affected by climate change directly through changes in thermal stress in animals, and indirectly by significant difficulties gaining access to food and water (Nuttall *et al.* 2005). Arctic communities located on coastlines may be affected by rising sea levels, increased coastal erosion, and severe storms. The fortunes of subsistence fisheries will depend on marine fish stocks and their climate related variations (Lange 2000). As the amount of sea ice decreases, seals, walrus, polar bears and other species that depend on it would suffer drastically.

Recent observations have demonstrated that there has been a distinct warming trend in lowland permafrost of 2-4°C over the last 100 hundred years (Fitzharris *et al.* 1996, Lange 2000), leading to disturbances of animal and human activities due to thawing, thermokarst formation and severe erosion. Further warming is likely to continue this trend and increase the likelihood of natural hazards for people (particularly affecting hunting and herding), buildings, communication links and pipelines. The documentation of widespread thawing of discontinuous permafrost in Alaska illustrates some of these hazards and the implications for habitat change and the physical infrastructure of communities. In western Alaska several communities in low-lying areas, including Shishmaref, Kivalina and Little Diomede are affected by recent climate changes and face severe problems as a result of erosion and thawing of the discontinuous permafrost (Callaway *et al.* 1999).

Unstable sea ice could make ice-edge hunting more difficult and dangerous. Temperature and precipitation changes could affect migration patterns of terrestrial mammals like caribou and alter breeding and moulting areas for birds. Salmon, herring, walrus, seals whales, caribou, moose and various species of waterfowl are expected to undergo shifts in range and abundance (ACIA 2005). Changes in snow cover could affect the growth and distribution of plants essential for survival of caribou and reindeer. Changes in snow cover could also make accessing hunting, fishing and herding areas more difficult by dogsled, snow-machine or other vehicles, making local adjustments in hunting practices and harvesting strategies necessary.

In many parts of the Arctic indigenous peoples are reporting that they are already experiencing the effects of climate change. In Canada's Nunavut Territory, Inuit hunters have noticed the thinning of sea ice and the appearance of birds not usually found in their region; Inupiat hunters in Alaska report that ice cellars are too warm to keep food frozen; Inuvialuit in the western Canadian Arctic report thunderstorms and lightning (a rare occurrence in the region); Gwich'in in Alaska have witnessed dramatic changes in weather, vegetation, and animal distribution patterns over the last 50 years or so; Saami reindeer herders in Norway have observed that prevailing winds relied on for navigation have shifted or that snow cannot be relied on to travel over on trails that people have always used and considered safe (Nuttall *et al.* 2005).

In the Kitikmeot region of Nunavut, Inuit observations of climate change have been recently documented (Thorpe *et al.* 2002). People have spoken of a changed climate in the 1990s compared with previous decades: increasing temperatures with earlier spring melts and later freeze-ups in autumn have meant periods of longer summer-like conditions, while weather has become variable and unpredictable. This change and variability has had many significant impacts on caribou. Migration routes and the location of calving grounds have shifted and food sources have sometimes become inaccessible. Inuit have recently noticed more frequent short-term changes in temperature, especially in freeze-thaw cycles, which, because these cycles help form an icy layer on the top of snow or tundra, prevent access to vegetation (Thorpe *et al.* 2002).

2.4 Concerns Over Irreversible Impacts

Indigenous peoples live with fluctuations in weather and climatic conditions. Experiencing year-to-year changes in weather, ice and snow patterns, animal behaviour and movement, and in hunting and fishing conditions is part of life in the Arctic. Yet the trends currently being observed give concern over major, irreversible impacts on indigenous communities and livelihoods. For example, since the late 1970s Alaska Natives in communities along the coast of the northern Bering and Chukchi Seas have noticed substantial changes in the ocean and the animals that live there, particularly in the patterns of wind, temperature, ice, and currents (Nuttall *et al.* 2005).

A significant collection of indigenous environmental observations was recorded during a study of environment changes in Canada's Hudson Bay region. The results are published in *Voices from the Bay* by the Canadian Arctic Resources Committee and the Municipality of Sanikiluaq, a small Inuit community on the Belcher Islands in the midst of Hudson Bay. Completed in 1996 and published in 1997 (McDonald *et al.* 1997), the study brought together 78 Inuit and Cree hunters and elders from 28 communities on the shores of Hudson and James bays in a series of workshops over three years to describe, record and verify ecological changes in the region, including but not limited to climate change. Observations include wholesale changes in location, number and duration of polynas—open water areas in winter—in eastern Hudson Bay, and changing routes of Canada and snow geese, but the study indicates that alterations in weather and climate are by no means uniform within the bioregion. *Voices from the Bay* and other recently reported observations from indigenous peoples (e.g. Krupnik and Jolly 2002, Nunavut Tunngavik 2001) illustrates an important and inescapable fact: much of the impact of climate change on northern indigenous peoples will be channelled through ecological changes to which they will have to respond, cope with and adapt to.

As indigenous peoples perceive and experience it, the Arctic is becoming an environment of risk and an environment at risk (Nuttall 1998) in the sense that sea ice is now unstable where hunters previously knew it to be safe, more dramatic weather patterns such as floods are occurring, vegetation cover is changing, and particular animals are no longer found in traditional hunting areas during specific seasons. The weather is becoming increasingly unpredictable and local landscapes, seascapes and icescapes are becoming unfamiliar.

2.5 Responding to Climate Change: Flexibility, Adaptation, Barriers and Opportunities

The archaeological record, ethnohistorical accounts and the memories of elders provide detailed accounts of how human life in the Arctic has always been dominated and influenced by periodic, irregular and often dramatic ecosystem changes, triggered by periods of warming and cooling, extreme weather events and fluctuations in animal populations (Krupnik 2000). The successful long-term occupation of the Arctic by indigenous peoples has been possible, in part, because of their adaptive capacity (in social, economic and cultural practices) to adjust to climate variation and change. Hundreds and even thousands of years ago, Arctic populations adapted to gradual or even rapid environmental change by settling amid favourable climate conditions and along the paths of animal migration routes. Many of these adaptations bore the hallmark of what is the essence of successful indigenous resource use systems throughout the Arctic -- flexibility in technology and social organisation and an ability to cope with climate change, responding both to its associated risks and seizing its opportunities. Cultural and ecological diversity required flexibility and resilient coping strategies during periods of extreme change. Subsistence diversity was the outcome of a successful cultural and social response to climatic variation and the resource instability of the Arctic environment (Krupnik 1993).

Today, Arctic peoples are not in a position to adapt, relocate or change resource use activities in the same ways as in the past, because most now live in permanent communities and have to negotiate greatly circumscribed social and economic situations. The majority of indigenous peoples live in planned settlements with elaborate infrastructures, and their hunting and herding activities are determined to a large extent by resource management regimes, land use and land ownership regulations and local and global markets. The mobility that indigenous peoples once possessed to move in response to shifts in the pattern and state of their resource base is no longer possible.

Changes to settlement patterns and the ecological relations between humans and animals often arise from government attempts to introduce new economic activities or to sedentarize indigenous peoples. In northern Russia and Siberia, for example, the Soviet authorities 'industrialised' reindeer herding as a way of facilitating the development of the Soviet North. The new settlements and industries in Siberia came to depend on reindeer herders to supply them with meat. Today, in post-Soviet Russia, privatisation and the transition to a market economy bring new challenges

Fig. 2.2. Greenlandic seal hunter. Ilulissat, Greenland, Spring 2005. Photograph by Niels Einarsson.

to reindeer herding peoples in Siberia and the Russian Far East, highlighting the dependence of Arctic reindeer systems on the complex interlinkages between local, regional and global economies.

Strict regulatory regimes, management practices and quota systems imposed by states and federal and provincial agencies increasingly affect hunting, fishing and herding (Anderson and Nuttall 2004). Some, while aiming, in principle, to protect and conserve wildlife also restrict access to resources. In Alaska, for example, state and federal policies make subsistence issues extremely complex. State and federal law define subsistence as the customary and traditional non-commercial use of wild resources and regulations limit the prospects of finding markets for caribou meat. Earning money through more commercial channels is not an option for Alaskan subsistence hunters. In northern Fennoscandia, Saami reindeer herders have traditionally ranged far and wide, crossing national borders as they follow their reindeer herds between winter and summer pastures. In modern times, political developments have restricted migration routes over the last one hundred years or so. Economic development in the nineteenth and twentieth centuries, such as mining, forestry, railways, roads, hydroelectric power and tourism have all had their impact on traditional Saami livelihoods. In Greenland, threats to the cultural and economic viability of hunting livelihoods in small communities come from transformations in resource management regimes and Home Rule government regulations, which conflict with local customary practices and knowledge systems (Dahl 2000; Nuttall 1992). Caribou, whales, seals, and fish, which have traditionally been subject to common use rights vested in members of a local community, are becoming national and privately-owned divisible commodities subject to rational management regimes defined by the state and the interest groups of hunters and fishers, rather than to locally understood and worked out rights, obligations, and practices.

Hunters and herders are increasingly constrained by institutional frameworks and management structures, as well as the legal recognition to resource use rights. They are commonly experiencing a transition from herding and hunting from what we may call a 'way of life' to an occupation and industry (Anderson and Nuttall 2004). The contemporary reality for many hunters and herders is that they are placed in very inflexible situations. Faced with climate change they are not necessarily in a position to respond appropriately. How indigenous peoples are in a position to take advantage of the opportunities that may arise, as well as being able to modify or change their mode of production in response to climatic variability, for example by switching hunting and fishing activities, is a critical research need.

For some peoples of the Arctic, the political and management systems are already in place that could assess the impacts of climate change, allow local and regional governments to act on policy recommendations to deal with the consequences, and improve the chances for indigenous peoples to deal successfully with climate change. Although complex, solutions to environmental problems are potentially realistic

Significant political changes since the 1970s have included land claims in Alaska and Canada and the formation of regional governments in Greenland and Nunavut. Settlements include the Alaska Native Claims Settlement Act (1971), Greenland Home Rule (1979), the James Bay and Northern Quebec Agreement (1975-77), the Inuvialuit Final Agreement (1984) and the Nunavut Agreement of 1992 (the Territory of Nunavut was inaugurated in 1999). These political changes often include changes in the ways that living and non-living resources are managed. A greater degree of local involvement in resource use management decisions has been introduced, including in some cases the actual transfer of decision-making authority to the local or regional level (AHDR 2004, CAFF 2001).

In addition, significant steps have been taken with innovative co-management regimes that allow for the sharing of responsibility for resource management between indigenous and other uses and the state

(Caulfield 2004, Huntington 1992). Examples include the Alaska Eskimo Whaling Commission, the Kola Saami Reindeer Breeding Project, the Inuvialuit Game Council and the North Atlantic Marine Mammal Commission (NAMMCO). Self-government is about being able to practise autonomy. The devolution of authority and the introduction of co-management allow indigenous peoples opportunities to improve the degree to which management and the regulation of resource use considers and incorporates indigenous views and traditional resource use systems (Huntington 1992).
Co-management projects involve greater recognition of indigenous rights to resource use and emphasise the importance of decentralised, non-hierarchical institutions and consensus decision-making. This presents tremendous opportunities for collaboration between indigenous peoples, scientists and policy-makers concerned with the sustainable use of living resources (Caulfield 2000, 2004). And it is within this new political and scientific environment of power sharing and dialogue that indigenous communities, scientists and policy-makers can work together to find solutions (such as building flexibility into otherwise constraining wildlife management regimes) to the pressing problems climate change may bring to the Arctic.

2.6 Conclusions

Indigenous communities across the Arctic are culturally and economically diverse and are affected by environmental change in different ways. Such diversity also means that local experiences of climate impacts and responses to climate variability and change may not be universal. As Scheraga and Grambsch (1998: 87) put it, there 'is a regional texture to changes in climate, and therefore to the effects of climate change,' as well as a regional texture to the risks and opportunities that climate change presents. For the Arctic, the Arctic Climate Impact Assessment points to significant regional variation in climate changes. This raises a question, therefore, about how communities differ in the ways they perceive risk, how they differ in the ways they utilise strategies for mitigating negative change and in the effectiveness of local adaptive capacity. This points to the importance for research on localised, regional and circumpolar studies of socio-economic impacts of recent climate change. Such work is necessary for a thorough understanding of how potential impacts of climate change are distributed across different regions and populations (e.g. see Scheraga and Grambsch 1998).

The emphasis of scientific research on climate change is to assess the impacts on the environment, ecosystem processes and wildlife. One gap in knowledge is how climate change affects social relations in indigenous communities. Nuttall *et al.*(2005) highlight this as a critical aspect of climate change research, arguing that a change in the ability of indigenous peoples to access traditional food resources can have a corresponding impact on the social fabric of their communities. In a very real sense, therefore, the discussion of climate impacts on traditional resource use activities by indigenous peoples is about sustaining human/food resource relationships and activities in indigenous societies, as well as being aware that climate change impacts pose a threat of severe and irreversible social changes. Arguing that research should place emphasis on understanding climate change impacts within the broader context of rapid social and economic change, Nuttall *et al.* (2005) illustrate the complexity of problems faced by indigenous peoples today and underscore the reality that climate change is but one of several problems affecting their livelihoods.

Is an ability to respond and cope with climate change, mainly by adjusting subsistence activities, a reliable indication of an ability to adapt in the future? Research is needed on understanding how much change can be accommodated by the existing ways of life of indigenous peoples. Case studies in the human dimensions chapters of the Arctic Climate Impact Assessment have pointed to the *resilience*, or the amount of perturbation that the resource use systems of indigenous peoples can absorb, and how they can adapt by learning and self-organization. The question of resilience (both social and ecological) is important, as it is a crucial aspect of the sustainability of local livelihoods and resource utilization (Adger 2000), and further research is needed in the Arctic because little is known about building adaptive capacity in the face of climate change.

As resilience not only depends on ecosystem diversity, but on the institutional rules which govern social and economic systems (Adger 2000), further research is also needed on co-management and the effectiveness of governance institutions and whether they can create additional opportunities to increase resilience, flexibility and the ability to deal with change (Nuttall *et al.* 2005, Young and Einarsson 2004). How can, for example, new governance mechanisms help indigenous peoples negotiate and manage the impacts of climate change? The answer to this will depend on a range of factors, including the importance of understanding the nature of the relationships between people, communities and institutions if effective policy responses are to be developed (O'Riordan and Jordan 1999). With a capacity-building strategy now being a key objective for the Arctic Council, tremendous opportunities exist for co-operation and constructive dialogue on dealing with climate change between communities, organisations, institutions and governments at circumpolar and wider international levels.

Acknowledgements

This chapter is based on work carried out largely within the framework of the Arctic Climate Impact Assessment (ACIA), my participation in which was made possible by the generous support of the Polar Regions section of the UK's Foreign and Commonwealth Office

References

ACIA 2005 *Arctic Climate Impact Assessment.* Cambridge: Cambridge University Press

AHDR 2004 *Arctic Human Development Report* Akureyri: Stefansson Arctic Institute

Adger, W.N. 2000 'Social and ecological resilience: are the related?' *Progress in Human Geography* 24(3): 347-364

Anderson, D.G. and M. Nuttall (eds.) 2004 *Cultivating Arctic Landscapes: knowing and managing animals in the circumpolar North* Oxford: Berghahn

CAFF 2001 *Arctic Flora and Fauna: status and conservation* Helsinki: Edita

Callaway, D., J. Eamer, E. Edwardsen, C. Jack, S. Marcy, A. Olrun, M. Patkotak, D. Rexford and A. Whiting, 1999. 'Effects of climate change on subsistence communities in Alaska' in G. Weller and P. Anderson (eds.). *Assessing the Consequences of Climate Change for Alaska and the Bering Sea Region* Center for Global Change and Arctic System Research, University of Alaska Fairbanks, pp. 59-73..

Caulfield, R.A 2000 'The political economy of renewable resource harvesting in the Arctic' in M. Nuttall and T.V. Callaghan (eds.) *The Arctic: environment, people, policy* Amsterdam: Harwood Academic Publishers

Caulfield, R.A 2004 'Resource governance' in AHDR 2004 *Arctic Human Development Report* Akureyri: Stefansson Arctic Institute

Chapin, S, M. Berman, T. V. Callaghan, A. Crepin, K. Danell, B.Forbes, G. Kofinas, D. McGuire, M. Nuttall, O.R. Young, and S. Zimov, in press 'Polar Systems' In *Millenium Ecosystem Assessment: conditions and trends*, Washington DC: Island Press

Chapman, W.L. and Walsh, J.E. 1993. 'Recent variations of sea ice and air temperatures in high latitudes' *Bulletin of the Meteorological Society of America* 73: 34-47

Dahl, J. 2000 *Saqqaq: an Inuit hunting community in the modern world* Toronto: University of Toronto Press

Fitzharris, B.B., I. Allison, R.J. Braithwaite, J. Brown, P.M.B. Foehn, W.

Haeberli, K. Higuchi, V.M. Kotlyakov, T.D. Prowse, C.A. Rinaldi, P. Wadhams, M-K Woo, Y. Xie, O. Anisimov, A. Aristarain, R.A. Assel, R.G. Barry, R.D. Brown, F. Dramis, S. Hastenrath, A.G. Lewkovicz, E.C. Malagnino, S. Neale, F.E. Nelson, D.A. Robinson, P. Skvarca, A.E. Taylor and A. Weidick, 1996. 'The cryosphere: changes and their impacts' in R.T. Watson, M.C. Zinyovera, R.H. Moss and D.J. Dokken (eds.). *Climate Change 1995: The Science of Climate Change*. Contribution of Working Group II to the Second Assessment Report of the Intergovernmental Panel on Climate Change. Cambridhe: Cambridge University Press.

Freeman, M.M.R. (ed.) 2000 *Endangered Peoples of the Arctic* Westport: Greenwood Press

Gray, A. 1995. 'The indigenous movement in Asia' in R.H. Barnes, A. Gray and B. Kingsbury (eds.). *Indigenous Peoples of Asia* Ann Arbor, Michigan: Association for Asian Studies, Inc.

Huntington, H.P. 1992 *Wildlife Management and Subsistence Hunting in Alaska* Seattle: University of Washington Press

Huntington, H.P., S. Fox, I. Krupnik and F. Berkes 2005 'The changing Arctic: indigenous perspectives' in ACIA, *Arctic Climate Impact Assessment* Cambridge: Cambridge University Press

Intergovernmental Panel on Climate Change (IPCC) (2001) *Climate Change 2001: Impacts, Adaptation, and Vulnerability*. Contribution of Working Group II to the Third Assessment Report of the Intergovernmental Panel on Climate Change. Cambridge: Cambridge University Press

Johanessen, O.M., Shalina, E.V. and Miles, M.W. 1999. 'Satellite evidence for an arctic sea ice cover in transformation' *Science* 286: 1937-1939

Krupnik, I. 1993. *Arctic adaptations: native whalers and reindeer herders of northern Eurasia*. University Press of New England, Hanover.

Krupnik, I. 2000. 'Reindeer pastoralism in modern Siberia: research and survival in the time of crash' *Polar Research* 19:49-56.

Krupnik, I. and D. Jolly (eds.) 2002. *The Earth is Faster Now: indigenous observations of Arctic environmental change*. Fairbanks: ARCUS

Lange, M. 2000 'Integrated global change impact assessments' in M. Nuttall and T.V. Callaghan (eds.) *The Arctic: environment, people, policy* Amsterdam: Harwood Academic Publishers

McDonald M., Arragutainaq L. and Z. Novalinga, 1997. *Voices from the Bay*, Ottawa: Canadian Arctic Resources Committee and Municipality of Sanikiluaq.

Maslanik, J.A., Serreze, M.C. and Agnew, T. 1999. 'On the record reduction in 1998 western Arctic sea ice cover' *Geophysical Research Letter* 26(13): 1905

Nunavut Tunngavik 2001 *Elders' Conference on Climate Change: final report* Cambridge Bay, Nunavut: Nunavut Tunngavik Inc.

Nuttall, M. 1992 *Arctic Homeland: kinship, community and development in Northwest Greenland* Toronto: University of Toronto Press

Nuttall, M. 1998 *Protecting the Arctic: indigenous peoples and cultural survival* New York: RoutledgeHarwood

Nuttall, M, F. Berkes, B. Forbes, G. Kofinas, T. Vlassova and G. Wenzel, 2005 'Hunting, herding, fishing and gathering: indigenous peoples and renewable resource use in the Arctic' in ACIA *Arctic Climate Impact Assessment* Cambridge: Cambridge University Press

O'Riordan, T. and A. Jordan 1999 'Institutions, climate change and cultural theory: towards a common analytical framework' *Global Environmental Change* 9: 81-93

Sakshaug, E. and D. Slagstad 1992 'Sea ice and wind: effects on primary productivity in the Barents Sea' *Atmosphere Ocean* 30: 579-591

Scheraga, J. and A. Grambsch 1998 'Risks, opportunities, and adaptation to climate change' *Climate Research* Vol. 10: 85-95

Siegert, M and J. Dowdeswell 2000 'Glaciology' in M. Nuttall and T.V. Callaghan (eds.) *The Arctic: environment, people, policy* Amsterdam: Harwood Academic Publishers

Thorpe, T., S. Eyegeot, N. Hakongak and K. Elders 2002 'Nowadays it is not the Same: Inuit Qaumjimajatuqangit, Climate and Caribou in the Kitikmeot Region of Nunavut, Canada' in I. Krupnik and D. Jolly (ed.s) *The Earth is Faster Now: Indigenous Observations of Arctic Environmental Change.* Fairbanks, ARCUS: 198-239.

Weller, G. 2000 'The weather and climate of the Arctic' in M. Nuttall and T.V. Callaghan (eds.) *The Arctic: environment, people, policy* Amsterdam: Harwood Academic Publishers

Weller, G., and M. Lange, (eds.) 1999. *Impacts of global climate change in the Arctic regions.* Report from a Workshop on the Impacts of Global Climate Change, 25-26 April 1999, Tromsø, Norway. Oslo: International Arctic Science Committee.

Wenzel, G.W. 1991 *Animal Rights, Human Rights: ecology, economy and ideology in the Canadian Arctic.* Toronto: University of Toronto Press.

Young, O.R. and N. Einarsson 2004 'A human development agenda for the Arctic: major findings and emerging issues' in AHDR *Arctic Human Development Report* Akureyri: Stefansson Arctic Institute

Climate change and ecosystem response

3 Climate variation in the European sector of the Arctic: Observations and scenarios

Inger Hanssen-Bauer

Norwegian Meteorological Institute, P.O.Box 43 Blindern, NO-0313 Oslo, Norway, e-mail: i.hanssen-bauer@met.no

3.1 Introduction

Global climate models typically indicate that increased concentrations of atmospheric greenhouse gases will lead to a larger temperature increase at high northern latitudes than anywhere else in the world (Cubasch et al. 2001, Räisänen 2003). On the other hand, a majority of the models indicate an area of minimum temperature response around southern Greenland. Large gradients in warming rates are thus projected in the Arctic.

The mean temperature in the Arctic did increase during the 20th century (e.g. Polyakov et al. 2003), but there are regional differences within the area. This paper is focused on the observed and modeled atmospheric climate in the European sector of the Arctic from 1900 to 2100. Are the observed changes in accordance with results from climate models? And what are the prospects for the future climate?

3.2 Observed climate variability in the European Arctic

The following climatic description is based upon observations from the European Arctic (Figure 3.1). Temperature and precipitation series from the NARP-dataset (http://projects.met.no/~narp/data) are analysed and related to sea-ice cover and atmospheric circulation indices. The climate in the European Arctic is also discussed in an Arctic and a global context.

3.2.1 Temperature

There are large spatial differences in surface air temperature (SAT) in the European sector of the Arctic. Annual mean temperature measured at the weather stations in Figure 3.1 span from -12.3 °C in Danmarkshavn

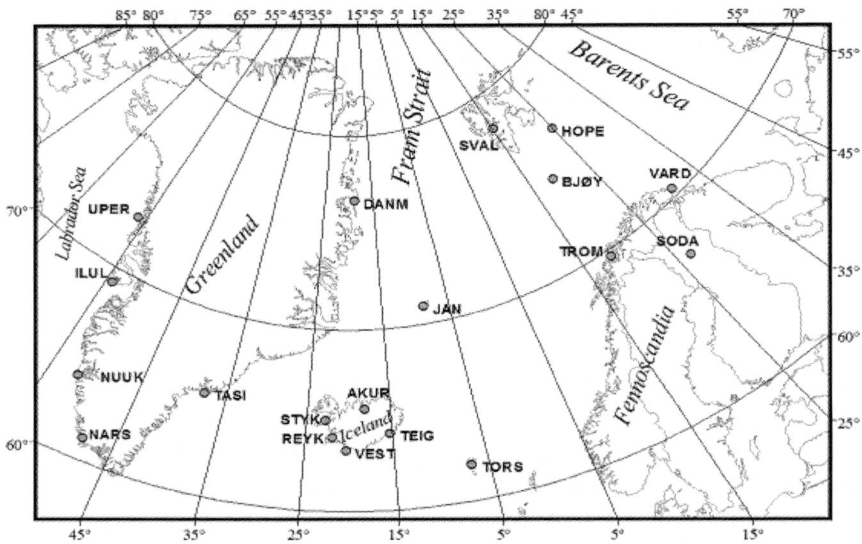

Fig. 3.1. Map showing the climate stations applied in the present analysis. The letter codes give the first letters in names applied in the text.

to +6.5 °C in Torshavn (Førland et al. 2002). Nevertheless, when the temperature series are standardised (by subtracting the 1961-1990 average and dividing by the standard deviation), they show similar long-term variability during the 20th century (Figure 3.2). Two periods with mainly positive temperature trends, the so-called "early 20th century warming" and a period towards the end of the century, are separated by a period with cooling. At most stations in the area, the thermal optimum of the century was experienced in the 1930s. In the north-eastern part of the area, the following period of cooling culminated in the1960s, followed by three decades with a warming trend. In Iceland and the Faroe Islands, there was a more gradual cooling to a minimum around 1980. At the westernmost stations in Greenland, the cooling continued even to around 1990. According to Mann-Kendalls non-parametric test, only series from northern Iceland and from Fennoscandia show a statistically significant warming (5 % level or better) from 1900 to 2002.

The multi-decadal temperature variation in the European Arctic during the 20th century is generally in accordance with the SAT series averaged over the whole Arctic north of 62°N (Polyakov et al. 2003). Also this time series shows a cold period around 1900, a period of warming up to about 1940, a cold spell in the 1960s and warming during the latest decades. The Arctic SAT series, however, shows a statistically significant long-term

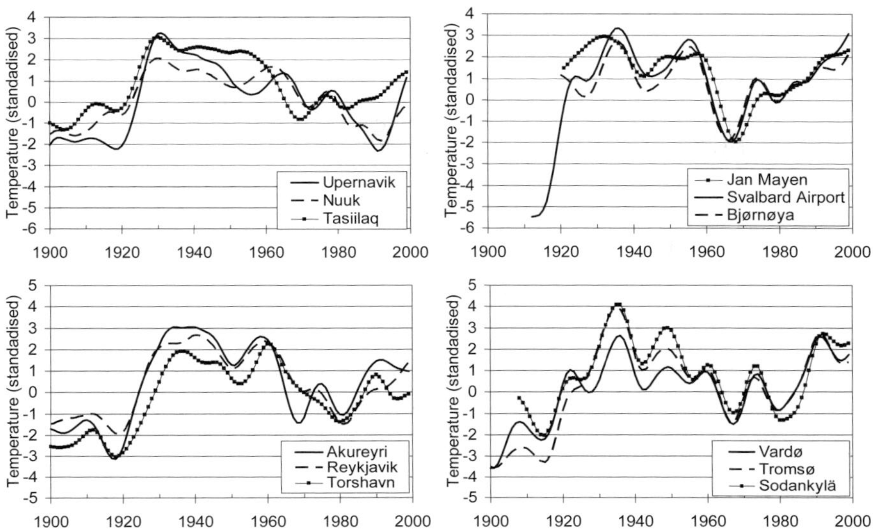

Fig. 3.2. Decadal scale variability in standardised series of annual mean temperature from climate stations in the European sector of the Arctic.

trend of +0.094 °C per decade from 1875 to 2000 (Polyakov et al. 2003).

The global SAT curve also shows one early and one recent period of warming during the 20th century, but there is no statistically significant cooling in the period between them (Folland et al. 2001). Johannessen et al. (2004) show that the early 20th century warming and the following cooling mainly are restricted to high northern latitudes. The recent warming is of more globally homogeneous character, though also this is enhanced in the Arctic (Serreze et al. 2000), and especially over the continents (Moritz et al. 2002). Whether or not the long-term positive temperature trend shows an Arctic amplification depend critically on the exact choice of period and dataset. Polyakov et al. (2003) conclude that there is no significant difference between the Arctic and global SAT trends during the 20th century. From 1875 to 2000 on the other hand, the temperature increase was larger in the Arctic.

3.2.2 Precipitation

The measured annual precipitation in the European Arctic spans from about 200 mm at some of the northernmost stations to more than 1500 mm at some stations in Iceland and the Faroe Islands (Førland et al. 2002). The decadal scale variability is less spatially consistent for precipitation than

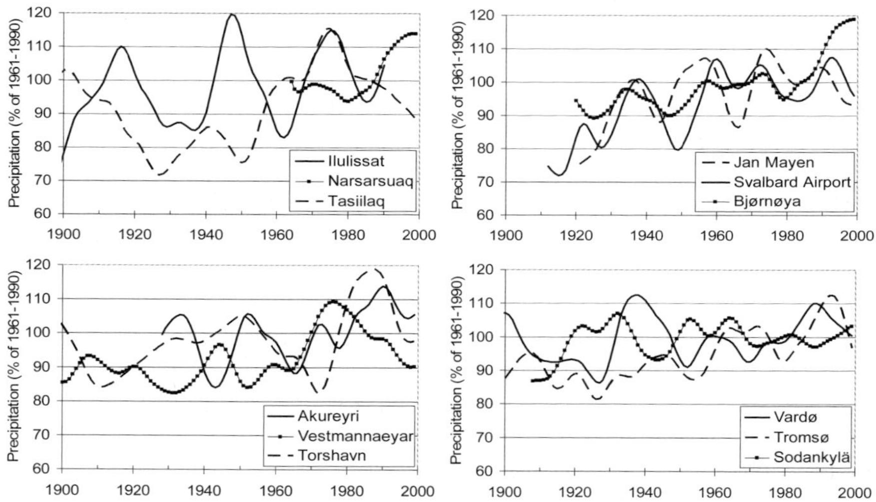

Fig. 3.3. Decadal scale variability in annual precipitation (in %) from climate stations in the European sector of the Arctic.

for temperature (Figure 3.3). However, Førland et al. (2002) conclude that none of the reliable series in the area show negative precipitation trends during the 20th century. According to Mann-Kendall non-parametric test, there are statistically significant positive trends in a belt from eastern parts north in the area to central parts in the south, while there tend to be no significant trend in eastern parts of Fennoscandia and at the westernmost stations. Positive trends of more than 2.0 % per decade were found in Svalbard, Bjørnøya and Tromsø, while trends from 1.5 to 2.0 % per decade were found at Jan Mayen, and parts of Iceland and the Faroe Islands.

The relative precipitation increase in northeastern and central parts of the area is considerably higher than the increase of 0.8 % per decade over the 20th century reported by New et al. (2001) for the 60-80°N land area.

3.2.3 Sea-ice

According to an estimate of Vinje (2001), the April sea-ice extent in the Arctic sector from 30°W to 70°E was reduced by 33 % of its 1864 value from 1864 to 1998. Vinje analysed the ice extent in sub-sectors west and east of 10°E, and found that the larger reduction (40 %) took place in the western sector. The reduction was not linear in time, and according to Vinje (2001) it was larger in the last decades of the 19th century than during the 20th century. Trend studies in the periods 1900-30, 1930-70 and

1970-2000 show that the April sea-ice extent tend to decrease in both sub-sectors during the first and the last decades of the 20th century (periods of warming), while they tend to increase slightly in the period around the middle of the century (period of cooling). All trends are stronger in the western area. In the eastern area, only the long-term trend is statistically significant.

Vinnikov et al. (1999) analysed in situ sea-ice data from the 20th century for the entire Arctic. The dataset is incomplete during the earlier decades, but shows a negative trend since about 1950, mainly caused by a 15 % decrease in summer sea-ice extent. Satellite-based sea-ice observations confirm a noticeable decrease during the last two decades of the 20th century, especially during summer (Johannessen et al. 1999; Cavalieri et al. 1997). However, Parkinson (2000) found that the trends vary by region. Deser et al. (2000) showed that the dominant mode of Northern Hemisphere winter sea-ice variability in the period 1958-1997 is an anti-phase decadal-scale fluctuation between the Labrador Sea and the Greenland/Barents Seas, which may be linked to atmospheric circulation.

3.2.4 Atmospheric circulation vs. temperature and sea-ice

Regional climate varies not only directly in response to variation in local radiative forcing, but also because of variations in the dominating atmospheric (and ocean) circulation patterns, which again may or may not be triggered by external forcing. The North Atlantic Oscillation (NAO, Hurrell 1995, 1996) and the Arctic Oscillation (AO, Thomson & Wallace 1998) winter indices are two correlated indicators of the mode of the northern hemisphere atmospheric circulation. Positive (negative) values of the AO index indicate increased (reduced) advection of cold Arctic air into western Greenland, and of warm humid air towards north-western Europe and the eastern Arctic. Thompson et al. (2000) conclude that a positive AO trend can account for half the observed winter warming over Eurasia from 1968-1997, while Rigor et al. (2000) show that more than half the warming over the eastern Arctic Ocean and the cooling over the Labrador Sea from the 1970s to the 1990s is accounted for by the AO. Also trends in extreme temperatures in western Greenland and Fennoscandia during the last five decades are in accordance with the positive NAO/AO trend (Tuomenvirta et al. 2000), and so is the antiphase fluctuation between the sea-ice conditions in the Labrador and Greenland/Barents Seas during the latest 4 decades (Deser et al. 2000).

Correlation analysis of the present dataset confirms that temperatures in the eastern part of the European Arctic are positively correlated with the

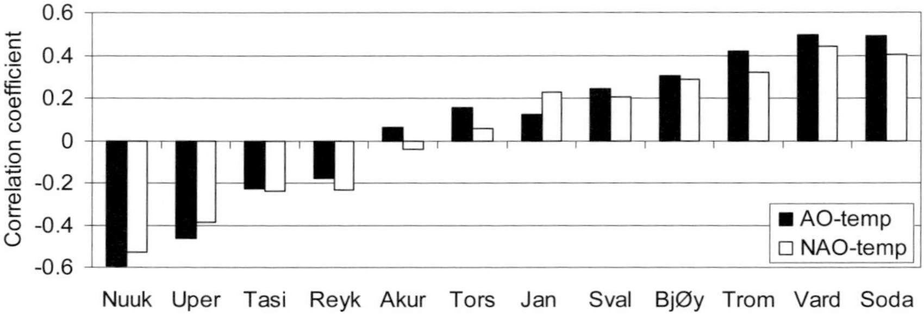

Fig. 3.4. Correlation coefficients between the annual mean temperature at different stations (Fig. 3.1) and the AO-index (black bars) and NAO-index (white bars).

NAO/AO indices, while the tendency is opposite at the westerly stations (Figure 3.4). Still, the AO's ability to account for decadal and multi-decadal variability, is definitely not stationary. While a positive trend in AO is consistent with the warming in Vardø and cooling in Nuuk from the 1960s to the 1990s, there is no positive AO trend which can account for the early 20th century warming (e.g. Moritz et al. 2002).

The early 20th century warming might still be connected to atmospheric circulation variability, as AO explains only a fraction of this variability. Further, the hemispheric circulation variability is highly nonlinear, and the AO index is only a linear approximation (Fyfe 2002).

3.2.5 Possible causes for the observed climate variation

Local and regional climate may change in response to variation in external forcing (natural or antropogenic), or because of internal variations in the climate system. The Arctic warming during the later decades is part of a global warming which is seen also in lower latitudes (Jones et al. 1999, Johannessen et al. 2002). Its signature agrees well with results from climate model runs where increasing concentrations of greenhouse gases and aerosols are included, but not with control simulations (Stott et al. 2000, Johannessen et al. 2002). It is thus concluded that the warming during the later decades most likely are induced mainly by antropogenic forcing. A matter of debate is still that, though several climate models produce a certain positive AO trend in response to increased greenhouse forcing, only one model produce an increase of the magnitude we have seen lately (Moritz et al. 2002).

Climate model experiments including historic variations in natural forcings (solar radiation and volcano activity) show realistic temperature variation during the early 20th century warming (Stott et al. 2000). It may thus be argued that this warming was caused by natural external forcing. Recent modelling studies (Delworth and Knutson 2000, Johannessen et al. 2004) showed, however, that climate models can produce similar features, just as a result of internal variability in the climate system. Bengtsson et al. (2003) argue that the early 20th century warming and the following cooling is an internal low frequency fluctuation involving atmospheric and ocean circulation and their effect on sea-ice extent. There are thus at least 2 different hypotheses concerning the early 20th century warming, but none of them include antropogenic forcing as a main factor.

3.3 Climate scenarios for the European Arctic

Atmosphere-ocean general circulation models (AOGCMs) are used to produce scenarios for future climate variation caused by increased atmospheric concentrations of "greenhouse-gases" and aerosols. Scenarios for future climate in the Arctic produced by different AOGCMs show substantial differences on local and regional scales. Still, climate scenarios at doubled CO_2 concentration from 19 AOGCMs run with the same emission scenario ("CMIP2 runs") show similarities (Räisänen 2003). Sections 3.1 and 3.2 are mainly based upon results from Räisänen (2003). The projected climate changes in these scenarios are defined as the differences between a 20-year period centred when the atmospheric CO_2 concentration is doubled and a 20-year period in the control simulation.

3.3.1 Temperature

Of the 19 CMIP2 models, 18 project a temperature increase in the Arctic area (north of 60°N) which exceeds the global mean warming. While the projected average global warming at doubled CO_2 is 1.75 °C, the Arctic area mean change is 3.4 °C (Räisänen 2003). The largest warming is projected in autumn and winter; the smallest warming in summer. A majority of the models show an area of reduced warming in the northern North Atlantic, which is also reflected in the average of the 19 scenarios (Figure 3.5, left panel). The reduced warming (which is qualitatively in accordance with observations during the latest decades) is mainly due to a projected reduction in the Atlantic meridional overturning circulation (Cubasch et al. 2001). The gradient in projected warming rates from 60 to 90°N is larger in the European sector of the Arctic than in any other sector. The inter-model variability is also considerable in the area, and has a

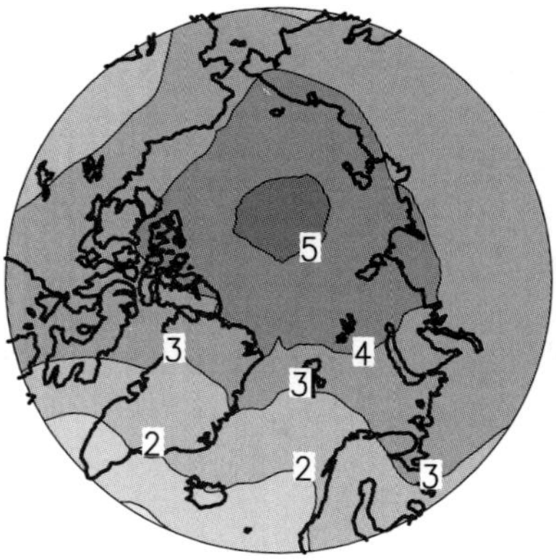

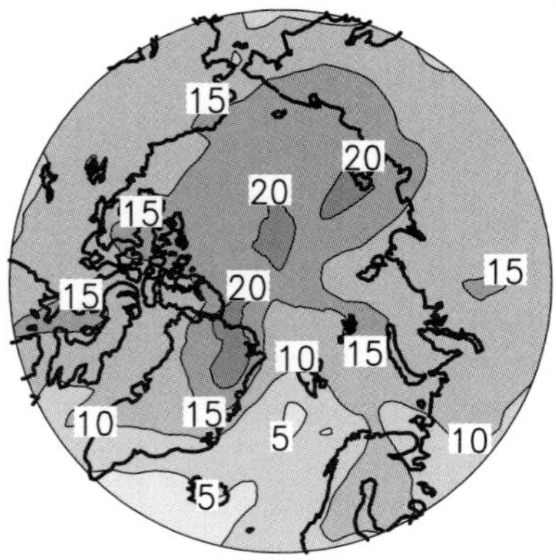

Fig. 3.5. Average from the 19 CMIP2 models of projected climate change at doubled CO_2 concentration north of 60°N (after Räisänen 2003). Left panel: Increased annual mean temperature (°C). Right panel: Increased precipitation (%).

maximum east of Svalbard. Some models show maximum warming and large sea-ice retreat in this area, while others indicate that the area of reduced warming includes the whole Barents Sea.

The global climate models have too coarse spatial resolution to give good regional climate scenarios. Downscaling is required to produce more detailed regional and local climate scenarios. Few such studies have been performed in the Arctic. However, Benestad et al. (2002) showed that local temperature scenarios in the Svalbard area are closely related to the projected changes in the regional sea-ice cover in the AOGCM. If the present sea-ice extent or the future melting in the region is unrealistic, the local temperature scenarios will thus be dubious. Downscaling studies from Fennoscandia (Christensen et al. 2001, Hanssen-Bauer et al. 2003) confirm that we can expect local spatial variation which is not resolved by the global models. The projected warming is larger in the inland than along the coast, especially during winter. Further, at high latitudes larger winter warming rates are projected in valleys than in mountain areas. This differs from projections from the Alps (Beniston et al. 2003) where projected warming rates tend to increase with altitude.

3.3.2 Precipitation

The ratio between projected increase in precipitation (in %) and temperature (in °C) in the Arctic is between 2 and 4 in all the 19 CMIP2 models (Räisänen 2003). On local scale, the scatter between the models concerning precipitation changes is larger than for temperature. A general feature is still that the projected precipitation increase averaged over the area north of 60°N is larger than the global average. Further, the Arctic is one of the few regions in the world where the 19 model average projected precipitation increase exceeds the inter-model standard deviation (Räisänen 2003). Figure 3.5 (right panel) illustrates that the average projected increase in annual precipitation at doubled CO_2 is smaller in the northern North Atlantic (5-10 %) than in other parts of the Arctic (10-20 %). This differs somewhat from the patterns of observed change during the 20th century. The projected increase in precipitation is largest in autumn and winter, and smallest in summer.

3.3.3 Other climate variables

With one exception, the 19 CMIP2 models project substantial reduction in the sea-ice extent. The spatial pattern varies a lot, partly because the simulations of the present-day sea-ice extent vary (e.g. Benestad et al. 2002). Sea-ice simulations from the ECHAM4 climate model reproduce the variability of the 20th century satisfactorily (Johannessen et al. 2004). For the 21st century, under the "old" emission scenario IS92a (IPCC 2001, p 14), the model projects a modest reduction during winter, but a radical (~80%) reduction in the summer ice extent. These results agree well with results from the HadCM3 model (Gregory et al. 2002), under the "new" B2 emission scenario (IPCC 2001, p 14).

Concerning changes in sea level pressure and atmospheric circulation, the models show a large spread in projections. Several (but not all) models project a modest positive trend in the NAO/AO indices. Several models also tend to show an eastward relocation of the Icelandic low. This would affect the southward transportation of cold air in western parts of the European sector as well as the advection of mild humid air into the eastern Arctic in a way that is not necessarily captured by the NAO/AO indices.

3.4 Summary and conclusions

The annual mean precipitation increased significantly in large parts of the European Arctic sector during the 20th century. Statistically significant warming is found in parts of Iceland and in Fennoscandia, while temperature series from other parts of the European Arctic show no significant long-term trend. "The early 20th century warming", which culminated around the 1930s and was followed by a period with cooling, is a dominating feature in all temperature series from the area. This feature may have been triggered by variation in solar radiation and volcano activity, or it may result from internal variability in the climate system. Positive temperature trends during the last decades of the 20th century are found in central and especially in eastern parts of the European Arctic. This recent warming, which is associated with a positive trend in the NAO index, is probably at least partly triggered by antropogenic forcing of the climate system. The sea-ice extent in the European sector of the Arctic has decreased during the 20th century. The April sea-ice extent shows negative trends during the first and last 30-year periods of the century, but a small (insignificant) increase in the period 1930-1970.

Though the global climate models (AOGCMs) certainly have their limitations, they are the best tools we have for projecting future climate. Comparison of results from several models may indicate the level of uncertainty. Projected changes in temperature, sea-ice extent and large-scale precipitation seem to be qualitatively robust. Generally, the AOGCMs project larger warming in the Arctic than the global average. However, a majority of the AOGCMs project an area of minimum warming near southern Greenland. In the European sector of the Arctic, there is thus a considerable gradient in the projected warming rates from 60°N to the high Arctic. As for temperature, the projected increase in precipitation is generally larger in the Arctic than globally averaged. Likewise for precipitation, most models project a smaller response in southwestern parts of the European sector than elsewhere in the Arctic. Projected changes in sea-ice extent for the 21st century show moderate reduction during winter, but large reduction during summer.

References

Benestad R, Førland EJ, Hanssen-Bauer I (2002) Empirically downscaled temperature scenarios for Svalbard. Atmos Science Lett (www.idealibrary.com): doi.10.1006/ asle.2002.0051

Beniston M, Keller F, Goyette S (2003) Snow pack in the Swiss Alps under changing climatic conditions: an empirical approach for climate impacts studies. Theor Appl Climatol 74: 19-31

Bengtsson L, Semenov VA, Johannessen OM (2003) The early century warming in the Arctic – a possible mechanism. Report No 345, Max-Planck-Institut für Meteorologie, Hamburg

Cavalieri DJ, Gloersen P, Parkinson CL, Comiso JC and Zwally HJ (1997) Observed hemispheric asymmetry in global sea ice changes. Science 278: 1104-1106

Christensen JH, Räisänen J, Iversen T, Bjørge D, Christensen OB, Rummukainen M (2001) A synthesis of regional climate change simulations – A Scandinavian perspective. Geophys Res Lett 28: 1003-1006

Cubasch U., Mehl GA, Boer GJ, Stouffer RJ, Dix M, Noda A, Senior CA, Raper S, Yap KS (2001) Projections of future climate change. In: Houghton JT, Ding Y, Griggs DJ, Noguer M, van der Linden PJ, Dai X, Maskell K, Johnson CA (eds) Climate Change 2001: The Scientific Basis. Contribution of Working Group I to the Third Assessment Report of the Intergovernmental Panel on Climate Change. Cambridge University Press, Cambridge, p 525-582

Delworth TL, Knutson TR (2000) Simulation of the early 20th century warming. Science 287: 2246-2250

Deser C, Walsh JE, Timlin MS (2000) Arctic sea ice variability in the context of recent atmospheric circulation trends. J Clim 13: 617-633

Folland CK, Karl TR, Christy JR, Clarke RA, Gruza GV, Jouzel J, Mann ME, Oerlemans J, Salinger MJ, Wang S-W (2001) Observed climate variability and change. In: Houghton JT, Ding Y, Griggs DJ, Noguer M, van der Linden PJ, Dai X, Maskell K, Johnson CA (eds) Climate Change 2001: The Scientific Basis. Contribution of Working Group I to the Third Assessment Report of the Intergovernmental Panel on Climate Change. Cambridge University Press, Cambridge, p 99-181

Fyfe JC (2002) Separating extratropical zonal wind variability and mean change. J Clim 16: 863-874

Førland EJ, Hanssen-Bauer I, Jónsson T, Kern-Hansen C, Nordli PØ, Tveito OE, Laursen EV (2002) Twentieth Century Variations in Temperature and Precipitation in the Nordic Arctic. Polar Record 38: 203-210

Gregory JM, Stott PA, Cresswell DJ, Rayner NA, Gordon C, Sexton DMH (2002) Recent and future changes in Arctic sea ice simulated by the HadCM3 AOGCM. Geophys Res Lett 29: 2175, doi:10.1029/2001GL014575.

Hanssen-Bauer I, Førland EJ, Haugen JE, Tveito OE (2003) Temperature and precipitation scenarios for Norway: Comparison of results from dynamical and empirical downscaling. Clim Res 25: 15-27

Hurrell JW (1995) Decadal trends in the North Atlantic Oscillation: Regional temperatures and precipitation. Science 269: 676-679

Hurrell JW (1996) Influence of variations in extratropical wintertime teleconnections on Northern Hemisphere temperature. Geophys Res Lett 23: 665-668

IPCC (2001) Summary for Policymakers In: Houghton JT, Ding Y, Griggs DJ, Noguer M, van der Linden PJ, Dai X, Maskell K, Johnson CA (eds) Climate Change 2001: The Scientific Basis. Contribution of Working Group I to the Third Assessment Report of the Intergovernmental Panel on Climate Change. Cambridge University Press, Cambridge, p 1-20

Johannessen OM, Shalina EV, Miles MW (1999) Satellite evidence for an Arctic sea ice cover in transformation. Science 286: 1937-1939

Johannessen OM, Bengtsson L, Miles MW, Kuzmina SI, Semenov VA, Alekseev GV, Nagurny AP, Zakharov VF, Bobylev LP, Pettersson LH, Hasselmann K, Cattle HP (2004) Arctic climate change: Observed and modelled temperature and sea ice variability. Tellus 56A: 328-341

Jones P, New M, Parker DE, Martin S, Rigor IG (1999) Surface air temperature and its changes over the past 150 years. Rev in Geophys 37: 173-199

Moritz RE, Blitz CM, Steig EJ (2002) Dynamics of recent climate change in the Arctic. Science 297: 1497-1502

New M, Todd M, Hulme M, Jones P (2001) Precipitation measurements and trends in the twentieth century. Int J Climatol 21: 1899-1922

Parkinson CL (2000) Recent trend reversals in Arctic sea ice extents: Possible connections to the North Atlantic Oscillation. Polar Geography 24: 1-12

Polyakov IV, Bekryaev RV, Alekseev GV, Bhatt U, Colony RL, Johnson MA, Makshtas AP (2003) Variability and trends of air temperature and pressure in the maritime Arctic, 1875-2000. J Clim 16: 2067-2077

Räisänen J (2003) CO2-induced climate change in the Arctic area in the CMIP2 experiments. In: Rummukainen M (ed) SweClim Newsletter 11, Rossby Centre, Stockholm, p 23-28

Rigor IG, Colony RL, Martin S (2000) Variations in surface air temperature observations in the Arctic, 1979-1997. J Clim 15: 2648-2663

Serreze MC, Walsh JE, Chapin III FS, Osterkamp T, Dyurgerov M, Romanovsky V, Oechel WC, Morison J, Zhang T, Barry RG (2000) Observational evidence of recent change in the northern high-latitude environment. Clim Change 46: 159-207

Stott PA, Tett SBF, Jones GS, Allen MR, Mitchell JFB, Jenkins GJ (2000) External control of twentieth century temperature variations by natural and anthropogenic forcings. Science 15: 2133-2137

Thompson DWJ, Wallace JM (1998) The Arctic Oscillation signature in the wintertime geopotential height and temperature fields. Geophys Res Lett 25: 1297-1300

Thompson DWJ, Wallace JM, Hegerl (2000) Annular modes in the extratropical circulation. Part II: Trends. J Clim 13:1018-1036

Tuomenvirta H, Alexandersson H, Drebs A, Frich P, Nordli PØ (2000): Trends in Nordic and Arctic Temperature Extremes and Ranges. J Clim 13: 977-990

Vinnikov KYa, Robock A, Stouffer RJ, Walsh JE, Parkinson CL (1999) Global warming and Northern Hemisphere sea ice extent. Science 286: 1934-1936

Vinje T (2001) Anomalies and Trends of Sea-Ice Extent and Atmospheric Circulation in the Nordic Seas during the Period 1864-1998. J Clim 14: 255-267

4 Impact of climate change on arctic and alpine lakes: Effects on phenology and community dynamics

R. Primicerio[1], G. Rossetti[2], P.-A. Amundsen[1] and A. Klemetsen[1]

[1]University of Tromsø, N-9037 Breivika, Tromsø, Norway, [2]University of Parma, I-43100 Parma, Italy, e-mail: raulp@nfh.uit.no

4.1 Introduction

The ecology of Arctic and Alpine lake communities is heavily influenced by ice and snow, two factors governed by climatic conditions (Rouse et al. 1997). Climate change generates variation in ice and snow cover phenology, thickness and texture that triggers a broad spectrum of physical, chemical and biotic responses (Quayle et al. 2002). Changes in community structure and dynamics caused by these environmental responses are early symptoms of an ecological impact of climate change. The simple character of Arctic and Alpine lake communities, with a small number of species arranged along few trophic levels, facilitates detection of such structural changes and permits detailed studies of dynamics. The sharp influence of climate and simple community structure, combined with the opportunity of tracking past ecological change using paleoecological records (Smol et al. 2005), make Arctic and Alpine lakes suitable ecosystems for an early, reliable assessment of ecological impacts of global climate change.

Global climate change is expected to modify the range and phenology of many terrestrial and aquatic species (Parmesan and Yohe 2003). Projections of the future impact of climate change on biodiversity, based on species-area relationships (Rosenzweig 1995) and the assumption that climate determines present distributions, stress how climate induced range alteration will greatly enhance the risk of extinction for many species (Thomas et al. 2004). Meanwhile, range expansion and changes in phenology consistent with global warming scenarios are being documented for an increasing number of taxa (Root et al. 2003). The above assessments of ecological impact of climate change are mainly based on comparative studies from tropical and temperate regions, the focus of past research efforts.

Evaluation of an impact of recent climate change on Arctic and Alpine communities must rely on evidence that is more sparse or indirect. Information needed for the extrapolation of climate impact scenarios in Arctic and Alpine systems is richer, for it includes data from lower latitudes and altitudes that may be assumed to reproduce relevant aspects of future environmental conditions.

Arctic and Alpine lakes are characterized by an extended period of ice cover, cold water and low production. Warming results in anticipated ice melt and delayed ice formation (Magnuson et al. 2000), and in increased water temperature (Schindler 1997). The resulting prolonged ice-free season, stronger thermal stratification and enhanced nutrient resuspension are expected to increase lake production (Rouse et al. 1997). Further increases in production may derive from indirect effects of warming on external inputs of nutrients and organic carbon (Quayle et al. 2002). The above changes in environmental conditions may also shift the main contribution to lake production from benthic to pelagic (Korhola et al. 2002). Increased temperature and production will facilitate successful invasion and introduction of species adapted to warmer and more productive waters. The ecological implications of the above environmental changes, however, reach beyond range alteration driven by physiological limitations. Within a lake, a longer productive season protracts the active period of community members increasing the scope for ecological interactions and ecosystem engineering. By relaxing the physical and energetic limitations on demographic and interaction parameters, and by modifying the nature and structure of interactions, warming resets the internal machinery that drives community dynamics (Harrington et al. 1999, Schmitz et al. 2003).

The above qualitative generalization is intended as a reference scenario for the following discussion of climatic impact on phenology and community dynamics of plankton, benthos and fish. The scenario invites some general considerations about impact on biodiversity. Medium term projections (year 2100) rank climate change as a main driver of future biodiversity loss in Arctic and Alpine freshwaters (Sala et al. 2000). Many of the cold stenothermal species populating these regions are relictual endemics that are forced into deep, cool waters of lakes in order to survive. Under warming, the deep refuges would gradually become unbearable due to the higher temperatures and lower oxygen concentrations. Loss of even a few endemic species will have a strong impact on regional diversity, considering that high altitudes and latitudes have relatively few species to begin with (Lomolino 2001, Willig et al. 2003). Empirical, hump shaped relations between productivity and species richness in lakes suggest that the expected increase in productivity will be followed by increased species richness (Rosenzweig 1995, Dodson et al. 2000). This does not contradict the expectation of higher exctinction rates given that recipient communities will experience substitutions by exotic species that are expanding their range. The rates of change in species composition and richness may also increase, due to higher propagule pressure. The latter effect is most likely in low latitude Alpine lakes, where the distance between different ecoclimatic regions is short and the species pool at lower altitudes is large.

The scenario of a warmer, wetter climate that will increase ice-free season duration, water temperature and productivity of Alpine and Arctic lakes must be treated with caution. The magnitude and even direction of ecological response to climate change will depend on a variety of environmental conditions that were not addressed. Basin morphometry, hydrology, catchment and geographic position are all likely to affect the character of climatic impact and will be considered in more detail below. The literature reviewed includes assessment and prognostic studies of climatic impact on Arctic and Alpine lakes. As mentioned earlier, the evidence of recent ecological changes reported in these studies is seldom direct. Reviews of assessment studies dealing with climatic impact on lakes mainly refer to data collected in temperate regions (Schindler 1997, Straile et al. 2003). In the absence of direct observations, microfossil accumulations in lake sediments provide compelling indication of climate driven ecological change (Smol et al. 2005). Paleolimnological records may even allow the detection of changes in the intensity of ecological interactions, as in the case of fish predation on the planktonic prey *Daphnia* (Jeppesen et al. 2002).

4.2 Plankton

A comparable environment leads to similarities in the structure and dynamics of plankton inhabiting Arctic and Alpine lakes. In these lakes, light and nutrients are available in measures adequate to support life for only a short period of the year. This productive, ice-free period has a limited temperature excursion, and summer thermal stratification, when present, lasts only a few weeks (Fig. 4.1, upper panels). The result is a relatively brief and environmentally homogeneous productive season with low food production sufficient to sustain only a small number of trophic levels. The limited scope for differentiation along resource and safety axes constrains the diversity of consumers. Accordingly, zooplankton species richness rapidly declines along the upper interval of latitudinal and altitudinal gradients (Patalas 1990, Rautio 1998). The species richness of algae is higher, but there is lower equitability, with a dominance of chrysophyceans and

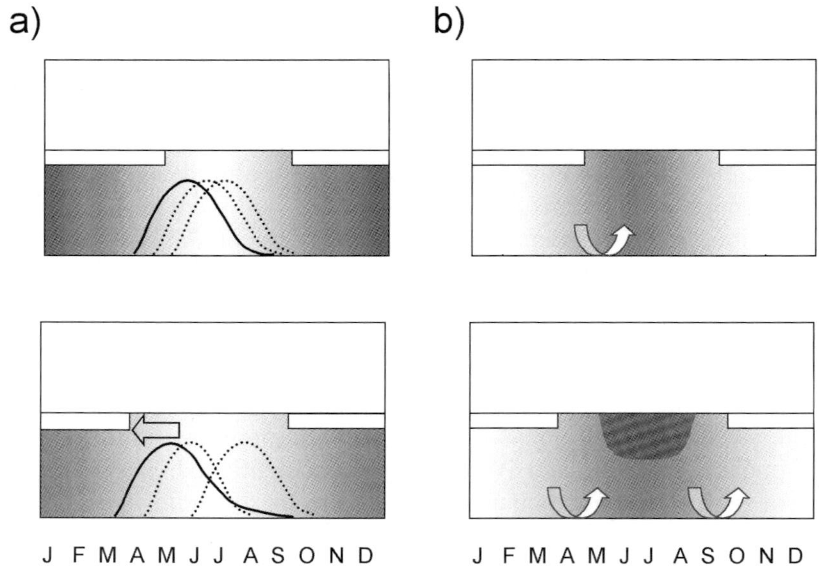

Fig. 4.1. Climate change affects both a) light, and b) temperature regimes in Arctic and Alpine lakes (darker shades depict low light and high temperature, respectively). A warmer climate (lower panels) anticipates ice melt increasing the amount of light available for photosynthesis during spring. The long photoperiod of high latitudes strengthens this effect. The growing season of algae (solid line) and herbivores (dotted lines) is anticipated and prolonged, increasing the scope for temporal segregation of competing species. Increased temperature promotes thermal stratification, boosting productivity in the epilimnion and producing an additional mixing event at the end of summer when temperature drops (bottom arrows). Mixing increases nutrients availability via sediment resuspension.

diatoms, typical of cool, unproductive waters. More surprising are the similarities in taxonomic composition. Some of the dominant planktonic species found in Alpine lakes are also present across the Arctic. The broad biogeographic distributions are obtained via passive dispersal powered by winds and animals (Kristiansen 1996, Bilton et al. 2001, Bohnak and Jenkins 2003). Migratory birds play an important role in this respect (Santamaria and Klaassen 2002), and the observed migration range expansions associated with warming will intensify plankton propagule pressure at the higher latitudes. The planktonic propagules that tolerate long-range dispersal are adaptive, dormant stages that have evolved to survive harsh conditions within lakes (Hairston 1998).

4.2.1 Phenology

Many planktonic species overwinter in a dormant state (Rengefors et al. 1998, Williams-Howze 1997, Brendonck and DeMeester 2003). The phenology of the onset and end of dormancy is bound to the timing of ice formation and melt by environmental changes that function as triggering cues for the planktonic organism. The environmental cues include physical stimuli connected with light and temperature regime shifts, and chemical stimuli related to biotic conditions (Rengefors et al. 1998, Brendonck and DeMeester 2003). Ice melt is accompanied by changes in light and temperature conditions that operate in concert to terminate winter dormancy. Earlier ice-melt, as brought about by warming (Fig. 4.1, lower panels), anticipates hatching and activation of dormant stages. The temporal relation between ice melt and the emergence of plankton is documented by variation in the timing of algal blooms and zooplankton outbursts in long-term series under variable interannual weather (Straile et al. 2003), and along altitudinal and latitudinal gradients (Reynolds 1989). The phenology of dormancy expected on the basis of the abiotic regime shifts can be modified by chemical cues associated with predation risk and crowding that inhibit emergence or induce dormancy (Hansson 2000, Lass and Spaak 2003). Control mechanisms triggered by chemicals, documented for both winter and summer dormancy (Rengefors et al. 1998, Williams-Howze 1997), imply that indirect effects of climate on timing and magnitude of predation risk and population peaks, discussed below, will contribute to further modify the phenology of dormancy.

Warming will also affect breeding phenology and the timing of ontogenetic stage shifts, with implications for population growth rates and stage-structured interactions. The dramatic impact of climate on breeding phenology and growth is best illustrated by copepods, which have complex and relatively long life cycles (Santer 1998). Copepods time reproduction with specific seasonal events to minimize the risk of predation and offspring starvation (DeMott 1989). In cyclopoids, hatching typically occurs during the algal bloom, minimizing the risk of starvation of the herbivorous larvae. But life cycles are flexible, as shown by *Cyclops scutifer*, a species that can vary life cycle duration from three years to one year depending on temperature and food conditions (Williams-Howze 1997). Increased temperature and food availability accelerate growth and reduce the need for summer dormancy, promoting short generation time (Maier 1989, Williams-Howze 1997). Calanoids often breed before the spring algal bloom, and their earliest larval stages rely on the rich egg reserves to survive and grow under the ice. The widespread genus *Eudiaptomus*

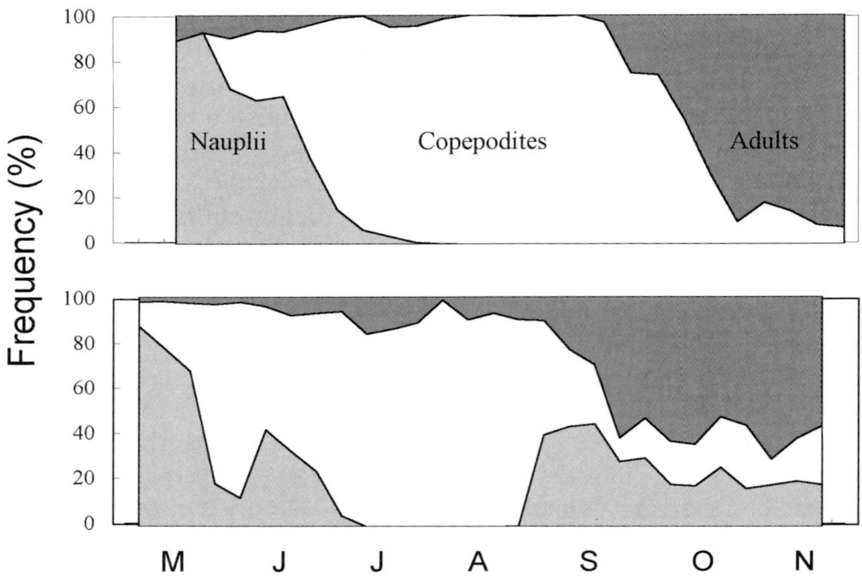

Fig. 4.2. The calanoid *Eudiaptomus intermedius* has typically a yearly life cycle in the subalpine lake Scuro (1527 m a.s.l., Italy), with nauplii being born in spring, under the ice (upper panel - 1992). But an early ice-free season, with relatively high water temperature, shortens the life cycle of the calanoid, allowing it to anticipate reproduction, and to double the number of generations per year (lower panel - 1993).

illustrates how such life history strategies may be drastically modified in response to warming (Fig. 4.2).

4.2.2 Community dynamics

Plankton communities are particularly sensitive to spring warming because the main dynamic events take place during the productive, ice-free season, when light is readily available (Sommer et al. 1986, Straile et al. 2003). After ice melt, egg banks and dormant stages seed the population and community dynamics in the pelagic (Rengefors 1998, Hairston et al. 2000, Brendonck and DeMeester 2003). Spring recruitment from dormant stages is particularly important in Arctic and Alpine lakes, as shown by demographic (Edmondson 1955, DeStasio 1990) and genetic studies of natural populations (Rossi et al. 1998). This is due to a limited scope for direct recruitment given the low density of individuals actively overwintering. After activation, population growth is constrained by long generation time

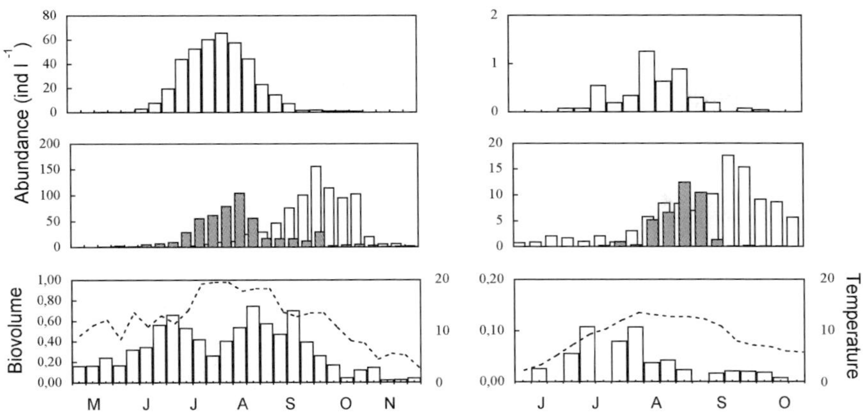

Fig. 4.3. Seasonal dynamics of the rotifer predator *Asplanchna priodonta* (top panels) and of its competing prey (middle panels), *Keratella cochlearis* and *Conochilus unicornis* (gray bars), in the subalpine lake Scuro (left panels) and subarctic lake Takvatn (right panels). Lower panels, stippled line, temperature (°C) at 1 m depth, bars, algae biovolume (mm3 l-1). The longer ice-free season of lake Scuro allows a greater temporal segregation among competing prey. In lake Takvatn, the high temporal overlap of competitors is permitted by their segregation along the water column, with the vulnerable, but competitively superior *Keratella* hiding in deep water layers.

and, in zooplankton, by relatively small clutch sizes. Even slight increases in temperature and production early in the ice-free season can burst population growth reducing the time from activation to peak density. The timing, duration and magnitude of the spring phytoplankton bloom are subject to climate forcing mediated by ice phenology, mixing, water temperature, and thermal stratification, which affect the light, nutrient and temperature environment experienced by the algae (Reynolds 1989). Climatic events also affect zooplankton grazing on algae (Scheffer et al. 2001). An early, warm productive season favours the early build up of zooplankton numbers resulting in a rapid decline of algae abundance (Straile et al. 2003). Prolonged and intensified herbivory, and sharper epipelagic gradients of nutrient availability due to increased thermal stratification, generate clearer patterns of succession in phytoplankton, with seasonal shifts to species with lower nutrient requirements or greater resistance to grazing (Sterner 1989, Reynolds 1997). Zooplankton may also experience new opportunities for temporal segregation under climate warming, as illustrated by the seasonal dynamics of rotifers (Fig. 4.3). In size structured populations (e.g. *Daphnia*), the number of synchronized generations between activation and dormancy, limited by ice phenology and water temperature, determines the seasonal change in size distribution and, consequently, in size dependent interactions (Primicerio 2003).

4.3 Benthos

Benthos is a key component of Arctic and Alpine lake food webs being responsible for a large portion of total production (Korhola et al. 2002). The composition and structure of benthic communities is affected by structural characteristics of the substrate on, or in, which they live (Moss 1998). This condition distinguishes the ecology of plankton and benthos and adds new sources of heterogeneity in the environment experienced by benthic organisms. In particular, variation in substrate steadiness and texture, two properties that affect the opportunity for attachment and hiding, promotes diversity. Stones, pebbles, sand, and lime establish a gradient of stability and interstitial grain often associated with increasing depth (Moss 1998). In the littoral zone, the character and heterogeneity of substrate depends also on the presence and composition of macrovegetation, used by other organisms to attach, hide and feed. The abundance and diversity of macrophytes decreases with increasing altitude and latitude; warming is expected to favour their expansion. Several benthic organisms have dispersal stages, such as adults of chironomids or resting propagules of algae and crustaceans (Kristiansen 1996, Bilton et al. 2001), that will ensure their rapid range expansion under favourable conditions promoted by warming.

4.3.1 Phenology

Timing of emergence of dormant benthic organisms is associated with physico-chemical regime shifts following ice melt. Data on the relation between phenology of emergence and interannual variation in ice phenology and water temperature are less rich for benthos than for plankton (Straile et al. 2003), but the available information suggests similar conclusions. Warming will anticipate emergence, prolonging the active season of benthic organisms. Breeding phenology, and seasonality and duration of life cycles will also change as a consequence of increased temperature and production. Long-term studies of aquatic insects illustrate how climatic events can modify the phenology of stage transition, as shown by Elliott (1996) for adult emergence of alder-flies living in the littoral zone of lake Windermere. Another example of climate driven change in timing of stage transition is provided by larvae of chironomids, which are often the numerically dominant component of benthic macroinvertebrates. Larvae of chironomids are likely to pupate earlier in warmer waters. Pupation is associated with a massive migration towards surface during summer, an event with important energetic implications for fish predators.

4.3.2 Community dynamics

In the recent past (~150 years), compositional shifts of algae in Arctic and Alpine lakes have occurred concomitantly with warming events (Douglas et al. 1994, Lotter and Bigler 2000). These taxonomic changes, traced in the paleolimnological record, can be explained by climate change impact (Sorvari et al. 2002). Similar changes in benthic algae are expected to occur in the near future as a consequence of temperature increases of similar magnitude (about 2 °C – Sorvari et al. 2002). In particular, diatoms, which play a central ecological role via their strong contribution to lake primary production, are sensitive to climate forcing and are likely to switch from a dominance of benthic species (e.g. members of *Fragilaria*) to mainly pelagic forms (e.g. species of *Cyclotella*). The change is expected due to the prolonged ice-free period and increased thermal stratification favouring small, lightly silicified planktonic diatoms that shade deep dwelling, benthic forms. Such changes from benthic to pelagic primary production may affect benthic consumers, although increased sedimentation may provide alternative food sources. The expected changes in phenology discussed above will influence demography and interactions among benthic species, but may also reverberate across the entire lake food web, as exemplified by the effects of chironomid larvae pupation. By diverting the attention of fish, chironomid pupation temporarily releases from predation other vulnerable prey, indirectly influencing their population dynamics. In turn, fish affect the structure of chironomid communities via predation (Mousavi et al. 2002).

4.4 Fish

Freshwater fish need water corridors to colonize new aquatic environments. They have therefore greater dispersal limitations than many planktonic and benthic species. As a result, the geographical range of a fish species maintains the historical pattern of colonization (Rosenfield 2002), unless this has been altered by human introductions. Introductions are common for recreational fish like salmonids, the dominant taxon in Arctic and Alpine lakes. Fish species dominance and composition changes with latitude and altitude, suggesting an influence of climate on the outcome of ecological interactions. Trophic interactions vary with season and fish ontogeny, although different fish species are not equally flexible in their habitat and diet use (Persson et al 1997). High flexibility characterizes Arctic charr, *Salvelinus alpinus*, the freshwater fish with the northernmost range limits, common in Arctic and Alpine lakes (Maitland 1995, Klemetsen et al. 2003). A distribution restricted to cool waters, as observed in Arctic charr, suggests vulnerability to global warming due to physiological stress (Lehtonen 1998). However, thermal stress need not be the driving factor behind cool water distributions. In Arctic charr, for example, the upper critical limits and optimum temperatures are relatively high (Lyytikäinen et al. 1997, Thyrel et al. 1999), and vary little with latitude (Elliott and Klemetsen 2002). Also, cold stenothermal species may be able to adapt to warmer temperatures, as suggested for lake trout, *Salvelinus namaycush* (Sellers et al. 1998). Thus, thermal stress may pose serious threats only under rapid change and extreme water temperatures, as could be experienced in shallow lakes. Changes in timing of life history events and seasonal resource availability, and new regimes of biotic interactions are likely to play a more decisive role in mediating the impact of climate change on fish (Davis et al. 1998).

4.4.1 Phenology

Seasonal changes in prey availability modify diet and habitat choice by fish (Persson et al 1997). After ice melt, higher zooplankton abundance increases the degree of planktivory and use of the pelagic habitat. In salmonids, this seasonal dietary shift is sharp (Klemetsen et al. 2003). Earlier zooplankton availability anticipates and prolongs the period of planktivory, as seen along latitudinal and elevational gradients (Fig. 4.4). Seasonal shifts in diet are not restricted to increased planktivory after ice melt, but include changes to insectivory, benthivory, and piscivory at different times of the year (Klemetsen et al. 2003). The timing of such foraging events is subject to indirect climatic influences via regulation of prey phenology. In addition to prey availability, the seasonal succession in diet composition depends on fish size and taxonomic identity. Higher specialization of given fish stages or taxa reduces the number and extent of their seasonal changes in diet. Because of the strong seasonality and low levels of prey availability, high latitude and altitude lakes are unfavourable environments for specialists, whereas flexible species like Arctic charr are well adapted to such conditions.

Warming will favour specialists, such as planktivorous coregonids, by prolonging and increasing the availability of their preferred prey group. In

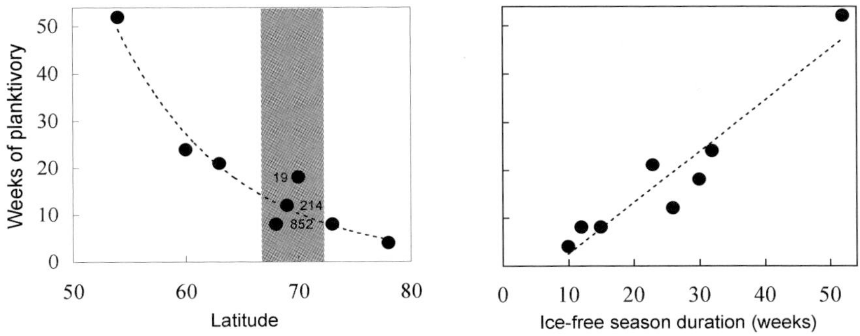

Fig. 4.4. Duration of planktivory in Arctic charr populations decreases with increasing latitude and altitude (left panel – labels within gray band mark altitude in m a.s.l. for lakes along elevation gradient) and is positively related to the duration of the ice-free season (right panel). The threshold for planktivory is set at 20 % in frequency of occurrence or stomach contents volume. Data are from lakes located in Western Europe, Bear Island and Spitzbergen. See Table 4.1 for a summary of lakes physiography. Data obtained from: Klemetsen & Grotnes (1975, 1980); Frost (1977); Hindar & Jonsson (1982); Klemetsen et al. (1985); Amundsen (1989); Christensen (1994); Forseth et al. (1994); and Elliott & Baroudy (1995).

most fish species, newly born individuals are specialist planktivores, highly vulnerable to seasonal changes in prey availability (Wootton 1999). Emergence must therefore be timed with zooplankton peak availability to avoid starvation after exhaustion of yolk reserves. Fish that are not able to synchronize emergence with earlier zooplankton outbursts driven by climate warming will risk extinction (McDonald et al. 1996). Fish that anticipate emergence will experience a prolonged, warmer and more productive growth season. Prolonged and accelerated growth allows earlier access to alternative prey not readily available to small juveniles due to size constraints.

4.4.2 Community dynamics

Indirect effects of warming, such as increased plankton production, generate new opportunities for resource specialization and growth of fish. The effects of increased plankton production on fish diet and growth are illustrated by nutrient enrichment studies at high latitudes. One such study from Norway (69°N) registered an increase in plankton contribution to Arctic charr stomach contents from less than 40% to more than 80% of the bio-volume, with *Daphnia* becoming the most important prey. Increased

Table 4.1. Physiography of lakes used to compare seasonal duration of planktivory in Arctic charr along gradients of latitude and altitude (Fig. 4.4). Area is lake area in km^2, Zmax is maximum depth in m, Lat is latitude in °N, Alt is altitude in m a.s.l., Open is the duration of the ice-free period in weeks, Tmax is maximum temperature in °C. Data obtained from: Klemetsen & Grotnes (1975, 1980); Frost (1977); Bøyum & Kjensmo (1978); Hindar & Jonsson (1982); Klemetsen et al. (1985); Amundsen (1989); Christensen (1994); and Elliott & Baroudy (1995).

Lake	Area	Z_{max}	Lat	Alt	Open	T_{max}
Windermere (UK)	15	64	54	39	52	24
Vangsvatn (Norway)	8	60	60	46	32	20
Høysjøen (Norway)	1	27	63	222	23	na
Takvatn (Norway)	15	80	69	214	26	14
Ellasjøen (Bear Island)	1	43	73	21	12	7
Linnévatn (Spitsbergen)	5	37	78	4	10	3
Gautelisvatn (Norway)	6	40	68	852	15	12
Skogsfjord-vatn (Norway)	14	80	70	19	30	15

availability of large, easy to catch zooplankton produced a dramatic acceleration in fish somatic growth within three years since the beginning of the nutrient enrichment program (Dahl-Hansen et al. 2001). Enhanced growth rates influence the size distribution of the whole fish population, affecting size dependent interactions. Similar effects driven by climate change are expected to modify the outcome of interactions among fish species. Competition between Arctic charr and whitefish (*Coregonus lavaretus*) illustrates the mechanisms involved. The whitefish is often regarded as the competitively superior species (Svärdson 1976, Lehtonen 1998), but ranking seems to depend on environmental conditions influenced by climate. A recurrent pattern is Arctic charr substituting whitefish, and other species, as latitude and elevation increase (Svärdson 1976, Heino 2001). These changes in composition and dominance are likely to be mediated by climatic effects on duration and productivity of the ice-free season. Greater availability of zooplanktonic prey, experienced at lower altitude and latitude, and promoted by warming, favours the coregonids, which are more efficient planktivores. Intensive, long-term studies of succesfull invasions of subarctic freshwaters by planktivorous fish species (e.g. Bøhn and Amundsen 2001) document the ecological mechanisms that may promote colonization of Arctic and Alpine lakes by exotics under warming (Shuter and Post 1990). The accelerated spread of non-native aquatic organisms (Lodge 1993), boosted by global warming, will result in a drastic change in the diversity and structure of Arctic and Alpine fish communities.

4.5 Higher-order effects of climate change and lake communities

Increased plankton production exemplifies a higher order effect of climate change that may reach the top of the lake food web altering fish assemblages. In turn, changes in fish composition, relative abundance, and size distribution will have top-down effects on plankton and benthos. These top-down, higher order effects of climate change are likely to generate some of the main structural alterations in lake communities (Schmitz et al. 2003). General predictions concerning the direction of such higher order effects are not easy due to the variation generated by the interplay between taxonomic composition, adaptation, and degree of environmental heterogeneity. This is illustrated by how size-structured interactions, anti-predator adaptation, and lake morphometry may modulate climate impact. The effect of predation by fish has a strong size-dependent component (Persson et al. 1997). In the absence of fish, invertebrates become the main top predators in Arctic and Alpine lakes. The top-down influence of invertebrates also depends on prey size, but the character of size dependence differs from that of fish (Hall et al 1976). Different configurations of size-structured interactions in absence vs. presence of fish will produce different ecological responses to climate change.

The indirect influence of top-predators on abundance and composition of planktonic and benthic species is mediated by changes in prey density and phenotypic traits. Inducible anti-predator adaptations responsible for trait-mediated interactions in plankton and benthos are predator specific (Tollrian and Harvell 1999). The importance of indirect interactions mediated by adaptive traits, such as risk sensitive foraging behaviour, is likely to increase with increasing productivity brought about by warming (Luttbeg et al. 2003). In lakes, a common anti-predator response that structures prey communities is the adaptive use of deep habitat refuges (Fig. 4.3 - Primicerio 2005). Shallow lakes lack deep habitat refuges, a simple consideration that stresses the influence of lake morphometry, and associated degree of environmental heterogeneity, on the effects of climatic variation.

Clearly the mechanisms by which climate change may affect lake communities, either directly or via an interplay with other sources of environmental stress (Schindler 2001), are sufficiently complicated to beg for a carefully designed, experimental research program with regional implementation (Harrington 1999). For Arctic and Alpine lakes, this research program is presently under way and will soon provide the information needed for more quantitative prediction.

4.6 Conclusions

Available data from extensive surveys, paleolimnological records, and long-term studies of natural and whole lake experiments confirm the sensitivity to climate change of these vulnerable ecosystems. Under the warming scenario, production is expected to increase and the seasonal dynamics of plankton and benthos should change, in accordance with observed alterations associated with temporal variation in temperature regimes. Plankton and benthos may also face compositional and structural changes resulting from range expansion of exotic species and local changes in nature and intensity of ecological interactions. When present, fish can play a central role in driving such structural changes by modifying the phenology and ontogeny of habitat and dietary shifts, and the character of resource specialization. In turn, the above phenotypic responses will affect fish population and community structure. Phenology of emergence and other frequency-dependent adaptations affected by climate generate nonlinear relations among interacting populations that amplify the effects of small change and influence the impact of environmental variation on population and community dynamics (Chesson 2003). The complex causal structure leading to higher order effects of climate change challenges assessments and predictions of ecological impact and has prompted carefully planned experimental investigations that will soon allow more quantitative projections.

References

Adrian R (1998) Effects of ice duration on the phyto- and zooplankton succession in a shallow lake. EcoSys Beiträge zur Ökosystemforschung BD 7: 43-49

Amundsen P-A (1989) Effects of intensive fishing on food consumption and growth of stunted Arctic charr (*Salvelinus alpinus* L.) in Takvatn, northern Norway. Physiology and Ecology Japan, Special Volume 1: 265-278

Bilton DT, Freeland JR, Okamura B (2001) Dispersal in aquatic invertebrates. Ann. Rev Ecol Sys 32: 159-181

Bohnak AJ and Jenkins DJ (2003) Ecological and evolutionary significance of dispersal by freshwater invertebrates. Ecol Lett 6: 783-796

Brendonck L and DeMeester L (2003) Egg banks in freshwater zooplankton: evolutionary and ecological archives in the sediment. Hydrobiologia 491: 65-84

Bøhn T and P-A Amundsen (2001) The competitive edge of an invading specialist. Ecology 82: 2150-2163

Bøyum A and Kjensmo J (1978) Physiography of Linnévatn, Western Spitsbergen. Verh Internat Verein Limnol 20: 609-614

Chesson P (2003) Understanding the role of environmental variation in population and community dynamics. Theor Pop Biol 64: 253-254

Christensen GN (1994) Selective predation by pelagic Arctic charr, *Salvelinus alpinus* (L.), in the oligotrophic lake Skogsfjordvatnet (Northern Norway). Master thesis, University of Tromsø, 69 p (in Norwegian)

Dahl-Hansen GAP, Berge D, Brettum P, Lindstrøm EA, Mjelde , Nygaard KA, Solheim L, Aanes KJ (2001) LIMNOCULT 1996-99. Nutrient enrichment of an ultra-oligotrophic watercourse. The results after three years of nutrient enrichment. APN report 534.1662 (in Norwegian)

Davis AJ, Jenkinson LS, Lawton JH, Shorrocks B, Wood S (1998) Making mistakes when predicting shifts in species range in response to global warming. Nature 391: 783-786

DeMott WR (1989) The role of competition in zooplankton succession. In: Sommer U (ed) Plankton ecology. Succession in plankton communities. Brock-Springer, Madison pp 195-252

DeStasio BT Jr (1990) The role of dormancy and emergence patterns in the dynamics of a freshwater zooplankton community. Limnol Oceanogr 35: 1079-1090

Dodson SI, Arnott SE and Cottingham K (2000) The relationship in lake communities between primary productivity and species richness. Ecology 81: 2662-2679

Douglas MSV, Smol JP, Blake W Jr (1994) Marked post-18th century environmental change in High-Arctic ecosystems. Science 266: 416-419

Edmondson WT (1955) The seasonal life history of *Daphnia* in an Arctic lake. Ecology 36: 439-455

Elliott JM (1996) Temperature-related fluctuations in the timing of emergence and pupation of Windermere alder-flies over 30 years. Ecol Entomol 21: 241-247

Elliott JM and Baroudy E (1995) The ecology of Arctic charr, *Salvelinus alpinus*, and brown trout, *Salmo trutta*, in Windermere (northwest England). Nordic J Freshwater Res 71: 33-48

Elliott JM and Klemetsen A (2002) The upper critical thermal limits for alevins of Arctic charr from a Norwegian lake north of the Arctic circle. J Fish Biol 60: 1338-1341

Forseth T, Ugedal O, Jonsson B (1994) The energy budget, niche shift, reproduction and growth in a population of Arctic charr, *Salvelinus alpinus*. J Animal Ecol 63: 116-126

Frost WE (1977) The food of charr, *Salvelinus willughbii* (Günther), in Widermere. J Fish Biol 11: 531-547
George DG (2000) The impact of regional-scale changes in the weather on the long-term dynamics of *Eudiaptomus* and *Daphnia* in Estwaithe Water, Cumbria. Freshwater Biol 45: 111-121
Gerten D and Adrian R (2002) Effects of climate warming, North Atlantic Oscillation, and El Nino Southern Oscillation on thermal conditions and plankton dynamics in Northern hemispheric lakes. The Scientific World Journal 2: 586-606
Hairston NG Jr (1998) Time travelers: what's timely in diapause research? Archiv Hydrobiol Spec Issues Adv Limnol 52: 1-15
Hairston NG Jr, Hansen A-M, Schaffner WR (2000) The effect of diapause emergence on the seasonal dynamics of a zooplankton assemblage. Freshwater Biol 45: 133-145
Hall DJ, Threlkeld ST, Burns CW, Crowley PH (1976) The size-efficiency hypothesis and the size structure of zooplankton communities. Ann Rev Ecol Sys 7: 177-208
Hansson L-A (2000) Synergistic effects of food chain dynamics and induced behavioral responses in aquatic ecosystems. Ecology 81: 842-851
Harrington R, Woiwod I, Sparks T (1999) Climate change and trophic interactions TREE 14: 146-150
Heino J (2001) Regional gradient analysis of freshwater biota: do similar biogeographic patterns exist among multiple taxonomic groups? J Biogeography 28: 69-76
Hindar K and Jonsson B (1982) Habitat and food segregation of dwarf and normal Arctic charr (*Salvelinus alpinus*) from Vangsvatnet lake, western Norway. Can J Fish Aquat Sci 39: 1030-1045
Jeppesen E, Jensen JP, Amsinck S, Landkildehus F, Lauridsen T, Mitchell SF (2002) Reconstructing the historical changes in *Daphnia* mean size and planktivorous fish abundance in lakes from the size of *Daphnia* ephippia in the sediment. J Paleolimnology 27: 133-143
Klemetsen A and Grotnes PE (1975) Food and habitat segregation of two sympatric Arctic char populations. Verh Internat Verein Limnol 19: 2521-2528
Klemetsen A and Grotnes PE (1980) Coexistence and immigration of two sympatric Arctic charr. In: Balon EK (ed) Charrs, Salmonid fishes of the genus *Salvelinus*. The Hague, Junk, pp. 757-763
Klemetsen A, Amundsen P-A, Dempson B, Jonsson B, Jonsson N, O'Connell MF, Mortensen E (2003) Atlantic salmon *Salmo salar* L., brown trout *Salmo trutta* L. and Arctic charr *Salvelinus alpinus* (L.): a review of aspects of their life histories. Ecol Freshw Fish 12: 1-59
Korhola A, Sorvari S, Rautio M, Appleby PG, Dearing JA, Hu Y, Rose N, Lami A, Cameron NG (2002) A multi-proxy analysis of climate impacts on the recent development of subarctic Lake Saanajarvi in Finnish Lapland. J. Paleolimnology 28: 59-77
Kristiansen J (1996) Dispersal of freshwater algae - a review. Hydrobiologia 336: 151-157

Lass S and Spaak P (2003) Chemically induced anti-predator defences in plankton: a review. Hydrobiologia 491: 221-239

Lehtonen H (1998) Does global warming threat the existence of Arctic charr, *Salvelinus alpinus* (Salmonidae), in northern Finland? Ital J Zool 65 (Suppl): 471-474

Lodge DM (1993) Species invasions and deletions: community effects and response to climatic change and human activity, Clearwater Lake, Saskatchewan. Geol Surv Can Bull No 534, pp. 125-138

Lomolino MV (2001) Elevation gradients of species-density: historical and prospective views. Global Ecology & Biogeography 10: 3-13

Lotter AF and Bigler C (2000) Do diatoms in the Swiss Alps reflect the length of ice-cover? Aquatic Sciences 62: 125-141

Luttbeg B, Rowe L, Mangel M (2003) Prey state and experimental design affect relative size of trait- and density-mediated indirect effects. Ecology 84: 1140-1150

Lyytikäinen T, Koskela J, Rissanen I (1997) Thermal resistance and upper lethal temperatures of underyearling Lake Inare Arctic charr. J Fish Biol 51: 515-525

Magnuson JJ et al (2000) Historical trends in lake and river ice cover in the Northern Hemisphere. Science 289: 1743-1746

Maier G (1989) The effect of temperature on the development times of eggs, naupliar and copepodite stages of five species of cyclopoid copepods. Hydrobiologia 184: 79-88

Maitland P (1995) World status and conservation of the Arctic charr *Salvelinus alpinus* (L.). Nordic J Freshwater Res 71: 113-127

McDonald ME, Hershey AE, Miller MC (1996) Global warming impacts on lake trout in arctic lakes. Limnol Oceanogr 41: 1102-1108

Moss B (1998) Ecology of freshwaters. Man and medium, past to future. 3rd ed. Blackwell, Oxford 557 p

Mousavi SK, Sandring S, Amundsen P-A (2002) Diversity of chironomid assemblages in contrasting subarctic lakes - impact of fish predation and lakes size. Arch Hydrobiol 154: 461-484

Parmesan C and Yohe G (2003) A globally coherent fingerprint of climate change impacts across natural systems. Nature 421:37-42

Patalas K (1990) Diversity of the zooplankton communities in Canadian lakes as a function of climate. Verh Internat Verein Limnol 24: 360-368

Persson L, Diehl S, Eklov P, Christensen B (1997) Flexibility in fish behaviour: consequences at the population and community levels. In: Godin J-G (ed) Behavioural ecology of teleost fishes. Oxford University Press, Oxford pp 316-343

Primicerio R (2003) Size dependent habitat choice in *Daphnia galeata* Sars and size-structured interactions among zooplankton in a subarctic lake (lake Lombola, Norway). Aquatic Ecol 37: 107-122

Primicerio R (2005) Predator mediated coexistence in zooplankton: the importance of behavioural mechanisms. Verh Internat Verein Limnol 29: 158-162

Quayle WC, Peck LS, Peat H, Ellis-Evans JC, Harrigan PR (2002) Extreme responses to climate change in Antarctic lakes. Science 295: 645

Rautio M (1998) Community structure of crustacean zooplankton in subarctic ponds – effects of altitude and physical heterogeneity. Ecography 21: 327-335

Rengefors K (1998) Seasonal succession of dinoflagellates coupled to the benthic cyst dynamics in lake Erken, Sweden. Archiv Hydrobiol Spec Issues Adv Limnol 52: 123-141

Rengefors K (1998) Algal cyst dormancy: a temporal escape from herbivory. Proc R Soc Lond B 265: 1353-1358

Reynolds CS (1989) Physical determinants of phytoplankton succession. In: Sommer U (ed) Plankton ecology. Succession in plankton communities. Brock-Springer, Madison pp 9-56

Reynolds CS (1997) Vegetation processes in the pelagic: a model for ecosystem theory. Excellence in Ecology, vol. 9. Ecology Institute, Oldendorf, Germany 371 p

Root TL, Price JT, Hall KR, Schneider SH, Rosenzweig C, Pounds JA (2003) Fingerprints of global warming on wild animals and plants. Nature 421: 57-60

Rosenzweig ML (1995) Species diversity in space and time. Cambridge University Press 436 p

Rossi V, Rossetti G, Benatti M, Menozzi P, Ferrari I (1998) Ephippial eggs and dynamics of the clonal structure of *Daphnia longispina* (Crustacea: Cladocera) in a mountain lake (Lago Scuro Parmense, Northern Italy). Archiv Hydrobiol Spec Issues Adv Limnol 52: 195-206

Rouse WR et al (1997) Effects of climate change on the freshwaters of arctic and subarctic North America. Hydrol Process 11: 873-902

Sala OE et al (2000) Global biodiversity scenarios for the year 2100. Science 287: 1770-1774

Santamaria L and Klaassen M (2002) Waterbird-mediated dispersal of aquatic organisms: an introduction. Acta Oecologica 23: 115-119

Santer B (1998) Life cycle strategies of free-living copepods in fresh waters. J Marine Systems 15: 327-336

Scheffer M, Straile D, van Nes EH, Hosper H (2001) Climate warming causes regime shifts in lake food webs. Limnol Oceanogr 46: 1780-1783

Schindler DW (1997) Widespread effects of climatic warming on freshwater ecosystems in North America. Hydrol Process 11: 1043-1067

Schindler DW (2001) The cumulative effects of climate warming and other human stresses on Canadian freshwaters in the new millenium. Can J Fish Aquat Sci 58: 1-12

Schmitz OJ, Post E, Burns CE, Johnston KM (2003) Ecosystem responses to global climate change: moving beyond color mapping. BioScience 53: 1199-1205

Sellers TJ, Parker BR, Schindler DW, Tonn WM (1998) The pelagic distribution of lake trout (*Salvelinus namaycush*) in small Canadian Shield lakes with respect to temperature, dissolved oxygen, and light. Can J Fish Aquat Sci 55: 170-179

Shuter BJ and Post JR (1990) Climate, population viability, and the zoogeography of temperate fishes. Trans Am Fish Soc 119: 314-336

Smol JP et al. (2005) Climate-driven regime shifts in the biological communities of arctic lakes. PNAS 102: 4397-4402

Sommer U, Gliwicz ZM, Lampert W, Duncan A (1986) The PEG-model of seasonal succession of planktonic events in fresh waters. Archiv Hydrobiol 106: 433-471

Sorvari S, Korhola A, Thompson R (2002) Lake diatom response to recent Arctic warming in Finnish Lapland. Global Change Biol 8: 171-181

Sterner RW (1989) The role of grazers in phytoplankton succession. In: Sommer U (ed) Plankton ecology. Succession in plankton communities. Brock-Springer, Madison pp 107-170

Straile D, Livingstone DM, Weyhenmeyer GA, George DG (2003) The response of freshwater ecosystems to climate variability associated with the North Atlantic oscillation. Geophysical monograph 134: 263-279

Svärdson G (1976) Interspecific population dominance in fish communities of Scandinavian lakes. Rep Inst Freshw Res, Drottningholm 55: 144-171

Thomas CD et al. (2004) Extinction risk from climate change. Nature 427: 145-148

Thyrel M, Berglund I, Larsson S, Näslund I (1999) Upper thermal limits for feeding and growth of 0+Arctic charr. J Fish Biol 55: 199-210

Tollrian R and Harvell CD (1999) The ecology and evolution of inducible defenses. Princeton University Press, Princeton, 383 p

Williams-Howze J (1997) Dormancy in the free-living copepod orders Cyclopoida, Calanoida, and Harpacticoida. Oceanogr Marine Biol Ann Rev 35: 257-321

Willig MR, Kaufman DM, Stevens RD (2003) Latitudinal gradients of biodiversity: pattern, process, scale and synthesis. Ann Rev Ecol Evol Syst 34: 273-309

Wootton RJ (1999) The ecology of teleost fishes. 2nd ed. Kluwer Academic Publishers, London, 392 p

5 Changes in growing season in Fennoscandia 1982-1999

Kjell Arild Høgda[1], Stein Rune Karlsen[1], and Hans Tømmervik[2]

[1]Norut IT, P.O. Box 6434, N-9294 Tromsø, Norway, [2]NINA, Polar Environmental Centre, N-9296 Tromsø, Norway, e-mail: kjell-arild.hogda@itek.norut.no

5.1 Introduction

Phenology is the science dealing with the influence of climate on the recurrence of annual phenomena of animal and plant life such as budburst and leaf fall etc. Changes in the phenological events are early indications of effects of changes in the climate. The distribution of plant species and vegetation types tends to be controlled by the climate (Moen 1999, Woodward 1987). Changes in the timing of phenological events are often the first indications of transitions of ecosystems, as e.g. the rise of the tree line and transitions from tundra to boreal forest. Accordingly, it is of interest to monitor the timing of important phenological events like the start of spring and autumn.

There is now ample evidence of the ecological impacts of recent climate change, from polar terrestrial to tropical marine environments (Walther et al. 2002). The responses of both flora and fauna span an array of ecosystems and organizational hierarchies, from the species to the community levels. Although we are only at an early stage in the projected trends of global warming, ecological responses to recent climate change are already clearly visible.

Traditional phenological studies in Europe indicate that the average annual growing season has lengthened during the past decades in western and central Europe (Menzel & Fabian 1999, Chmielewski and Rötzer 2002, Sparks and Menzel 2002, Ahas et al. 2002). Most of this lengthening of the growing season is due to an earlier start of spring, while the effect of a later end of the growing season is less. The earlier onset of spring is most likely due to increased winter/spring temperature. However, there are indications of regional differences in Europe (Ahas et al. 2002). A delay trend in onset of birch-pollen seasons in northern Finland for the last two decades has been reported (Emberlin et al. 2002). From Kola Peninsula in north-western Russia, Kozlov and Berlina (2002) showed a decline in length of growing season for the last 70-years. However, we should be aware that a delay in spring is not necessarily a result of colder climate in northern areas, it could be the result of more snow in the winter, and thereby later snowmelt. We should also be aware of that the phenological response to changes in climate would change with the altitude, latitude and longitude gradients.

Use of satellite data is a way to examine these geographical differences in climatic change and corresponding phenological responses. Several studies show the possibilities to combine Normalized Difference Vegetation Index images (NDVI) derived form satellite images with phenological ground registrations (i.e. Reed et al. 1994, Schwartz and Reed 1999, Chen et al. 2001). Reed et al. (1994) estimated twelve metrics linked to phenological events based on NDVI data from National Oceanic and Atmospheric Administration Advanced Very High Resolution Radiometer (NOAA AVHRR) data. These were compared to certain agricultural crops, grasslands, shrub-lands and forests for the United States. Chen et al. (2001) used phenological data from the Chinese Yearbooks of Animal and Plant Phenological Observations from the period 1982-1993 and compared these with NDVI data estimated form NOAA AVHRR images. Using threshold NDVI values, they found a high correlation between the end of the growing season, but a low correlation for the beginning of the growing season. Schwartz and Reed (1999) found a good agreement between NOAA AVHRR NDVI data and modelled phenology based on phenological registrations across eastern United States. However, in Europe, and in particular in Fennoscandia, these types of comparative studies have not been published.

On a global scale, the use of satellite data support the general trend against earlier onset of spring and prolonged growing season in most of Eurasia towards the end of the 20th century (Myneni et al. 1997, Zhou et al. 2001, Tucker et al. 2001, Bogaert et al. 2002, Shabanov et al. 2002). Tucker et al. (2001) analyzed a time series from 1982 to 1999 of NDVI data from the NOAA AVHRR series of satellites. They found that there where significant variations in photosynthetic activity and growing season length at northern latitudes above 35°N. They especially noticed that two distinct periods of increasing plant growth where apparent, 1982 to 1991 and 1992 to 1999. The decline from 1991 to 1992 where explained by global cooling caused by the eruption of Mt. Pinatubo in 1991. Tucker et al. (2001) found that average May to September NDVI values from 45°N to 75°N increased by 9% from 1982 to 1991, decreased 5% from 1991 to 1992 and again increased 8% from 1992 to 1999. In the same time periods, the start of the growing season varied with –5.6, +3.9 and –1.7 days, respectively.

Zhou et al. (2001) estimated that about 61% of the total vegetated area between 40ºN and 70ºN in Eurasia shows a persistent increase in the growing season from central Europe through Siberia to the Aldan plateau. The increase in the length of the growing season in the period 1981-1999 was estimated to be in the range 18-24 days, depending on a NDVI threshold value set to define the growing season. In Zhou et al. (2002), the coastal and northern Fennoscandia was determined too sparsely vegetated to be included. In North America they found a fragmented pattern of change in smaller areas, mostly in the forests of southeast and grasslands of the upper Midwest. They also found a larger increase in growing season NDVI magnitude (12% versus 8%) and a longer active growing season (18 versus 12 days) caused by an earlier spring and delayed autumn in Eurasia relative to North America. They observed NDVI decrease in parts of Alaska, boreal Canada and northeastern Asia. They explained this decrease by temperature-induced drought caused by pronounced warming without a concurrent increase in rainfall.

We used the NOAA AVHRR NDVI satellite dataset, produced by the Global Inventory Modeling and Mapping Studies (GIMMS) group, to investigate regional climate change impact on the length of the growing season in Fennoscandia during the last two decades. The study area is characterized by large regional climatic differences from south to north, from west to east, and from lowland to mountains. The aim of this study is to reveal the regional pattern of changes in start of spring and autumn. In order to achieve our aim, we developed a method using individual threshold NDVI values for each pixel for defining the start and end of the growing season. The individual pixel specific threshold value is tuned to give the highest correlation with selected phenological, climate and pollen in-situ data for the period 1982-1999.

The estimated onset of spring is shown to be well correlated with ground data on onset of leafing of birch. The results show high regional differences in trends in the onset of spring. In the southern part of Fennoscandia, and on the oceanic west coast of Norway, the spring starts considerably earlier in the late nineties compared to the early eighties. This earlier trend fits with the pattern from western and central Europe, and is likely to be related to increased spring temperature. On the other hand, the results indicate no or even a slightly delayed trend in onset of spring in the mountain areas in southern Norway and in the continental parts of northern Fennoscandia. At the same time the autumn is delayed in the whole area except an area in the continental part of northern Fennoscandia. This also means that the growing season is prolonged for the whole area, except the northern continental parts of Fennoscandia.

5.2 Data and Methods

5.2.1 Satellite Data

We have used the NOAA AVHRR NDVI satellite dataset produced by the Global Inventory Monitoring and Modelling Studies (GIMMS) group for the period January 1982 to December 1999, with 8x8 km^2 spatial resolution (Tucker et al. 2001, Zhou et al. 2001, Bogart et al. 2002). The GIMMS dataset is composed of the maximum value NDVI for 15-day periods. The highest NDVI value for each composite is chosen as the NDVI for the compositing period as it corresponds to lower aerosol and cloud contamination. This procedure ensures the highest possible degree of data coverage. It is also sufficiently accurate for detecting phenology change, as absolute NDVI values throughout the growing season can be compared, and not only the date for specific events.

The processing for the particular dataset used in this work include improved navigation (Rosborough, 1994), calibration of the four different sensors by the method of Vermote and Kaufman (1995), and corrections for sensor degradation and partial atmospheric correction of the data (Los 1998). The GIMMS data from April 1982 to December 1984 and from June 1991 to December 1993 are corrected to remove the effects of stratospheric aerosols from El Chichon and Mt. Pinatubo eruptions applying the method of Vermote and El Saleous (1994). No explicit atmospheric correction except for stratospheric aerosols is applied to this GIMMS NDVI dataset (Zhou et al. 2001). However, the 15-days maximum value NDVI composition minimizes residual atmospheric and bidirectional effects.

5.2.2 Ground Data

In this study phenological data on birch (*Betula* spp.) collected at four stations are used (see Table 5.1). Due to the lack of phenological stations in southern parts of Scandinavia, the dataset is supplemented by pollen data from birch in this area. Birch has a strong correlation between budburst and appearance of first pollen. However, the Scandinavian countries have different methods for defining the onset of the pollen season. In Sweden

(Stockholm) the first of five continuous days with pollen are used as the onset criteria. In Denmark (Copenhagen) the start of the pollen season is defined when the accumulated level reaches 2.5% of the seasonal level. In Norway (Tromsø, Trondheim, and Oslo) the criteria is equal to Denmark, but the accumulation level is defined at 5%. In this study we use the first day of recorded pollen as a start of the onset period in Norway. However, this date could be influenced by long-distance pollen transport, and in Tromsø in 1984 and 1995 the second day of recorded pollen is used due to obvious long-distance pollen transport at the first recorded pollen. In addition to the biological data, air temperature and snow cover data from Karesuando in northern Sweden and Karlstad in southern Sweden are used as reference data in this study. The time of a 21 days moving average mean temperature 0°C and 5°C crossing at spring and autumn, and snow melt at spring is compared with the NDVI value.

5.2.3 Growing Season Analysis

A common method is to define e.g. the start of the growing-season as the point in time when the NDVI value exceeds a certain threshold (Zhou et al. 2001). This method works well in areas with a rich and homogenous vegetation cover. However, in areas, like the cost of Norway where each pixel most likely will represent a mixture of water, farmland, forest, and alpine vegetation, this method does not work. We have therefore used a method to set the threshold individually for each pixel. Approaches where the thresholds are estimated individually for each pixel, a geographic cell, or land cover types can be found in Reed et al. (1994), Chen et al. (2001), and Tucker et al. (2001). This is an objective method that works well over Fennoscandia.

For each pixel an 18 years mean NDVI value (NDVI > 0) for the period from 1982 to 1999 was computed. The timing for the 15 days composite NDVI value upward and downward passing of this mean value was used as a threshold for determining the start and end of the growing season at each pixel for each year. This threshold value was found to best correlate with ground data. The procedure ensures sensitivity for changes in timing during the 18 years period, independent of the actual ground cover present at each pixel. A standard linear regression model was applied to the timing data. The difference of the start and the end of the model was used as an estimate of changes in timing of the growing season. Correlation analysis was carried out between NDVI based events and in-situ observed dates for metrics such as "leaf onset" and start of pollen season.

5.3 Results

5.3.1 Onset of Growing Season

Table 5.1 shows the relationship between the timing of spring measured from the NDVI images and the ground based data. Difference is the mean difference in number of days from the ground based and the NDVI based onset/end of growing season, meaning that negative values indicate that the ground measurements were before the NDVI based date. The 0 °C and 5 °C are crossing upward (spring) and downward (autumn), and is based on 21 days moving average. Positions of the stations are shown in Fig. 5.2.

For the onset of spring, data from the four phenological stations and the five pollen stations show a correlation with the NDVI based measurements in the range of r=0.58 to r=0.86, with a mean correlation of r=0.68. The correlations were significant (< 0.05) or highly significant (< 0.01) (Table 5.1) for the two of the phenological stations and all the pollen stations. While the timing of onset of leafing of birch is tuned to have almost exactly the same date as onset of spring measured from NDVI, the onset of birch pollen season is about 10 days earlier (Fig 5.1a).

Table 5.1. Correlation between onset of growing season measured from NDVI and as defined from birch-pollen, phenology on birch, or climatic data. *Correlation significant (< 0.05), **correlation highly significant (< 0.01). Trend measured in (day year-1) and difference measured in (days).

Station	Type of data	Period	r	Trend	Diff. ground– sat.
Svanhovd	Onset of leafing	1994-99	0.63	1.9	-1
Abisko	Onset of leafing	1982-99	0.58*	0.7	1
Kevo	Onset of leafing	1982-99	0.86**	0.1	0
Kilpisjärvi	Onset of leafing	1989-99	0.60	1.0	1
Tromsø	Pollen	1984-99	0.72**	0.9	-14
Trondheim	Pollen	1982-99	0.59*	-0.4	-15
Oslo	Pollen	1984-99	0.85**	-0.6	-10
Stockholm	Pollen	1982-99	0.61**	-0.5	-10
Copenhagen	Pollen	1982-99	0.64**	-0.7	-12
Karesuando	Last day snow	1982-98	0.39	0.2	-4
Karesuando	0°C	1982-98	0.48	0.2	-29
Karesuando	5°C	1982-98	0.36	0.3	-2
Karlstad	Last day snow	1982-97	-0.19	-0.8	-54
Karlstad	5°C	1982-98	0.24	-0.1	-40

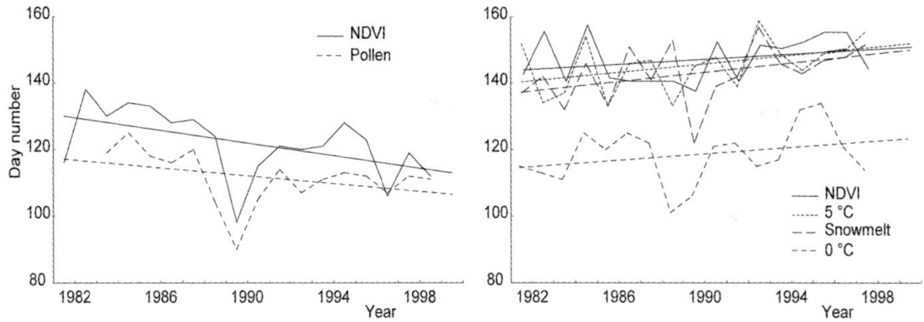

Fig. 5.1ab. a) Data from Oslo. Date of first recorded birch pollen and onset of spring measured from NDVI. b) Data from Karesuando in northern Sweden. Onset of spring measured from temperature data and from NDVI. Temperature data are 21 days smoothed average.

According to climatic data at spring, in Karesuando (Fig. 5.1b) the NDVI values show good fit with the temperature passing 5 °C and staying above (2 days mean difference). However, in southern Sweden (Karlstad), the mean difference is 40 days.

In Fig. 5.2 the changes in time of onset of spring are illustrated. In general, the results show a pattern according to the north-south, the oceanic-continental, and the altitude gradient. In the southern part of Fennoscandia the spring starts considerably earlier. The most significant change is within southern Sweden, with changes up to one month. In the most oceanic parts of Fennoscandia, the coastline of western Norway, the spring starts 2-3 weeks earlier. A slightly delay trend is observed in the alpine and in the northern continental parts.

In general the ground data support the trends in onset of spring as measured from NDVI. All the data from the southern ground stations in this study have a trend of earlier onset of spring, and all the corresponding NDVI values have a similar trend. On the other hand, ground data from the northern parts show a delayed trend. Using linear regression, the onset of birch pollen season in Tromsø in northern Norway, show a delay trend on 0.9 day year^{-1}, and the corresponding NDVI values show a delay trend of 0.8 day year^{-1}. In the year 1990 in Tromsø the onset of birch-pollen season was fourteen days later than the normal for the period 1984 to 1999.

When we relate the different trends in north and south to changes in climate, the climatic station in southern Sweden has a trend of considerably earlier disappearing of the snow cover. In 1998 there was no snow cover at all. During the period 1982 to 1998 annual mean temperature has increased 1.8 °C, March mean temperature has increased 1.8 °C

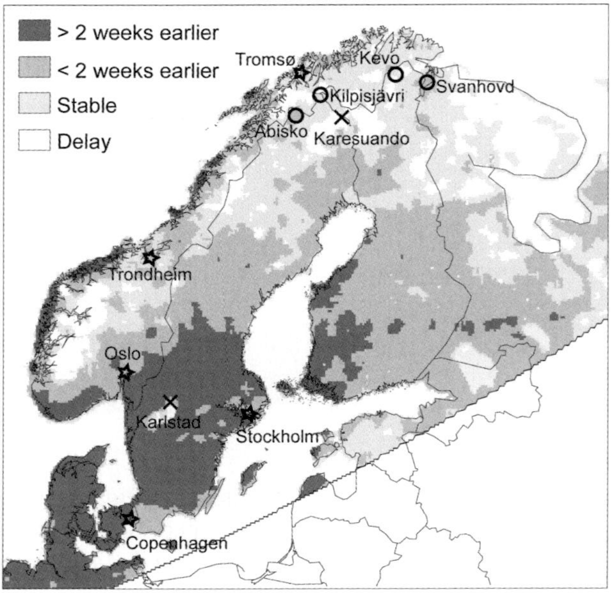

Fig. 5.2. Change in onset of spring during the period 1982 to 1999. Position of the pollen stations (star), phenological observation points (circle), and climatic stations (cross) used in this study.

(from -0.6 to 1.2 °C), and April mean temperature has increased 0.9 °C (from 3.7 to 4.6 °C).

Concerning the climatic station in Karesuando in northern Sweden, this station has a trend towards warmer yearly average, but colder springs. We estimate that during the period 1982 to 1998 annual mean temperature has increased with 1.5 °C. April mean temperature has decreased with 1.4 °C (from –2.5 to –3.9 °C) and May mean temperature has decreased with 2.4 °C (from 4.9 to 2.5 °C).

5.3.2 End of Growing Season

As in-situ data from autumn, data from two phenological stations and one climatic station were used, all located in the northern part of Fennoscandia (Fig. 5.2 and Table 5.2). The six-year series of data on more than 50 % yellowing of birch leaves from Svanhovd in north-easternmost Norway show a correlation on 0.50 with the NDVI based measurements. However, a 12 years series from Kilpisjärvi, in northernmost Finland, of similar ground data show low correlation with NDVI values. Concerning climatic

Table 5.2. Correlation between end of growing season measured from NDVI and as defined from phenology on birch. *Correlation significant (< 0.05), **correlation higly significant (< 0.01). Trend measured in (day year^{-1}) and difference measured in (days).

Station	Type of data	Period	r	Trend	Diff. ground– sat.
Svanhovd	Yellowing >50 %	1994-99	0.50	-2.9	-11
Kilpisjärvi	Yellowing >50 %	1988-99	0.15	-0.2	3
Karesuando	5°C	1982-98	0.60	0.1	-32
Karesuando	0°C	1982-98	-0.03	0.1	-8

data, there is a correlation of 0.60 between passing 5 °C at autumn and NDVI values. However, the climatic event happens more than a month before in mean.

The map of changes in onset of autumn (Fig. 5.3) indicates that the autumn is delayed 1-3 weeks in the whole area. There is no clear pattern, but in the continental parts in the north, in the border area between Norway, Sweden, and Finland, the situation is stable. This area partly overlaps with the area that has a delay trend in onset of spring (Fig. 5.2).

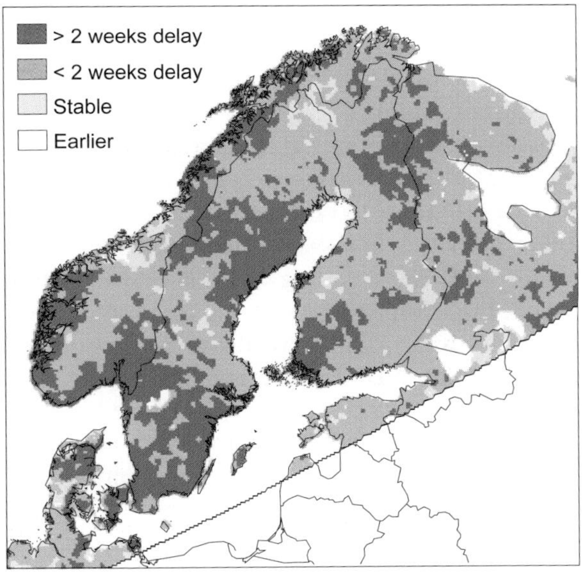

Fig. 5.3. Change in onset of autumn during the period 1982 to 1999.

5.4 Discussion

Due to the high relationship between NDVI and photosynthesis activity, and the method with a threshold value for each individual pixel, it is likely we measure real variations in nature. At spring, the correlation between phenological data on birch (bud burst and pollen) and NDVI values (interpreted as start of growing) are high and significant. Birch *(Betula pubescens)* is common in most of Fennoscandia and dominating in large parts of the oceanic north (Hulten 1971, Påhlsson 1998). In areas where birch is frequent but do not dominate, it is reasonable that the NDVI based measurements shows the timing of the phenophases of the photosynthetic dominating species that correspond with the timing of the nearby onset of leafing of birch. However, in mountain areas and farming land, where birch not even occurs, a calibration of the NDVI images with surface observations remains to be investigated.

In this study, less attention was given the autumn phenophases, and the correlations with the few observation points are weak and not significant. A botanical description of the changes on the ground in relation to the measured NDVI values at autumn remain to be investigated. For autumn, and for spring in areas where birch not occur, the nearest biological description of the phenological events is a threshold passing of the mean photosynthesis activity from all the plants in a 8x8 km^2 large area, upward passing in spring and downward in autumn.

There are several possible error sources in the measurements. Extended changes in forestry management, in the grazing pressure by reindeer and sheep, in long-transported air pollution (e.g. nitrogen in precipitation), in agricultural practice, or other activities that causes large changes in land cover, would affect the results. Due to an ending of subsidies in 1994 extended reduction in cleaning of birch and willows on the clear cut areas has occurred in Sweden (Anonymous 2003). This may affect the proportion of deciduous forest and the thereby the NDVI value and the timing of the growing season. Denmark and large parts of southwestern Sweden are totally dominated by agriculture. Extended changes in crop type, for instance from spring wheat to autumn wheat, with earlier green up in spring would be misinterpret as measurements of earlier spring. Possible influence of changing land cover remains to be investigated. However, due to the large pixel size of 8x8 km^2, and since in most cases the agricultural areas shows the same trend as the nearby natural vegetation not heavily influenced by human impact, we believe the main pattern is due to real changes in onset of spring and autumn. This is even believed to be the case in southernmost Fennoscandia much more influenced of human impact than in the north.

The NDVI based spring trend observed in southern Sweden is statistically significant on the 5% level. More than two weeks earlier onset of spring occurs in Denmark, southern Sweden, southernmost Norway, and south-westernmost Finland. These areas belong to the Nemoral and Boreonemoral zone, and are characterized by broad-leaved deciduous woodland that requires warm summers (Moen 1999). The oceanic coast of west and central Norway also have considerably earlier onset of spring. Climatically the oceanic area is characterized by having few frosty days, high frequency of precipitation, and long growing season (Moen 1999). In both these areas, warmer winters and warmer spring often leads to precipitation as rain, and the snow cover is disappearing either totally or earlier in the spring. The climatic station in Karlstad is situated in the middle of southern Fennoscandia with indications of one-month earlier onset of spring. The climate station has an increase of mean May and April temperature and significant earlier disappearing of the snow cover consistent with an earlier spring/end of winter. The pattern of earlier onset of spring in southern parts during the last decades fits with phenological studies in western and central Europe, and has been linked to increased early spring temperature (Chmielewski and Rötzer 2002, Menzel and Fabian 1999, Ahas et al. 2002, Menzel 2002).

In the continental part of northern Fennoscandia and mountains areas in southern Norway, the spring is stable or there are indications of a slightly delayed trend (Fig. 5.2). However, the delay trend in Fig. 5.2 is not statistically significant. On the other hand the phenological observation data (bud burst and pollen) showed a significant correlation with the NDVI-based phenological metrics. Climatically the continental north is characterized by short growing season and very cold winters with mean January temperature of less than -15 °C (Moen 1999, Tuhkanen 1980, Tveito et al. 2000). In such areas increased precipitation at winter will lead to thicker snow cover, and could result in delay in snow melt, even if the temperature increase. The delayed onset of spring in Karesuando in northernmost Sweden is correlated with colder April and May and delayed temperature passing of 0 °C and 5 °C and more snow in May. The few studies in the north within the stable/slightly delayed areas somewhat support our results. During the period 1930 to 1998, at one phenological station on Kola Peninsula, Kozlov and Berlina (2002) show a slightly delayed trend in snow melt, and a stable situation for onset of leafing on birch. The leaf unrolling showed highest correlation with temperature sum from March to May ($r=0.52$, $p< 0.001$). In Kevo in northern Finland (Fig. 5.2), Emberline et al. (2002) have shown distinct trends towards later onset of birch pollen seasons during 1982-1999, and in the same period the March temperature has increased and May temperature decreased. However, the delayed trend in the continental north could be a result of a more complicated pattern of changes in both temperature and precipitation, and the full explanation remains unknown.

For autumn, there is much less information available. The phenological gardens throughout Europe indicate that the beginning of autumn is gradually delayed (Menzel and Fabian 1999, Menzel 2000). Our results indicate a delay in most areas. Kozlov and Berlina (2002) found indications of earlier onset of autumn phases, however, they investigated the last 70 years, and in this study we only investigate 18 years. On the other hand they used the metric "beginning of leaf fall in mountain birch", which is not a good phenological metric to detect using earth observation. The "beginning of leaf fall" metric is very dependent of the wind regime and a little local gale can wipe out the leaves on the observation site leaving the surrounding areas untouched. By using GIMMS data with a spatial resolution of 8x8 km^2, it is very difficult to detect such local events. We applied the metric, "yellowing of leaves more than 50 %", which fit better with NDVI measurements done at Svanhovd, although we will consider other metrics or a combination of several metrics to detect and monitor the autumn in the future. Experimentation with several different threshold values applied on each individual pixel for the autumn season is also necessary to perform in order to obtain a better fit between earth observations and ground data.

5.5 Conclusion

This study has shown large regional differences in the changes of the length of the growing season, during the period 1982-1999 in Fennoscandia. It clearly indicates that the impact on length of growing season of global warming will be different depending on the regional climate. Whether the measured changes in spring and autumn events are caused by climate change or are of a temporal and stochastic nature remains to be seen. This study also demonstrates that NOAA AVHRR data, with a method using a different threshold value applied on each individual pixel, are well suited for studying the regional impact of a changing climate even in heterogeneous and sparsely vegetated areas.

Acknowledgements

We thank Dr. Compton Tucker at NASA Goddard and Prof. Ranga Myneni at Boston University for providing us with the GIMMS dataset.

We thank Professor E. Kubin at Finnish Forest Research Institute, Finland, Professor S. Neuvonen at Kevo Subarctic Research Institute/University of Turku, Finland, Professor P.S. Karlsson at Abisko Scientific Research Station/University of Uppsala, Sweden, and Scientific Leader P.E. Aspholm at Svanhovd Environmental Centre, Norway, for providing us with phenology data for birch. We also thank Scientist H. Ramfjord at the Norwegian University of Science and Technology, Norway, A. Ekebom, Palynological laboratory/Swedish Museum of Natural History, Sweden, A. Frøsig, Astma-Allergi Forbundet (AAF), Denmark, and A. Rasmussen, Danish Meteorological Institute, Denmark, for providing us with birch-pollen data.

References

Ahas R, Aasa A, Menzel A Fedotova, VG, Scheifinger H (2002) Changes in European spring phenology. Int. J. Climatol. 22:1727-1738

Bogaert J, Zhou L, Tucker CJ, Myneni RB, Ceulemans R (2002) Evidence for a persistent and extensive greening trend in Eurasia inferred from satellite vegetation index data. J Geophys Res 107(D11):10.1029/2001JD001075

Chen X, Xu C, Tan Z (2001) An analysis of relationships among plant community phenology and seasonal metric of normalized difference vegetation index in the northern part of monsoon region of China. Int. J. Biometrol. 45:170-177

Chmielewski F-M, Rötzer T (2002) Annual and spatial variability of the beginning of growing season in Europe in relation to air temperature changes. Clim Res 19: 257-264

Emberlin J, Detandt M, Gehrig R, Jaeger S, Nolard N, Rantio-Lehtimäki A (2002) Responses in the start of *Betula* (birch) pollen seasons to recent changes in spring temperatures across Europe. Int J Biometeorol 46: 159-170

Hulten E (1971) Atlas over växternas utbredning i norden, Fanerogamer och ormbunksväxter. Ed. 2. LVI +531 pp. Generalstabens litografiska anstalts förlag, Stockholm

Kozlov MV, Berlina NG (2002) Decline in Length of the Summer Season on the Kola Peninsula, Russia. Climate Change 54 (4): 387-398

Los, S. O. (1998). "Estimation of the Ratio of Sensor Degradation between NOAA AVHRR Channels 1 and 2 from Monthly NDVI Composites." IEEE Transactions on Geoscience and Remote Sensing 36(1): 206-213

Menzel A, (2002) Plant phenological anomalies in Germany and their relation to air temperature and NAO. Climatic Chang. 57: 243-263

Menzel A, Fabian P (1999) Growing season extended in Europe. Nature 397:659

Moen A (1999) National Atlas of Norway: Vegetation. Norwegian Mapping Authority, Hønefoss. 200 pp.

Myneni RB, Keeling CD, Tucker CJ, Asrar G, Nemani RR (1997) Increased plant growth in the northern high latitudes from 1981 to 1991. Nature 386:698-702

Påhlsson L (ed.) (1998) Vegetationstyper i Norden. TemaNord 1998:510. Nordisk Ministerråd, København 1998

Reed BC, Brown JF, VanderZee D, Loveland TR, Merchant JW, Ohlen DO (1994) Measuring phenological variability from satellite imagery. J of Vegetation Science. 5:703-714

Rosborough, G. W., D. Baldwin and W. J. Emory (1994) Precise AVHRR Navigation. IEEE Transactions Geoscience and Remote Sensing **32**: 644-657

Shabanov NV, Zhou L, Knyazikhin Y, Myneni RB, Tucker J (2002) Analysis of interannual change in northern vegetation activity observed in AVHRR data from 1981-1994. IEEE Trans Geosci Remote Sens 40(1):115-130

Schwartz MD, Reed BC (1999) Surface phenology and satellite sensor-derived onset of greenness: an initial comparison. Int J Remote Sens 20(17):3451-3457

Sparks TH, Menzel A (2002) Observed changes in seasons: an overview. Int J Climatol 22: 1715-1725

Tuhkanen S (1980) Climatic parameters and indices in plant geography. Acta Phytogeogr. Suecica, 67, 1-105

Tveito OE, Førland E, Heino R, Hanssen-Bauer I, Alexandersson H, Dahlström B, Drebs A, Kern-Hansen C, Jónsson T, Vaarby Laursen E, Westman Y (2000) Nordic temperature maps. DNMI Report 09/00. 54 pp.

Tucker CT, Slayback DA, Pinzon JE, Los SO, Myneni RB, Taylor MG (2001). Higher Northern Latitude NDVI and Growing Season Trends from1982 to 1999. Int J Biometrol 45:184-190

Vermote, EE. and El. Saleous, N (1994) Statospheric aerosol perturbing effect on remote sensing of vegetation: operational method for the correction of AVHRR composite NDVI, SPIE Atmospheric Sens. and Mod., 29, 2311-2319

Vermote EF, Kaufman YJ (1995) Absolute calibration of AVHRR visible and near infrared channels using ocean and cloud views. Int J Remote Sens 16:2317-2340

Walther GR, Post E, Convey P, Menzel A, Parmesan C, Beebee TJC, Frometin JM, Hoegh-Guldberg O and Bairlen F (2002) Ecological responses to resent climate change. Nature Vol 416, March 28

Woodward FI (1987) Climate and plant distribution. Cambridge University Press.

Zhou L, Tucker CJ, Kaufmann RK, Slayback D, Shabanov NV, Myneni RB (2001) Variations in northern vegetation activity inferred from satellite data of vegetation index during 1981 to 1999. J Geophys Res 106(D17):20069-20083

6 Northern climates and woody plant distribution

R.M.M. Crawford[1] and C.E. Jeffree[2]

[1]Plant Science Laboratory, Sir Harold Mitchell Building, St Andrews University, St Andrews, KY16 9AL, UK. [2]Edinburgh University Institute of Cell and Molecular Biology, Daniel Rutherford Building, King's Buildings, Mayfield Road Edinburgh, EH9 3JH, UK, e-mail: rmmc@st-andrews.ac.uk

6.1 Introduction

Northern regions have always engendered enquiry in relation to how far temperate plants can migrate north or arctic plants migrate south before they reach some climatic limitation either from a failure to grow or from an inability to reproduce. Consequently, any attempt to predict the consequences of current climatic change for northern plant species has to give particular attention to marginal regions and consider the biological causes for the failure of plants to survive outside any given area. Limits to plant distribution in the Arctic invite investigation as they represent the ultimate peripheral locations for high-latitude survival. Biologically, plants are ideally suited for the study of peripheral situations as their sedentary nature facilitates mapping. Many atlases record limits to plant distribution both past and present at high latitudes (Hultén and Fries 1986b; Huntley and Birks 1983; Löve and Löve 1975; Meusel and Jäger 1992). These circumpolar plant distribution maps are unique, for in no other area of the World is it possible to examine across adjacent continents both the north-south and east-west extensions of species ranges. The ease with which temperature records can now be collated and compared with past and present plant distribution maps can create an impression that cartographic representation combined with mathematical modeling is all that is needed to relate climatic limits with plant distribution. Unfortunately, as pointed by the pioneering Norwegian eco-physiologist Eilif Dahl "the many indices that have been used for correlation between the northern and alpine limits of plants and meteorological factors are rarely based on any eco-physiological foundation" (Dahl 1998). As he expressed the problem "the question is in which way does temperature affect plant performance?"

Despite the attractive convenience of the concept of a generalised temperature limit for plant growth it is important to take note of the fact that mean temperatures do not exist in nature and therefore should be considered only as indicators and not causal factors (Holtmeier 2003). Maps which make some accommodation for seasonality and the differences between oceanic and continental climates can begin to answer some of the problems raised by Dahl in relation to matching climatic data with plant performance.

6.2 Interpreting distribution maps

Presenting data in the form of maps is intellectually satisfying. Maps create a global perception of distribution that permits comparisons to be made readily with physical factors such as climate and soils. Unfortunately, this simplicity of representation may be misleading in relating cause to effect as the factors which control distribution along any particular temperature boundary may differ from one region to another. The southern limit of the Tundra is generally determined by the northern limits of the boreal forest. However, the northern limit of the boreal forest may be controlled by differing ecological circumstances in diverse regions (Callaghan et al. 2002). Temperature, as it affects the length of the growing season, can be related to the position of the boreal treeline over large land-masses in regions with marked continental climates. However, over extensive regions of western Siberia the northern limit of the boreal forest is displaced southwards by the development of vast bogs. It follows therefore that although the boreal treeline (the tundra-taiga interface) is a global phenomenon that can be mapped and even observed by remote sensing (Fig. 6.1a-b), the reasons for its position from one area to another may differ (Crawford et al. 2003). Consequently, if the factors controlling distribution are not understood then information that may be derived from maps is open to misinterpretation.

Given these uncertainties about the physiological impact of temperature change it is relevant to examine in closer detail the natural distribution of plants in relation to seasonal variations in temperature. Predicting plant migration in terms of temperature change demands correlative observations from a range of variables. Iversen's classical studies pioneered the matching of plant distribution to summer and winter isotherms (Iversen 1944). In such studies the choice of isotherm is often arbitrary and the coincidence of a species boundary and an isotherm is not necessarily causal

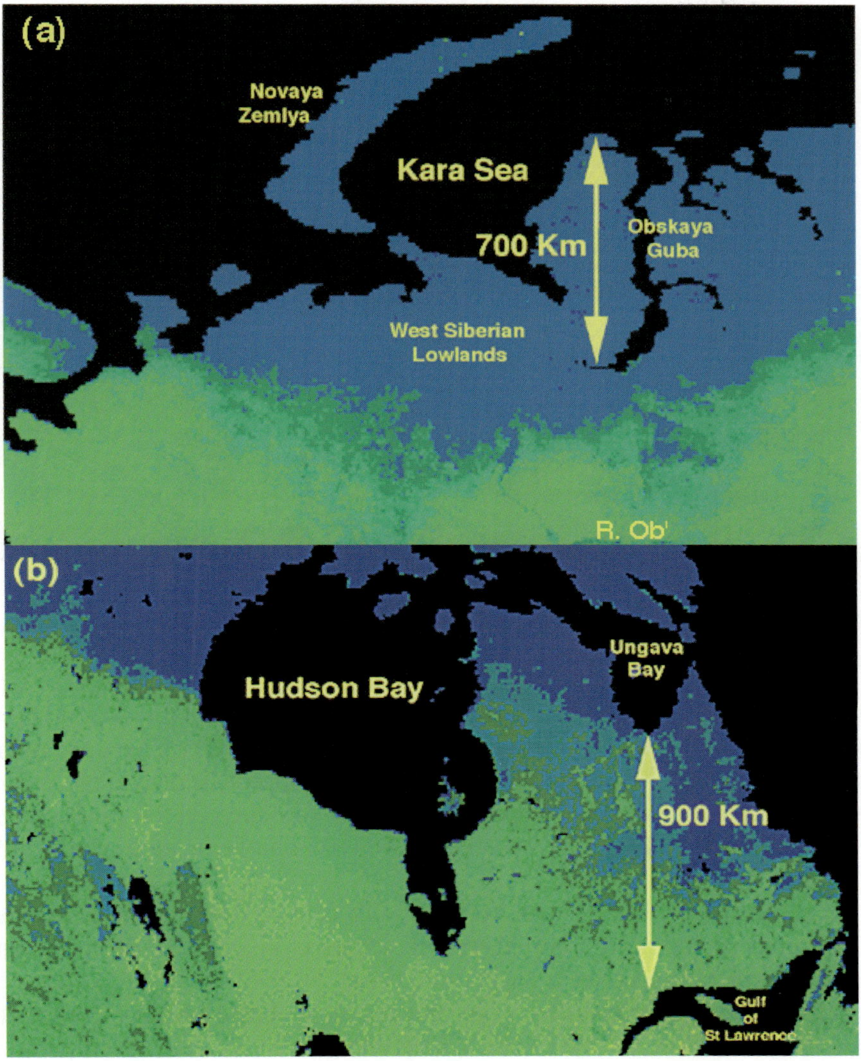

Fig. 6.1. Transition zone between Forest and Tundra as seen at the northern limits of (a) Western Siberian Lowlands and (b) the eastern section of the North American Boreal Forest boundary in normalised vegetation index (NDVI) images recorded in May 1999. Colour scale; blue = 0 as the Tundra is still covered by snow and ice. Other colours from green towards yellow represent increasing positive values in the NDVI index denoting the presence of photosynthetically active woody evergreen species emerging above the snow cover (images by courtesy of Goddard Space Flight Center).

Fig. 6.2. Photograph taken at the southernmost occurrence of lichen–spruce (*Picea mariana*) forest in Québec at 47°N (photo R.M.M. Crawford).

(Crawford 1989; Hengeveld 1990) and may over-estimate the geographical area occupied by species (Jeffree and Jeffree, 1994, 1996). There are many situations where a demographic or probability approach provides a better assessment of plant distribution, as can be seen below in the positions of the northern boreal treeline which cannot always be delineated by a simple line on a map.

Satellite NDVI images can be used with care to detect various vegetation zones including the boreal treeline (Rees et al., 2002). The problems of distinguishing between vegetation zones in relation to the evergreen treeline is greatly simplified when images are collected before the tundra ice and snow have melted as the woody evergreen species emerging above the snow cover stand out against the photosynthetically inert tundra surface. Images taken in the month of May in western Siberia reveal in many areas a marked transition zones between continuous tree cover and total absence of emergent evergreen vegetation.

Examination of Figs. 6.1a-b shows that in some regions the treeline is not a distinct boundary but a mosaic. This condition is found particularly in areas where bog formation occurs as in the West Siberian Lowlands where the transition extends North-South for approximately 600 km from the South-eastern shore of Obskaya Guba to the east-west flowing section of the River Ob'. This mosaic is considered by Russian ecologists to be a self-renewing cyclic process taking place over hundreds of years. Cryoperturbation causes the soil surface in localised areas to rise above the general level of the bog (Chernov and Matveyeva 1997). In some areas this permits the active layer to dry out sufficiently to allow the re-establishment of trees for a period until they shade the ground and cause the permafrost to rise and favour once again the growth of mosses as opposed to trees (Crawford et al. 2003). In Northern Quebec, a similar situation occurs under different circumstances. Here, the impact of combined insect and fire disturbances translates into a dramatic decrease in post-fire tree regeneration of the forest inducing the shift to lichen woodland (Payette et al. 2000). The result is a transition zone which begins at 47°N (Fig. 6.2) and extends over several hundred kilometers in a north–south direction (Fig. 6.1b).

6.3 Mapping species occurrence probability in relation to temperature

In terms of relating temperature conditions to the presence or absence of particular species it would appear more realistic to use a method that related probability of occurrence to geographical distribution rather than any supposed absolute temperate limit. Attention therefore has to be given to the nature of isotherms and what they represent. In essence, winter (x) and summer (y) isotherms are equivalent to straight lines in x, y cartesian space. Therefore, sets of 4 isotherms (upper and lower, x and y temperature limits) map out rectangles of temperature space. Species however, occupy scatters which have the form of inclined ellipses. Consequently, the fit of an isotherm to an ellipse is good only locally, and any isotherm-box overestimates species distributions in environmental space by adding unoccupied corners (Jeffree and Jeffree 1994).

A more objective manner of comparing species distribution to the interaction of winter and summer temperatures has been described and examined for a number of European species (Jeffree and Jeffree 1994, 1996), in which climatic temperature preferences of species are described in terms of the temperatures of the coldest (t_X) and warmest (t_Y) months of the year recorded at locations within a species' geographical distribution. Plotted on Cartesian co-ordinates for t_X and t_Y, the temperature data for locations occupied by a species form a bell-shaped probability density distribution described about the bivariate mean t_X bar, t_Y bar. Elliptical contours on this surface, representing defined limits on the continuously diminishing scale of probability at which a species may occur with increasing distance from the bivariate mean, may be calculated using the equations given by Jeffree & Jeffree (1994) for any desired level of probability. Consequently, ellipses calculated from temperatures within species distributions may be used to define and map those parts of the world's surface in which temperatures are potentially suitable for the species at a specified level of probability. Thus, a temperature ellipsoid calculated for a species distribution under current climatic conditions may be used as a template for identifying those geographical areas which may fall within the temperature range suitable for the species in any scenario of past or future climate change (Jeffree and Jeffree 1996). The use of ellipsoids derived from t_x and t_y for a number of woody species (Figs 6.3-6) allows a probability approach to relating temperature to the presence or absence of species. This approach enables the preparation of maps which not only show the probability of occurrence of the species in relation to temperature, but also where the species are absent even although the temperature regime is suitable for their occurrence. Such areas of non-occurrence may indicate that the absence of the species may be due to some factor other than temperature and deserve careful study, especially if they are contiguous with the main area of distribution.

Two types of map can be produced. The first, which are referred to as the *normal* (**N maps**), record the areas actually occupied by the species and compare this with the potential distribution as represented by the calculated ellipse. Logically, any location on Earth could be either inside or outside the species distribution and the temperatures associated with that location could be either inside or outside the calculated species temperature ellipse. Consequently, any one site can have one of four possible states;

1. Inside the ellipse with the species present (IESP - shown in maps as green) - here the species distribution agrees with the prediction based on the calculated ellipse.
2. Inside ellipse with the species absent (IESA – shown in maps as blue) - indicating an unoccupied potential environmental space for the species in question.
3. Outside the ellipse with the species present, (OESP – shown in maps as red) indicating that the temperature conditions may be marginal for the species.
4. Outside the ellipse with the species absent - indicating an unoccupied space probably with unsuitable temperatures for survival (OESA - shown in maps as grey).

In the second type of map, probability or (**P maps),** the colours represent bands of increasing probability that specified winter and summer temperature combinations are suitable for the species with the most favourable zones being shown in red, and the least favorable as purple and grey.

6.4 Woody shrub case histories

The following case studies are based on distribution maps specifically prepared for Northern Europe (Hultén and Fries 1986). The data representing the baseline temperature regime are taken from the University of East Anglia 0.5°gridded monthly temperature data for the years 1961 to 1990 (New et al. 1999; New et al. 2000). Hypothetical changes of +/- 1C° or +/- 4C° are imposed on each species as separate and opposing changes to both summer and winter temperatures in order to model the potential impact of mean annual temperature changes of + or – 2,5C° with contrasting seasonality on the species distributions. An important aspect of these maps is that they also represent areas where the thermal regime is suitable for the species in question but where the species is nevertheless absent.

6.4.1 Possible migration anomalies and case histories

From the results of these models it can be deduced that the following phenomena occur in relation to the effects of temperature on species distribution.

1. Species migration is sensitive to existing temperature *seasonality* and the superimposed *seasonality* of temperature change.
2. Migration cannot be predicted from annual mean temperature alone.
3. For the same patterns of change, in different parts of their ranges, species could migrate in *opposite* directions.
4. Seasonality gradients may present barriers to migration notwithstanding overall warming, particularly near coastal or continental margins.

(a) *Cassiope hypnoides* The Dwarf Arctic Heath (*Cassiope hypnoides*) is a species of transatlantic distribution which under the present climatic regime extends from Eastern Canada northwards to 75°N in Labrador. It is widespread latitudinaly in Greenland to 75°N on the west coast and 80°N on the east coast. The species also occurs in central and northern Iceland and the more northern montane regions of Scandinavia, the Kola Peninsula, and N. Siberia extending to the Southern part of Novaya Zemlya. The most northerly currently known location is on the West Coast of Svalbard

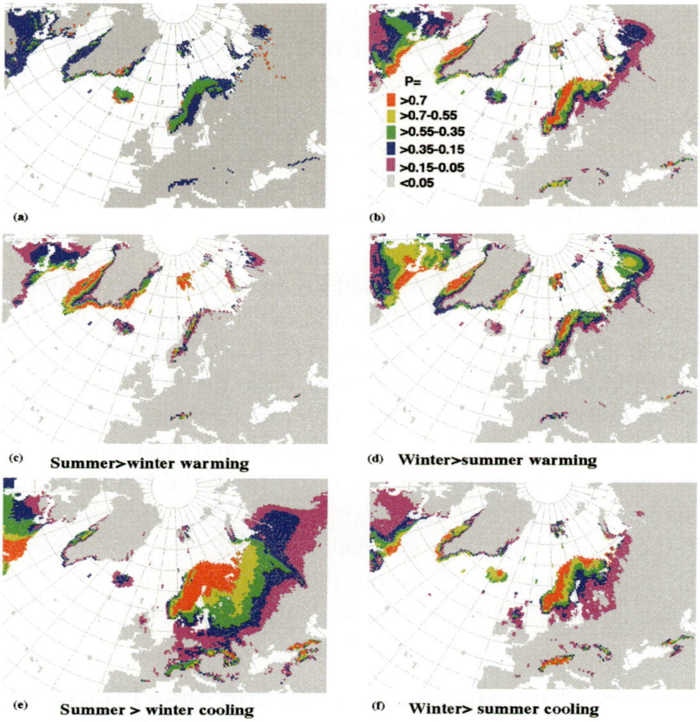

Fig. 6.3. Probability density plots of the possible ranges of *Cassiope hypnoides* in relation to temperature. **(a)** Distribution based on the temperature data from the CRU 1961-1990 Global Climate Dataset (New et al., 1999, 2000) relative to an ellipse enclosing 75 per cent of the observations. Green sites are within this "optimum" ellipse with the species present (IESP). Red sites (OESP) are outside the ellipse, but with the species present. These red sites can be regarded as current locations in which the species is at the low-probability margins of its temperature range. The blue sites (IESA) are inside the ellipse but where the species is currently absent and may be seen as representing potentially-suitable locations, where temperatures in summer and winter are suitable for the species. In plots **(b-f)** the colours represent bands of increasing probability of the (x), winter and (y) summer temperatures being suitable for the species. Red is most suitable (see inserted scale); **(b) P map** zones of increasing probability for winter and summer temperatures being suitable for the species at 1961-1990 temperatures; **(c)** 1961-1990 temperatures + 4C° in summer and + 1C° I winter; **(d)** 1961-1990 temperatures + 1C° in summer and + 4C° in winter; **(e)** 1961-1990 temperature - 4C° in summer and - 1C° in winter; **(f)** 1961-1990 temperature - 1C° in summer and –4 C° in winter.

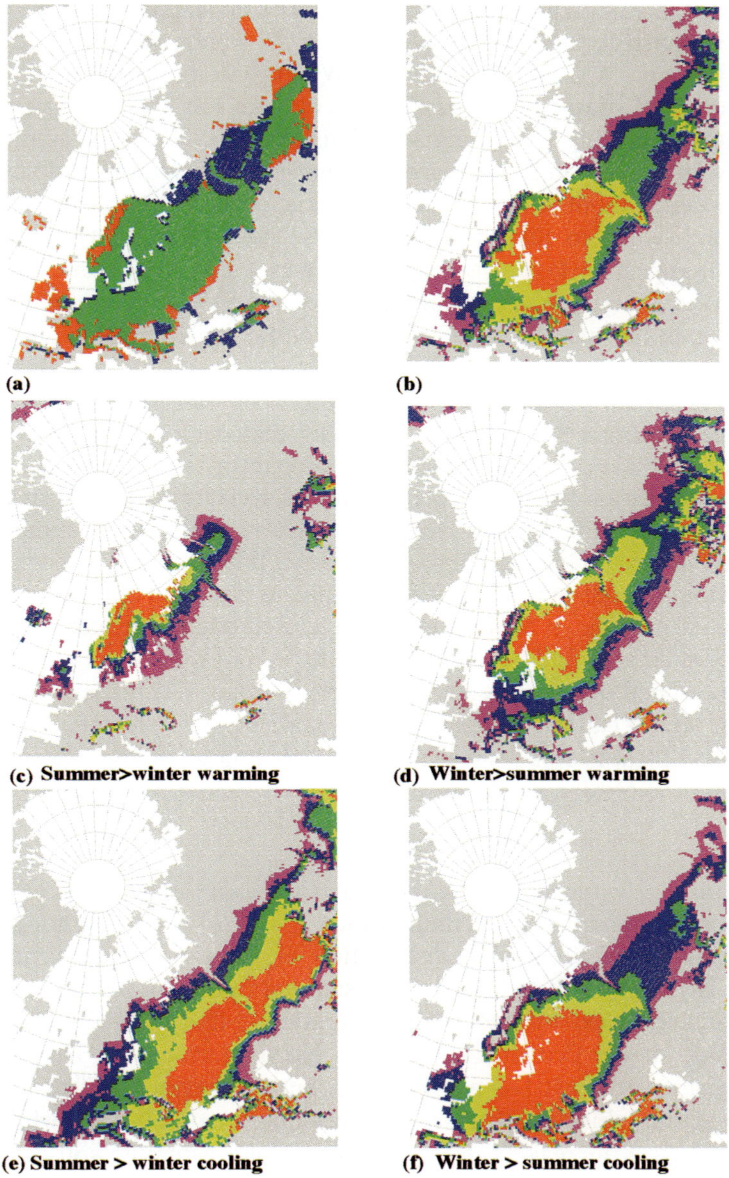

Fig. 6.4. Probability density plots of the possible ranges of *Vaccinium myrtillus* in relation to temperature. See legend Fig. 6.3 for details of images (a-f).

at 79°N in Kongsfjord (Rønning 1996). The present temperature conditions appear suitable for a wider extension in the region to the East of the Hudson Bay and in Western Greenland. When examined as a probability of occurrence (**P maps**) it is possible to compare the effects of winter versus summer warming or cooling on its distribution. Increased summer warming would result in an increased presence in Greenland, Svalbard and Novaya Zemlya, but would reduce the occurrence of the species in Scandinavia and Northern Siberia. The converse conditions with warmer winters would appear to favour an expansion westwards in N. America and eastwards in Siberia. Winter cooling would be likely to favour a southern extension of the species in Europe and a retreat in Eastern Siberia.

(b) *Vaccinium myrtillus* The bilberry (*Vaccinium myrtillus*) is a plant of northern Europe but with a much wider distribution southwards and eastwards than *Cassiope hypnoides*. The probability map for its present distribution based on temperature classifies its western extension in the British Isles Western France and Northern Spain as marginal. Summer warming would appear to lead to a major decline in this species while winter warming would result in a western retreat and an eastward expansion. However, climatic cooling, if it took place largely through a reduction in summer temperatures, would have a different effect and would be likely to increase the presence of the species in western Europe. Colder summer temperatures, would cause a retreat from the northernmost habitats and an expansion southwards and eastwards. Winter cooling would be likely to favour a southern extension of the species in Europe and a retreat in Eastern Siberia.

(c) *Calluna vulgaris* Heather (*Calluna vulgaris*) under present conditions is more oceanic in its distribution than *V. myrtillus* even although the temperature–based probability distribution suggests that its presence in Scotland is marginal. It has been proposed that optimal conditions for heather establishment exist in those regions bordering the North Sea from the eastern side of the British Isles through North Germany to Denmark (Gimmingham, 1972). It has also been found that heather suffers a decline in seed-setting ability with increasing oceanicity and altitude (Miller, 2001). Further west, increasing oceanicity and winter-saturated soil saturation favours *Erica tetralix* rather than *Calluna vulgaris*. Here it should be recognized that the current prevalence of heather in Scotland and Ireland has been accentuated by the late Holocene deforestation that has taken place in these regions (Fossitt 1996). Summer warmth appears from these predictions to be a limiting factor as summer warming would expand the range while winter warming would cause a further retreat from the west an and advance eastwards. As might therefore be expected, summer cooling by 4 C° below the present temperature levels would cause a south-western migration which would not take place if it were the winter temperatures that had been reduced.

(d) *Salix polaris* The polar willow (*Salix polaris*) is at present mainly restricted to the Arctic. These maps suggest that summer warming would probably lead to a retreat from its present Scandinavian occurrences, which agrees with a physiological study of this species carried out in Spitsbergen (Muraoka, et al., 2002), where a large increase in respiration with rising temperatures during the growing season is predicted as likely to cause a marked decrease in net primary productivity. However, in contrast to the other species examined in the maps presented here i.e., *Vaccinium myrtillus, Calluna vulgaris* and *Pinus sylvestris* (see below), it is probable that winter-warming, would result in a less marked retreat from oceanic areas in the distribution of *S. polaris* than that which is predicted for these other species.

6.5 Ecological limitations for the survival of woody plants

The most thoroughly examined aspects of the ecology of woody plants in relation to temperature are the altitudinal and latitudinal limits for the survival of trees. Many studies have sought to determine whether or not the low temperature regimes of high latitudes and altitudes cause trees to come into a carbon balance deficit. Intuitively, this would appear a simple and logical explanation. It might be expected that woody plants which devote a considerable part of their resources into the formation of non-productive trunks and stems may be unable to support such a growth strategy when growing seasons are cool and short. However, an extensive world-wide study of the carbon balance in trees at their upper altitudinal boundaries has shown the converse, namely, that tree growth near the timberline is not limited by carbon supply and that is more probable that it is sink activity and its direct control by the environment that restricts biomass production of trees under current ambient carbon dioxide concentrations (Körner 2003). Although carbon limitation is not a feature of woody plants at their upper limits of distribution it is nevertheless possible that certain organs can be seriously carbon-deficient under specific conditions. Overwintering tree roots in oceanic climates, if not fully dormant when flooded can through anaerobic respiration rapidly deplete their carbohydrate supplies and thus become vulnerable to post-anoxic injury when aeration is

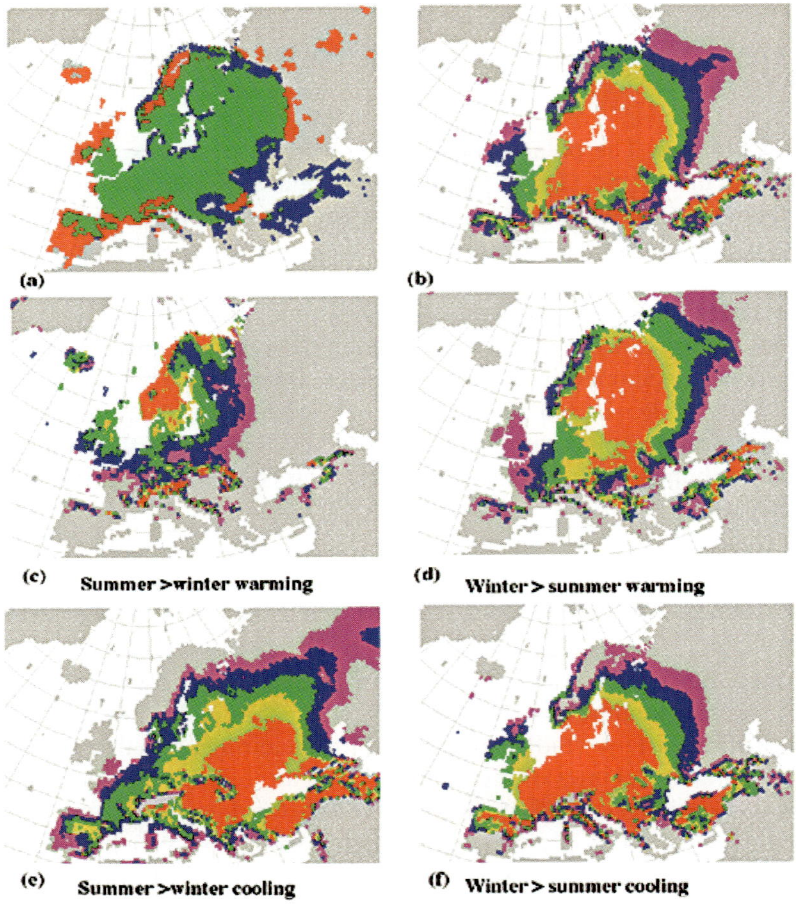

Fig. 6.5. Probability density plots of the possible ranges of *Calluna vulgaris* in relation to temperature. See legend Fig. 6.3 for details of images (a-f).

restored in spring (Crawford et al. 2003). Similarly, intermittent mild winter periods can deplete carbohydrate levels and render over-wintering shoots sensitive to frost injury (see below).

Oceanic conditions have long been known to reduce both the altitudinal and latitudinal position of the treeline (Fig. 6.1a-b). It would appear (see above) that this is also the case for a number of shrubby woody species. The blanket bogs of Scotland and Ireland stand out among the habitats that are significantly correlated with regions of high oceanicity. The basis on which this community is distinguished from less oceanic associations is

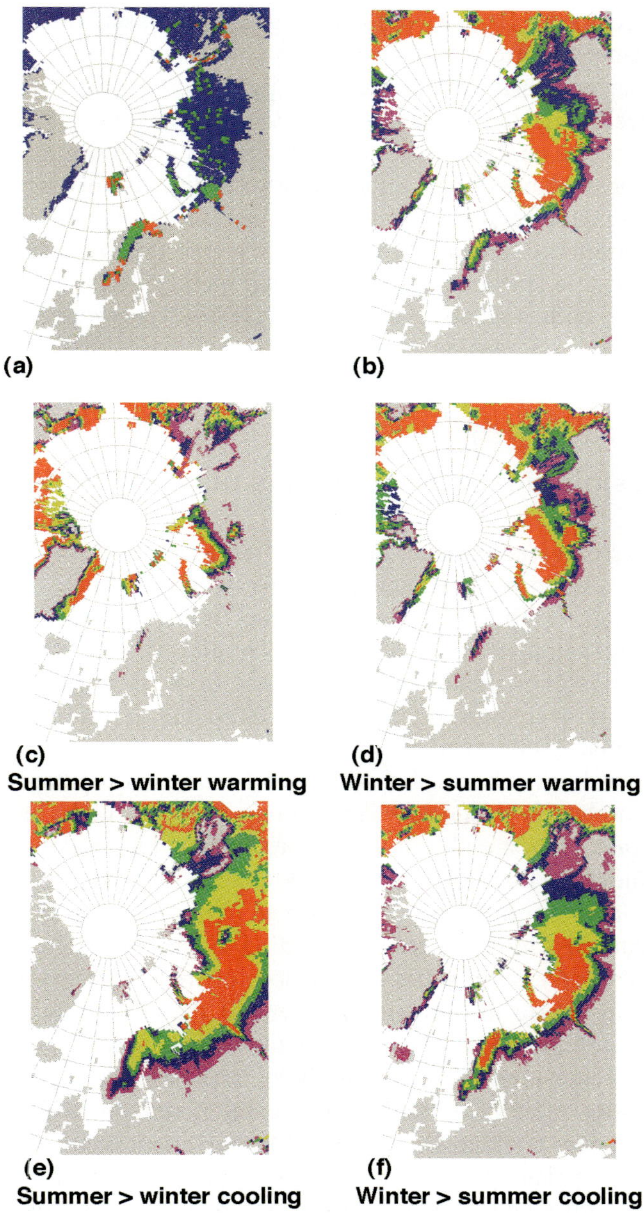

Fig. 6.6. Probability density plots of the possible ranges of *Salix polaris* in relation to temperature. See legend Fig. 6.3 for details of images (a-f).

determined in the British National Vegetation Classification not so much by the plants that are in this community, but more in negative terms by the species that are absent. In particular it is marked by the absence of woody shrub species such as *Vaccinium vitis-idaea*, *V. uliginosum* and *Empetrum nigrum* ssp. *hermaphroditum*, although there is a high frequency of *Myrica gale*. In these blanket bogs, although bryophytes flourish, oceanicity exerts a predominantly negative influence on many higher plants in that it gives rise to communities that are species poor, particularly in woody species, with the ericoid species being represented mainly by the flood-tolerant *Erica tetralix* with a relative paucity of *Calluna vulgaris*. This latter case agrees with the increasingly marginal position that is predicted for *Vaccinium myrtillus* and *C. vulgaris* should winters become warmer (Fig. 6.4-5).

6.6 Physiological disadvantages of warm winters

Some physiological explanation is required to explain why milder oceanic conditions should be disadvantageous for woody species. The Norwegian plant ecologist Eilif Dahl was one of the first to distinguish between the positive and negative effects of oceanic conditions on mountains. The relatively species-poor mountain floras of the Scottish Highlands and south-west Norway, were considered by Dahl to be due to mild periods of winter weather that encouraged premature spring growth causing severe die-back of non-hardy shoots. He described Norwegian montane species such as the woody *Rhododendron lapponicum* and the herbaceous *Aconitum septentrionale* that are absent from more oceanic mountains as "south-west coast avoiders" (Dahl 1951; Dahl 1990).

Vaccinium myrtillus has been observed in north-eastern Sweden to show a significant loss in frost hardiness leading to lethal injuries after a warmer than usual winter (Ögren 1996). This loss of frost tolerance was accompanied by a decrease in the solute content of the shoots, suggesting a progressive respiratory loss of cryoprotective sugars. Gas exchange measurements estimated that the initial carbohydrate reserves would have lasted for only four months if tissue water content remained high. When thin snow cover was coupled with clear skies then shoot dehydration could improve cold tolerance by 5-10 C°. However, in mild winters, with recurrent periods of mist and rain, increased metabolic activity resulted in shoot damage that was twice as frequent as a result of long-term de-hardening. It has also been demonstrated in Finland that a small elevation in air temperature can accelerate de-hardening in *V. myrtillus* (Taulavuori, et al., 1997).

Depending on habitat type the impact of oceanicity on woody species is likely to vary. On the sides of Norwegian fjords and Scottish sea-lochs, the steep inclination of the terrain encourages the survival of trees. However, in level areas there is frequently a tendency for bog formation to replace forest. In many northern maritime regions, with large areas of level ground, bog may be the climax vegetation and not forest (Klinger 1996). Once bog vegetation becomes established forest regeneration becomes problematic. Palynological studies show that the paludification of north-west Europe has a long history. A marked climatic deterioration (Klitgaard-Kristensen et al. 1998), commonly termed 'the 8200 BP event' (probably due to freshwater fluxes in the final de-glaciation of the Laurentide ice sheet), appears to have been accompanied by a temporary reduction in tree cover throughout western Europe, which in the oceanic conditions of Scotland set in train the development of blanket peat. In the Outer Hebrides blanket peat began its appearance between 9000 and 8000 BP, most probably as a consequence of this climatic episode, before there was any significant human settlement (Fossitt 1996).

It may seem counter intuitive, but there is even an argument for suggesting that in oceanic areas climatic warming may lead to a retreat rather than an advance of the treeline. Examination of temperature variations over the past century for Europe and the Arctic from Northern Norway to Siberia suggests that variations in the North Atlantic Oscillation are associated with an increase in oceanicity in certain maritime regions. A southward depression of the treeline in favour of wet heaths, bogs and wetland tundra communities is also observed in several northern oceanic environments. The heightened values currently detected in the North Atlantic Oscillation Index, together with rising winter temperatures, and increased rainfall in many areas in Northern Europe, present an increasing risk of paludification with adverse consequences for forest regeneration, particularly in areas with oceanic climates. Climatic warming in oceanic areas may increase the area covered by bogs and thus, contrary to general expectations, may lead to a retreat rather than an advance in the northern limit of the boreal forest (Crawford et al. 2003).

It is also to be expected at high latitudes, as woody plants approach their most northerly limits, that the distinction between oceanic and continental climates will become more critical for plant survival. A detailed dendroclimatological study of Scots Pine (*Pinus sylvestris*) in northern Norway (69°N) has shown that high winter temperatures represent a stress factor at the limit of pine in oceanic habitats. Consequently, a period around 1920, with low winter temperatures, co-coincided with a marked rise in growth (Kirchhefer, 2001). As with the studies on *Vaccinium myrtillus* discussed above (Ögren, 1996) mild periods in winter, causing bud dormancy to

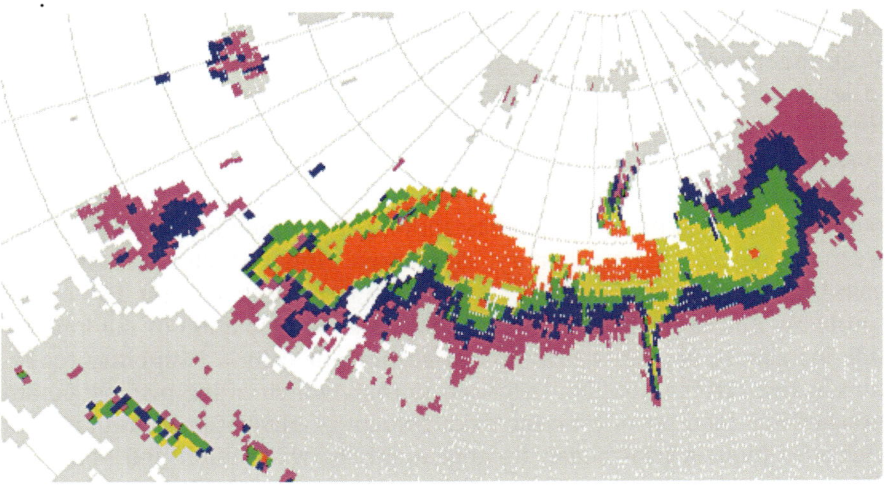

(a)

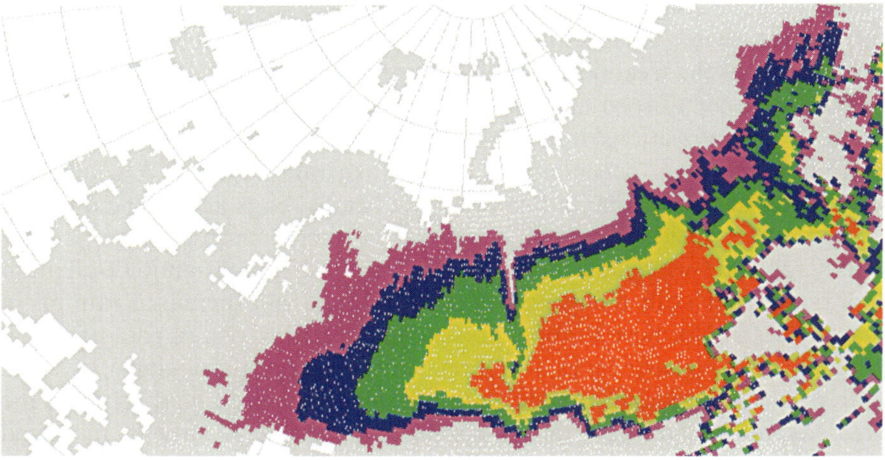

(b)

Fig. 6.7. Comparison of possible contrasting changes in distribution of *Pinus sylvestris* with varying climate conditions as compared with 1961-1990. **(a)** Winter 4C° colder, summer 4C°warmer; **(b)** winter 4C° warmer, summer 4C° colder. Note the retreat from western Europe with the imposition of warmer winters.

break and increasing subsequent frost damage may be implicated, as may also heavy needle loss after mild winters (Kirchhefer, 1999). Fig. 6.7(a-b) shows the contrasting effects that might be produced on the distribution of *Pinus sylvestris* should there be significant climatic warming. In the event of warm winters prevailing there could be a marked retreat of Scots pine from western Europe.

There may be many reasons as to why oceanic conditions are detrimental for the survival of Scots Pine and other woody species. Among these is the effect of mild winters combined with wet soil conditions. Warm weather in autumn and early winter delays the onset of dormancy in tree roots which if followed by soil saturation or flooding can induce anaerobic conditions which result in a rapid winter consumption of winter carbohydrate reserves Crawford, 2003).

Studies which seek to match species range against changing environmental conditions may be geographically illuminating, but are nevertheless only a first step in understanding ecological limitations. Just as there are many causes of damage to plants, so there are many variations in local populations which have evolved to withstand particular local situations. Further mapping studies are needed at the sub-species level to determine if intra-specific differences exist between populations in their responses to oceanic and continental climates.

6.7 Conclusions

Climatic warming in northern latitudes will undoubtedly create new ecological opportunities for vegetation advance as ice sheets retreat and permanent snow cover is reduced. Depending on proximity to the oceans, and this includes the Arctic Ocean, the degree of winter versus summer warming is likely to differ. In northern regions the influence of oceanicity on plant distribution is very marked. The models described in this paper show that among some of the commonest woody species winter warming may in some cases cause a significant retreat in areas that where the climate is influenced by the ocean, while in other areas the species may make significant advances. Due to seasonal differentiation in responses to temperature the nature of the migration of species into these vacated areas or changing communities is unlikely to be a mere latitudinal shift northwards of existing species assemblages. It appears therefore highly probable that species migration will be strongly influenced by *seasonality* of temperature change, and for any one species similar patterns of climatic change in different parts of its range, could result in migration in *opposite* directions. Consequently, changing thermal conditions will be likely to alter species composition of existing northern plant communities. Attention will therefore have to be given to the roles of seasonality gradients as potential barriers to migration notwithstanding overall warming.

Acknowledgements

Figures 6.1a and b were generated from the 8-km resolution Pathfinder data set. The authors thank Dr. W.G. Rees for assistance in the use of satellite images and the Distributed Active Archive Centre (Code 902.2) at the Goddard Space Flight Center, Greenbelt, MD. 20771, for producing and distributing the data in their present form. We are also greatly indebted to Professor Christian Körner for careful and helpful reading of earlier versions of this chapter.

References

Callaghan TV, Werkman BR, Crawford RMM (2002) The tundra-taiga interface and its dynamics: Concepts and applications. Ambio: 6-14.

Chernov YI, Matveyeva NV (1997) Arctic ecosystems in Russia. In: Wielgolaski FE (ed) Polar and alpine tundra, Amsterdam, Elsevier, pp 361-507.

Crawford RMM (1989) Studies in Plant Survival. Blackwell Scientific Publications, Oxford.

Crawford RMM, Jeffree CE, Rees WG (2003) Paludification and forest retreat in northern oceanic environments. Annals of Botany 91: 213-226.

Crawford RMM (2003) Seasonal differences in plant responses to flooding and anoxia. Canadian Journal of Botany 81: 1224-1246.

Dahl E (1951) On the relation between summer temperatures and the distribution of alpine plants in the lowlands of Fennoscandinavia. Oikos 3: 22-52.

Dahl E (1990) Probable effects of climatic change due to the greenhouse effect on plant productivity and survival in North Europe. Effects of climate change on terrestrial ecosystems 4: 7-18.

Dahl E (1998) The Phytogeography of Northern Europe. Cambridge University Press, Cambridge.

Fossitt JA (1996) Late quaternary vegetation history of the Western Isles of Scotland. New Phytologist 132: 171-196.

Gimingham CH (1972) Ecology of heathlands. Chapman and Hall, London.

Hengeveld R (1990) Dynamic biogeography. Cambridge University Press, Cambridge.

Holtmeier F-K (2003) Mountain timberlines. Kluwer, Dordrecht.

Hultén E, Fries M (1986) Atlas of North European vascular plants north of the tropic of Cancer. Koeltz Scientific Books, Königstein.

Huntley B, Birks HJB (1983) An Atlas of past and present pollen maps for Europe: 0-13000 years ago. Cambridge University Press.

Iversen J (1944) *Viscum*, *Hedera* and *Ilex* as climatic indicators. Geologiska Foereningens i Stockholm Foerhandlingar 66: 463-483.

Jeffree CE, Jeffree EP (1996) Redistribution of the potential geographical ranges of Mistletoe and Colorado beetle in Europe in response to the temperature component of climate change. Functional Ecology 10: 562-577.

Jeffree EP, Jeffree CE (1994) Temperature and the biogeographical distribution of species. Functional Ecology 8: 640-650.

Kirchhefer AJ (1999) PhD thesis, Dendroclimatology on Scots pine (*Pinus sylvestris*) in northern Norway Department of Biology. Tromsø, University of Tromsø

Kirchhefer AJ (2001) Reconstruction of summer temperatures from tree-rings of Scots pine (*Pinus sylvestris* L.) in coastal northern Norway. Holocene 11: 41-52.

Klinger LF (1996) Coupling of soils and vegetation in peatland succession. Arctic and Alpine Research 28: 380-387.

Klitgaard-Kristensen D, Sejrup HP, Haflidason H, Johnsen S Spurk M (1998) A regional 8200 cal. yr BP cooling event in northwest Europe, induced by final stages of the Laurentide ice-sheet deglaciation? Journal of Quaternary Science 13: 165-169.

Körner C (1999) Alpine Plant Life. Springer Verlag, Berlin.

Körner, C. (2003) Carbon limitation in trees. Journal of Ecology 91: 4-1.

Löve A, Löve D (1975) Cytotaxonomical atlas of the arctic flora. J. Cramer, Lehre.

Meusel H, Jäger EJ (1992) Vergleichende Chorologie der Zentraleuropäischen Flora. Gustav Fischer Verlag, Jena.

Miller GR, Cummins RP (2001) Geographic variation in seed-setting by heather (Calluna vulgaris (L.) Hull) in the Scottish Highlands. Journal of Biogeography 28: 1023-1031.

Muraoka H, Uchida M, Mishio M, Nakatsubo T, Kanda H Koizumi H (2002) Leaf photosynthetic characteristics and net primary production of the polar willow (*Salix polaris*) in a high arctic polar semi-desert, Ny-Alesund, Svalbard. Canadian Journal of Botany-Revue 80: 1193-1202.

New M, Hulme M, Jones P (1999) Representing twentieth-century space-time climate variability. Part I. Development of a 1961-1990 mean monthly terrestrial climatology. Journal of Climate 12: 829-856.

New M, Hulme M, Jones P (2000) Representing twentieth-century space-time climate variability. Part II. Development of 1901-1996 monthly grids of terrestrial climate. Journal of Climate 13: 2217-2238.

Ögren E (1996) Premature dehardening in *Vaccinium myrtillus* during a mild winter: A cause for winter dieback? Functional Ecology 10: 724-732.

Payette S, Bhiry N, Delwaide A Simard M (2000) Origin of the lichen woodland at its southern range limit in eastern Canada: the catastrophic impact of insect defoliators and fire on the spruce-moss forest. Canadian Journal of Forest Research 30: 288-305.

Rees G, Brown I, Mikkola K, Virtanen T Werkman BR (2002) How can the dynamics of the Tundra-Taiga boundary be remotely monitored? Ambio Special Report 12: 56-62.

Rønning OI (1996) The flora of Svaldbard. Norwegian Polar Institute, Oslo.

Taulavuori K, Laine K, Taulavuori E, Pakonen T Saari E (1997) Accelerated dehardening in bilberry (Vaccinium myrtillus L.) induced by a small elevation in air temperature. Environmental Pollution 98: 91-95.

7 Topographic complexity and terrestrial biotic response to high-latitude climate change: Variance is as important as the mean

W. Scott Armbruster[1,3,4], David A. Rae[1], and Mary E. Edwards[2,3]

[1]Department of Biology, Norwegian University of Science and Technology, N-7491 Trondheim, Norway, [2]School of Geography, University of Southampton, Highfield, Southampton S017 1BJ, UK, [3]Institute of Arctic Biology, University of Alaska, Fairbanks, AK 99775 USA, [4]School of Biological Sciences, University of Portsmouth, Portsmouth PO1 2DY, UK, e-mail: scott.armbruster@bio.ntnu.no

7.1 Introduction

Climate changes that occur as a result of the increased greenhouse effect are expected to be greatest at high latitudes (Cattle and Crossley 1995; Watson et al. 2001), and resultant earth-system responses are likely to be dramatic (Weller 1998). For example, even the small increase in temperature witnessed over the past few decades has melted permafrost in many places, causing major changes in soil drainage, chemistry, and biology, as well as damage to roads, buildings, airports, and pipelines (Williams 1995; Weller 1998; Osterkamp and Romanovsky 1999). The expectation of dramatic environmental change has stimulated numerous empirical and model-based studies addressing both ecosystem responses to predicted changes in temperature and CO_2 concentration and how these responses feed back to the climate system (e.g. Sellers et al 1997; Kane and Reeburgh 1998; Arft et al. 1999; Chapin et al. 2000; Eugster et al. 2000; Rupp et al. 2001; Levis et al. 1999, 2000; Friedlingstein et al 2003; Ichii et al. 2003; Jones et al. 2003; Kaplan et al. 2003; McGuire et al. 2002, 2003).

Nearly all studies of environmental change are based on modelling and scaling-up activities that use mean values or properties as input. Yet it is well known that the values of climatic parameters vary locally due to topographic influences (e.g. Hungerford et al. 1989; Young et al. 1997). Elevation has well documented effects on temperature and precipitation. Microclimate is also influenced by the radiation load, which varies in response to variation in aspect of slopes ("aspect") and steepness of slopes ("slope"). Hence, rough (mountainous or hilly) landscapes generate a mosaic of diverse microclimates, with direct effects on organism physiology and ecology and biotic community structure and function (Knapp 1985; Young et al. 1997, Porter et al. 2002). Implicit in this variation is variation in ecosystem processes. For example, an increase in temperature will probably lead to a release of CO_2 from many high-latitude soils, at least transiently, generating a positive feedback and further warming (Chapin et al. 2000; Jones et al. 2003; Knorr et al. 2005). However, carbon sources and sinks are finely balanced, and the sign of the projected change may be sensitive to scaling-up procedures (Chapin et al. 2000).

While analysis of means is practical and necessary in current modelling, ignoring variation about the mean may leave out complexity of potentially great ecological importance. Most studies of biotic response to climate depend on the mean environmental parameters because these are the only data that are widely available. However, all terrestrial biotic response to climate change is mediated by the local microclimate actually experienced by the organism. Some of the ecological complexity that exists on the landscape due to small-scale variation in topographic setting has been studied using dense arrays of micro-loggers; results indicate the potential importance of local microclimate variation for both modelling and empirical studies of terrestrial biotic responses to environmental change (Edwards and Armbruster 1989; Wesser and Armbruster 1991; Lloyd et al 1995; Rae 2003). Large-scale patterns in elevation-generated microclimate variation are determined by the location of mountain ranges. In contrast, patterns in the amount of variation induced by aspect and slope appear to have a latitudinal trend. For reasons explained below, aspect/slope-induced variation in microclimate appears to peak near the Arctic and Antarctic Circles. Thus, the sensitivities of vegetation and ecological processes to small-scale climate variation may be especially great in the Arctic and Antarctic.

7.2 Variation and climate models

Models that simulate global- and continental-scale patterns of interactions between climate, vegetation, and biogeochemical processes, e.g. VEMAP (Vegetation/Ecosystem Modeling and Analysis Project; Pan et al. 1998) and LPJC (Lund-Potsdam-Jena dynamic global vegetation model; Sitch et al. 2003), have focused on large-scale patterns and processes. The significance of small-scale heterogeneity to such large-scale modelling efforts lies in the need to scale down to, and up from, landscape-level processes when estimating the potential feedbacks inherent in the climate system

(Bonan et al. 1995). Error may result from non-additive and non-linear responses to variation, which may generate values that depart significantly from response to mean values ("aggregation error").

It is well known that microclimates vary significantly at high latitudes, because the angle of the sun is low during the growing season, and north- and south-facing slopes experience very different radiation loads. Theoretical micro- and regional-scale climate models can help assess potential microclimate variation (e.g. MTCLIM; Hungerford et al. 1989) and its effects on the distribution of species, communities, and ecosystem processes, but the sensitivity of microclimate to cloudiness and wind (e.g., Porter et al. 2002) argues for direct estimation and testing with empirical observations. There is surprisingly little detailed empirical information on the nature of microclimatic variation and its effects, perhaps because, until recently, it has been prohibitively expensive to install dense grids of weather stations (or other climatic instruments) on rough landscapes.

7.3 Biotic response to microclimatic variation

Organisms generally respond to the local climate and not regional means. This has many direct implications for the ecology and population biology of most organisms, especially those that are sedentary or disperse locally (see Porter et al. 2002). For example, the extinction of a plant species may be predicted from models of mean temperature change, whereas in reality the species may persist locally in favourable microsites, such as on cool, moist, north-facing slopes in scenarios of climate warming (e.g. Coker and Coker 1973; Sydes 1997), or on warm south-facing slopes in scenarios of climate cooling (another kind of aggregation error).

The biological importance of micro- and regional-scale climatic heterogeneity is implicit in the current shift in paleoecological thinking about the post-glacial re-colonization of landscapes adjacent to retreating ice-sheets. There is increasing recognition that many local refugia probably existed and played important roles in the recolonisation of post-glacial landscapes, rather than recolonisation occurring through waves of long-distance migration from lower latitudes (Stewart and Lister 2001, Schauffler & Jacobson 2002, Clark et al. 2003, Brubaker et al. 2005). With respect to biotic response to ongoing climate change, it is therefore important to focus on the local reorganisation of communities in addition to major latitudinal shifts in distribution (Rupp et al. 2001).

The possible scale of biotic response to past and ongoing climate change (by genes, populations, or species) has received insufficient attention (but see Davis and Shaw 2001). Increases and decreases in local abundances of species with different climatic tolerances are likely, with the extreme case being local extinction and replacement of species. This may come about through processes of long-distance migration or may result from the reshuffling of co-occurring (parapatric) species with diverse climatic tolerances, as a result of, in part, the local heterogeneity in microclimate (e.g. Brockway 1998; Goldblatt and Manning 2002; but also see Chapin et al. 1995; Chapin and Shaver 1996). In addition, a change in climate may induce a plastic response by genetic individuals (within genotypes; Crawford and Abbott 1994, Bret-Harte et al. 2001, Edwards et al. 2005), or there may be changes in the relative frequencies of genotypes and alleles as a result of natural selection and/or gene flow (i.e. evolution; Davis and Shaw 2001).

In this context it becomes important to know the degree to which microclimate varies on the local landscape and the degree to which organisms respond to this variation. Research challenges include assessing i) whether local spatial variation in microclimate is large or small relative to the kinds of regional temperature changes expected from the anthropogenically enhanced greenhouse effect over the next century, ii) how local variation compares with larger-scale temperature variation with altitude and latitude, iii) the extent of associated shifts in species composition of communities, if any. With respect to this last question, little species-composition change associated with significant microclimate variation suggests that plasticity and/or local evolution (ecotypic differentiation) are the biotic responses to climate changes of this scale. Alternatively, strong changes in species composition in response to microclimate variation suggest that replacement of some species by others will occur with climate change of this scale. Thus, additional work on biotic response to spatial variation in microclimate would have value for understanding biotic response to temporal climate change.

7.4 Latitudinal Trends in Variation in Radiation Load

Well established theory and empirical data show that radiation loads striking the earth's surface vary with latitude and season (Strahler and Strahler 1989). Furthermore, incident radiation varies with the slope angle of the surface, with a surface perpendicular to the incident radiation receiving the greatest load. Combining these two analytical results into a simple spreadsheet model (see Weiss et al. 1993) allows us to explore latitudinal patterns of variation in radiation environments experienced by rough landscapes.

For heuristic purposes, consider the difference in radiation loads between north- and south-facing surfaces at 35° inclination. On the summer solstice at 30° and ~90° latitude (N. Hemisphere), south-facing surfaces receive about the same radiation as north-facing surfaces, but between 30° and ~90° they receive up to 50% more radiation. On 1 September, south-facing surfaces at ~90° latitude receive the same radiation load as north-facing surfaces. Between 30° and 80° they receive 60-400% more radiation.

Integrating this over the high-latitude growing season results in minimal differences between south- and north-facing surfaces at 30° and 90° but about a doubling of radiation sums on south-facing surfaces relative to north-facing in the subarctic and low Arctic (60° -70°). Thus the contrast between radiation budgets on south- versus north-facing slopes reaches its theoretical maximum near the Arctic Circle, remains high throughout the subarctic and low-Arctic regions and declines sharply in the low temperate and extreme high-Arctic latitudes (Fig. 7.1).

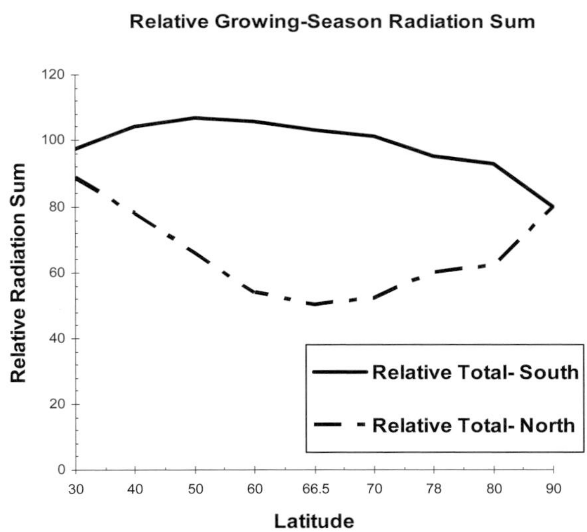

Fig. 7.1. Growing-season radiation sums on north-facing and south-facing surfaces (at 35° inclination) across a range of latitudes. Atmospheric transmissivity was assumed to be 83%.

7.5 Measuring Microclimate and Biotic Response to Variation in Slope and Aspect

The above arguments demonstrate the importance of obtaining better information on small- to medium-scale variation in climate, and the potential effects of this variation on the distribution of species, communities, and ecosystem processes in Arctic and Antarctic regions.

7.5.1 Effect of Topography on Seasonal Radiation Sums and Microclimate

Very few empirical studies of local variation in microclimate have been conducted, so we review here one conducted recently in Svalbard. The approach is to place in the field three thermal microloggers at each of a series of plots of different slope and/or aspect. The three microloggers are in: 1) silver film canisters, shielded from sun (measuring air temperature,

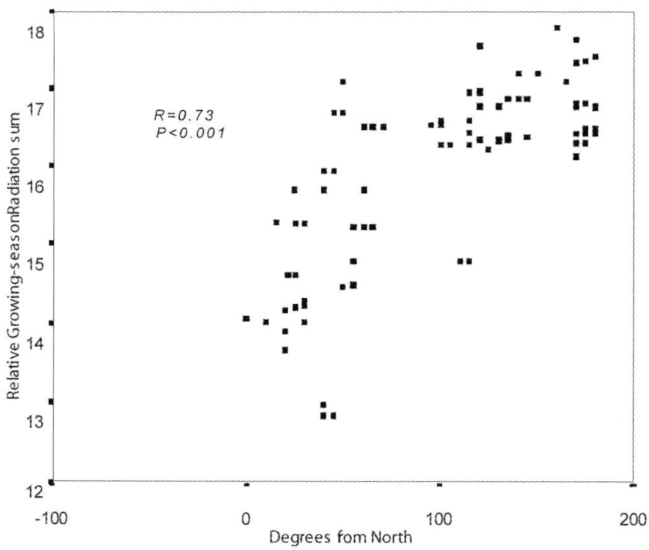

Fig. 7.2. Relative growing-season radiation sums measured on plots located on slopes across a range of aspects, from south facing (180°) to north facing (0°). Plots were located in Adventdalen, Spitsbergen and measurements made July-August 2000. Aspect explained 53% of the variance in radiation sum ($P < 0.001$).

T_a); 2) black film canisters, shielded from wind, but not the sun (= radiation load); and 3) black film canisters, exposed to sun and wind (operative environmental temperature, T_e; Bakken 1992; Bishop and Armbruster 1999).

Using these three measures of thermal parameters it is possible to estimate directly, or by calculation: air temperature (T_a), operative environmental temperature (T_e), radiation load, and wind speed (Moe 2000; Rae 2002; Armbruster et al. submitted). To check the accuracy of these estimates, we compared estimated values of radiation load and wind speed with directly measured values of radiation and wind at a subset of plots. We found reasonably close correspondence in both cases ($R = 0.97$, $P < 0.001$; $R = 0.64$, $P < 0.001$; respectively). Three thermal microloggers, as described above, were placed in each of 45 plots an area of complex micro-relief in Todalen, a side valley of Adventdalen, near Longyearbyen, Spitsbergen, Svalbard (78° N) in July-August 1999 and 2000.

The seasonal radiation sums at 45 micro-sites on Spitsbergen were estimated from the data recorded by the three loggers at each sites using the formula derived in Rae (2003). These radiation values were strongly influenced by the aspect of the slopes on which the loggers were placed (Fig. 7.2). South-facing slopes experienced about 50% higher radiation loads than

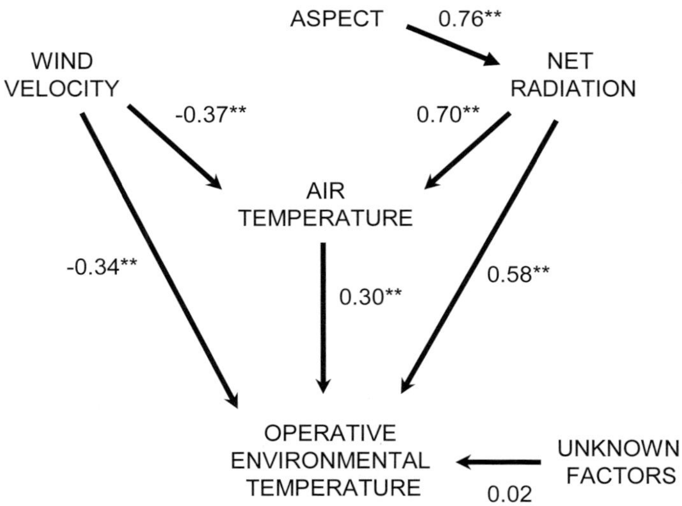

Fig. 7.3. Estimated influences of various measured or calculated microclimate variables on operative environmental temperature, Adventdalen, Spitsbergen, July 2001.

north-facing slopes during the study period. South-facing slopes also experienced up to 5.0° C higher maximum air temperatures and 1.0° C higher mean air temperatures than north-facing slopes during the study period. The overall thermal environment experienced by small surface-dwelling organisms ("operative environmental temperature") on south-facing slopes was up to 6.6° C higher (maximum T_e), and the mean was up to 2.6° C higher, than on north-facing slopes. This approach also allows one to estimate the relative contributions of variation in radiation load, wind, and air temperature to the variation in operative environmental temperature across plots. While aspect had a major influence on radiation load, radiation load positively affected thermal environment, both directly and indirectly via air temperature. Wind had a significant negative effect on operative environmental temperature (Fig. 7.3). A surprising 96% of the variance in operative environmental temperature was explained by the estimated wind speed, radiation load, and air temperature (Moe, 2000; Rae 2003).

7.5.2 Effects of Topography and Microclimate on Composition of the Biotic Communities and Soil Environment

We estimated the percent cover of each vascular plant species in each plot and sampled ground-dwelling insects with a small pitfall trap (150 ml plastic cup) in each plot. We then ordinated the plant (see Table 7.1) and invertebrate community-composition data using detrended correspondence

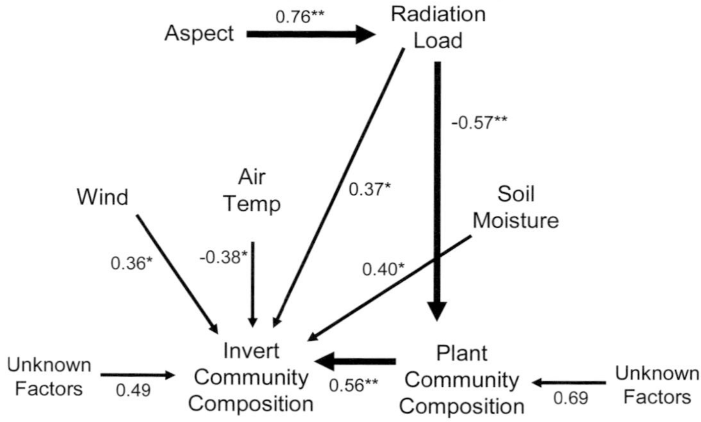

Fig. 7.4. Estimated influences of topography on microclimate, and microclimate on the composition of biotic communities, Adventdalen, Spitsbergen.

Table 7.1. Species and species scores on first axis of the detrended correspondence analysis (DCA) run on the plants recorded in each plot at the study site in Todalen, Adventdalen, Spitsbergen.

Species	DCA Score	Species	DCA Score
Ranunculus nivalis	432	Cerastium arcticum	169
Sphagnum spp.	413	Equisetum scirpoides	138
Trisetum spicatum	402	Salix polaris	126
Minuartia biflora	386	Poa arctica	118
Ranunculus pygmaeus	381	Alopecurus borealis	102
Arnica angustifolia	380	Luzula confusa	94
Equisetum arvense	379	Rhacomitrium spp.	77
Dicranum spp.	369	Papaver dahlianum	62
Oxyria digyna	331	Stellaria crassipes	61
Betula nana	311	Crustose lichens	60
Selene uralensis	305	Draba subcapitata	52
Saxifraga cernua	297	Draba micropetala	44
Festuca cryophila	293	Pedicularis hirsuta	27
Selene acaulis	283	Saxifraga cespitosa	25
Cardamine bellidifolia	245	Cassiope tetragona	-2
Saxifraga hieracifolia	224	Carex rupestris	-2
Polytrichum spp.	218	Saxifraga nivalis	-12
Taraxacum arcticum	204	Hierochloe alpina	-12
Polygonum viviparum	170	Dryas octopetala	-14
Potentilla rubricaulis	177	Saxifraga oppositifolia	-27

analysis (DCA), implemented on the PC-Ord 4.0 package (McCune and Mefford 1999) to reduce dimensionality and create a linear variable (Table 7.1). We related variation in community composition among plots to variation in measured or calculated microclimate variables using path analysis (Li 1975; Shipley 2000).

Microclimate and, indirectly, topography had profound effects on the composition of the plant community and to a lesser extent, the invertebrate community. Radiation load appeared to have the biggest "direct" influence on the composition of the plant community, while all measured environmental variables appeared to influence the invertebrate community directly and almost equally (Fig. 7.4). (Note that "direct" in this context must be qualified by the caveat that radiation and other effects could influence the communities indirectly via environmental variables not included in the model, e.g., soil temperature, snow accumulation and persistence, etc.) Aspect, slope, and microclimate also induced variation in soil temperature, (not shown in the figure). Although other soil-environment factors were not measured in this study, similar studies in Alaska showed that radiation loads and environmental temperatures also influence soil moisture, soil pH, soil-nutrient availability, and soil carbon storage (Wesser and Armbruster 1991; Lloyd et al. 1994).

7.5.3 Complex Indirect Effects of Microclimate

In addition to the direct (or apparently direct) effects of microclimate on the structure of the invertebrate community, indirect effects of microclimate were mediated via changes in the plant community (Fig. 7.4). At low-arctic and subarctic latitudes, grazing animals such as sheep (both wild and domestic) utilize south-facing slopes more intensively, because 1) they are snow-free for more of the year, 2) they offer more and/or better forage, and/or 3) the microclimate is more favourable and vertebrates experience reduced thermoregulatory costs (Pearson et al. 1995; Mysterud et al. 2001; Porter et al. 2002). Herbivore behaviour, in turn, can have major effects on the amount of disturbed and open ground, the species composition of the plant community, and hence —by extension— factors such as soil temperature, nutrient availability, and carbon storage in the soil. For example, at an alpine site near the Arctic Circle in Norway (Loftan), microclimate appeared to influence the grazing intensity of sheep, and hence the former indirectly affected plant community composition (Svengård Barre 2000).

7.6 Scaling Micro-scale Patterns up to "Real Space" Landscape Models of Ecosystem Response to Climate Change

Incorporating microclimatic variation, and the community and ecosystem variation it generates, into large-scale models requires scaling up from the micro-landscape to the regional landscape. One approach involves the use of high-resolution digital elevation models (DEM's). Modelling on the scale of real landscapes allows investigation of the consequences of making aggregation assumptions. Landscape-based outcomes can be compared to grid-cell means over a modest area to see just how much the two deviate for a variety of functions and conditions. In this way it is possible to assess the risk involved in running standard large-scale models, whether they are species-distribution models or ecosystem-function models.
Such an exercise was attempted by Wesser et al. (1994) for a small region (Kathul Mountain; 65° 21' N, 142° 17' W) in Yukon-Charley National Preserve, in subarctic Alaska. They explored the potential effects of increases and decreases in regional temperature on the response of vegetation on the actual mountainous landscape. The goal was to simulate the effects of past colder conditions (-2° C) and future climate warming (+2° C) on the distribution of plants on the slopes of Kathul Mountain. There were clear and dramatic rearrangements of communities on the landscape, with major increases and decreases in the abundance of extreme community types (Fig. 7.5). Notably, a decrease by 2° C caused warm steppe (red) to almost disappear from the landscape, and an increase by 2° C caused it to dominate the south-facing slopes of Kathul Mountain.

It thus seems probable that the climatic and biotic variation as documented by this and similar studies is indicative of similar magnitudes of variation in ecosystem properties and processes. Ecosystem responses to climate variation are likely non-linear, in which case ecological complexity generated by topography could pose difficulties in scaling from topographically complex local landscapes up to coarse grid-scale means. The latitudinal variation in patterns of radiation across diverse topographic settings suggests that such aggregation problems may be especially large at high latitudes. It is here that the modelling of ecosystem response to climate change may be the most vulnerable to aggregation errors, and further assessment of the potential magnitude and consequences of such errors is desirable.

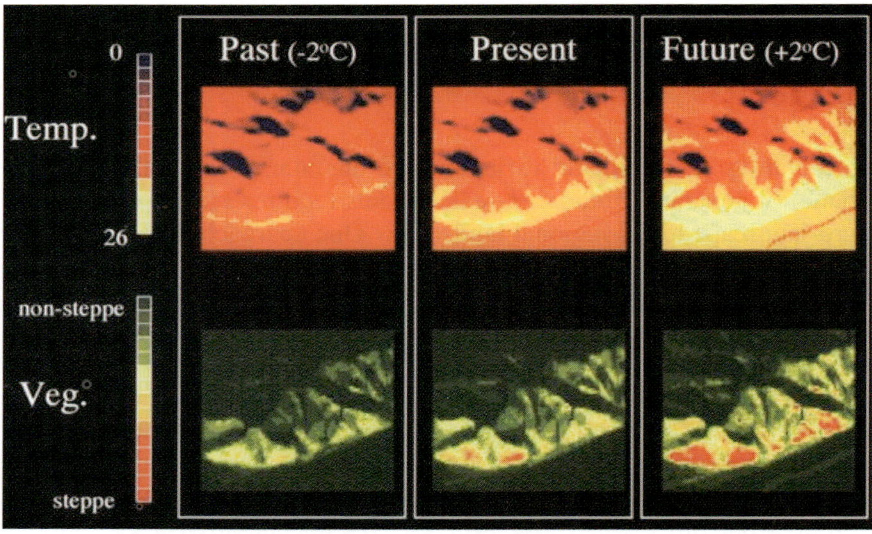

Fig. 7.5. Model outputs under scenarios of climate cooling (e.g. past climates) and warming (e.g. future climate). Codes for vegetation reconstruction: red = steppe, green = forest (low elevation) or tundra (high elevation). Figure modified from Wesser et al. 1994, with permission of S. D. Wesser.

7.7 Topographically induced variation in UV stress

Ongoing thinning of the ozone layer is causing significant increases in ultraviolet (UV) radiation in Polar regions, and this and projected changes may pose serious threats to terrestrial as well as aquatic/marine life (see chapters by Blumthaler; Vincent; Hanelt et al.; Hessen; and Nybakken; this volume). The variation in UV radiation loads on the terrestrial landscape and its consequences have received less attention than in the marine environment. UV radiation loads are distributed on rough landscapes in fundamentally the same way as are other radiation wavelengths. This suggests similar excesses in UV radiation load on south-facing slopes, potentially leading to significantly higher UV stress at these sites. Correspondingly, north-facing slopes could act as refugia from UV-stress, even with ozone thinning.

There is likely to be a genotype-environment correlation, however, in that organisms already living on south-facing slopes may have adaptations allowing them to cope with higher radiation loads (see Feder and Hofmann 1999). Thus, they may be pre-adapted to dealing with higher UV loads associated with ozone thinning (e.g. more phenolics, anthocyanins, or heat-shock proteins in leaves), perhaps in combination with migrating off the most stressful sites. This line of investigation deserves considerably more attention from researchers using UV-specific instrumentation, as well as those studying the genetics of UV tolerance.

7.8 Conclusions

Rough landscapes experience a mosaic of radiation budgets and microclimates, even when that roughness occurs at a small spatial scale. In our study area, growing-season means of environmental temperatures on slopes of different aspect separated by only a few tens of meters differed by up to 2.6 °C; this is near the middle of the range of projected temperature increases for CO_2-enhanced warming over the next 100 years (Watson et al. 2001). Microclimatic variation of this magnitude affects local community composition of both plants and animals. For plants in the study area, nearly 100% species replacement occurs between north-facing and south-facing slopes (Moe 2000). This suggests that the main biological response to future climate change may occur by species replacement through shifts in local distribution, rather than by phenotypic plasticity or ecotypic differentiation within taxa. Observations of dramatic variation in species composition in response to microclimatic variation suggest that landscapes have significant capacity to respond to large changes in temperature by local rearrangement of species already present. This is likely to occur more rapidly than long-distance migration of extra-limital species.

Acknowledgments

We thank Stuart and Andy Weiss for discussion of radiation-budget calculations and for sharing their spreadsheet program and Ørjan Totland and Jon Børre Ørbæk for comments on the manuscript, and the Norwegian Research Council for funding under the Arctic Light and Heat (ALV) program.

References

Arft AM, Alatalo J, Bret-Harte MS, Dale M, Deimer M, Gugerli F, Gurevitch J, Henry GHR, Hollister R, Jones MH, Jónsdóttir IS, Laine K, Lévesque E, Marion GM, Molau U, Molgaard P, Nordenhäll U, Raszhivin V, Robinson

CH, Starr G,. Stenström A, Stenström M, Totland Ø, Turner L, Walker L, Walker, MD, Webber PJ, Welker JM, Wookey PA (1999) Responses of tundra plants to experimental warming: meta-analysis of the International Tundra Experiment. Ecological Monographs 69: 491-511.

Armbruster WS, Rae DA, Edwards, ME Assessing biotic-community response to microclimatic change: measurement and analysis of variation in environmental temperature and plant and invertebrate communities in Arctic tundra. Ecography (submitted).

Bakken GS (1992) Measurement and application of operative and standard operative temperatures in ecology. American Zoologist 32:194-216.

Bishop JA, Armbruster WS (1999) Thermoregulatory abilities of Alaskan bees: effects of size, phylogeny, and ecology. Functional Ecology 13:711-724.

Bonan GB, Chapin FS, Thompson SL (1995) Boreal forest and tundra ecosystems as components of the climate system. Climatic Change 29:145-167.

Bret-Harte MS, Shaver GR, Zoerner JP, Johnstone JF, Wagner JL, Chavez AS, Gunkelman RF, Lippert SC, Laundre JA (2001) Developmental plasticity allows *Betula nana* to dominate tundra subjected to an altered environment. Ecology 82:18-32

Brockway DG (1998) Forest plant diversity at local and landscape scales in the Cascade Mountains of southwestern Washington. Forest Ecology and Management 109:323-341.

Brubaker LB, Anderson PM, Edwards ME, Lozhkin AV (2005) Beringia as a glacial refugium for boreal trees and shrubs: new perspectives from mapped pollen data. J. Biogeography 32:833-848.

Cattle H, Crossley J (1995) Modeling Arctic climate-change. Philosophical Transactions of the Royal Society of London Series A- Mathematical Physical and Engineering Sciences 352:201-213.

Chapin FS, Shaver GR (1996) Physiological and growth responses of arctic plants to a field experiment simulating climatic change. Ecology 77: 822-840.

Chapin FS, Shaver GR, Giblin AE, Nadelhoffer KJ, Laundre JA (1995) Responses of arctic tundra to experimental and observed changes in climate. Ecology 76:694-711.

Chapin FS, McGuire AD, Randerson J, Pielke R, Baldocchi D, Hobbie SE, Roulet N, Eugster W, Kasischke E, Rastetter EB, Zimov SA, Running SW (2000) Arctic and boreal ecosystems of western North America as components of the climate system. Global Change Biology 6 (suppl. 1):211-223.

Clark JS, Lewis M, McLachlan JS, HilleRisLambers, J (2003) Estimating population spread: What can we forecast and how well? Ecology, 84, 1979-1988.

Coker PD, Coker AM (1973) *Phyllodoce caerulea.* Journal of Ecology 61:901-913.

Crawford, R. M. M. and Abbott, R. J. 1994. Pre-adaptation of arctic plants to climate-change. Botanica Acta 107: 271-278.

Davis MB, Shaw RG (2001) Range shifts and adaptive responses to Quaternary climate change. Science. 292: 673-679.

Edwards ME, Armbruster WS (1989) A steppe-tundra transition on Kathul Mountain, Alaska. Arctic and Alpine Research 21:296-304.

Edwards ME, Brubaker LB, Lozhkin AV, Anderson PM Structurally novel biomes: a response to past warming in Beringia. Ecology 86:1696-1703.

Eugster W, Rouse WR, Pielke RA, McFadden JP, Baldocchi DD, Kittel TGF, Chapin FS, Liston GE, Vidale PL, Vaganov E, Chambers S (2000) Land–atmosphere energy exchange in Arctic tundra and boreal forest: available data and feedbacks to climate. Global Change Biology 6 (suppl. 1):84-115.

Friedlingstein P, Dufresne JL, Cox PM, Rayner P. (2003) How positive is the feedback between climate change and the carbon cycle? Telus B. Chemical and Physical Meteorology 55:692-700.

Hamrick JL, Allard, RW (1972) Microgeographical variation in allozymes frequencies in *Avena barbata*. Proc. National Acad. Sci. USA 69:2100-2104.

Hungerford RD, Nemani RR, Running SW, Coughlan JC (1989) MtClim - A mountain microclimate simulation-model. USDA Forest Service Intermountain Research Station Research Paper 414:1-52.

Goldblatt P, Manning JC (2002) Plant diversity of the Cape Region of southern Africa. Annals of the Missouri Botanical Garden 89:281-302.

Ichii K, Matsu Y, Murakami K, Mukai T, Yamaguchi Y, Ogawa K (2003) A simple global carbon and energy coupled cycle model for global warming simulation: sensitivity to the light saturation effect. Tellus B- Chemical and Physical Meteorology 55:676-691.

Jones CD, Cox P, Huntingford C. (2003) Uncertainty in climate-carbon-cycle projections associated with the sensitivity of soil respiration to temperature. Tellus B-Chemical and Physical Meteorology 55: 642-648.

Kane DL, Reeburgh WS (1998) Introduction to special section: Land-Air-Ice Interactions (LAII) Flux Study. Journal of Geophysical Research-Atmospheres 103 (D22): 28913-28915.

Kaplan JO, Bigelow NH, Bartlein PJ, Christiansen TR, Cramer W, Harrison SP, Matveyeva NV, McGuire AD, Murray DF, Prentice IC, Razzhivin VY, Smith B, Walker DA, Anderson PM, Andreev AA, Brubaker LB, Edwards ME, Lozhkin AV, Ritchie JC (2003) Climate change and arctic ecosystems II: Modeling paleodata-model comparisons, and future projections. Journal of Geophysical Research 108 (D19): Art. No. 8171. doi: 10.1029/2002JD002559

Knapp AK (1985) Early season production and microclimate associated with topography in a C-4 dominated grassland. Acta Oecologica-Oecologia Plantarum 6:337-346.

Knorr W, Prentice IC, House JI, Holland EA (2005) Long-term sensitivity of soil carbon turnover to warming. Nature 433:298-301.

Levis S, Foley JA, Pollard D. (2000) Large-scale vegetation feedbacks on a doubled CO2 climate. Journal of Climate 13, 1313-1325.

Levis S, Foley JA, Pollard D. (1999) Potential high-latitude vegetation feedbacks on CO2-induced climate change. Geophysical Research Letters 26, 747-750.

Li CC (1975) Path analysis-- A primer. Boxwood Press, Pacific Grove, CA, USA.

Lloyd AH, Armbruster WS, Edwards ME (1994) Ecology of a steppe-tundra gradient in interior Alaska. J. Vegetation Sci. 5:897-912.

McCune B, Mefford MJ (1999) PC-Ord. Multivariate analysis of ecological data, Version 4.0. MjM Software Development, Gleneden Beach, OR, USA.

McGuire AD, Wirth C, Apps M, Beringer J, Clein J, Epstein H, Kicklighter DW, Bhatti J, Chapin FS, de Groot B, Efremov D, Eugster W, Fukuda M, Gower T, Hinzman L, Huntley B, Jia GJ, Kasischke E, Melillo J, Romanovsky V, Shvidenko A, Vaganov E, Walker D (2002) Environmental variation, vegetation distribution, carbon dynamics and water/energy exchange at high latitudes. Journal of Vegetation Science 13:301-314.

McGuire AD, Sturm M, Chapin FS (2003) Arctic transitions in the land-atmosphere system (ATLAS): Background, objectives, results, and future directions
Journal of Geophysical Research-Atmospheres 108 (D2): art. no. 8166.

Moe A (2000) Patterns of insect and plant distribution in relation to radiation loads and microclimate on Spitsbergen. Hovedfag thesis, Botany Dept., NTNU, Trondheim, Norway.

Mysterud A, Langvatn R, Yoccoz NG, Stenseth NC (2001) Plant phenology, migration and geographical variation in body weight of a large herbivore: the effect of a variable topography. Journal of Animal Ecology 70:915-923.

Osterkamp TE, Romanovsky VE. (1999) Evidence for warming and thawing of discontinuous permafrost in Alaska. Permafrost Periglac. 10:17-37.

Pan YD, Melillo JM, McGuire AD, Kicklighter DW, Pitelka LF, Hibbard K, Pierce LL, Running SW, Ojima DS, Parton WJ, Schimel DS (1998) Modeled responses of terrestrial ecosystems to elevated atmospheric CO2: a comparison of simulations by the biogeochemistry models of the Vegetation/Ecosystem Modeling and Analysis Project (VEMAP). Oecologia 114:389-404.

Pearson SM, Turner MG, Wallace LL, Romme WH (1995) Winter habitat use by large ungulates following fire in northern Yellowstone National Park. Ecological Applications 5:744-755.

Porter WP, Sabo JL, Tracy CR, Reichman OJ, Ramankutty N (2002) Physiology on a landscape scale: Plant-animal interactions. Integrative and Comparative Biology 42:431-453.

Rae DA (2003) Plant- and invertebrate-community responses to species interactions and microclimate gradients in alpine and arctic environments. Ph.D. thesis, Biology Dept., NTNU, Trondheim, Norway.

Rupp TS, Chapin FS, Starfield AM (2001) Modeling the influence of topographic barriers on treeline advance at the forest-tundra ecotone in northwestern Alaska. Climatic Change 48: 399-416.

Schauffler M, Jacobson, GL Jr (2002) Persistence of coastal refugia during the Holocene in northern New England, USA, detected by stand-scale pollen stratigraphies. Journal of Ecology, 90:235-250.

Sellers PJ, Hall FG, Kelly RD, Black A, Baldocchi D, Berry J, Ryan M, Ranson KJ, Crill PM, Lettenmaier DP, Margolis H, Cihlar J, Newcomer J, Fitzjarrald D, Jarvis PG, Gower ST, Halliwell D, Williams D, Goodison B, Wickland DE, Guertin FE (1997) BOREAS in 1997: Experiment overview, scientific results, and future directions. Journal of Geophysical Research-Atmospheres 102 (D24): 28731-28769.

Sitch S, Smith B, Prentice IC, Arneth A, Bondeau A, Cramer W, Kaplan JO, Levis S, Lucht W, Sykes MT, Thonicke K, Venevsky S (2003) Evaluation of ecosystem dynamics, plant geography and terrestrial carbon cycling in the LPJ dynamic global vegetation model. Global Change Biology 9:161-185.

Stewart, JR, Lister, AM (2001) Cryptic northern refugia and the origins of the modern biota. Trends in Ecology and Evolution16:608-613.

Strahler AN, Strahler AH (1989) Elements of physical geography, 4th ed. Wiley, New York, 562 pp.

Sveingård Barre M (2000) Tundra plant community composition in relation to radiation budgets and disturbance, North Norway. Hovedfag thesis, Geography Dept. NTNU, Trondheim, Norway.

Sydes, C (1997) Scotland's rare plants. An overview of distribution and threats. Scottish Natural Heritage, Edinburgh, UK.

Watson RT and the Core Writing Team (2001) IPCC Third Assessment Report: Climate Change 2001: Synthesis report. IPCC, Geneva, Switzerland. pp 184

Weiss SB, Murphy DD, Ehrlich PR, Metzler CF (1993) Adult emergence phenology in checkerspot butterflies - The effects of macroclimate, topoclimate, and population history. Oecologia 96:261-270.

Weller G (1998) Regional impacts of climate change in the Arctic and Antarctic. Ann. Glaciol. 27:543-552.

Wesser S D, Armbruster WS (1991) Controls over species distribution across a forest-steppe transition: a causal model and experimental test. Ecological Monographs 61:323-342.

Wesser SD, Armbruster WS, Debevec EM, Edwards ME (1994) Subarctic steppe distribution through space and time: a microclimate modeling approach. In R. H. Meehan, V. Sergienka, and G. Weller (eds.), Bridges of Science between North America and the Russian Far East. Arctic Division, AAAS, Fairbanks, AK, pp 63-69.

Williams PJ (1995) Permafrost and climate-change - Geotechnical implications. Philosophical Transactions of the Royal Society of London Series A - Mathematical Physical and Engineering Sciences 352:347-358.

Young KL, Woo MK, Edlund SA (1997) Influence of local topography, soils, and vegetation on microclimate and hydrology at a high Arctic site, Ellesmere Island, Canada. Arctic and Alpine Research 29:270-284.

8 The flow of Atlantic water to the Nordic Seas and Arctic Ocean

Tore Furevik[1,2], Cecilie Mauritzen[3], and Randi Ingvaldsen[4,2]

[1]Geophysical Institute, University of Bergen, Norway, [2]Bjerknes Centre for Climate Research, Bergen, Norway, [3]Norwegian Meteorological Institute, Oslo, Norway, [4]Institute of Marine Research, Bergen, Norway, e-mail: tore@gfi.uib.no

8.1 Introduction

The Nordic Seas, encompassing the Greenland, Iceland, Norwegian and Barents Seas, act as a buffer zone between the warm and saline waters of the North Atlantic Ocean, and the cold and fresh waters of the Arctic Ocean. The region, with its area of 4.1×10^6 km^2 and a water volume of 4.5×10^6 km^3, covers 1% and 0.3% of the World Ocean area and volume respectively (Jakobsson, 2002). Despite these modest numbers, the Nordic Seas with their complex topography, water mass distribution and flow regimes, is recognized as a key area for the conversion from light surface water to dense deep waters, having climatic impacts over large parts of the globe.

The most important factor for the Nordic Seas marine climate is the amount of warm and saline Atlantic Water (AW) entering over the Greenland-Scotland Ridge, and flowing north in the eastern part of the Norwegian Sea eventually reaching the Arctic Ocean via the Barents Sea or through the Fram Strait between Greenland and Svalbard (Fig. 8.1). Its vital importance for the marine climate, water mass transformation and biomass production was recognized almost a century ago (Helland-Hansen and Nansen, 1909). Despite this, there is still no consensus on what is actually driving the inflow of AW to the Nordic Seas, or what are the main mechanisms for transforming the AW into the dense overflow water that is returning south over the Greenland-Scotland Ridge. This overflow water ventilates the world oceans as the deepest part of the Atlantic meridional overturning circulation (AMOC), the Atlantic part of what is commonly referred to as the "great conveyor belt" (Broecker, 1991).

Besides the heat transported by the ocean currents, the Nordic Seas and northwest Europe owe a very mild and favourable climate to the heat

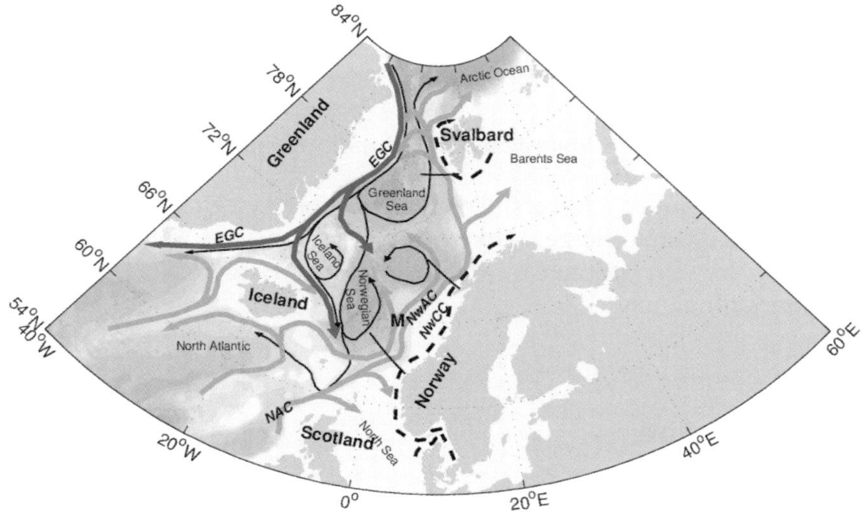

Fig. 8.1. Schematic flow map of the Nordic Seas showing the northward flow of Atlantic Water (light arrows), the southward flow of Polar Water (dark arrows), the coastal currents (dashed arrows), and finally the intermediate and deep circulation (black arrows). The thin straight lines show from south to north the position of the Svinøy, Gimsøy and Sørkapp sections (Fig. 8.2), and M marks the position of the OWSM. Depths are shaded at 500m intervals. Abbreviations are explained in the main text.

associated with the North Atlantic storm track. As a consequence of the dependency on heat transported into the region, the North Atlantic region has been particularly sensitive to climate changes in the past. Abrupt climate changes with ~10°C temperature jumps over just a few decades occurred in the wake of the last glacial maximum. The last major climate perturbation in the region was the Younger Dryas termination some 11 600 years ago (Dansgaard et al., 1989). Observational evidence suggests that the abrupt changes in climate have been driven by sudden switches in the strength or positioning of the Gulf Stream and its extension into the Nordic Seas. These switches may have been caused by large ice discharges from inland glaciers or fresh water flushing from dammed lakes that have made the surface waters fresh enough to inhibit the deepwater formation (Clark et al., 2001).

The findings in paleo records and early results based on relatively simple numerical climate models, have generated much debate on what may happen to the inflow of AW in a global warming scenario, and in particular if unpleasant surprises are laying ahead of us (Broecker, 1997;

Rahmstorf, 1995). Despite the fact that our climate has been very stable for more than 10 000 years, and abrupt climate changes therefore may seem unlikely in present day climate, the release of greenhouse gases and the associated global warming may perturb the balance. Since the late '60s there has been no deep-reaching convection in the Greenland Sea (Dickson et al., 1996), the deep waters of the Nordic Seas have become warmer (Østerhus and Gammelsrød, 1999), the intermediate waters have become fresher (Blindheim et al., 2000), and there are indications of a reduced southward flow of cold, dense water over the Greenland-Scotland Ridge (Hansen et al., 2001). Most state-of-the-art climate models participating in the third assessment report of the Intergovernmental Panel of Climate Change (IPCC-TAR) suggested a 30-40% reduction in the strength of the AMOC during this century (Houghton et al., 2001). It is therefore tempting to ask if the recent observations in the Nordic Seas are indicating that such a reduction has already started.

In this review we will focus on the inflow of AW to the Nordic Seas, the pathways and transformation of the AW within the Nordic Seas and Arctic Ocean, and the expected consequences of global warming. In section 2, we outline the main bathymetric features, the circulation patterns and the hydrographical distribution. In section 3, some of the large-scale changes that have taken place in the Nordic Seas climate during the last couple of decades will be reviewed. In section 4, different forcing mechanisms for inflow and conversion of water masses are discussed in the context of the changes that have taken place. Finally the review is concluded and summarized in section 5.

8.2 Nordic Seas bathymetry, circulation, and water mass transformation

8.2.1 Bathymetry

The Nordic Seas are characterized by a complex topography, with deep basins, generally shallow shelves and steep continental slopes (Fig. 8.1). The deepest areas are found in the Greenland, Norwegian and Lofoten basins. They have depths exceeding 3000 m, and are separated by the Jan Mayen fracture zone and the Mohn Ridge. In the east, the continental shelf extends from west of Ireland to the Fram Strait, with the North Sea and the Barents Sea regions being part of the shelf. Off the western coast of Norway, and in particular off Lofoten, the continental shelf becomes narrow, having a sharp shelf break and a very steep slope towards the deep basins.

In the west, the continental shelf becomes more narrow towards south, from approximately ~300 km in the north to only ~100 km in the Denmark Strait. The Greenland-Scotland Ridge, which separates the Nordic Seas from the Atlantic Ocean, is shallow. The sill depths of the three main passages are, from west to east, 620 m, 480 m, and 840 m, where the deepest sill is found in the narrow Faroe-Bank Channel. Thus waters at depths between 620 m and 840 m can only escape through the Faroese Channels, while waters below 840 m cannot directly cross the ridge. The exchange with the Arctic Ocean is much less restricted; the sill depth of the Fram Strait is ~ 2600 meters.

8.2.2 Circulation

The AW that eventually enters the Nordic Seas originates in the tropical Atlantic from where it is transported north by the Gulf Stream and its northward extension known as the North Atlantic Current (NAC). By the time the AW reaches the Nordic Seas it has been significantly cooled and freshened compared to its tropical origin.

The AW enters the Nordic Seas along three relatively distinct paths; west of Shetland, west of the Faroes and west of Iceland (Fig. 8.1). Given the heat losses over the Nordic Seas and Arctic Ocean, simple heat and volume budget calculations demand that the exchange between those seas and the Atlantic Ocean ought to be around 8-9 Sv (Worthington 1970). Systematic measurements during the last decade confirm this estimate:

The eastern branch of the NAC enters the Nordic Seas at the eastern slope of the Faroe-Shetland Channel. The mean volume and heat transports (relative to 0°C) based on direct measurements from summer 1998 through 2002 are 3.2 Sv (1 Sv=10^6 m^3s^{-1}) and 123 TW (1 TW=10^{12} W) respectively (Turrell et al., 2003). The annual cycle is found to be only 0.2 Sv.

The middle branch of the NAC enters the Nordic Seas between Iceland and the Faroe Islands. This branch veers to the right upon crossing the ridge, and can be seen as a sharply defined warm current north of the Faroe Islands. East of the Faroes some of this water flows south and partly mixes with the eastern branch of the NAC. From summer 1997 to summer 2001 the mean volume and heat transports north of the Faroes were 3.5 Sv and 124 TW respectively (Hansen et al., 2003), with insignificant annual or interannual variability. The average volume flux is thus slightly higher than the volume flux of the eastern branch, while the heat flux remains the same due to lower temperatures of the inflowing water.

The third and westernmost branch of the NAC enters the Nordic Seas along the eastern rim of the Denmark Strait, where it is known as the North

Icelandic Irminger Current. The flow continues to the east and southeast around the periphery of the Iceland shelf, where it looses most of its heat through atmospheric forcing and mixing with local water masses. Based on measurements from summer 1997 to summer 2000 the amount of water and heat carried by this current are roughly 1 Sv and 25 TW respectively (Jónsson and Briem, 2003).

The former two branches constitute the Norwegian Atlantic Current (NwAC), which flows towards the Barents Sea and the Fram Strait (Fig. 8.1). Both branches follow topography; the inner branch follows roughly the 500-meter isobath whereas the outer branch follows roughly the 2000-meter isobath. The warm waters of the current are found not only in the high-velocity cores but also in the broad region between the cores. This gives the current a broad, sluggish river-like appearance in hydrography data.

During the 2000 km long transit northwards from the southern Norwegian Sea the warm current continues to loose heat and gain freshwater. It has also lost some volume to the surrounding water masses; of the roughly 6.5 Sv entering the southern Norwegian Sea an estimated 1.5 Sv exits to the Barents Sea (Ingvaldsen et al., 2004) and 3-5 Sv exits through the Fram Strait (Schauer et al., 2004). Assuming that the changes in heat and salinity only require entrainment of small volumes of the surrounding water masses, there can only be of the order of 1 Sv (between 0 and 2) of AW that can be lost to the surroundings within the Nordic Seas.

There are however large uncertainties in both entrainment and transport rates that make this a rather crude estimate.

8.2.3 Water mass transformation

Fresher and mostly colder waters surround the NwAC during the transit through the Norwegian Sea; inshore is the fresh water in the Norwegian Coastal Current (NwCC), and offshore is the Polar Water (often called Arctic Water). The NwCC originates as a roughly 1 Sv branch (Furnes et al., 1986) of the Faroe-Shetland inflow of warm AW takes a detour into the North Sea and gains freshwater through runoff from the continents, export of brackish water from the Baltic Sea, and mixing with ambient water masses. The low-salinity current continues from the North Sea towards the Barents Sea along the coast of Norway. Offshore, the Polar Water is brought south with the East Greenland Current (EGC in fig. 8.1) and into the interior basin with the Jan Mayen and East Icelandic Currents.

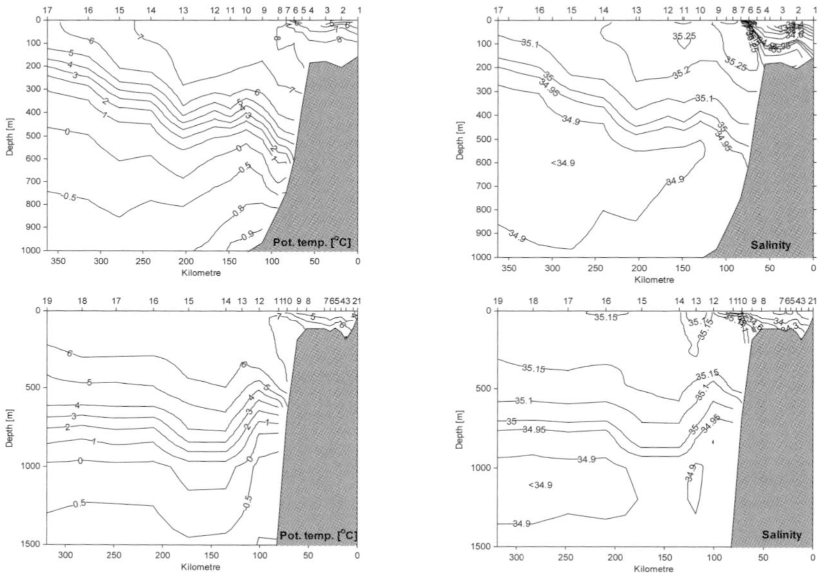

Fig. 8.2. Distribution of potential temperature (left fig.) and salinity (right fig.) in the Svinøy and Gimsøy sections (locations shown in fig. 8.1). The data are based on respectively 17 and 19 regular CTD-stations taken by the R/V Johan Hjort in the period 2-18 March 2001. The numbers over each plot indicate the standard stations (Courtesy of Kjell Arne Mork, Institute of Marine Research, Bergen, Norway).

The NwAC interacts with the surrounding waters; satellite images show vigorous eddy activity on both sides of the warm current (not shown). This results in mixing of water masses and a lowering of temperature and salinity of the AW as it flows northwards. The other main agent of transformation is air-sea heat loss (the freshwater fluxes here do not make a large direct impact on density). Near the inflow region the AW typically has temperature and salinity (TS) values in the range 7-9°C and 35.10-35.40 (Blindheim and Østerhus, 2004), with TS values close to 8°C and 35.3 when averaged over the 50-200 m depth interval (Furevik et al., 2002). At the Svinøy section located at approximately 63°N, typical TS values are 7-8°C and 35.2, decreasing to 6-7°C and 35.15 at the Gimsøy section crossing the Lofoten Basin some 400 km further to the north (Fig. 8.2). From Svinøy to Gimsøy a significant deepening of the Atlantic layer is seen both in the temperature and salinity structures, recently described to be a result of the AW meeting a southward flowing deep current in this region (Orvik, 2004). This feature is also seen in fig. 8.3, where climatological mean

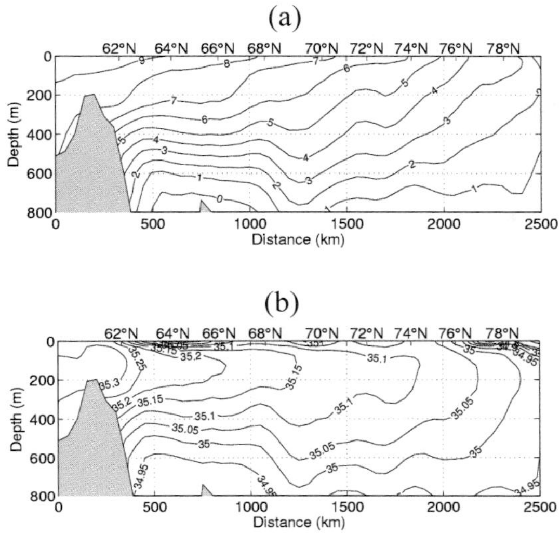

Fig. 8.3: Mean temperature and salinity profiles along the pathway of the AW inflow (defined by the positions of the salinity maximum) from the Faroe-Shetland inflow to the Fram Strait. The plots are based on a climatology compiled from ~100 000 hydrographical station (Furevik et al., 2002).

temperature and salinities are plotted for a section following the core of the maximum salinity through the Nordic Seas.

Further to the north, where the Barents Sea and the West Spitsbergen branches separate, the Atlantic layer shallows, and typical TS values are 3-5°C and 34.90-35.15 (Fig. 8.3). By the time the AW reaches the 79°N latitude off northwest Spitsbergen, the TS values have dropped below 2°C and 35 respectively. North of Spitsbergen the AW continues east as a subsurface current insulated from the atmospheric forcing by fresh Polar Water above. In the Arctic Ocean maximum temperature of the Atlantic layer decreases from 2°C near the inflowing region in the Nansen Basin, to 0°C in the Canadian Basin (Carmack, 2000), while the mean depth descends from about 200 m to about 400 m along this path.

In the Barents Sea the AW completely looses its TS characteristics during its passage to the Arctic Ocean due to a very strong cooling by the atmosphere. When the water eventually leaves the Barents Sea between Novaja Zemlja and Frantz Josef Land, the TS characteristics are close to 0°C and 34.85 respectively (Schauer et al., 2002). Thus the AW flowing into the Arctic Ocean through the Barents Sea is much colder than the water entering through the Fram Strait, and is not contributing to the warmest parts of the AW layer in the Arctic Ocean.

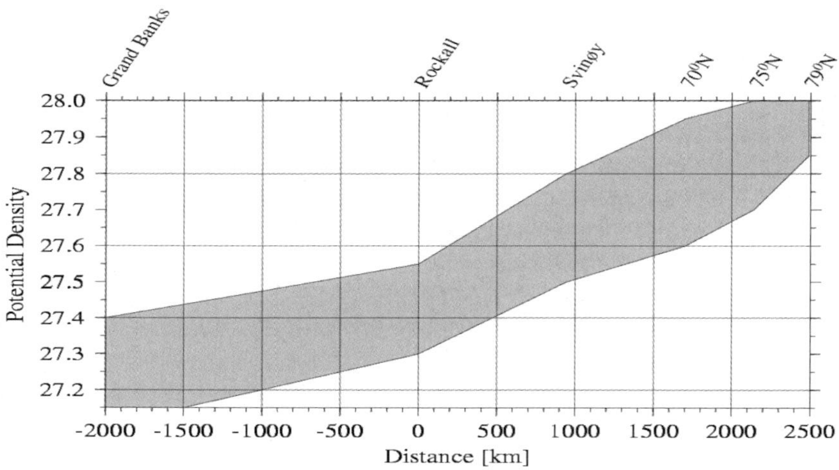

Fig. 8.4: Schematic changes in AW density from Grand Banks (east of New Foundland, Canada) via Rockall (west of Ireland) to the Svinøy section and eventually into the Fram Strait. The shading represents the spread of the measurements within the core of the AW at each location. The numbers on the y-axis are in σ_θ units, thus 28.0 equals 1028 kgm^{-3}. The plot is based on the World Ocean Atlas 1998 data provided by the NOAA-CIRES Climate Diagnostics Center, Boulder, Colorado, USA, from their Web site at http://www.cdc.noaa.gov/.

The buoyancy evolution along the path of the NAC and the NwAC is visualized in fig. 8.4. The largest buoyancy losses of the AW are experienced in the Nordic Seas, and the region from the Lofoten Basin to Spitsbergen and the Barents Sea is an area of particularly large change in density. This is also the region that experiences the highest mean annual heat losses of the Nordic Seas (Fig. 8.5a), so a large change in density is not a surprising result. According to the NCEP-NCAR reanalysis data, winter mean heat loss reaches close to 300 Wm-2 in the northern Fram Strait and Barents Sea (Fig. 8.5b). The sensible heat flux is found to be the dominant flux component (Furevik and Nilsen, 2004).

The return flow to the North Atlantic, which also adds up to between 6 and 8 Sv (the Bering Strait inflow to the Arctic is roughly balanced by the outflow through the Canadian Archipelago), is still colder and fresher than the transformed AW at 75-80°N. Some of it exits the Denmark Strait with the East Greenland Current as relatively light water; this is the very fresh Polar Water, which does not sink upon entering the North Atlantic. But the bulk of the return flow (roughly 5 Sv, see Dickson and Brown, 1994) is dense and thus constitutes "overflow water". Many processes have been mentioned as contributing to that final transformation; the most classical

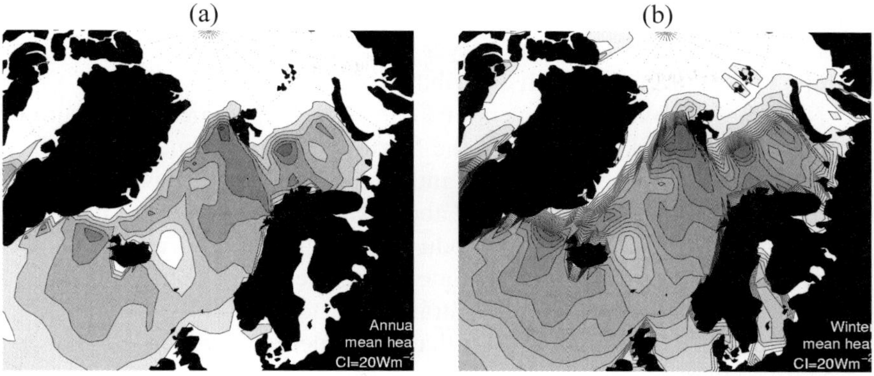

Fig. 8.5: Annual and winter (Dec-Mar) mean total heat flux calculated from the NCEP/NCAR re-analysis data 1948-2003, provided by the NOAA-CIRES Climate Diagnostic Center, Boulder, Colorado, USA from their Web site at http://www.cdc.noaa.gov/. Contour intervals are 20Wm-2. The fluxes are everywhere directed out of the ocean.

explanation, going back to Helland-Hansen and Nansen (1909), is that of open-ocean convection in the Greenland Sea, where cooling of warm, saline AW makes the upper waters dense enough to sink. Another process discussed is dense water formation, by cooling and/or brine rejection, on the Arctic shelves (see e.g. Aagaard et al., 1985; Schauer, 1995).

Mauritzen (1996a,b) aims to quantify the various processes of transformation, and finds that 1) it is unrealistic that the production rate through open-ocean convection is larger than 1 Sv (consistent with the recent transport estimates discussed in the previous section, where the implicit volume loss to the Nordic Seas interior is 0 – 2 Sv); 2) the main product, by volume, of the Greenland and Iceland Seas open-ocean gyres is Arctic Intermediate Water (the water discussed e.g. in Blindheim, 1990); 3) transformation to dense water within the NwAC itself actually is the most important process for the overturning (roughly 7 Sv of water is made denser by roughly 0.5 kgm^{-3} from Rockall, and 0.7 kgm^{-3} from the Grand Banks; see fig. 8.4). Finally, Mauritzen (1996a,b) finds that since the branch that enters the Barents Sea (which has a strength of roughly 1.5 Sv, see above) is transformed to truly dense water by the time it reaches the Arctic Ocean (as discussed above), shelf production in that sea is of first-order importance for the transformation process.

To summarize, many steps are involved in completing the overturn. The largest steps are taken within the NAC and NwAC where the warm water is being made cooler and fresher. Our best quantitative estimates suggest that the most important final steps are open-ocean convection in the Greenland and Iceland Seas, shelf production in the Barents Sea and interior mixing within the Atlantic layer under the ice and the halocline in the Arctic Ocean, each of which contribute roughly equally.

Note that the overturn does not take the shape of a vertical overturn until the return flow sinks to depth in the North Atlantic Ocean; within the Nordic Seas light northward-flowing and dense southward-flowing waters are both found at shallow depths (and above sill levels), above the densest waters. Note also that the main product of the Greenland and Iceland Sea convection, the Arctic Intermediate Water, exits primarily through the Faroese Channels (the Denmark Strait being too shallow). As we shall see in the next section, it is this water that has undergone the largest changes in recent decades.

8.3 Observed changes in the Nordic Seas marine climate

8.3.1 North Atlantic Oscillation

The leading mode of atmospheric variability in the Atlantic sector is the well-known North Atlantic Oscillation (NAO), representing a shift of atmospheric mass between the Icelandic low and the Azores high, directly influencing the strength of the westerlies (Hurrell, 1995). Several indices in use are based on the pressure difference across the westerlies. Measurements from Gibraltar and Iceland going back to 1824 show strong year-to-year variability in the strength of the westerlies, and also indications of more decadal-scale variability (Fig. 8.6a). A pronounces shift occurred from the relatively weak westerlies in the '60s to the strong westerlies in the '90s. During positive NAO winters (high index years), a strong Azores high and a deep Icelandic low make the meridional pressure gradient and thus the westerlies stronger than normal (Fig. 8.6b). This results in higher temperatures and more precipitation over northern Europe, and correspondingly colder and dryer conditions over southern Europe (see Marshall et al., 2001, Hurrell et al., 2003 for reviews on the NAO and its many impacts on several physical, ecological, and biological parameters).

While the NAO is the dominant mode for the North Atlantic region as a whole, it should be noted that for more regional studies the NAO index has often proved insufficient in describing the main climatic variations. A reason for this is that the spatial pattern of the NAO has shifted throughout the years, as more low pressure systems have gone further northeast into the Nordic Seas. As a consequence sea-level pressure (SLP) variability in

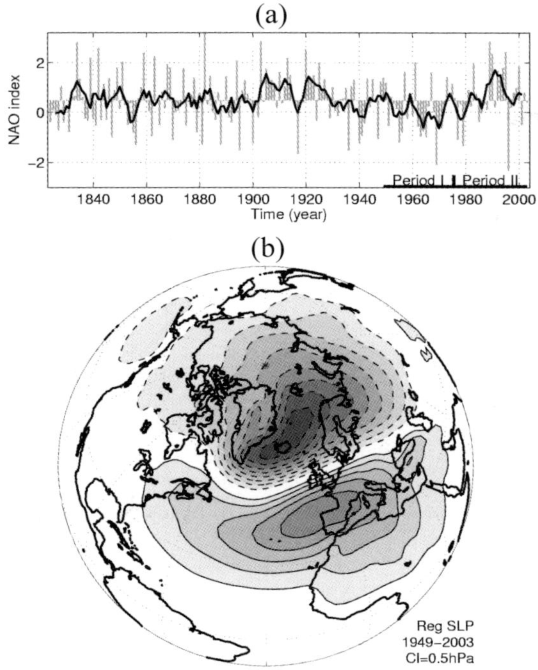

Fig. 8.6: a) The NAO index based on sea-level pressure at Gibraltar and Stykkisholmur, Iceland (updated from Jones et al., 1997). Thin line shows the winter (Dec-Mar) means, thick line shows the data filtered with a 5 years moving average. Periods used for making the regression maps in fig. 8.7, are marked on the time axis. b) The winter-mean sea-level pressure regressed on the NAO index in (a). Contour intervals are 0.5 hPa, with negative values dashed. The plotted values correspond to changes in the sea-level pressure associated with a unit increase in the index. Data are from the NCEP/NCAR reanalysis product.

the Barents Sea and Fram Strait not captured by the NAO index in the '50s and '60s, show a strong correlation with the NAO index during the '80s and '90s. The regression plots between SLP and NAO index in fig. 8.7 demonstrate this.

The NAO can influence the ocean circulation by a variety of mechanisms and on a wide range of time scales. The fastest responses are due to anomalous momentum forcing (wind), where divergences in the Ekman transports change the surface elevation and accelerate the flows due to horizontal pressure gradients. Time scales are here on the order of days. Slower responses involve changes of the internal density structures that can be due to anomalous momentum forcing, heat fluxes, or fresh water fluxes. These mechanisms can work locally in the Nordic Seas, or they can

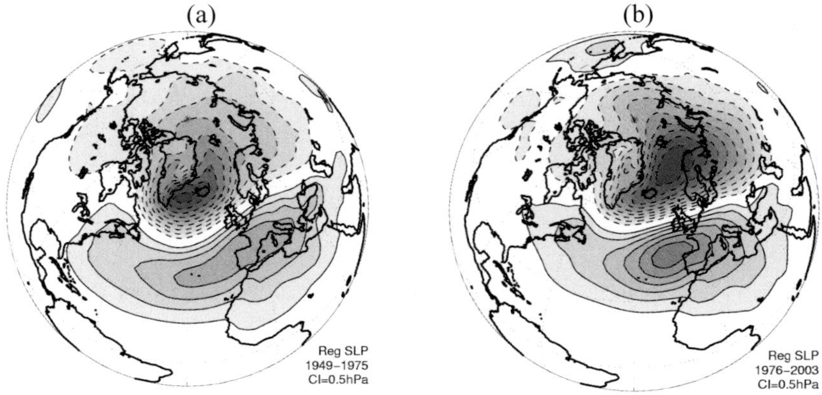

Fig. 8.7: Winter-mean sea-level pressure regressed on the NAO index shown in fig. 8.6a for the periods 1949-1975 (a) and 1976-2003 (b). Contour intervals are 0.5 hPa, with negative values dashed. The plotted values correspond to changes in the sea-level pressure associated with a unit increase in the index. Data are from the NCEP/NCAR reanalysis product.

invoke signals in the ocean that are either advected with the mean flow into the region, or propagating as slow baroclinic waves. Time scales can here be from weeks to years (see papers by Marshall et al., 2001; Visbeck et al., 2003; Furevik and Nilsen, 2004 for reviews).

8.3.2 Atlantic inflow

Due to the baroclinic structure of the Gulf Stream and its NAC extension, it is not surprising that the flow seems to be susceptible to atmospheric forcing on a variety of time scales. Drifter data from the '90s combined with sea surface temperature and sea level height data suggest that during positive NAO years, the branch going towards the Irish-Scottish Shelf is strengthened, and there is also a stronger cyclonic circulation in the Iceland and Irminger Basins (Flatau et al., 2003). The latter is in concordance with estimates based on changes in hydrography (Bersch, 2002), and from advection rates for various salinity anomalies observed during the last three decades (Belkin, 2004). Also observations at the Svinøy section indicate that the eastward branch of the NwAC is stronger during winter than during summer, and that there is a positive correlation between the NAO and the flow (Orvik et al., 2001). Similar results are indicated by altimeter data (Furevik and Nilsen, 2004; Skagseth et al., 2004), and also model simulations by Karcher et al. (2003) showed an increased inflow of AW through the Faroe-Shetland Channel during the strongly positive NAO years of the early '90s.

In contrast, direct measurements from the three branches of AW inflow to the Nordic Seas, fail to show any significant seasonal signal in the total inflow, nor do they show any correlation with the strength of the NAO (Hansen et al., 2003). A way to explain the seasonality and NAO correlation downstream at the Svinøy section, is that with enhanced cyclonic wind forcing (more lows), a larger fraction of the waters that originally enter the Nordic Seas over the Iceland-Faroe Ridge, recirculates and mixes with the Faroe-Shetland branch east of the Faroes before eventually flowing north in the eastern branch of the NwAC (Hansen et al., 2003). Thus the eastern branch of the NwAC is strengthened and the western branch weakened during strong cyclonic wind forcing (positive NAO years). It should nonetheless be noted that there are still discrepancies between this explanation and the altimeter and drifter data.

Considering the variability in the Fram Strait and the Barents Sea branches, direct measurements show that both branches have strong variability on a variety of time scales, including interannual. The AW transport in the West Spitsbergen Current in the Fram Strait increased between 1997 and 2000 from about 3 to 4 Sv (Schauer et al., 2004). The stronger flow, in combination with higher temperatures, resulted in a more than a doubling of the annual mean net heat transport into the Arctic Ocean (Schauer et al., 2004). A similar increase could explain the warming of the intermediate layers of the Eurasian Basin in the early '90s. During the same period as the Fram Strait branch increased (1997 to 2000), the Barents Sea branch decreased from about 2 to 1.5 Sv (Ingvaldsen et al., 2002; 2004). Considering that the AW flowing into the Arctic Ocean from the Barents Sea is much colder than the water entering through the Fram Strait, a reduction in the Barents Sea branch may enhance an eventual warming signal in the Arctic Ocean.

While direct evidence for long-term changes in the flow regimes are sparse due to lack of long-term homogeneous measurements, indirect evidence in the form of changes in hydrography may be found. Blindheim et al. (2000) reported a 2-3 year time lag between the phase of the NAO and the western extension of the AW in the Nordic Seas (defined as the 35 salinity contour), where positive NAO years were associated with a more eastern position of the front. Furevik et al. (2002) showed that the Iceland-Faroe Front migrated further south during positive NAO years, and finally geostrophic calculations from the Svinøy Section since 1955 have indicated an out-of-phase relation between the inner and outer branch of the NwAC (Mork and Blindheim, 2000). Without being conclusive, it is likely that the effect of strengthened westerlies over the North Atlantic is to increase the AW inflow through the Faroe-Shetland Channel, and correspondingly reduce the inflow over the Iceland-Faroe Ridge. Less is known about the total inflow, although Dickson et al., (2000) suggested that the total inflow of AW increases during high-NAO years, and Nilsen et al., (2003) find a slight increase in their model simulations of the total inflow.

During the generally positive NAO period of the last 40 years, the volume and temperature flux of the NwAC that eventually flows into the Arctic Ocean is thus likely to have increased.

8.3.3 Atlantic Water temperatures

Thanks to traditions going back to the latter part of the 19th century (see Blindheim and Østerhus, 2004, for a review), the NwAC is among the best-sampled ocean currents in the world. The many repeated hydrographical sections with multi-decadal long temperature and salinity records provide us with a well of information to be used to study natural climate variability. Among the well-known sections are the Scotland-Faroe section conducted by the Marine Laboratory in Aberdeen, Scotland, the Svinøy, Gimsøy, Bjørnøya-Fugløya, and Sørkapp sections conducted by the Institute of Marine Research in Bergen, Norway, and the 78°N and 79°N Fram Strait sections conducted by the Norwegian Polar Institute in Tromsø, Norway and the Alfred Wegener Institute for Polar and Marine Research in Bremerhaven, Germany.

The temperature measurements show high year-to-year correlations, as anomalous high or low temperatures have a tendency to persist for some years. Furevik (2000, 2001) combined satellite measurements of sea-surface temperature with in situ observations to look at the temperature anomalies that propagated through the system in the '80s and '90s. In contrast to a warm anomaly that propagated from the Faroe-Shetland Channel to the Barents Sea and Fram Strait in the first half of the '80s, the warm period of the early '90s was much larger in magnitude, lasted for a longer period, and was generated inside the Nordic Seas rather than being advected with the flow of AW from the south. The main reason for the warming was higher air temperatures over the Nordic Seas, possibly a more narrow flow of AW, and possibly an increased inflow of AW through the Faroe-Shetland Channel as discussed above. Together these forcing factors, all probably related to the extreme positive phase of the NAO from the 1989 to 1995, contributed to very high AW temperatures in the Barents Sea and Fram Strait, and subsequently in the Arctic Ocean. And numerous observations from icebreaker and submarine cruises in the Arctic Ocean during the early '90s, revealed that the Atlantic layer in the

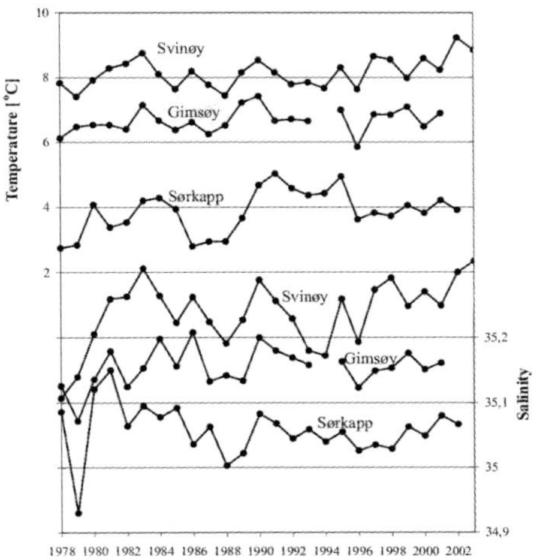

Fig. 8.8: Temperature and salinity in the core of the Atlantic Water in the Svinøy, Gimsøy and Sørkapp sections, averaged between 50 and 200 m depth. Updated from Mork and Blindheim (2003).

Arctic was warmer than ever before (Carmack et al., 1995; Grotefendt et al., 1998). In addition, the front between the waters of Atlantic and Pacific origins, shifted from the Lomonosov Ridge to the Mendeleyev Ridge (McLaughlin et al., 1996; Morison et al., 1998), implying that also the total volume of the Atlantic layer had increased. A new warm anomaly was observed in the Barents Sea (Ingvaldsen et al., 2003) and the Fram Strait (Schauer et al., 2004) in the late '90s, and it is likely that this water has later spread into the central Arctic (Karcher et al., 2003; Schauer et al., 2004).

The warm and cold episodes are clearly shown in fig. 8.8, where temperatures and salinities in the core of the AW are shown. The plots reveal that the signals generally propagate from south (Svinøy) to north (Sørkapp), and also the strong amplification of the warm signal of the '90s. The cold and fresh episodes occurring at the Svinøy section in 1978, 1987, and 1994 are known as the great salinity anomalies of the '70s, '80s, and '90s. They have been shown to originate in the northwestern North Atlantic, probably as a result of strong outflow of Polar Water. From here they have propagated south with the sub-polar gyre, and east and north with the North Atlantic Current before entering the Nordic Seas after 5-7 years (see Belkin, 2004, and the references therein). A new strong warm and saline anomaly is seen in the southernmost section in the last year. This can be traced back to the Faroe-Shetland inflowing region, which in 2003 had the warmest and most saline AW inflow since the measurements started in

1902 (Mork, 2004). It is probable that the warm and saline water will be advected downstream into the other sections and into the Fram Strait and Barents Sea in the years to come.

8.3.4 Intermediate Waters and Overflows

As discussed above, the overflows do not originate from the deepest parts of the Nordic Seas, but rather from intermediate levels. At least one of these intermediate water masses (Arctic Intermediate Water) has seen extreme changes in salinity since the late '60s or early '70s (Turrell et al., 1999, Blindheim et al., 2000, Dickson et al., 2002). As readily seen in fig. 8.9a, there has been a pronounced freshening during the last 30 years. Further on, the signal can be followed south through the gaps in the Greenland-Scotland Ridge, from where the Labrador Sea and the rest of the North Atlantic Ocean are being diluted (see Dickson et al. 2002, Visbeck et al., 2003). The freshening signal has continued equatorward in the Deep Western Boundary Current, the deepest part of the AMOC, and can now be found even south of the Equator (Dickson et al., 2002).

Different sources for this extra fresh water, which correspond to mixing several meters of fresh water into the ocean, have been proposed; increased fresh water and ice export from the Arctic, more precipitation over the region, or internal redistribution of the water masses. In a recent study, Curry et al. (2003) suggest that global warming is the cause. In an analysis of salinity data from the entire Atlantic Ocean, they found clear indications of a freshening at depth also in the southernmost part of the Atlantic, showing that the freshening was not only a North Atlantic feature. In addition, the upper water of the tropical Atlantic was found to become more saline, which as they state clearly suggest a spin-up of the global hydrological cycle. Warmer air can contain more moisture, and a global warming should therefore be accompanied by increased atmospheric fresh water transport from the tropical oceans towards the high latitudes. If they are right the Nordic Seas freshening is a strong indication of global warming.

At the same time there is indirect evidence that the dense overflows may be decreasing in strength: Hansen et al. (2001) combined density measurements at ocean weather station M (OWSM) in the Norwegian Sea (Fig. 8.1) with direct measurements of volume transports through the Faroe-Shetland Channel, and suggest a 20% reduction in that transport since the '50s. Their simple argument is that less deep-water formation in the Nordic Seas has made the intermediate and deep waters less dense, resulting in a less north-south pressure gradient to drive the flow. In Hansen et

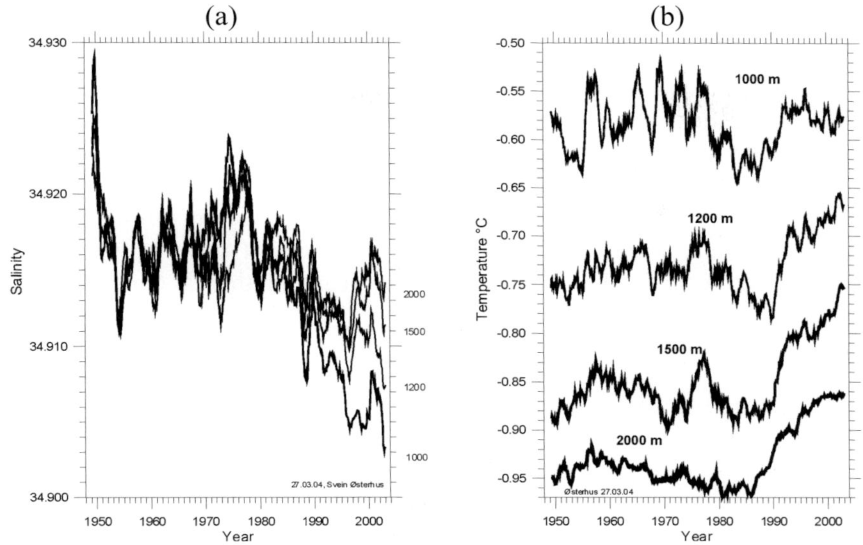

Fig. 8.9: Observed salinity (a) and temperature (b) at 1000, 1200, 1500, and 2000 m depths at OWSM. The position is shown on map in fig. 8.1 (courtesy of Svein Østerhus, Bjerknes Centre for Climate Research, Norway).

al. (2003), they draw the same conclusion for the cold component of the overflow across the southern part of the Iceland-Faroe Ridge and for the cold component of the overflow across the Wyville-Thomson Ridge.

8.3.5 Deep Waters

The densest waters of the Nordic Seas consist of Greenland Sea Deep Water (the product of convection reaching below ~ 2000 m in the Greenland Sea) and Arctic Ocean Deep Water (which enters the Nordic Seas in the western Fram Strait). The open ocean convection in the Greenland Sea was at its maximum depth in the '60s, resulting in very cold and dense deep waters in the region (Dickson et al., 1996). Since then, convection has in general not reached as deep, due to mild winters and fresher surface waters, and the warmer Arctic Ocean Deep Water influence has increased (Meincke et al., 1997). With a few years time lag, the warming signal continued into the Nordic Seas, where the long deep-water record from the OWSM provides an excellent historical record of the change (Østerhus and Gammelsrød, 1999), as seen in fig. 8.9b. In the Norwegian Sea, below the AW and the Intermediate Waters, a mixture of these two deep-water masses is found.

8.4 Expected impacts of anthropogenic climate change

Substantial changes in the Nordic Seas climate have been observed during the last decades. The increased hydrological cycle and increased melting of Arctic land ice under global warming is of particular concern because both fresher surface waters and milder winters will act to reduce the upper water density and thus stabilise the water column. Based on simple model experiments (e.g. Rahmstorf, 1995) and paleo data (Alley et al., 1993), a worst-case scenario has been proposed. That is that a consequence of global warming may be a total shutdown of the AMOC with a substantial cooling of the North Atlantic region.

There are several reasons for why we believe that is a highly unlikely scenario: The northward flow of AW does not depend on a single factor that easily can be switched on or off. The large-scale oceanic circulation in the North Atlantic is mainly wind driven. Prolonged wind forcing results in enhanced horizontal gradients of surface elevation and internal pressure fields that maintain a relatively steady circulation. The driving forces for the inflow to the Nordic Seas are subtler. They probably consist of wind forcing that pushes the waters against the northwest European coasts, buoyancy forcing that produces dense southward flowing water which has to be compensated by an equal inflow, and the estuarine forcing which is simply that water entrained by the southward flowing light waters from the Arctic, also requires a compensating inflow.

Global warming will likely affect these forcing mechanisms in different ways. While the response in wind differs between the climate models, there is a general tendency for most models to respond to increased greenhouse gas concentrations by producing stronger westerlies and more cyclonic circulations over the Nordic Seas (see Furevik and Nilsen, 2004, for a review). Stronger or more persistent westerlies are likely to enhance the general oceanic circulation in the Atlantic, and also to increase the inflow to the Nordic Seas, in particular through the Faro-Shetland Channel.

Another response to global warming is a warming and freshening of the surface waters in the high-latitude convection regions in the ocean, both acting to reduce the surface density and stabilise the water column. While the latter is usually blamed to be responsible for the rapid (abrupt) climate changes in previous climate, reduction in the rate of deep-water formation in present day climate models seem to be a combined result of warming and freshening, and there is no consensus on what is the most important of the two. While there has been a substantial reduction in the production of deep waters by open-ocean convection in the Nordic Seas, less is known about the shelf convection that is mainly driven by brine release during ice-freezing. As ice freezing and thus the rate of salt release are negatively correlated with the ice thickness (thick ice insulates the waters below and prohibits the freezing), a thinner ice cover is likely to enhance the ice production. Shelf processes and down slope sinking of waters is generally poorly described in climate models, and there is therefore little knowledge on how this will change in a future climate. It may nevertheless happen to be of increasing importance in a future climate. Thus, despite the fact that most of the climate models participating in the third assessment report of the IPCC show a modest decline in the strength of the AMOC, unresolved processes may modify these results.

Finally, we have strong reasons to believe that global warming will also affect the third driving force for the Atlantic Water inflow to the Nordic Seas, the estuarine forcing. An enhanced atmospheric fresh water flux to the Arctic, with more precipitation and river runoff, will likely increase the sea-level difference between the Arctic and the Atlantic Ocean, and thus the southward transport of freshwater through the Nordic Seas. In addition, a thinner sea-ice cover will reduce the fraction of the freshwater that is exported in solid form. The net result of this will probably be more entrainment and a stronger estuarine forcing, again a factor that will work to maintain the inflow of AW to the Nordic Seas. In a recent paper, de Boer and Nof (2004) argued that the present day stable AMOC conditions started when sea-level rise opened the Bering Strait 10 400 years ago, and fresh-water anomalies in the convection regions could be flushed south and replaced by denser AW. With their interpretation, a substantial or complete shutdown of the AMOC is unlikely to appear in present day climate, and global warming will not change this.

8.5 Summary and conclusion

In this paper we have discussed the flow of AW through the Nordic Seas, an important factor for ocean climate and biomass production in the area. The average volume and heat transports from the North Atlantic to the Nordic Seas are slightly below $8 \times 10^6 m^3 s^{-1}$ and $300 \times 10^{12} W$ respectively, with the main passages being the Faroe-Shetland Channel and the Iceland-Faroe Ridge. Observations suggest that more than ¾ of this water is carried into the Arctic through the Fram Strait and Barents Sea. Due to atmospheric forcing and mixing with ambient fresh water masses, a substantial cooling and freshening of the water masses occurs within the Nordic Seas and Arctic Ocean. Most of the heat is released in the Nordic Seas, mainly in the Lofoten Basin, Fram Strait and Barents Sea, where the annual mean heat loss exceeds 100 Wm^{-2}. Within the Nordic Seas and Arctic Ocean the AW is transformed into a denser water mass that is known as the over-flow water. This water spills over the ridges to the south, and forms the deepest parts of the AMOC.

During the last decades, substantial changes in the Nordic Seas ocean climate have been observed. The primary forcing mechanism seems to have been the low-frequency change from weak westerlies in the '60s to record strong westerlies in the '90s. The result of the changed atmospheric forcing seems to have been a stronger flow of AW through the Faroe-Shetland Channel, and possibly a weaker inflow over the Iceland-Faroe Ridge. Less is known about the total inflow. Other, better-documented changes have been that the AW has become much warmer and also more saline during the last decades, the intermediate waters that end up as over-flow water has freshened, and the deep waters of the Nordic Seas have become warmer.

Paleo-records have shown that the North Atlantic region has undergone periods characterised by rapid changes in climate, probably linked to abrupt changes in the AMOC. Furthermore, most climate models indicate a significant reduction in the strength of the AMOC in a global warming scenario. We here review the mechanisms responsible for producing dense water and maintaining the AW inflow to the Nordic Seas. Observations and model runs indicate a reduction in the open-ocean convections of the north. The result may be a weaker AMOC and a reduced inflow of AW to the Nordic Seas. We believe, however, that other forcing mechanisms at least partly will compensate this: stronger wind forcing, increased shelf convections, or strengthened estuarine forcing. The effect of global warming will likely affect all forcing mechanisms in various ways. We therefore believe that a total shutdown of the AMOC under global warming is unlikely, and that the inflow of AW to a large degree will be maintained. We may nevertheless expect that the trends we have seen in AW temperatures and salinity can be sustained in the decades to come, and that this can have significant physical and ecological impacts in various parts of the region.

Acknowledgements

We are grateful to Bogi Hansen, Faroese Fisheries Laboratory, and Igor M. Belkin, University of Rhode Island, for their valuable comments and suggestions that improved the manuscript. TF and CM have been supported by the Norwegian Research Council project NOClim II (155972/720). This work was completed during a visit of TF at the Danish Meteorological Institute, Copenhagen, whom he will thank for hospitality and good working environment.

References

Aagaard K, JH Swift, and EC Carmack (1985). Thermohaline circulation in the Arctic Mediterranean Seas. J Geophys Res, 90, 4833-4846.

Alley, RB., DA Meese, CA Shuman, AJ Gow, KD Taylor, PM Grootes, JC White, M Ram, ED Waddington, PA Mayewski, and GA Zielinksi (1993). Abrupt increase in Greenland snow accumulation at the end of the Younger Dryas event. Nature 362, 527–29.

Belkin I. M. (2004), Propagation of the "Great Salinity Anomaly" of the 1990s around the northern North Atlantic, Geophys. Res. Lett., 31, L08306, doi:10.1029/2003GL019334.

Bersch M., North Atlantic Oscillation–induced changes of the upper layer circulation in the northern North Atlantic Ocean, J. Geophys. Res., 107 (C10), 3156, doi:10.1029/2001JC000901, 2002

Blindheim J, and S Østerhus (2005). The Nordic Seas main oceanographic features. Climate Variability in the Nordic Seas, Geophysical Monograph Series, AGU, accepted.

Blindheim J, V Borovkov, B Hansen, SA Malmberg, WR Turrell, and S Østerhus (2000). Upper Layer Cooling and Freshening in the Norwegian Seas in Relation to Atmospheric Forcing. Deep-Sea Res. I, 47(4), 655-680.

Blindheim, J. (1990) Arctic Intermediate Water in the Norwegian Sea. Deep-Sea res., 37(9), 1475-1489.

Broecker WS (1991). The great conveyor belt. Oceanography, 4, 79-89.

Broecker WS (1997). Thermohaline circulation, the Achilles heel of our climate system: will manmade CO2 upset the current balance? Science, 278, 1582-1588.

Carmack EC (2000). The Arctic Ocean's freshwater budget: sources, storage and export. In The Fresh Water Budget of the Arctic Ocean (Ed by L. Lewis) Nato Science Ser. 2 Environmental security. Vol.70, 91- 126. Kluwer Academic Press.

Carmack EC, RW Macdonald, RG Perkin, FA McLaughlin, and RJ Pearson (1995). Evidence for warming of Atlantic Water in the southern Canadian Basin of the Arctic Ocean: Results from the Larsen-93 expedition. Geophys. Res. Lett., 22(9), 1061--1064, doi:10.1029/95GL00808.

Clark P, S Marshall, G Clarke, S Hostetler, J Licciardi, and J Teller, Freshwater forcing of abrupt climate change during the last glaciation (2001). Science, V. 293, p. 283-286.

Curry R, RR Dickson, and I Yashayaev (2003). A change in the freshwater balance of the Atlantic Ocean over the past four decades. Nature, 426(6968), 826--829, doi:10.1038/nature02206.

Dansgaard W, JWC White, and SJ Johnsen (1989). The abrupt termination of the Younger Dryas climate event. Nature, 339, 532-534.

de Boer AM and D Nof (2004). The Exhaust Valve of the North Atlantic, J. Climate, 17(3), 417-422.

Dickson RR, I Yashayaev, J Meincke, B Turrell, S Dye, and J Holfort (2002). Rapid freshening of the deep North Atlantic Ocean over the past four decades. Nature, 416(6883), 832--837, doi:10.1038/416832a.

Dickson RR, J Lazier, J Meincke, P Rhines, and J Swift(1996). Longterm coordinated changes in the convective activity of the North Atlantic. Prog. Oceanog., 38(3), 241--295, doi:10.1016/S0079-6611(97)00002-5

Dickson RR, R Osborn, TJ Hurrell, JW Meincke, J Blindheim, B Ådlandsvik, T Vinje, G Alekseev and W Maslowski (2000). The Arctic Ocean response to the North Atlantic Oscillation. J. Climate, 13, 2671-2696.

Dickson, R. R., and J. Brown (1994). The production of North Atlantic Deep Water: sources, rates, and pathways, J. Geophys. Res., 99, 12,319–12,341. doi:10.1006/jmsc.1997.0229.

Flatau MK., L Talley, and PP Niiler(2003). The North Atlantic Oscillation, surface current velocities, and SST change in the Subpolar North Atlantic. J. Climate, 16(14), 2355--2369.

Furevik T, and JEØ Nilsen (2005). Large-Scale Atmospheric Circulation Variability and its Impacts on the Nordic Seas Ocean Climate – a review. Climate Variability in the Nordic Seas, Geophysical Monograph Series, AGU, accepted.

Furevik T, M Bentsen, H Drange, JA Johannessen and A Korablev (2002). Temporal and spatial variability of the sea surface salinity in the Nordic Seas. J Geophys Res., 107(C12): 8009, doi:10.1029/2001JC001118.

Furevik T. (2000). On anomalous sea surface temperatures in the Nordic Seas. J. Climate, 13(5), 1044--1053.

Furevik T. (2001). Annual and interannual variability of Atlantic Water temperatures in the Norwegian and Barents Seas: 1980--1996. Deep-Sea Res. I, 48(2), 383--404, doi:10.1016/S0967-0637(00)00050-9.

Furnes GK, B Hackett, and R Sætre (1986). Retroflection of Atlantic Water in the Norwegian Trench. Deep-Sea Res, 33 (2), 247-265.

Grotenfendt K, K Logemann, D Quadfasel, and S Ronski (1998). Is the Arctic Ocean warming?. J. Geophys. Res. 103(C12) 27,679-27,687, 1998.

Hansen B, S Østerhus, H Hatun, R Kristiansen, and KMH Larsen (2003). The Iceland-Faroe inflow of Atlantic water to the Nordic Seas. Progress in Oceanography, 59. pp 443-474.

Hansen, B, WR Turrell and S Østerhus (2001). Decreasing overflow from the Nordic Seas into the Atlantic Ocean through the Faroe Bank Channel since 1950, Nature, 411 (6840), 927-930, doi:10.1038/35082034.

Helland-Hansen B and F Nansen (1909). The Norwegian Sea, its physical oceanography. Based on the Norwegian researches 1900-1904. In Report on Norwegian fishery and marine-investigations, vol 2, Bergen, Norway, 390 pp.

Houghton JT, Y Ding, DJ Griggs, M Noguer, PJ van~der Linden, X Dai, K Maskell, and CA Johnson, editors (2001). Climate Change 2001: The Scientific

Basis. Contribution of Working Group I to the Third Assessment Report of the Intergovernmental Panel on Climate Change. Cambridge University Press. 801 pp.

Hurrell J (1995). Decadal trends in the North Atlantic Oscillation: Regional temperatures and precipitation. Science, 269(5224), 676--679.

Hurrell JW, Y Kushnir, G Ottersen, and M Visbeck (2003). An Overview of the North Atlantic Oscillation. In The North Atlantic Oscillation: Climatic Significance and Environmental Impact, volume 134 of Geophysical Monograph Series. AGU. ISBN 0875909949.

Ingvaldsen R, H Loeng, and L Asplin (2002). Variability in the Atlantic Inflow to the Barents Sea based on a one-year time series from moored current meters. Cont. Shelf. Res 22, 505-519.

Ingvaldsen R, H Loeng, G Ottersen and B Ådlandsvik (2003). Climate variability in the Barents Sea during the 20th century with a focus on the 1990s. ICES Marine Science Symposia, 219, 160-168.

Ingvaldsen R, L Asplin and H Loeng (2004). The seasonal cycle in the Atlantic transport to the Barents Sea during the years 1997-2001. Cont. Shelf Res., 24 (2004), 1015-1032.

Jakobsson PK, MH Ribergaard, D Quadfasel, T Schmith, and CW Hughes (2003). Near-surface circulation in the North Atlantic as inferred from Lagrangian drifters: Variability from the mesoscale to interannual. J. Geophys. Res., 108(C8), 3251, doi: 10.1029/2002JC001554.

Jónsson, S, and J Briem (2003). Flow of Atlantic water west of Iceland and onto the north Icelandic shelf. ICES Marine Science Symposia, 219, 326-328.

Karcher, MJ, R Gerdes, F Kauker, and C Kõberle (2003). Arctic warming: Evolution and spreading of the 1990s warm event in the Nordic Seas and the Arctic Ocean, J. Geophys. Res., 108 (C2), 3034, doi: 10.1029/2001JC001265.

Marshall J, Y Kushnir, D Battisti, P Chang, A Czaja, R Dickson, M McCartney, R Saravanan, and M Visbeck (2001). North Atlantic climate variability: Phenomena, impacts and mechanisms. J. Climate, 21(15), 1863--1898.

Mauritzen C (1996a). Production of dense overflow waters feeding the North Atlantic across the Greenland-Scotland Ridge. Part 1: Evidence for a revised circulation scheme. Deep-Sea Res, 43, 769-806.

Mauritzen C (1996b). Production of dense overflow waters feeding the North Atlantic across the Greenland-Scotland Ridge. Part 2: An inverse model. Deep-Sea Res, 43, 807-835.

McLaughlin, FA, EC Carmack, RW Macdonald, and JKB Bishop (1996). Physical and geochemical properties across the Atlantic/Pacific water mass front in the southern Canadian Basin. J. Geophys. Res. 101, 1183-1197.

Meincke J, B Rudels, and H Friedrich (1997). The Arctic Ocean - Nordic Seas thermohaline system. ICES J. Mar. Sci., 54(3), 283--299,

Morison J, M Steele, and R Andersen (1998). Hydrography of the upper Arctic Ocean measured from the nuclear submarine U.S.S. Pargo. Deep-Sea Res. I, 45(1), 15-38, doi:10.1016/S0967-0637(97)00025-3.

Mork KA and J Blindheim (2000). Variation in the Atlantic Inflow to the Nordic Seas, 1955-1996. Deep-Sea Res. I, 47(6), 1035-1057, doi:10.1016/S0967-0637(99)00091-6.

Mork KA and J Blindheim (2003). Heat loss of the Norwegian Atlantic Current toward the Arctic. ICES Marine Science Symposia, 219, 144-149.

Mork, KA (2004). The Norwegian Sea ecosystem, ocean climate (in Norwegian), in Havets Miljø 2004, Fisken og Havet, spesial issue 2-2004, Institute of Marine Research Bergen, Norway, edited by K Sjøtun.

Nilsen J. E. Ø., Y. Gao, H. Drange, T. Furevik, M. Bentsen, Simulated North Atlantic-Nordic Seas water mass exchanges in an isopycnic coordinate OGCM, Geophys. Res. Lett., 30 (10), 1536, doi:10.1029/2002GL016597, 2003.

Orvik KA (2004). The deepening of the Atlantic Water in the Lofoten Basin of the Norwegian Sea, demonstrated by using an active reduced gravity model. Geophys. Res. Lett, 31, L01306, doi:10.1029/2003GL018687.

Orvik KA, Ø Skagseth, and M Mork (2001). Atlantic inflow to the Nordic Seas. Current structure and volume fluxes from moored current meters, VM-ADCM and SeasSoar-CTD observations. Deep-Sea Res. I, 48, 937-957.

Østerhus S and T Gammelsrød (1999). The abyss of the Nordic Seas is warming. J Climate, 12(11), 3297-3304.

Rahmstorf, S., 1995: Bifurcations of the Atlantic thermohaline circulation in response to changes in the hydrological cycle. Nature, 378, 145-149.

Schauer U (1995). The release of brine-enriched shelf water from Storfjord into the Norwegian Sea, J. Geophys. Res., 100 (C8): 16,015-16,028.

Schauer U, E Fahrbach, S Østerhus and G Rohardt (2004). Arctic warming through the Fram Strait – Oceanic heat transport from three years of measurements. Subm to J. Geophys Res.

Schauer U, H Loeng, B Rudels, VK Ozhigin, and W Dick (2002). Atlantic Water flow through the Barents and Kara Seas. Deep-Sea Res, 49, 2281-2298.

Skagseth Ø, KA Orvik, and T Furevik (2004). Coherent variability of the Norwegian Atlantic slope current determined by using TOPEX/ERS altimeter data, Geophys. Res. Lett., 31, L14304, doi:10.1029/2004GL020057.

Turrell WR, B Hansen, S Hughes and S Østerhus (2003). Hydrographic variability during the decades of the 1990s in the Northeast Atlantic and southern Norwegian Sea. ICES Marine Science Symposia, 219, 111-120.

Turrell WR, G Slesser, RD Adams, R Payne, and PA Gillibrand (1999). Decadal variability in the composition of Faroe Shetland Channel Bottom Water. Deep-Sea Research, 46, 1-25.

Visbeck M, EP Chassignet, R Curry, T Delworth, B Dickson, and G Krahmann (2003). The ocean's response to North Atlantic Oscillation variability. In The North Atlantic Oscillation: Climatic Significance and Environmental Impact, volume 134 of Geophysical Monograph Series. AGU. ISBN 0875909949.

Worthington, LV (1970). The Norwegian Sea as a Mediterranean basin. Deep-Sea Res., 17, 77-84.

9 Climate variability and possible effects on arctic food chains: The role of *Calanus*

Stig Falk-Petersen[1], Vladimir Pavlov[1], Sergey Timofeev[2] and John R. Sargent[3]

[1]Norwegian Polar Institute, N-9296 Tromsø, Norway, [2]Murmansk Marine Biological Institute, Russian Academy of Science, Murmansk 183010, Russia, [3]University of Stirling, Stirling, FK9 4LA, Scotland, UK, e-mail: stig@npolar.no

9.1 Introduction

The large oscillations of abiotic factors in the Arctic is critical in structuring its marine biota and the biodiversity of its indigenous populations and communities. The seasonal light cycle is modified by the sea ice cover, creating a situation dominated by phytoplankton blooms, that follow the receding ice edge, and in leads as the sea ice opens during the Arctic summer (Sakshaug 1997, 2003; Hegseth 1998; Falk-Petersen et al. 2000a; Engelsen et al. 2002). Blooms of phytoplankton propagate through Arctic waters (Zenkevich 1963) and carbon fixed through photosynthesis is rapidly converted into large, specialised lipid (marine fat) stores by the herbivorous *Calanus* species (Lee 1975; Sargent and Henderson 1986). These high-energy lipids are then rapidly transferred upwards through the food chain in large amounts (Falk-Petersen et al. 1990). The increase in lipid level from 10-20% of dry mass in phytoplankton to 50 - 70% in herbivorous zooplankton is probably one of the most fundamental specialisations in polar bioproduction. The lipid – based energy flux is one of the primary reasons for the large stocks of fish and mammals in Arctic waters.

The importance of the diatom => *Calanus* food chain in the Arctic pelagic food has been demonstrated by Falk-Petersen et al. (1986; 2002) and Scott et al. (2002). A wide spectrum of predators from zooplankton to fish and sea birds has also been analysed by using fatty acid trophic markers in Arctic waters. In all of these studies the *Calanus* C20 and C22 lipid trophic markers were strikingly dominant, demonstrating the importance of the

Calanus species in the Arctic pelagic ecosystem (Falk-Petersen et al. 2001, 2002, 2004; Dahl et al. 2003.

The population size spectrum and energy content of the key *Calanus* species, being potential prey for zooplankton-eating fish and sea birds, is therefore instrumental in structuring the biodiversity of Arctic ecosystems. We believe that understanding the climate variability is a key to understand the biology of Arctic animals and the biodiversity of Arctic systems. In this paper we discuss how different climate regimes in the Nordic Seas can influence *Calanus*- based Arctic food chains.

9.2 Climate variability and species distribution

9.2.1 The distribution of the *Calanus* species and the current system.

The circulation of water in the Nordic Seas, the modification of water masses and changes of ice conditions are important for a) the local climate and b) the larger-scale systems of oceanic and atmospheric circulation affecting the climate of northern Europe and c) the marine food resources.

The distribution of the three main *Calanus* species in the Nordic Seas and the Eurasian Basin of the Arctic Ocean is strongly influenced by two large, interconnected current systems (Fig. 9.1). The current system is mainly governed by the topography (Rudels et al. 1999; Hansen and Østerhus 2000). In the south the North Atlantic Current enters the Norwegian Sea on both sides of the Faroe Islands and flows northwards parallel to the Norwegian Coast (Hansen and Østerhus 2000). North of the Lofoten Islands at approx. 72 °N one branch turns northwest and flows along the Barents Sea margin as the West Spitsbergen Current. Northwest of Spitsbergen, a part of the current branch westward and southwards and meets the East Greenland Currents flowing southwards before part of the current turns eastwards first into the Iceland Sea and then into the Norwegian Sea north of Iceland.

In the north the Atlantic water enters the Arctic north of Spitsbergen, Svalbard where it submerges under the Arctic water masses (Nansen 1902). It then circulates the Eurasian basin and returns into the Fram Strait on the Greenland side as Arctic Intermediate Waters at depths of 500-900 m. This deep and cold-water mass turns east north of Jan Mayen, and follows the same circulation system as surface water along the Spitsbergen

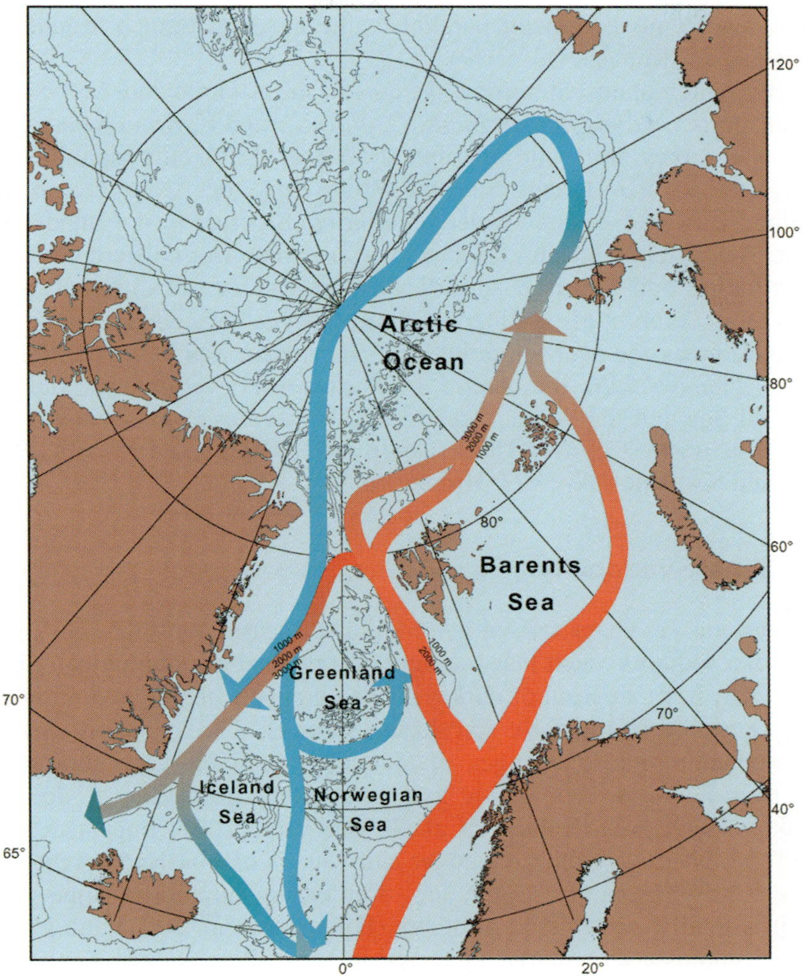

Fig. 9.1. The current system in the European Arctic. The red arrow showing the North Atlantic current entering the Arctic and the blue indicating gradually cooling and sinking (modified after Hansen and Østerhus 2000, Jones et al. 1995).

Shelf slope. In the Fram Strait, therefore, there is an exchange of the waters between the Polar Ocean and the Nordic Seas, with a southward flow on the western side and a northward flow on the eastern side (Jones et al. 1995). The North Cape Branch of the North Atlantic Current enters Barents Sea through the Bear Island Trough (The Barents sea opening) before it flows into the eastern Barents Sea. It then flows northwards along Novaya Zemla before it enters the Arctic Ocean between Franz Josef Land and Novaya Zemlya, flowing parallel to the northern branch around the Eurasian basin (Rudels et al. 1999).

The two interconnected current systems in the Atlantic and Arctic (Fig. 9.1) transports *Calanus finmarchicus, C. glacialis* and *C. hyperboreus* long distances and they are found distributed all over the Arctic, including the Norwegian Sea, the Barents Sea, the White Sea, the Arctic Ocean, the Greenland Sea and coastal waters bordering Siberia, East Canada and Alaska. The different species do, however, originate from different centres of distribution and are used as indicator species for the different water masses (Van Aken et al. 1991). The three *Calanus* species also have different core areas for over-wintering, the Norwegian Sea being central for *Calanus finmarchicus,* the Arctic shelf area for *C. glacialis* and the Greenland Sea and the Arctic Ocean for *C. hyperboreus* (Jaschnov 1970; Runge et al. 1986; Conover 1988; Tande 1991; Hirche and Mumm 1992; Hirche and Kwasniewski 1997; Hirche 1997).

9.2.2 Climate variability.

Arctic systems are characterised by pronounced seasonal oscillations in incident solar radiation such that the light regime changes dramatically during the year, from a period of darkness lasting approximately 4.5 month at 80°N to a period of midnight sun during summer. The ice cover and sea water temperature also changes dramatically, both on very short time scales from hours to days and on long time scales from years to decades (Polyakov 1999, Falk-Petersen et al. 2000a; Vinje 2001; Engelsen et al. 2002). The extent to which ice cover can vary is emphasised by the fact that the North Pole can be largely ice - free during summer in some years while it is totally ice covered in others (Polyakov et al. 1999) and phytoplankton blooms are regularly registered in North Pole waters (Zenkevich 1963). Starting and continuation of the bloom in the different parts of the Arctic Ocean corresponds to features of the annual cycle of the ice concentration (Fig. 9.2).

Analysis of the inter-annual water temperature variations along a section across the Atlantic water inflow in the Norwegian, Barents and Greenland seas (Tereshchenko 1997; Furevik 2001) have shown several cold and warm periods during last five decades, varying in periods of 2 to 7 years. Inter-annual and decadal variability of the summer water temperature (Fig. 9.3) was studied between 1950 and 2000 in the core Atlantic inflow in the West Spitsbergen Current (data extracted from a box delineated by 78-79N, 8-10E) and the Barents Sea Opening (box delineated by 72.5-73.5N,

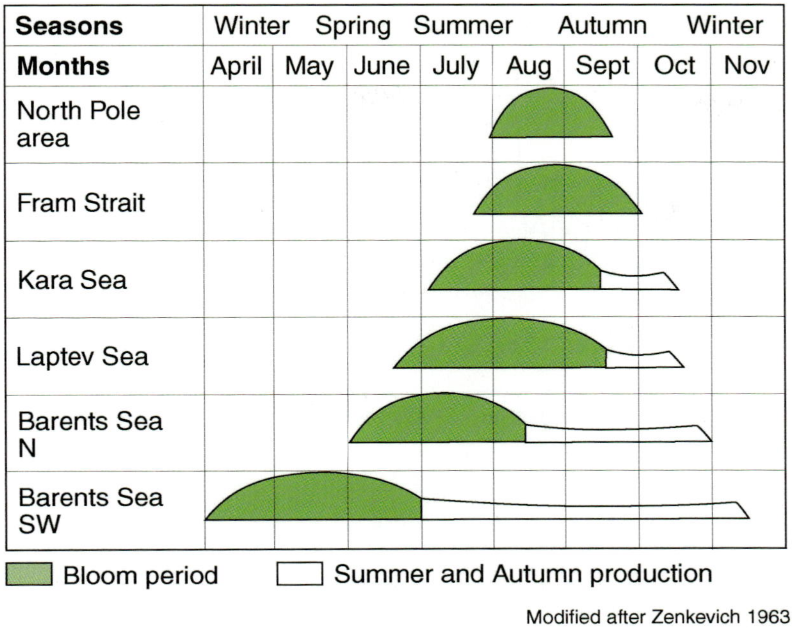

Fig. 9.2. Time related plankton blooms in the Arctic Oceans (modified after Zenkevitch 1963).

19.5-21.5E). For this study all available data from the National Oceanographic Data Center archive (NODC-02, period 1900-1990), Greenland Sea Project dataset (period 1987-1993), ESOP project I and II dataset (period 1993-1998), VEINS project dataset (period 1997-2000), NPI dataset (period 1959-2003), Russian data collected at the NPI (period 1959-1992) and Climatic atlas of the Barents Sea (Matishov et al. 1998) were used. The temperature decrease in the 50's, 60's and 70's and increased in the 80's and 90's in both areas. The regime shift, between the two periods took place between 1978 and 1979. The surface water of the West Spitsbergen Current had a maximum values at the start of the 50s and the end of the 90s of 4 to 6.5°C and a minimum at the end of the 70s of 1.5 to 3 °C. The Barents Sea Opening had higher temperatures with a maximum at the start of the 50ties and the end of the 90s of 6 to 7 °C and a minimum at the end of the 80s of 4.5 to 5.5°C at 50 m. Periods with warming is related to high North Atlantic Oscillation (NAO) modus and increased Atlantic water inflow (Furevik et al. 2005).

Analysis of historical data of sea ice (Vinje 2001) shows that the ice extent has been reduced monotonically over a period 135 years (Fig. 9.4).

April ice extent has decreased by ~33% over the past 135 years and reached

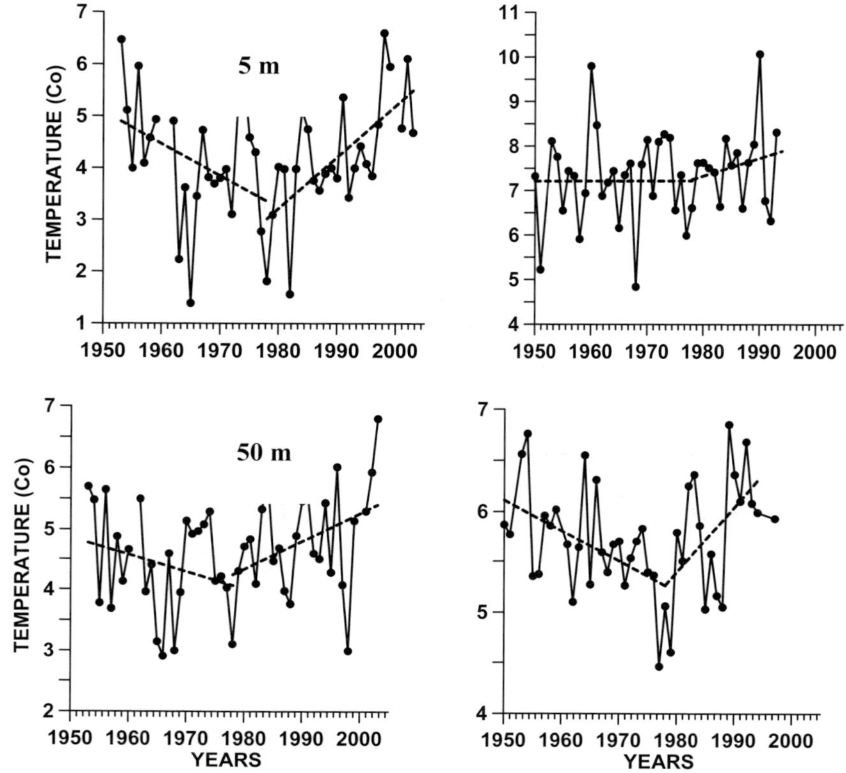

Fig. 9.3. The time series of the summer (July-September) water temperature (solid line and points) along the path of Atlantic water in the West Spitsbergen current for box: 78-79N, 8-10E (left panels) and in the Barents Sea Opening for box: 72.5-73.5N, 19.5-21.5E (right panels) at the depths 5 and 50 m. Linear trends (dashed lines) are shown for two periods 1953-1978 and 1978 – 2003 in West Spitsbergen current and 1950-1978 and 1978-1995 in the Barents Sea Opening.

minima in the 1990s. A strong negative correlation is found between NAO winter index and the Nordic Seas April ice extent (Fig. 9.4, Vinje 2001). For example most of the Barents Sea, parts of the Greenland Sea and the Norwegian Sea were ice covered in the spring of 1966 when NAO index was low while in 1995 (high NAO) most of the same areas were ice free.

9.2.3 Phytoplankton bloom.

The large and spatial changes in ice concentration directly influence the light available for primary production. Blooms of phytoplankton develop as the ice opens (Engelsen et al. 2002) and Zenkevich (1963) described the

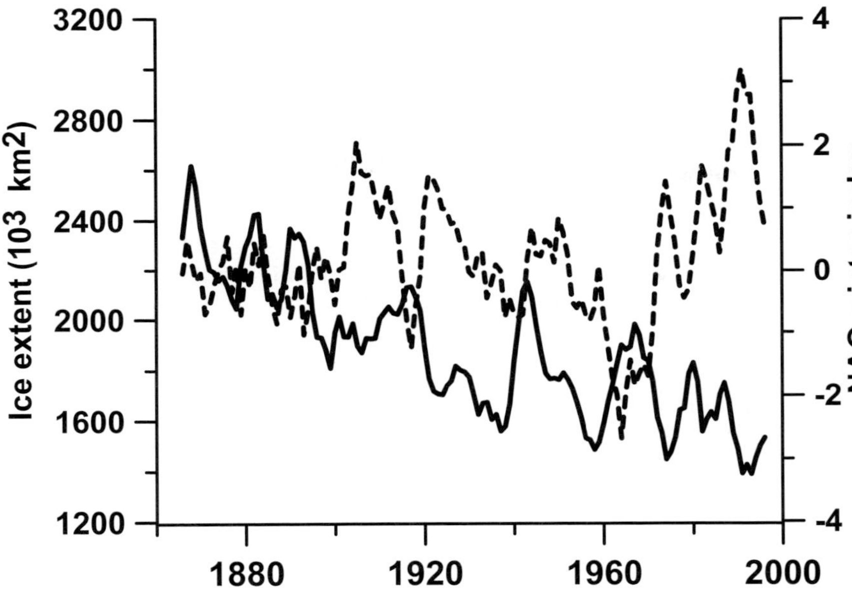

Fig. 9.4. 5-yr running means April ice extent in the Nordic Seas (solid line) and NAO winter index (dashed line). Modified after Vinje (2001).

propagation of the algal bloom along a latitudinal gradient from the southern Barents Sea to the North Pole (Fig. 9.2). The algal bloom is often found in the southern Barents Sea as early as March – April, during the summer in the northern Barents Sea and in August – September in the Fram Strait and in the Arctic Ocean. Thus, herbivorous zooplankton in high latitude, ice-covered seas is exposed to large variations in food availability, not only between seasons (Lee and Hirota 1973; Falk-Petersen et al. 2000b) but also between years, decades and longer periods. The pelagic *Calanus* species being one of the major components of the Arctic marine ecosystem must, therefore, be adapted to an environment changing markedly on different time scales. This readily accounts for the biodiversity of the *Calanus* complex in terms of the species' different life strategies, different ecological niches and different centres of distribution.

9.3 The *Calanus* species

Table 9.1 shows the impressive plasticity of the Arctic *Calanus* species. In the North Sea, *C. finmarchicus* can have a life span of less than a year (Wiborg 1954; Marshall and Orr 1955) while in the Norwegian Sea, along the coasts of north Norway, Greenland and east Canada and the Nordic Seas, the life span is mainly one year (MacLellan 1967, Lie 1968, Sekerak et al. 1976, Tande 1991, Falk-Petersen et al. 1999). *C. glacialis* has a life span of 1 to 3 years but for most areas a life span of 2 years is reported (Conover and Huntly 1991, Kosobokova 1999). *C. hyperboreus* shows the most impressive plasticity, with a life span from two to five years (Dawsen 1978, Conover and Huntly 1991, Hirche 1997, Falk-Petersen et al. 1999). Such plasticity in the different species reflects differences in their environments. The central Arctic Ocean has the highest variability in sea climate and, therefore, the lowest predictability of food available for herbivorous zooplankton. The ability of the different developmental stages of calanoids to over-winter depends on their abilities to store energy. It is interesting to note that all three species have similar energy content (dry weight 0.14, lipid 0.05 mg pr individual), at their first main over-wintering stage (Fig. 9.5). The similarity of the energy content of the first over - wintering copepodite stages, V, IV and III for *C. finmarchicus*, *C. glacialis* and *C. hyperboreus,* respectively, is consistent with the energy demand of the first over - wintering being the same for the three species.

Table 9.1. The Arctic *Calanus* complex. Life span (years), maximum size (mm), lipid weight (LW) as mg • ind-1, and % total lipid (TL) of DW, % wax ester (WE) of TL. IV and V = copepodite stages, F = females.

Species	Life span	Stage	Max. size	LW	% TL	% WE
C.finmarchicus.	1	IV	2.2	0.02	53	63
.		V	3.0	0.04	34	68
		F	3.2	0.08	31	62
C. glacialis	2-3	IV	2.8	0.05	56	68
		V	4.0	0.38	61	72
		F	4.6	0.45	70	68
C.hyperboreus.	2-5	IV	4.0	0.28	54	72
		V	6.0	1.03	65	75
		F	7.0	1.81	62	75

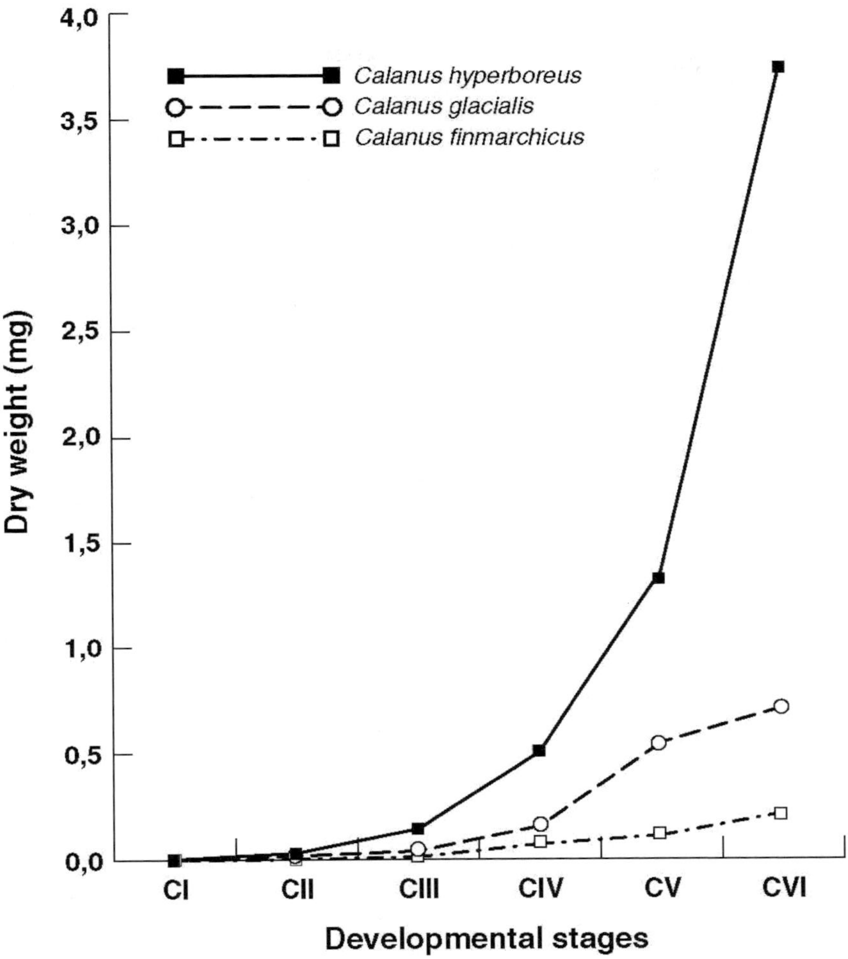

Fig. 9.5. Growth in terms of dry weights of the different copepodite stages (I to VI) of Calanus finmarchicus, Calanus glacialis and Calanus hyperboreus. Data from Mauchline 1998.

9.3.1 *Calanus hyperboreus*

C. hyperboreus, the most polar species, has its centre of distribution in the current system connecting the deep-sea areas such as the Greenland Sea, the Fram Strait and the Arctic Ocean (Fig. 9.1). These are areas with marked annual and inter-annual variations in ice conditions. In favourable seasons, when the ice cover opens up for longer periods, food is plenty, while in years with permanent ice cover food is scarce. *C. hyperboreus* has adapted to this condition by developing a large plasticity in its life strategy (Table 9.1). It is a large copepod and can over - winter, as early as stage III, after accumulating large energy reserves in the form of high - energy wax esters (Hirche 1997; Scott et al. 2000). Under favourable conditions when primary production is high, the animal has a life span of 2 years, while under extensive ice cover it has a life span from 3 to 5 years. *C. hyperboreus* females store large quantities of wax esters which are consumed in the spawning process during winter. Spawning takes place independent of the algal bloom and the fertilised eggs will normally develop rapidly via the naupliar stages to stage III or IV copepodites during the short bloom in the following summer/ autumn. When *C. hyperboreus* has accumulated sufficient lipid reserves it migrates down to deep waters (500-2000 m) and enters diapause (Vinogradov 1997). Under these conditions it has sufficient high-energy reserves to meet its minimal metabolic requirements for long periods, exceeding a single season if need be. It is logical that the greater the variation in ice cover and especially the greater the likelihood of prolonged ice cover, the greater the advantage for an herbivorous copepod to arrest development at any stage. This is more likely to occur in large copepods with large reserves of high-energy lipids in early as well as late developmental stages.

9.3.2 *Calanus glacialis*

C. glacialis is a typical shelf species which spawns in waters all around the Arctic shelf and in the White Sea. Although the shelf areas experience large variations in ice cover, the ice nearly always opens up for shorter or longer periods during the summer or autumn allowing an annual algal bloom to occur. *C. glacialis* has a life span of 1-3 years. It spawns before or during the yearly spring bloom (Niehoff et al. 2002) and it grows rapidly in the same year to reach its over-wintering stage IV, with large energy-rich wax ester reserves (Table 9.1). It then descends to deep areas on the shelf (200 to 300 m) to enter diapause and over-winter. Scott et al. (2000) suggested that this species is well capable of developing from eggs via the naupliar stages to wax ester-rich stage III and stage IV copepodites within a single year. However, the large increment in body mass and lipid reserves as it develops further from stage IV to stage V copepodites is unlikely to be achieved in a single year, except under very favourable conditions. Thus, development from stage IV to stage V probably occurs in two year-old *C. glacialis*. The resulting large, wax ester-rich stage V copepodites then over-winter and spawn immediately prior to or during the spring bloom in their subsequent third year of life.

9.3.3 *Calanus finmarchicus*

C. *finmarchicus*, the smallest of the three *Calanus* species, is an Atlantic boreal deep-water species. It has two main centres of distribution, one connected to the large gyre in the Norwegian Sea (Fig. 9.1) and the other in the Subarctic Gyre south of the Labrador Sea and east of Newfoundland. In addition there is a small sub-population south of Newfoundland (Matthews 1969; Aksnes and Blindheim 1996; Planque et al. 1997; Bucklin et al 2000). Breeding populations are also found along the Norwegian coast and in the southern Barents Sea. These are areas where a predictable annual spring bloom occurs between March and May. North of the Arctic Circle, *C. finmarchicus* develops from eggs, via 6 naupliar and several copepodite stages to their major over-wintering stage V within 6 to 10 weeks in favourable years (Sargent and Falk-Petersen 1988). They also accumulated high-energy wax esters as reserves and over-winter in diapause at depths of 500 to more than 2000m (Gislason and Astthorson 2000). The onset of reproductive maturity and egg production occur before the onset of the phytoplankton bloom and are mainly fuelled by internal reserves of wax esters (Ohman and Hirche 2001). The per capita egg production rate increases at the time of the bloom (Niehoff et al. 1999) and final gonad maturation followed by successful spawning depends on timing of the spring bloom (Niehoff et al. 2002).

9.4 Ecosystem effects of Arctic warming

According to our present understanding there are two possible scenarios of climate change in the region. The first is based on the hypotheses that the Arctic climate varies in cycles (Proshutinsky et al. 1999) and that we now are at the beginning of a cold period, leading to heavy ice conditions in the Barents, Greenland and parts of the Norwegian Seas (Vinje 2001). The second scenario is based on a hypothesis of a continuation of the present trend, resulting in a permanent warming of the Arctic climate (IPPC 2001) and a decrease of the ice extent of the Barents Sea and the Southern Arctic Ocean. The extent of the ice cover in the Nordic Seas in spring has decreased since 1860 due to the net thermal effect of the north-bound currents in the area (Vinje 2001). A warming of the water masses in the central Arctic Ocean has been observed in the 1990s, explained by increased

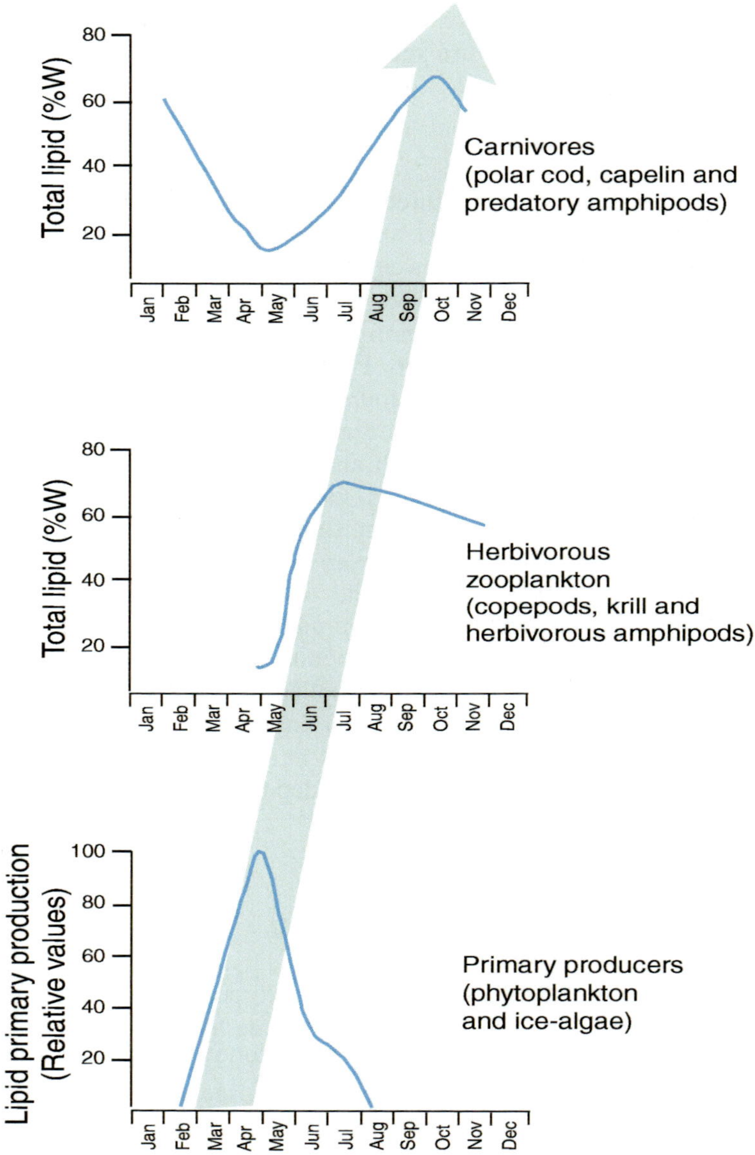

Fig. 9.6. The transfer of lipid from the spring bloom via Calanus and krill to capelin within six months (modified after Falk-Petersen et al. 1990).

Atlantic water influx in periods with high NAO modus (Furevik et al. 2005). Should Arctic warming be a long-term trend, as suggested by IPPC, and much of the southern part of the Arctic Ocean become ice - free during spring and summer, a yearly spring algal bloom will occur in this area, resulting in a large increase in "new" primary production. This will probably create an environment favourable to *C. finmarchicus* and detrimental to *C. glacialis* and especially to *C. hyperboreus*.

Polar pelagic systems are notoriously instable (Sakshaug 1997) and herbivorous zooplankton exposed to marked variations in food available has adapted by storing lipids as energy reserves. The lipid accumulated in large amounts by calanoid copepods is exploited as an energy source by large-stocks of fishes like herring (*Clupea harengus*) and the capelin (*Mallotus villosus*), which themselves can store large quantities of oil derived from zooplankton (Fig. 9.6). These energy rich fishes, together with *Calanus*, krill and amphipods species, occurring in dense schools and swarms, constitute large energy-packed food items for sea birds and marine mammals enabling them to sustain over-wintering populations (Welch et al. 1992). This lipid-based energy flux, from algae to marine mammals (Fig. 9.6), is likely to be a key determinant of the biodiversity and productivity of this environment.

The size spectrum and energy content of the major zooplankton species in Arctic ecosystems determine their value as food sources for the upper trophic levels. Both vary substantially among the three *Calanus* species. For example, *C. glacialis* and *C. hyperboreus* are considerably larger and adults contain approx. 10 and 25 times more energy (lipids), respectively, than *C. finmarchicus*. Fig. 9.5 and Table 9.1 summarise the important characteristics of the individual *Calanus* CV. Taken together this tells us that in the Arctic basin *C. hyperboreus* is the most highly adapted of the three species in accumulating large, energy-rich lipid reserves, followed by the Arctic shelf *C. glacialis* and the north Atlantic *C. finmarchicus* (Sargent & Falk-Petersen 1988; Scott et al. 2000). Perhaps more important for a zooplankton predator, is the difference in the food size spectrum between the three *Calanus* species, decreasing from the large energy-rich quanta provided by *C. hyperboreus* to the smaller less energy-rich quanta provided by *C. finmarchicus*.

9.5 Conclusions

From the above we conclude that *Calanus finmarchicus* is a deep-water species adapted to an environment with a regular yearly spring bloom, as occurs in the Norwegian Sea. *Calanus glacialis* is a shelf species adapted to an environment subjected to large variations in the timing and length of the annual bloom, as found in the northern Barents Sea and the Arctic

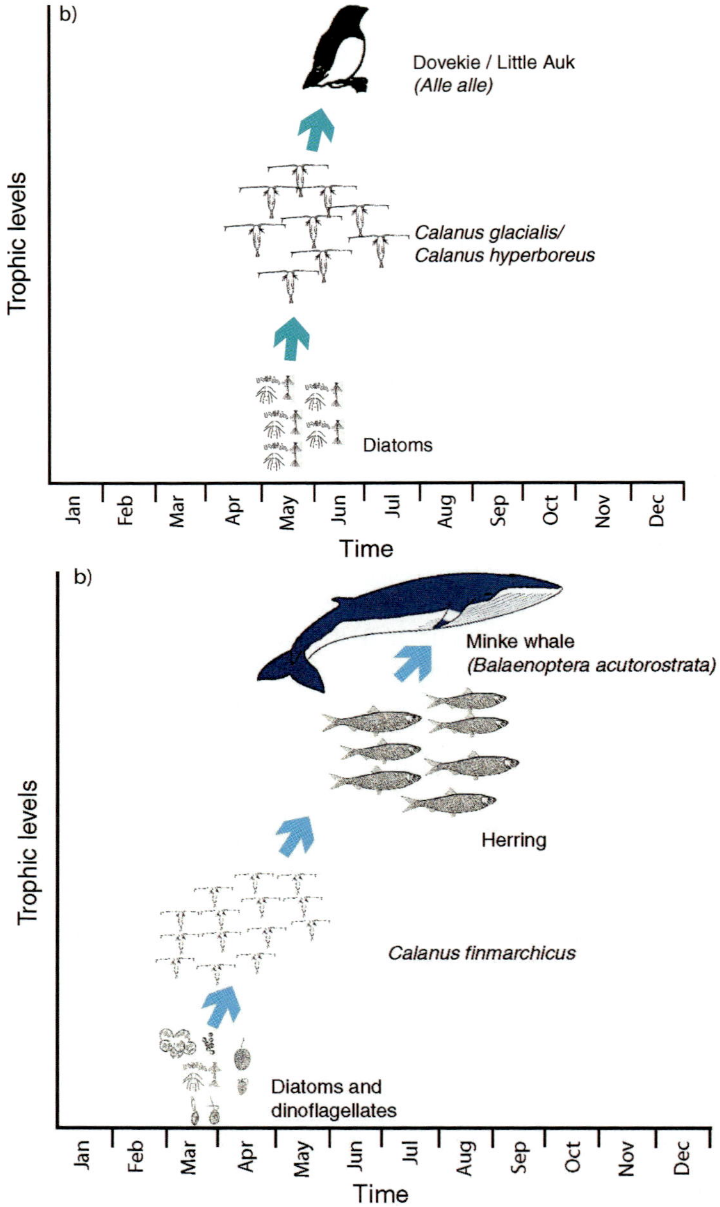

Fig. 9.7. *Calanus* based food web in the Nordic Seas. a) *C. glacialis / C. hyperboreus* based food web b) *C. finmarchicus* based food web.

shelves. *Calanus hyperboreus* is a deep-water species adapted to an environment with large inter-annual variations in ice cover and algal blooms, as found in the central Arctic Ocean and the Fram Strait.

We postulate that a warmer climate with reduced ice cover and new phytoplankton bloom regimes will shift zooplankton community structure towards a smaller zooplankton size spectrum (*C. finmarchicus*), with lower energy content per individual. The changes in zooplankton communities will be reflected in the diet composition and affect the general ecology of specialised seabirds. The little auk (*Alle alle*) requires access large energy-rich zooplankton (*C. glacialis and C. hyperboreus*) to successfully raise their chicks (Fig. 9.7a, Weslawski et al. 1994; Pedersen and Falk 2001). In Iceland, colonies of this high - arctic seabird were heavily reduced during the first half of the last century associated with the rapid warming in the Icelandic climate during the same period (Gudmundsson 1952). The rise in temperature probably triggered a shift in the zooplankton distribution favouring *C. finmarchicus* instead of *C. hyperboreus*, a situation having fatal impact on the little auk. It is more difficult to deduce consequences for generalised sea birds and sea mammals. However, Kitaysky and Golubova (2000) concluded, after 8 years of observation in the Okhotsk Sea, that internal oceanographic change impacted alcid reproduction performances by affecting food accessibility to planktivorous auklets and piscivorous puffins in opposite ways. A northward expansion of *C. finmarchicus* to the northern Barents Sea and into the Arctic Ocean north of Svalbard, due to a continuous warming of the water masses and new phytoplankton bloom regime, will create a prey base in the area that will favour the development of herring stocks. From this we postulate that Arctic warming will cause a ecosystem shift, from today's *C. glacialis / C. hyperboreus* based energy transfer favouring little auk (Fig. 9.7a) to a *C. finmarchicus* based food chain leading to minke whale (*Balaenoptera acutorostata*) via herring (Fig. 9.5b).

References

Aksnes DL, Blindheim J. 1996. Circulation patterns in the North Atlantic North Atlantic and possible impact on population dynamics of *Calanus finmarchicus*. Ophelia 44: 7-28.

Arctic Climatology Project 1998. Joint U.S.-Russian Atlas of the Arctic Ocean, Environmental Working Group joint U.S.–Russian atlas of the Arctic Ocean—1997, 1998 summer period. L. Timokhov & F. Tanis (eds.). Ann Arbor, MI: Environmental Research Institute of Michigan in association with the National Snow and Ice Data Center. CD-ROM.

Bucklin A, Assthorsson OS, Gislason A, Allen LD, Smolenack SB, Wiebe PH. 2000. Population genetic variation of *Calanus finmarchicus* in Icelandic waters: preliminary evidence of genetic differences between Atlantic and Arctic populations. ICES Journal of Marine Science 57 (6): 1592-1604.

Conover RJ 1988. Comparative life histories in the genera Calanus and Neocalanus in high latitudes of the Northern Hemisphere. Hydrobiologia 167/168:127-142.

Conover RJ Huntley M 1991. Copepods in ice-covered seas – Distribution, adaptations to seasonally limited food, metabolism, growth patterns and life cycle strategies in polar seas. Journal of Marine Systems 2:1-41.

Dahl TM, Falk-Petersen S, Gabrielsen GW, Sargent JR, Hop H, Millar R-M. 2003. Lipids and stable isotopes in common eider, black-legged kittiwake and northern fulmar - A trophic study from an Arctic fjord. Marine Ecology Progress Series 256:257-269.

Dawson JK 1978. Vertical distribution of Calanus hyperboreus in the Central Arctic Ocean. Limnology and Oceanography 23:950-957.

Dickson RR, Osborn TJ, Hurrel JW, Meincke J, Blindheim J, Ådlandsvik B, Vinje T, Alekseev G, Maslowski W. 2000. The Arctic Ocean response to the North Atlantic Oscillation. Journal of Climate 13:2671- 2696.

Engelsen O, Hegseth EN, Hop H, Hansen E, Falk-Petersen S. 2002: Spatial variability of chlorophyll-a in the Marginal Ice Zone of the Barents Sea, with relations to sea ice and oceanographic conditions. Journal of Marine Systems 35:79-97

Falk-Petersen, S., Sargent, J.R. and Middleton, C. 1986. Level and composition of tracylglycerols and wax esters in commercial capelin oils from the Barents Sea Fishery, 1983. Sarsia 71: 49-54.

Falk-Petersen S, Sargent JR, Tande KS. 1987. Lipid composition of zooplankton in relation to the Sub-Arctic food web. Polar Biology 8:115-120.

Falk-Petersen S, Hopkins CCE, Sargent JR. 1990. Trophic relationships in the pelagic food web. In: Barnes M, Gibson RN (Eds.), Trophic relationships in the marine environment. Aberdeen University Press, Aberdeen, pp. 315-333.

Falk-Petersen S, Pedersen G, Kwasniewski S, Hegseth EN, Hop H. 1999. Spatial distribution and life cycle timing of zooplankton in the marginal ice zone of the Barents Sea during the summer melt season in 1995. Journal of Plankton Research 21:1249-1264.

Falk-Petersen S, Hop H, Budgell WP, Hegseth EN, Korsnes R, Løyning TB, Ørbæk JB, Kawamura T, Shirasawa K. 2000a. Physical and ecological processes in the Marginal Ice Zone of the northern Barents Sea during the summer melt periods. Journal of Marine Systems 27:131-159.

Falk-Petersen S, Hagen W, Kattner G, Clarke A, Sargent JR. 2000b. Lipids, trophic relationships and biodiversity in Arctic and Antarctic krill. Canadian Journal of Fisheries and Aquatic Science 57 :178-191.

Falk-Petersen S, Sargent JR, Kwasniewski S, Gulliksen B, Millar R-M. 2001: Lipids and fatty acids in *Clione limacina* and *Limacina helicina* in Svalbard waters and the Arctic Ocean: Trophic implications. Polar Biology 24:163-170.

Falk-Petersen S, Dahl TM, Scott CL, Sargent JR, Gulliksen B, Kwasniewski S, Hop H, Millar, R-M. 2002: Lipid biomarkers and trophic linkages between the Arctic ctenophores and calanoid copepods in Svalbard waters. Marine Ecology Progress Series 227:187-194.

Falk-Petersen, S., T. Haug, Nilsen, K.T., Wold, A. and Dahl T.M. 2004. Lipids and trophic linkages in harp seal (*Phoca groenlandica*) from the eastern Barents Sea. Polar Research 23 (1): 43-50.

Furevik T. 2001. Annual and interannual variability of Atlantic water temperatures in the Norwegian and Barents Seas: 1980-1996. Deep-See Research I, 48: 383-404.

Furevik,T, Mauritzen, T, and Ingvaldsen, R. 2005: The Flow of Atlantic Water to the Nordic Seas and Arctic Ocean. In: Orbaek, J.B.,Tombre, T., Kallenborn,, R., Hegseth, E., Falk-Petersen, S. and Hoel A.H. (Eds.), Arctic-alpine Ecosystems and People in a Changing Environment, Springer Verlag, Berlin,

Gislason A, Astthorsson OS. 2000. Winter distribution, ontogenetic migration, and rates of egg production of Calanus finmarchicus southwest of Iceland. ICES Journal of Marine Science 57 (6): 1727-1739.

Gudmunsson F. (1952). The effects of recent climatic changes on bird life of Iceland. Proceedings of the Xth International Ornithological Congress pp. 502-514.

Hansen B, Østerhus S. 2000. North Atlantic- Nordic Sea exchanges. Progress in Oceanography 45:109-208.

Hegseth EN. 1998. Primary production of the northern Barents Sea. Polar Research 17:113-123.

Hirche H-J. 1997. Life cycle of the copepod *Calanus hyperboreus* in the Greenland Sea. Marine Biology 128:607-618.

Hirche H-J. 1998 ?Hirche H-J, Kwasniewski S. 1997. Distribution, reproduction and development of *Calanus* species in the Northeast Water in relation to environmental conditions. Journal of Marine Systems 10:299-317.

Hirche H-J, Mumm N. 1992. Distribution of dominant copepods in the Nansen Basin, Arctic Ocean, in summer. Deep Sea Research 39:485-505

IPPC 2001. Climate change 2001. The scientific basis. Contribution of working group I to the third assessment report of the intergovernmental panel on climate change. Houston JT, Ding Y, Griggs DJ, Nouger M, van der Linden PJ, Dai X, Johnson CA (eds). Cambridge University 881pp.

Jaschnov WA. 1970. Distribution of *Calanus* species in the seas of the northern hemisphere. Internationale Revue der gesamten Hydrobiologie und Hydrographie 55:197-212.

Jones EP, Rudels B, Anderson LG. 1995. Deep waters of the Arctic: origin and circulation. Deep-Sea Research 42:737-760.

Kitaysky AS, Golubova EG. 2000. Climate change causes contrasting trends in reproductive performance of planktivorous and piscivorous alcids. Journal of Animal Ecology 69:248-262

Kosobokova KN. 1999. The reproductive cycle and life history of the copepod *Calanus glacialis* in the White Sea. Polar Biology 22:254-263.

Lee RF. 1975. Lipids of zooplankton. Comparative Biochemistry and Physiology B51: 263-266.

Lee RF, Hirota J. 1973. Wax esters in tropical zooplankton and nekton and the geographical distribution of wax esters in marine copepods. Limnology and Oceanography 18:277-239.

Lie U. 1968. Variations in the quantity of zooplankton and the propagation of *Calanus finmarchicus* at Station M in the Norwegian Sea. Fiskeridirektorates Skrifter, Serie Havundersøkelser 14:121-128.

MacLellan DC. 1967. The annual cycle of certain calanoid species in west Greenland. Canadian Journal of Zoology 45:101-115.

Marshall SM, Orr AP. 1955. The biology of the marine copepod *Calanus finmarchicus* (Gunnerus). Oliver & Boyd, Edinburgh 188pp.

Matishov G., Zuyev A., Golubev V., Adrov N., Slobodin V., Levitus S., Smolyar I. 1998. Climatic Atlas of the Barents Sea, 1998: temperature, salinity, oxygen. Murmansk Marine Biological Institute (Russia), National Oceanographic Data Centre, Ocean Climate Laboratory (USA), Washington, 122 p. (CD)

Matthews JBL.1969. Continuous plankton records: the geographical and seasonal distribution of *Calanus finmarchicus* s.l. in the North Atlantic. Bulletin of Marine Ecology 6:251-273.

Mauchline J. 1998. The biology of Calanoid Copepods. Advances in Marine Biology 33: 1-710

Mehlum F, Gabrielsen GW 1993. The diet of high-arctic seabirds in coastal and ice-covered, pelagic areas near the Svalbard archipelago. Polar Research 12:1-20.

Nansen F. 1902. Oceanography of the North Polar basin. The Norwegian North Polar Expedition 1893-1896. Scientific results (3) 9, 427 pp. Greenwood Press, New York.

Niehoff B, Klenke U, Hirche HJ, Irigoien X, Head R, Harris R. 1999. A high frequency time series at Weathership M, Norwegian Sea, during the 1997 spring bloom: the reproductive biology of *Calanus finmarchicus*. Marine Ecology-Progress Series 176: 81-92.

Niehoff B, Madsen BW, Hansen BW and Nielsen TG 2002. Reproductive cycles of the three dominant *Calanus* species in Disko Bay, West Greenland. Marine Biology 140:567-576.

Ohman MD, Hirche H-J. 2001. Density-dependent mortality in an oceanic copepod population. Nature 412:638-641.

Pedersen CE, Falk K. 2001. Chick diet of dovekies *Alle alle* in Northwest Greenland. Polar Biology 24 :53-58.

Planque B, Hayes G, Ibanez F, Gamble JC. 1997. Large-scale spatial variations in the seasonal abundance of *Calanus finmarchicus*. Deep-Sea Research 44: 325-326.

Polyakov IV, Proshutinsky AY, Johnson MA. 1999. Seasonal cycles in two regimes of Arctic climate. Journal of Geophysical Research 104 (C11):761-788.

Proshutinsky AY, Polyakov IV, Johnson MA. 1999. Climate states and variability of Arctic ice and water dynamics during 1946-1997. Polar Research 18:135-142.

Rudels B, Friedrich HJ, Quadfasel D. 1999. The Arctic Circumpolar Boundary Current. Deep-Sea Research II 46:1023-1062.

Runge JA, McLaren IA, Corkett CJ, Bohrer RN, Koslow JA. 1986. Molting rates and cohort development of *Calanus finmarchicus* and *C. glacialis* in the sea off southwest Nova Scotia. Marine Biology 86:241-246.

Sakshaug E. 1997. Biomass and productivity distributions and their variability in the Barents Sea. ICES Journal of Marine Science 54:341-351.

Sakshaug E. 2003. Variability in primary production in Arctic Seas. In: Stein R. (ed). The Arctic Carbon Cycle Springer (in press)

Sargent JR, Henderson RJ, 1986. Lipids. In: Corner, E.D.S. (Ed.) The biological chemistry of marine copepods. Clarendon Press, Oxford, pp. 59-108.

Sargent JR, Falk-Petersen S. 1988. The lipid biochemistry of calanoid copepods. Hydrobiologia 167/168: 101-114.

Scott CL, Kwasniewski S, Falk-Petersen S, Sargent JR. 2000. Lipids and Life Strategies of *Calanus finmarchicus*, *Calanus glacialis* and *Calanus hyperboreus* in Late Autumn, Kongsfjorden, Svalbard. Polar Biology 23:510-516.

Scott CL, Kwasniewski S, Falk-Petersen S, Sargent JR. 2002. Species differences, origins and functions of fatty alcohols and fatty acids in the wax esters and phospholipids of *Calanus hyperboreus, Calanus glacialis* and *Calanus finmarchicus* from Arctic waters. Marine Ecology Progress Series 235:127-134.

Sekerak A.D, Thomson D, Bain H, Acreman J. 1976. Summer surveys of the marine ecology of Creswell Bay, Somerset Island, and Assistance Bay, Cornwallis Island, NWT. Unpublished Report by LGL Ltd, for Polar Gas. Ltd: 1-251

Tande KS. 1991. *Calanus* in North Norwegian fjords and in the Barents Sea. Polar Research 10:389-408.

Tereshchenko, V.V. 1997. Seasonal and interannual changes of temperature and salinity of waters of the main currents on the section "Kola Meridian" in the Barents Sea. Murmansk, publ. PINRO: 1-71.

Van Aken HM, Quadfasel D, Warpakowski A. 1991. The Arctic front in the Greenland Sea during February 1989: Hydrographic and biological observations. Journal of Geophysical Research 96, C3:4739-4750.

VEINS. Variability of exchanges in the Northern Seas (VEINS). EU Marine Science and Technology Programme (MAST-III), Project MAS3-CT96-0070, 1997-2000.

Vinje T. 2001. Anomalies and trends of ice extent and atmospheric circulation in the Nordic Seas during the period 1864-1998. Journal of Climate 14:255-267.

Vinogradov ME. 1997. Some problems of vertical distribution of meso-macroplankton in the ocean. Advances in Marine Biology 32:1-92.

Welch HE, Bergmann MA, Siferd TD, Martin KA, Curtis MF, Crawford RE, Conover RJ, Hop H. 1992. Energy flow through the marine ecosystem of the Lancaster Sound region, Canada. Arctic 45: 343-357.

Weslawski JM, Stempniewicz L, Galaktionov K (1994) Summer diet of seabirds from the Franz-Josef-Land Archipelago, Russian Arctic. Polar Research 13:173-181

Wiborg KF. 1954. Animal plankton of the Norwegian coast and offshore waters of west and north-western Norway. Report from the Norwegian Fisheries Investigations 11:1-246.

Zenkevitch L. 1963. Biology of the Seas of the USSR. Georg Allen and Unwin, London, 955 pp.

10 Adjustment to reality: Social responses to climate changes in Greenland

Rasmus Ole Rasmussen

NORS – North Atlantic Regional Studies, Roskilde University 21.2, Box 260, DK4000 Roskilde, Denmark, e-mail: rasmus@ruc.dk

10.1 Introduction

Greenland has experienced three major socio-economic shifts during the 19th and 20th century, all of them induced by the interactions between the natural system of climate change, and the socio-economic and socio-technical system of resource exploitation (Hamilton et al., 2003; Rasmussen and Tommasini 2003).

The first was the shift from a sea mammal based economy to fisheries during the latter part of the 19th century and up till around World War II (WWII). This had to do with a marked increase in sea temperature, resulting in a decrease in the sea mammal stock, combined with a dwindling world market for blubber and sealskin. The cod became the dominating species, and in general the changes resulted in the process of sedentarization as well as the starting point of the modernization of Greenland.

The second transformation was the shift from cod fisheries to a mono economy based on shrimp fisheries during the 1970's and the 1980's. The main reason for this shift was a reduction in sea temperature. This process of diminishing cod resources had already started in the 1960's and leading to a low in the 1970's. The decline – in combination with over-exploitation of the stock - basically eliminated the spawning possibilities of the cod stock. At the same time it gave way for a massive expansion of the shrimp fisheries, an expansion which was facilitated by a shift from inshore to off-shore fisheries. Parallel to the transition in resource base Greenland experienced a modernization process. This process created the present infrastructure as well as the production structures dominating Greenland today. Politically it lead to the creating of Home Rule in Greenland in 1979, and in relation to production it resulted in massive investment in the off-shore shrimp fleet during the 1980s.

The third transformation is the ongoing shift towards a more diversified focus of fisheries. This shift has definitely been influenced by the present temperature changes, and is based on shrimp fisheries as the backbone of the economy, in combination with Greenland Halibut, Snowcrab and other species offering substantial contributions to the economy. At the same time the economy seem to be directing towards a situation where major contributions from other business activities outside renewable resource extraction have become crucial. Besides the importance of the local environment on the resource situation in general, however, the characteristics of the world market are instrumental in this new transition process.

10.2 General patterns of resource usage

The marine ecosystem of Greenland waters are in terms of renewable resources among the most productive in the world. They are intermediate between the cold polar water masses of the Arctic region and the temperate water masses of the Atlantic Ocean (Pedersen and Zeller, 2001). The resource situation is characterized by three major water masses. First, the cold (–1.8 °C) polar water coming through the Nares Strait, Baffin Bay and Davis Strait. Secondly, the comparatively warm (3–4 °C) Irminger Current, a branch of the North Atlantic Current, which transports warm water from the Gulf of Mexico to Europe (Smidt 1989:13). On Greenland's east coast the Irminger Current meets the Polar Current moving south, and both flow south of Cape Farewell and then northward along the west coast of Greenland. The third water mass is the fresh water melting from Greenland's ice sheet (Nielsen et al 1970:57–58).

As the resource situation is determined by water from both the polar and temperate regions, the marine productivity in the Greenland shelf areas is highly influenced by changes in the North Atlantic circulation system. These changes are impacting the distribution of species and fisheries yield around Greenland (Pedersen and Rice, 2002; Buch et al., 2003). And as the distribution of several species are marginal to their core distribution, even a small variations in summer temperatures will highly affect the composition of the marine species.

10.3 Historic changes in resource usage patterns

Hunting, especially for sea mammals, had been the main focus of the traditional lifestyle of the Inuit in Greenland, and also the focus of the colonial

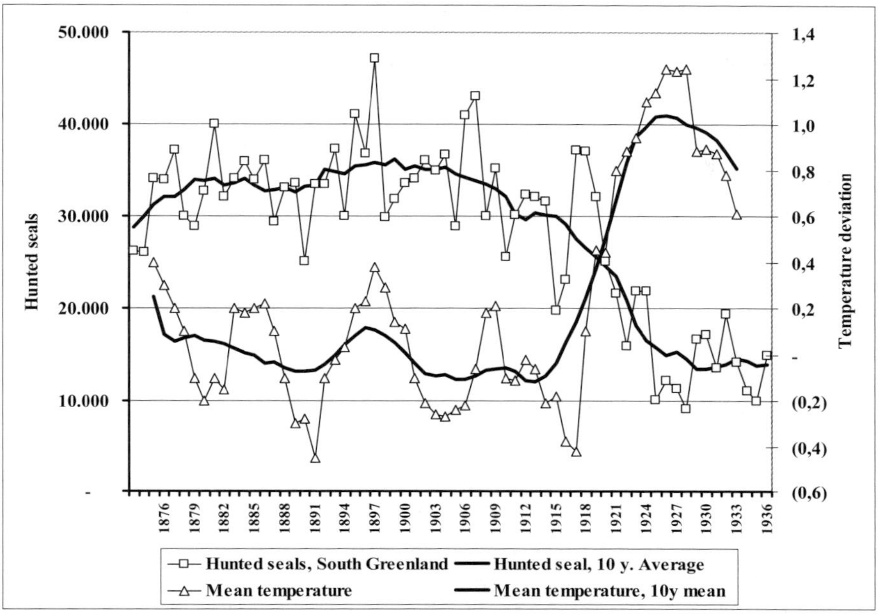

Fig. 10.1: Relationship between decline in seal hunting and temperature change in Greenland 1876-1936. The increase in seal hunting up till 1900 was due to an increase in population. The higher pressure on the resource caused a process of depreciation of the stock, and consequently a decline in number of hunted seals. The process of depreciation of the stock was accelerated by the rather sudden change in temperature from 1918. Partly based on Rasmussen, 2003b.

economic policy for Greenland well into the 20th century. An increase in the population, however, demanded more resources for survival, and as an increase in sea temperatures resulted in a natural decline of the stock, a general problem emerged. Simultaneously the European and North American markets for seal blubber based lamp oil was replaced by mineral oil, so a new basis for as well subsistence as commercial activities was highly needed. The result of the dramatic increase in sea temperature within a 10-year period shows clearly on fig. 10.1, and was followed by a change in resource base from sea mammals - mainly seals - toward fisheries.

10.3.1 The process of sedentarization

The first commercial fishery began in Ilulissat in the 1890s (Nørrevang et al. 1971), followed by an initiative in 1904 towards commercialization by enabling sale of Greenland halibut (Reinhardtius hippoglossoides) in

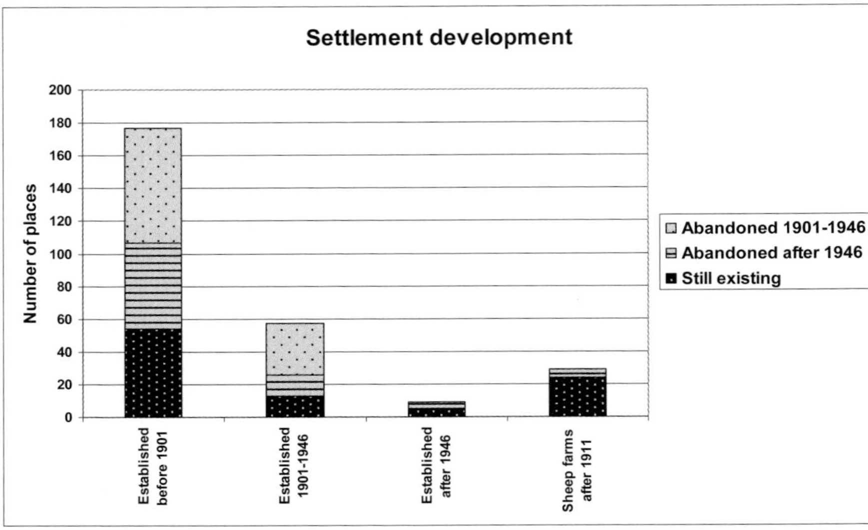

Fig. 10.2: The sedentarization process in Greenland. The major part of the settlements in Greenland have been established in the latter part of the 19th and the early years of the 20th century. Based on data from Rasmussen, 2003b.

Ilulissat, and Arctic char (Salvelinus alpinus) in other parts of Greenland. (Smidt 1989). The Tjalfe research expedition in 1908–1909 had shown that fish were available in several other places along the coast, so in 1910 foreign fisheries for halibut (Hippoglossus hippoglossus) started around Sisimiut, just as a fishery for Greenland halibut was established in the fjords around Qaqortoq. At the same time, the first processing plant for salted cod was established at Qeqertarsuatsiaat south of Nuuk.

The experiences from this plant were of great importance for the further development and expansion of fish processing activities, as several Greenlanders became experienced in the processing technique (Nørrevang et al. 1971). In the following years additional salting plants were established along the coast, and as waters warmed markedly the seals declined, and cod fishing replaced seal hunting (Mattox, 1973). But whereas seals had traditionally provided Greenlanders with almost everything they needed to survive, not only food but also basic elements of clothing, cooking and heating, the fisheries provided only food. This created an obvious demand for commercial activities (Smidt 1989), eventually leading to the sedentarization of the population.

As shown on fig. 10.2, the major part of settlements have been established in the late 19th century, and even a large number of settlements have been abandoned – most of them due to limited or just temporary resource access – the present settlement pattern is more or less reflecting this transition process from a semi-nomadic hunting society to a pattern of fisheries dependent settlements.

10.3.2 Industrialization

After World War II a new approach to development was initiated (Rasmussen and Hamilton, 2001). Its purpose was to reduce the backwardness of the economy, and to bring a modern standard of living to the population. The starting point was the recognition of Greenland as a county in Denmark, and the G50 plan which paved the way for new development of the economy based on fisheries, primarily based on private initiatives. After 10 years, however, it became obvious that the privatization had not succeeded, as there were not any private initiatives on which to build the economy. By 1964 the next plan – the G60-plan - was ready, stressing that public activities were necessary to create a take-off situation for the home economy.

The promotion of large scale industrial activities led to a situation where the major population increase took place in the larger settlements, and especially in settlements with a high potential for development in fisheries. The population growth was matched by massive investments in new infrastructure such as concrete housing on a large scale industrial base. And fisheries, which were considered to be the backbone of the economy, received special attention.

The cod resource was considered inexhaustible, and the world market open and available for new products (Poppel, 1997), so the development perspectives seemed to be unlimited. Consequently the capacity of the fleet was doubled from 1965 to 1975. The total volume of output from fisheries was 50.000 tons and increasing, as shown on fig. 10.3 below. The expansion of the fisheries was almost solely in cod fisheries, but slowly commercial fishing for other species was introduced as well.

The fishermen were aware of the fluctuations in the resources, but the fishing industry - and especially the Royal Greenland Trading Company - was very slow to respond to the ongoing changes. The industry was based on large-scale processing of cod, but it seemed difficult to react to the fact that the situation was changing. So it was not recognized by the authorities until during the 1970s that the changes in sea temperature on one hand contributed to the diminishing of the cod stock, and on the other hand gave way to a new shift in resource base, as indicated on fig. 10.4 below. The decline in sea temperature contributed to the decline in cod stock, which eventually opened up for a boom in shrimp fisheries. And when the

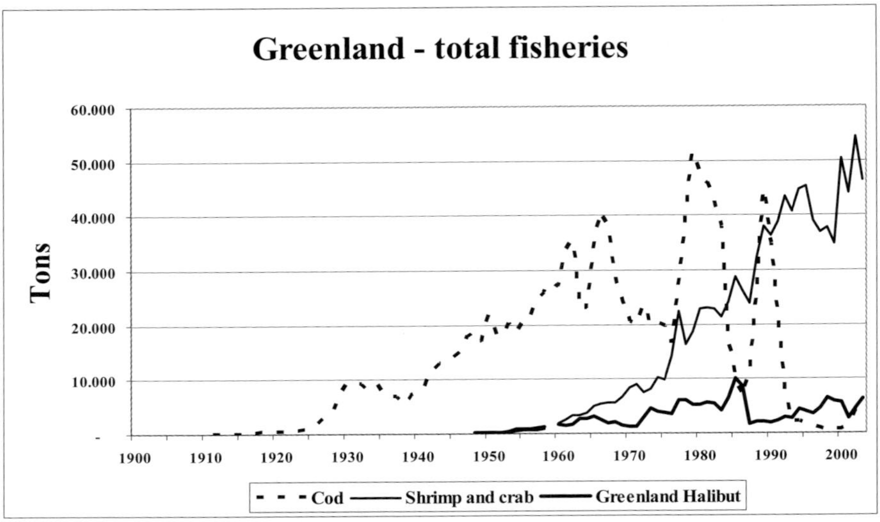

Fig. 10.3: Development of fisheries in Greenland during the 20th century. Cod was the dominating resource until it collapsed in the 1980s, when the dominating position was taken over by shrimp fisheries. During the last 10 years Greenland Halibut has become a major resource for fisheries. Based on data from Rasmussen and Hamilton (2001).

temperature again increased in the late 1980s and 1990s, the cod was not able to regenerate so other species consequently became dominant (Hamilton et al, 2003).

10.3.3 From a cod to a shrimp based economy

It was realized by the Home Rule Government in Greenland, when they took over in 1979, that the changes in the cod stock were permanent and could therefore not be a proper basis for an economic development. A 5-year development plan was introduced in 1984, aiming at investments that should create a massive change of focus from cod towards the much more promising shrimp fisheries.

Until the 1950s, shrimps were mainly pursued by inshore vessels on in-shore banks. But the combination of the cooling of water and a reduced cod stock led to a massive expansion of the shrimp stock. And consequently it caused a massive investment in new fishing gear and off-shore vessels. When the Home Rule took over the government of Greenland, the economic basis was partly transfer money from Denmark. The money were supposed to cover the continuation of both the social and the institutional structures inherited from the colonial days. Instead the Home Rule

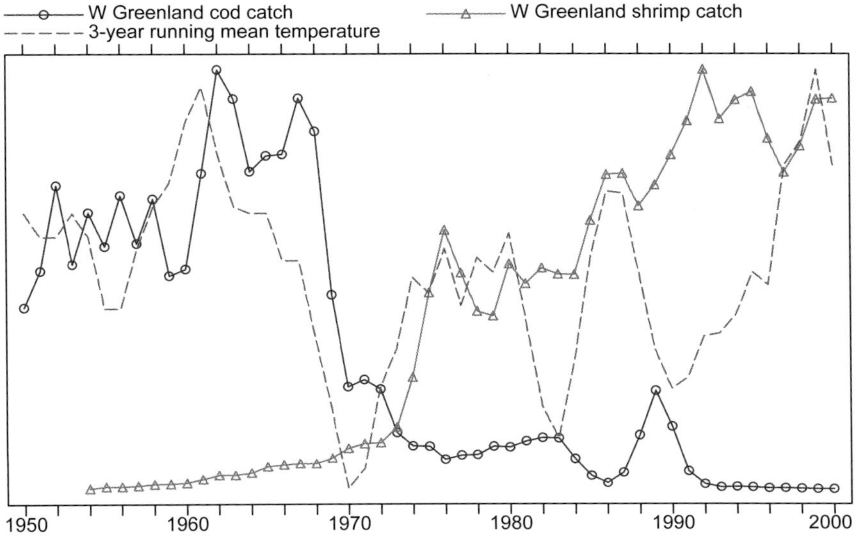

Fig. 10.4: The shift in resource base partly induced by changes in sea temperatures. The marked decline in temperature from 1960 to 1980 was paralleled by the collapse of the cod stock. The following temperature increase then was paralleled by the expansion of the shrimp fisheries. From Hamilton, Brown and Rasmussen, 2003.

Government used part of the money for new initiatives – primarily investments in connection with fisheries, and first of all in the larger west-coast towns. Consequently, a new era was established. The shrimp resource and shrimp related activities became the backbone of the economy, with the Home Rule owned processing plants and a very efficient off-shore fleet as the major players in the game. This shift is very evident when looking at fig. 10.3 and 10.4, eventually transforming Greenland into the worlds largest producer of coldwater shrimp.

10.3.4 The need for diversification

During the last decade the Home Rule government has been confronted with two major obstacles for the economic expansion based on the shrimp resource. First of all, the shrimp fisheries has now been developed to its maximum. Both national and international fisheries advisors are recommending maintenance of the fisheries at the present level, only with minimal adjustments, and this reality seem to be accepted generally by the trade (Det Rådgivende Udvalg, 2004). Secondly, the fishing industry is

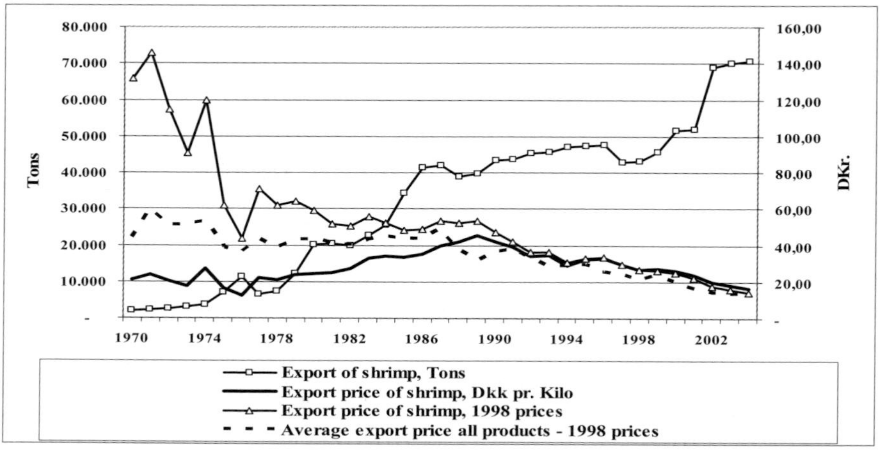

Fig. 10.5: Export volume and export price development of shrimp production in Greenland. Up till the late 1980s both volume and average prices were increasing, but during the last 15 years the average price has declined, while production volume has increased in order to compensate for this decline.

confronted with the harsh reality of the world market. This situation becomes clear when looking at fig. 10.5. The average fixed price of fish products has been more or less in a situation of constant decline. The average price per kilo of export products from Greenland, mainly cooked shrimp, has declined from a level around 30 DKK in the 1980s to the present level around 15 DKK per kilo (both in fixed 1998 prices), and when looking at the price development for shrimp the situation has been even worse.

With a continuation of this trend in world market prices, maintaining a future income at the present level would require a production increase of at least 4-5% per year. And this is not possible - as discussed above - with the present exploitation of the shrimp resource at its maximum.

The changes in the environment and the lack of influx of a new cod stock, however, have contributed to increased access to other renewable resources (fig. 10.6). First of all Greenland halibut has turned out to be an important asset for most of the northern communities in Northwest Greenland during the last 10 years. Settlements in the municipalities of Upernavik and Uummannaq which previously were depending on hunting of seals has become thriving fisheries communities (Det Rådgivende Udvalg, 2003, 2004, 2005). And during the recent years also an increased fishing for snow crab has turned out to be an important contributor to the total economy. Parallel to the development of fisheries, the hunting for both land- and sea-mammals are still contributing to both subsistence and

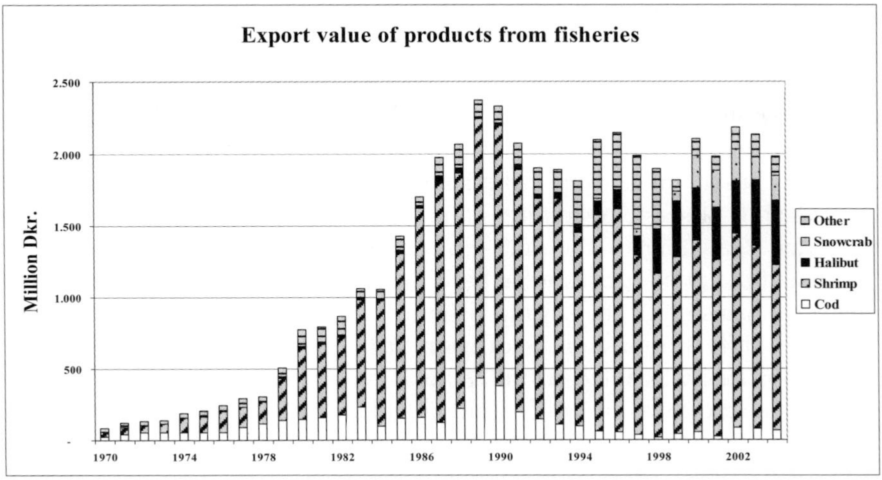

Fig. 10.6: Value of export of fish and fish products from Greenland. The export which was previously dominated by a mono-economy of shrimp products has emerged into a more diverse structure, with more than 1/3 of the total export covered by other species. Partly based on Rasmussen, 2003a.

commercial activities. As shown by Rasmussen (1993b), between 10 to 30 percent of the local economy is still due to contributions from the informal and subsistence sector!

Since the mid 1990s, however, the political aims for the Home Rule in Greenland have been towards consolidation and new business strategies (Poppel, 1997). It has become obvious that there are limitations to economic expansion based on renewable resources, and as a consequence the concept of " the three pillars" - Fishing, Minerals and Tourism – was developed as a strategy for further development. More recently the whole concept of “land based activities” has been added as a crucial cornerstone for future economic development, and if these steps succeed, substantial changes in the resource dependency may eventually appear.

10.4 Conclusions

As presented above, Greenland has experienced three major socio-economic changes during the last 150 years. The first was the shift from a sea mammal to fisheries based economy during the latter part of the 19th century and up till around WWII. It resulted in sedentarization as well as the first steps of modernization of Greenland. While an increase in sea temperature pushed towards a change in resource base, the dwindling world market for blubber and sealskin combined with a growing market for salted and dried fish pulled the economy towards fisheries. This, however, required more centralized processing facilities, and therefore also a more permanent settlement structure.

The second transformation was the shift from cod fisheries to a mono economy based on shrimp fisheries during the 1970's and the 1980's. Again a shift in sea temperature – this time a temperature reduction which prevented spawning of the cod – in combination with over-exploitation eliminated the previous resource base. Instead the changes gave way for a massive expansion of the shrimp fisheries, an expansion which was facilitated by a shift from inshore to offshore fisheries. This shift, however, was also rather easily adapted to the general modernization process, because the focus already was on concentration and centralization of production as well as of population.

The third transformation is the ongoing shift towards a more diversified focus of fisheries as well as of the economy. This shift is also influenced by the present temperature change, a temperature increase which eventually may lead to a re-introduction of the cod stock, but so far has pushed the Greenland Halibut-fisheries towards Northwest Greenland, and has opened up for a substantial Snowcrab fisheries in the southern and central parts of Greenland's Westcoast. An important pull in this process towards a more diversified fisheries has been the continuous decline in world market prices of shrimp products. Even a substantial increase in production has not been able to compensate the diminishing revenue, so the diversification has simply been required in order just to maintain a stable income from fisheries. And it has been still clearer that not even a diversified fisheries is enough for the future, but that a diversified economy would be needed.

A common characteristic of all three cases is that both the process and the outcome has been the result of an interaction between environmental changes and changes in the socio-economic structures. In none of the cases the result has been determined by only one of the factors, just as there has been no simple one-way causality. The implications of climate change have been the strengthening of ongoing socio-economic changes, just as it has contributed to the outlining of possible development paths. The socio-economic changes, on the other hand, have been reflecting a more or less pragmatic approach to both the material and the political reality – what could be characterized as an "adjustment to reality".

The differences between the cases have been the level of implications of the environmental changes. The shift from a sea mammal based economy to a fisheries economy had profound consequences on society, first of all through the sedentratization process. The shift from a cod to a shrimp based economy strengthened an already ongoing process of concentration and centralization, which made the villages more or less redundant, but did not totally re-structure the organization of the communities. And the present diversification of both renewable resource exploitation and the economy in general seems to have reversed parts of the former concentration policy, by making some of the villages close to the new resources as a more active part of the economy.

There is little doubt that the living conditions in Greenland also in the future will experience social changes in interaction with the environmental dynamics similar to what has been the case during the last 150 years.

References

Buch, E., Pedersen, S.A. and Ribergaard, M.H., 2003: Ecosystem variability in West Greenland waters. Journal of Northwest Atlantic Fishery Science, in press.

Det Rådgivende Udvalg, 2003, 2004, 2005: Den Økonomiske Udvikling i Grønland. Det Rådgivende Udvalg Vedrørende Grønlands Økonomi. Grønlands Hjemmestyre.

Hamilton, L.C. Brown, B.C. Rasmussen, R.O., 2003: West Greenland's cod-to-shrimp transition: local dimensions of climatic change. Arctic, v. 56, no. 3, Sept. 2003, p. 271-282, ill., maps

Mattox, W.G., 1973: Fishing in West Greenland 1910-1966. The Development of a new native industry. Meddelelser om Grønland bind 197, nr 1, København: C.A. Reitzels Forlag

Nielsen, N., Skautrup, P., Vibe, Chr, eds, 1970: J.P. Trap - Danmark. Vol. XIV: Grønland. G.E.C. Gads Forlag, København.

Nørrevang, A., Meyer, T., Christensen, S. (ed), 1971: Danmarks Natur bind 10. Grønland og Færøerne. Politikens forlag.

Pedersen, S.A., and Rice, J. 2002. Dynamics of fish larvae, zooplankton, and hydrographical characteristics in the West Greenland Large Marine Ecosystem 1950-1984. In: Large Marine Ecosystems of the North Atlantic. Changing States and Sustainability. K.S. Shermann, H.-R. and Skjoldal (eds). Chapter 5. Elsevier Science. p. 151-193

Pedersen, S.A. and Zeller, D., 2001. A mass balance model for the West Greenland marine ecosystem. In: Guenette, S., Christensen, V. and Pauly, D. (eds). Fisheries impacts on North Atlantic Ecosystems: Models and Analyses. Fisheries Centre Research Reports 9(4)., 111-127.

Poppel, Birger, 1997: Greenland's Road to Recovery and the Pattern of Settlement. in: North - The Nordic Journal of Regional Development and Territorial Policy. Vol. 8 no. 2, July 1997. Oslo.

Rasmussen, Rasmus Ole, 2003a: Havfiskeri/kystfiskeri – magt og afmagt I Grønlands hoveerhverv. I: Gorm Winther (red): Demokrati og Magt i Grønland. Aarhus Universitetsforlag.

Rasmussen, Rasmus Ole, 2003b: Social response to short and long term changes in sea temperatures in Greenland . Paper presented at "Arctic Alpine Ecosystems and People in a Changing Environment", Tromsø 24 February - 1 March 2003

Rasmussen, Rasmus Ole and Hamilton, Lawrence C., 2001: The Development of Fisheries in Greenland, with focus on Paamiut/Frederikshåb and Sisimiut/Holsteinsborg. Roskilde, NORS Research Report 53.

Rasmussen, Rasmus Ole and Tommasini, Daniela: Adjustment to Reality - Cases of detached, dependent, and sustained community development in Greenland. Paper and poster presentet at SEARCH Open Science Meeting, Seattle, 27-30 October 2003.

Smidt, Erik L. Balslev, 1989: Min tid i Grønland - Grønland i min tid. Fiskeri, Biologi, Samfund 1948-1985.

Statistical Yearbook, 2000, 2001, 2002, 2003. Nuuk: Home Rule Government.

UV radiation and biological effects

11 Factors, trends and scenarios of UV radiation in arctic-alpine environments

Mario Blumthaler

Innsbruck Medical University, Department of Medical Physics, Muellerstr. 44, A-6020 Innsbruck, Austria, e-mail: Mario.Blumthaler@uibk.ac.at

11.1 Introduction

The levels of UV radiation at the Earth's surface vary widely, depending on the atmospheric and environmental conditions of the observation site. The knowledge of the effects of the individual parameters affecting UV radiation as well as their interaction is a prerequisite for any analysis of possible long-term trends.

A special situation is found for UV radiation in the Alps and similarly in the Arctic. There the effects of parameters like aerosol amount, altitude above sea level and reflectivity of the ground (albedo) work together to yield the highest possible levels of UV radiation, in addition to the effects of solar zenith angle and atmospheric ozone. Detailed measurements of solar UV radiation levels in Arctic and Alpine regions have been carried out, allowing the estimation of the different effects of these parameters on future UV radiation scenarios.

The measurements of solar UV radiation discussed here are carried out with detectors with a horizontal detector surface, thus determining the irradiance, which assumes a weighting of the incoming radiation with the cosine of the zenith angle. One part of the irradiance comes in a straight line from the sun (direct component), the other part is scattered and arrives as the radiance of the upper hemisphere of the sky (diffuse component). Both parts together form the global irradiance, which is the usual quantity for any meteorological or biological discussion of UV effects.

The measurement results presented here stem from two types of instruments. Measurements with a spectroradiometer (spectral resolution of 0.5 to 1 nm) provide the most valuable data, allowing a range of analyses of the results. However, especially in the UV-B range this is still challenging and it requires highly experienced manpower to reach an absolute level of uncertainty of ±5% (Bais et al. 2001). Therefore, routine observations are often carried out with broadband detectors, which are less costly to acquire and operate, but are usually also less accurate. Still they need careful maintenance. The most widely used are broadband detectors which have a spectral sensitivity simulating the sensitivity of the human skin for erythema (McKinlay and Diffey 1987). With proper calibration, these detectors can directly measure biologically relevant irradiance.

11.2 Variability of solar UV radiation

Assuming cloudless conditions, the most important parameters for solar UV irradiance are solar zenith angle, ozone content of the atmosphere, amount of aerosols, albedo of the ground and altitude above sea level. Some other parameters are of less significance, i.e. further absorbing gases in the atmosphere besides ozone, type and optical characteristics of aerosols or vertical distribution of the constituents of the atmosphere. For Alpine and Arctic environments, albedo of the ground and altitude above sea level are of special significance. If clouds cover the sky, then the radiation levels are dominated by the type and distribution of the clouds.

11.2.1 Effect of solar zenith angle

The solar zenith angle (SZA) at the time of observation is the most significant parameter, determining the level of UV radiation. It is fixed by the latitude of the observation site and by the local time. The higher the SZA the smaller is the irradiance as a consequence of increasing absorption and backscattering in the atmosphere with increasing path length of the radiation from the sun. This effect is especially pronounced for UV-B irradiance (280 - 315 nm), as additionally these wavelengths are absorbed more efficiently in the ozone layer of the atmosphere when the SZA is higher. Therefore, the diurnal and annual course of UV-B irradiance is much steeper than the course of UV-A irradiance (315 - 400 nm) or visible irradiance. As an example, in Innsbruck (47.3°N, 577 m above sea level) the ratio between maximum values of daily sums in summer relative to winter is about 5 for UV-A irradiance and about 15 for erythemally weighted UV-B irradiance.

Another consequence of the effect of the SZA on irradiance is the decrease of daily maximum irradiance levels, when going from mid latitudes towards the North Pole (assuming that all other parameters remain constant). Again, this effect is more pronounced for UV-B radiation compared to UV-A radiation. However, also the length of the day in summer is increasing, so that daily totals are less reduced than daily maxima.

11.2.2 Effect of ozone

The total ozone content of the atmosphere is the second most important parameter, to influence UV-B radiation. The absorption cross section of ozone is strongly decreasing in the UV range. Almost no radiation with wavelengths below 285 nm can be measured at the earth's surface, whereas the effect of ozone can almost be neglected at wavelengths above 330 nm. About 9% of the total amount of ozone is in the troposphere, the maximum concentration is located at altitudes between about 20 to 30 km in the so-called ozone layer.

The total amount of ozone varies strongly both locally and temporally, generally being between about 250 and 400 Dobson Units. Maximum values in mid-latitudes are observed in spring-time, when the day to day variability is also the highest, minimum values occur in autumn.

The reduction of global UV irradiance (*I*) by ozone (*O*) can be quantified by complex radiative transfer models, but it can be simplified with a power-law: $I \approx O^{-RAF}$. The exponent *RAF* is the radiation amplification factor. For relatively small variations of O (ΔO) the power law can be transformed to a linear form:

$$\frac{\Delta I}{I} = -RAF * \frac{\Delta O}{O}$$

This law is valid for specific wavelengths as well as for broadband wavelength ranges. For erythemally weighted irradiance the RAF amounts to about 1.1, thus a reduction of ozone by 1% will result in an increase of erythemal irradiance by 1.1%. This relation has been verified by many measurements under the assumption that all other relevant parameters are constant (McKenzie et al. 1991, Blumthaler et al. 1995). The RAF is only slightly dependent on SZA, therefore it is a useful fast tool to estimate the effect of ozone changes.

11.2.3 Effect of aerosols

In many environments the amount of aerosols in the atmosphere is a further important parameter affecting UV radiation. The influence of aerosols depends only slightly on wavelength, with an increasing effect towards shorter wavelengths. In many cases, aerosol scattering is dominant, thus

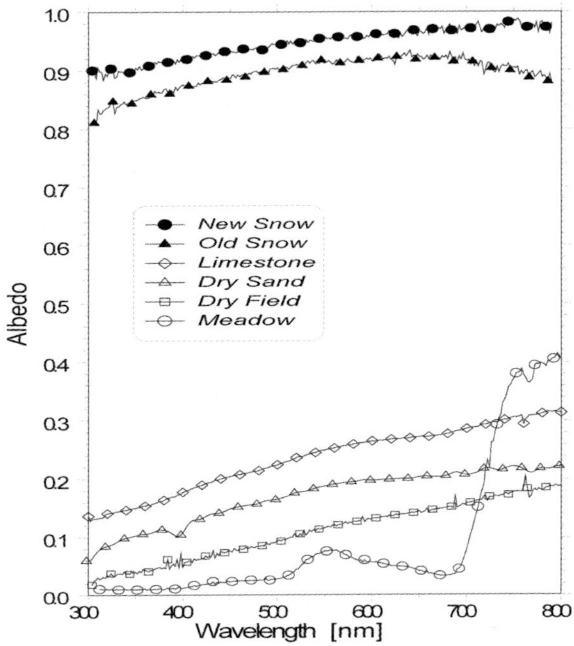

Fig. 11.1. Spectral albedo of various types of surfaces.

the amount of direct irradiance is reduced and the amount of diffuse irradiance is increased. However, if the absorbing component of the aerosols is high, then global irradiance can be reduced significantly. In urban areas, reductions of up to 30% for erythemal irradiance have been measured, compared to an aerosol-free atmosphere (Kylling et al. 1998).

In Alpine environments and especially at higher altitudes as well as in Arctic regions, the aerosol load is usually low. Only in basin areas layers of aerosols can appear when no exchange of the air mass by local winds occurs for certain time periods.

11.2.4 Effect of albedo

The albedo of a horizontal terrain is defined as the ratio between reflected radiation and incoming radiation. The reflected radiation enhances the diffuse irradiance due to multiple reflections between the ground and the atmosphere. Therefore the albedo effect is generally higher with a cloud layer, compared with a cloud-free situation.

The amount of reflected radiation depends on the type of surface, and furthermore it is dependent on the wavelength. Spectral measurements of albedo of various types of surfaces (Fig. 11.1) show a clear division into two regions: all snow-free surfaces have relatively small values of albedo, which are especially small in the UV range. The smallest values for albedo were found for green grassland, where in the UV-B the albedo was less than 1%. On the other hand, snow-covered surfaces show very high values of the albedo, which can be higher than 90%. However, these very high values are only valid for snow-covered surfaces with fresh, clean snow. Due to aging and deposition of pollution the snow can become less reflective, which may reduce the albedo down to approximately 60%, depending on the local conditions (Ambach and Eisner 1986).

Since the share of diffuse irradiance is higher in the UV than in the visible range, the effect of high albedo is especially pronounced in this wavelength range. The amplification of global irradiance due to increasing albedo has a maximum at around 320 nm (Fig. 11.2). The albedo effect decreases in the UV-B range due to absorption of the multiply scattered UV-B radiation by tropospheric ozone. The increase in global irradiance when changing from a snow-free surface (average albedo 7%) to a snow-covered surface (average albedo 75%) amounts to about 25% at 320 nm

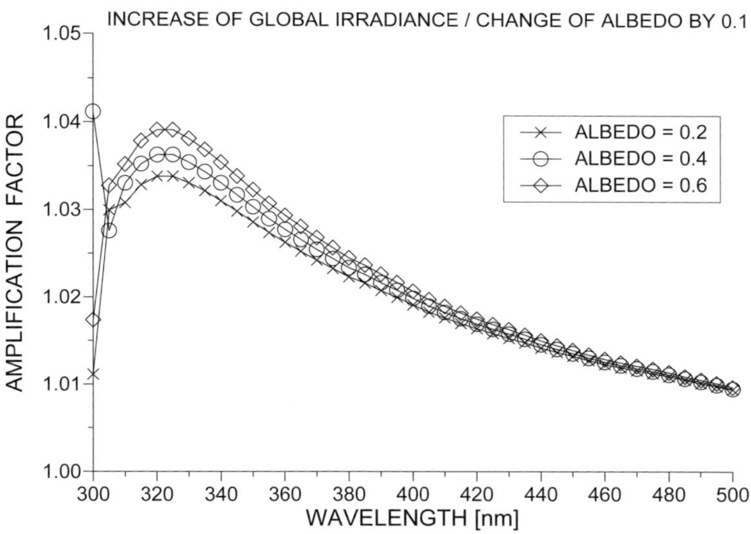

Fig. 11.2. Increase of global irradiance due to an increase of albedo by 0.1 for albedo values of 0.2, 0.4 and 0.6.

and 7% at 500 nm, which is a significant contribution to the high UV levels in Alpine and Arctic environments.

For estimating the effect of a surface with a certain albedo on UV irradiance an area up to 30 km around the measurement site is significant (Degünther et al. 1998). If the terrain is covered by snow it is usually not a homogenous surface. For example, trees usually lose their snow cover rapidly so that the regional albedo can be significantly lower than the albedo of a homogenous snow surface. This effect can be estimated when the 'effective' albedo is determined. This is the value of albedo for a homogenous, infinite surface, introduced into a one-dimensional radiative transfer model. The value of the albedo is then used to fit the model output to measurement results of spectral global irradiance (Weihs et al. 2001). By applying this method, temporal variations of the effective albedo can be followed (Fig. 11.3). During a measurement campaign at the High Mountain Station Jungfraujoch (Swiss Alps, 3576 m above sea level) in spring 1998, the effective albedo decreased from about 0.7 to 0.45. At the beginning, most of the surrounding was covered by snow, although several rocks and the trees at the lower altitudes were snow free. Later, when the snow at altitudes below about 2500 m (covering about 50% of the surrounding of the station) melted away, the overall effective albedo was reduced due to the average effect of snow-covered and snow-free terrain. During periods with cold weather and precipitation, snow returned for a short period at lower altitudes again (around days 30-35 in Fig. 11.3), and the effective albedo increased to values around 0.6. During a following period of good-weather, the snow melted again and the effective albedo became smaller again (days 35-40 in Fig. 11.3).

11.2.5 Effect of altitude

The increase of UV radiation with increasing altitude is a consequence of several factors working together. The basic effect is that at higher altitudes the air mass above the site is less and therefore the attenuation of radiation due to scattering and absorption is less too. This effect is especially pronounced for direct irradiance, whereas diffuse irradiance is only slightly reduced at higher altitudes. As a consequence, at higher altitudes a greater part of the global irradiance stems from the direct component. The scattering processes in the atmosphere are much more pronounced for shorter wavelengths and therefore the increase of irradiance with increasing altitude is greater for the shorter wavelengths. Additionally, in the UV-B range the altitude effect (increase of global irradiance for an increase of altitude of 1000 m) increases strongly due to the effect of absorption by

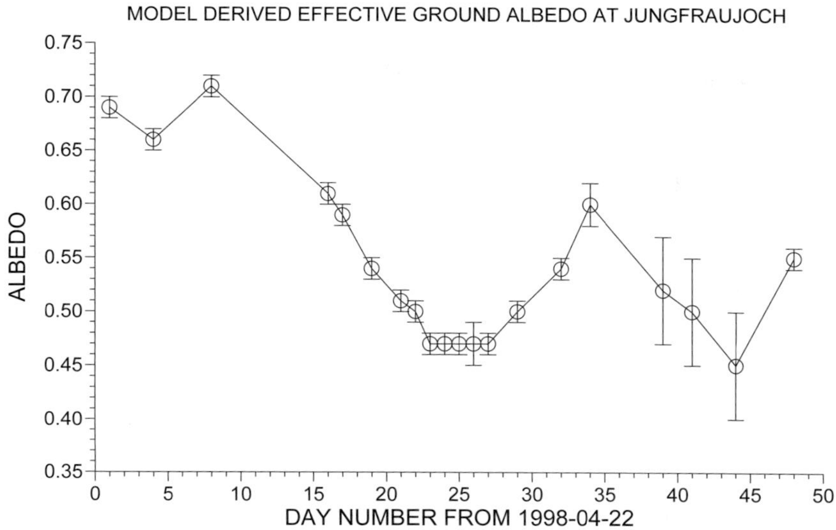

Fig. 11.3. Temporal variation of effective albedo at Jungfraujoch (3576 m above sea level).

tropospheric ozone. Furthermore, with increasing altitude usually the amount of aerosols above the observer is reduced, so that the irradiance is enhanced, only very slightly increasing with decreasing wavelength. Consequently, the altitude effect depends strongly on the local situation of the amount and vertical distribution of aerosols and ozone. Finally, albedo can increase the irradiance at higher altitudes, because there the surroundings are often covered by snow, whereas the lower altitudes are snow-free. This will increase the altitude effect (Gröbner et al. 2000).

All the afore-mentioned processes contribute together in a variable way to the final altitude effect. Therefore, it is not meaningful to mention only one number for the altitude effect but it is more realistic to consider a broad range of values. In any case, at shorter wavelengths the altitude effect becomes more pronounced.

From synchronous spectral measurements in the Alps at different altitudes the increase per 1000 m was found to be around 10% at 360 nm and 25% at 300 nm (Fig. 11.4) (Blumthaler et al. 1994a). Measurements of the altitude effect with broadband detectors in the Alps showed values of 15-25% for erythemal UV irradiance and about 10-15% for total irradiance (300 nm to 3000 nm) (Blumthaler et al. 1997). In contrast, measurements in the Andean mountains with small amounts of aerosol and no snow cover showed values for the altitude effect of about 9% in the UV-B (Piazena 1996).

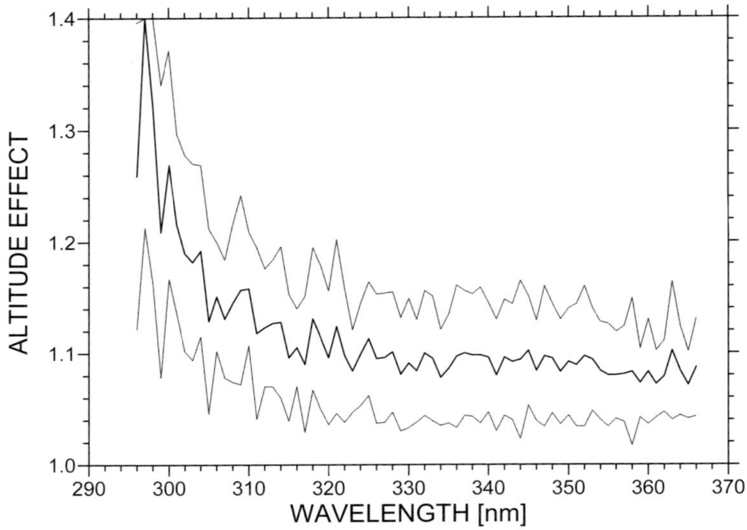

Fig. 11.4. Increase of spectral irradiance with altitude, average (solid line) and range of variation (dashed lines). From Blumthaler et al. 1994a.

11.2.6 Effect of clouds

Clouds in front of the sun attenuate the direct irradiance in proportion to their optical thickness. Clouds covering the sky usually cause an attenuation of global irradiance too. The overall reduction of global irradiance by clouds can vary over a broad range, on the average under completely overcast sky the UV irradiance is reduced by 50-80% (Josefsson and Landelius 2000). Measurements at higher altitudes show less attenuation because the optical depth of the clouds is on average smaller than at sea level. For very thick clouds, the reduction can be up to 90%. These numbers show that clouds represent the most dominant factor governing the level of global UV irradiance, and all other parameters, discussed previously, are less significant.

Short term intensification of UV irradiance of up to 20% can occur when clouds are close to the sun and the sun itself is not covered by clouds (i.e. Cede et al. 2002). Usually this happens only for short time periods of minutes, because clouds are moving with the wind and therefore this special situation with reflections is only transient.

The attenuation of global irradiance by clouds is also wavelength dependent. Simultaneous measurements of irradiance in the total wavelength range, in the UV-A and in the UV-B range show that the irradiance in the UV range is less attenuated than in the total wavelength range by up to 40% under completely overcast conditions (Blumthaler et al. 1994b). This is a consequence of the high diffuse component of UV irradiance compared to total irradiance. Also measurements of global irradiance above and below a homogeneous cloud layer in the Alps show that the transmission is higher at shorter wavelengths (Seckmeyer et al. 1996).

11.3 Long-term variations

Long-term variations of solar UV radiation at the Earth's surface can occur due to long-term variations of one or more of the parameters discussed above. Concerning specifically the Alpine and Arctic environment, the possible parameters responsible for any long-term variation of UV radiation are total ozone, snow-coverage and cloudiness.

Long-term variations of total ozone have been monitored world-wide for several decades, and significant changes have been observed, which are partly global and partly regional. For middle Northern latitudes, a slight ozone decrease was observed from beginning of the 80's to the middle of the 90's, since then the levels have been about constant at about 3% below the pre-1980 values. For high Northern latitudes, especially in spring-time a higher ozone reduction has been observed, with strong variations from year to year (WMO 2003).

The very small amount of aerosols at high altitudes allows investigation of the long-term effects of changing total ozone on UV irradiance on clear sky days without disturbance by changing aerosol amounts. Measurements of erythemally weighted UV irradiance over more than 20 years at the High Mountain Station Jungfraujoch (Swiss Alps, 3576 m above sea level) demonstrate these long-term variations (Fig. 11.5). The data are collected during measurement campaigns of about 8 weeks every year, at varying seasons from year to year. Cloudless conditions at 30° solar elevation are analysed. The relative variation of total radiation shows the effect of varying albedo, which gives about ±10% variability with highest values during measurements in winter time. The relative variation of erythemally weighted irradiance shows a variability of about -30% to +40%, which is generally in phase with the variation of total irradiance and therefore correlated with albedo variations. In addition, the variability of total ozone modulates the UV variations in the sense that during periods of low ozone

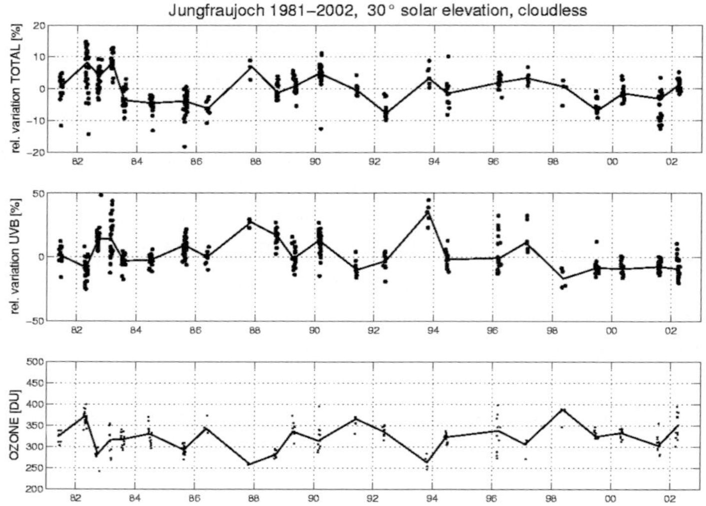

Fig. 11.5. Long-term variation of total irradiance, erythemal UV irradiance ('UV-B') and total ozone on cloudless days of the measurement campaigns at Jungfraujoch (Swiss Alps, 3576 m above sea level) between 1981 and 2002. For total and UV-B irradiance the relative variation from the mean over the whole period is shown. Total ozone data are from Arosa, Switzerland, taken from the World Ozone and UV Data Center, Canada (http://www.woudc.org).

and high albedo an especially high erythemal irradiance is observed, whereas with high ozone and low albedo especially low values of erythemal irradiance are observed. Besides the short-term variability of erythemal irradiance, no systematic long-term trend can be observed over the whole period from 1981 to 2002. For shorter time intervals in the 80's, a small but significant trend of increasing erythemal irradiance was observed (Blumthaler and Ambach 1990), in agreement with the ozone trend at that time.

Trends of erythemal UV irradiance between 5-10% per decade during the last 2 decades have been found by Kaurola et al. (2000) in Northern Europe, and are most significant during spring time.

Long-term spectral measurements in the Arctic (Sodankylä, Finland, 67° N) report spring season increases of irradiance at 310 nm of about 20% from 1990 to 1998 and of more than 40% at 305 nm, showing clearly an effect associated with ozone decrease (Masson and Kyrö 2001). In the same time period, irradiance at 325 nm increased by about 10%, which could be caused by albedo changes during that period and at that site. However, although this time series of spectroradiometric measurements at Sodankylä is among the longest in Europe, it is still relatively short, and the choice of the time interval may affect the results.

These examples of experimental results show how important long-term observations of solar UV radiation are for interpretation of any trend. Both analyses were carried out for almost clear sky conditions, because if all sky conditions would have been analysed, then the strong variability of cloudiness would require much longer time series (Weatherhead et al. 1997). For separation of any long-term variation of UV radiation into the contributing factors, spectral measurements would be preferable, because the different spectral behaviour of changes in ozone, in albedo and in cloudiness could be identified in the results. Such spectral time series are shorter and less frequent than broadband data, so that results cannot be expected soon.

11.4 Conclusions

The main parameters determining solar UV radiation in arctic-alpine environments, which might be changing over longer time periods, are ozone, aerosols, albedo and cloudiness. Based on measurements, quantifications of the effects of these parameters on solar UV radiation are derived. This allows estimating consequences for solar UV radiation in the future, if certain scenarios for the variations of these parameters are assumed. Some possible examples for future scenarios are:

UV radiation at middle and northern latitudes will decrease slightly, if stratospheric ozone is recovering due to the actions of the Montreal protocol and its successors.

UV radiation will increase due to a decrease of aerosols, if air pollution is reduced in industrial areas due to actions for protecting the environment.

UV radiation will decrease, if the area of snow-covered terrain is reduced due to increasing temperatures.

UV radiation will increase, if the average amount of cloudiness is decreasing.

In practice, all these possible changes are linked together and coupled by feedback mechanisms. Therefore, predictions about future levels of UV radiation are at the moment very uncertain. However, the knowledge of the quantitative relations between the significant parameters and the level of solar UV radiation at the Earth's surface is available and therefore any scenario for the future can be described. Nevertheless, also in the future, high quality measurements of UV radiation will be necessary to monitor any significant variation.

References

Ambach W, Eisner H (1986) Albedo verschiedener Oberflächen für erythemwirksame solare Strahlung. Wetter und Leben 38:1-4.

Bais AF, Gardiner BG, Slaper H, Blumthaler M, Bernhard G, McKenzie R, Webb AR, Seckmeyer G, Kjeldstad B, Koskela T, Kirsch P, Gröbner J, Kerr JB, Kazadzis S, Leszczynski K, Wardle D, Brogniez C, Josefsson W, Gillotay D, Reinen H, Weihs P, Svenoe T, Eriksen P, Kuik F, Redondas A (2001) The SUSPEN intercomparison of ultraviolet spectroradiometers. J Geophys Res 106:12509-12526.

Blumthaler M, Ambach W (1990) Indication of increasing solar ultraviolet-B radiation flux in Alpine regions. Science 248:206-208.

Blumthaler M, Webb AR, Seckmeyer G, Bais AF, Huber M, Mayer B (1994a) Simultaneous Spectroradiometry: A Study of Solar UV Irradiance at Two Altitudes. Geophys Res Let 21:2805-2808.

Blumthaler M, Ambach W, Salzgeber M (1994b) Effects of cloudiness on global and diffuse UV irradiance in a high-mountain area. Theor Appl Climatol 50:23-30.

Blumthaler M, Salzgeber M, Ambach W (1995) Ozone and ultraviolet-B irradiances: experimental determination of the radiation amplification factor. Photochem Photobiol 61:159-162.

Blumthaler M, Ambach W, Ellinger R (1997) Increase of solar UV radiation with altitude. J Photochem Photobiol B: Biol 39:130-134.

Cede A, Blumthaler M, Luccini E, Piacentini R, Nunez L (2002) Effects of clouds on erythemal and total irradiance as derived from data of the Argentine network. Geophys Res Lett 29:76/1-4.

Degünther M, Meerkötter R, Albold A, Seckmeyer G (1998) Case study on the influence of inhomogeneous surface albedo on UV irradiance. Geophys Res Lett 25:3587-3590.

Gröbner J, Albold A, Blumthaler M, Cabot T, De la Casinieri A, Lenoble J, Martin T, Masserot D, Müller M, Philipona R, Pichler T, Pougatch E, Rengarajan G, Schmucki D, Seckmeyer G, Sergent C, Touré ML, Weihs P (2000) Variability of spectral solar ultraviolet irradiance in an Alpine environment. J Geophys Res 105:26991-27003.

Josefsson W, Landelius T (2000) Effect of clouds on UV irradiance: As estimated from cloud amount, cloud type, precipitation, global radiation and sunshine duration. J Geophys Res 105:4927-4935.

Kaurola J, Taalas P, Koskela T, Borkowski J, Josefsson W (2000) Long-term variations of UV-B doses at three stations in northern Europe. J Geophys Res 105:20813-20820 .

Kylling A, Bais AF, Blumthaler M, Schreder J, Zerefos CS (1998) The effect of aerosols on solar UV irradiances during the PAUR campaign. J Geophys Res 103:26051-26060.

Masson K, Kyrö E (2001) A decade of spectral UV-B measurements in Sodankylä, Finland. European Geophysical Society 26. General Assembly 2001, EGS Newsletter 203.

McKenzie RL, Matthews WA, Johnston PV (1991) The relationship between erythemal UV and ozone, derived from spectral irradiance measurements. Geophys Res Lett 18:2269-2272.

McKinlay AF, Diffey BL (1987) A reference action spectrum for ultraviolet induced erythema in human skin. CIE Journal 6:17-22.

Piazena H (1996) The effect of altitude upon the solar UV-B and UV-A irradiantce in the tropical Chilean Andes. Sol Energy 57:133-140.

Seckmeyer G, Erb R, Albold A (1996) Transmittance of a cloud is wavelength-dependent in the UV-range. Geophys Res Lett 23:2753-2755.

Weatherhead EC, Reinsel GC, Taio G, Meng X, Choi D, Cheang W, Keller T, DeLuisi J, Wuebbles DJ, Kerr JB, Miller AJ, Oltmans SJ, Frederick JE (1997) Factors affecting the detection of trends: statistical considerations and applications to environmental data. J Geophys Res 103:17149-17161.

Weihs P, Lenoble J, Blumthaler M, Martin T, Seckmeyer G, Philipona R, De la Casinieri A, Sergent C, Gröbner J, Cabot T, Masserot D, Pichler T, Pougatch E, Rengarajan G, Schmucki D, Simic S (2001) Modeling the effect on an inhomogeneous surface albedo on incident UV radiation in mountainous terrain: determination of an effective surface albedo. Geophys Res Let 28:3111-3114.

WMO (World Meteorological Organisation) Scientific Assessment of Ozone Depletion: 2002. Global Ozone Research and Monitoring Project. Report Nr. 47, Geneva. 2003.

12 Effects of enhanced UV-B radiation and epidermal UV screening in arctic and alpine plants

Line Nybakken[1] and Wolfgang Bilger[2]

[1]Department of Ecology and Natural Resource Management, Agricultural University of Norway, Box 5003, NO-1432 Ås, Norway, [2] Botanisches Institut, Christian- Albrechts Universität zu Kiel, Olshausenstr. 40, 24098 Kiel, Germany, e-mail: line.nybakken@ina.nlh.no

12.1 Introduction

Plants, as sessile organisms dependent on sunlight for their very existence, are often expected to be vulnerable to increasing UV-B radiation. During the last three decades, more than 800 articles have been published concerning UV-effects on plants (ISI database). Studies have been performed at all levels, from cellular investigations to large projects concerning the effects on entire ecosystems. Of the large number, however, less than 5 % of UV-B response studies on plants have been conducted in the field, and even fewer with plants growing in their natural ecosystems (Caldwell et al. 1998). Only about ten studies have been published on the effects on plants from arctic and alpine areas.

This chapter focuses on possible negative effects of increased UV-B radiation and protection against UV-B offered by absorbing pigments. The UV-absorbing compounds in higher plants are mainly located in the epidermis of leaves, where they screen damaging radiation before reaching the mesophyll cells. Experiments with mutants of higher plants have shown that the UV-B sensitivity of leaves and whole plants is strongly correlated with the concentrations of constitutive and UV-B induced screening pigments in the epidermis (e.g. Lois and Buchanan 1994; Landry et al. 1995). UV-B screening compounds in higher plants are often phenolics, of which hydroxycinnamic acid esters (HCAs) and flavonoids are of particular importance. HCAs absorb mainly in the UV-B, while the flavonoids have a second major absorption band in the UV-A. An increased concentration of phenolic compounds in higher plants is a common response to

UV-B, documented in numerous field and growth chamber studies, (for a review, see Searles et al. 2001).

12.2 Research on UV-B effects

UV-B radiation not attenuated by the epidermis may have various effects on plants, direct and indirect, from cellular to ecosystem level. UV-B effects on higher plants have recently been reviewed with emphasis on natural conditions by e.g. Rozema et al. (1997b), Caldwell et al. (1998) and Day (2001); on arctic terrestrial organisms by Björn (2002). Few studies have been published from the arctic environment. Apart from some extensive studies including plants from the American Arctic published in the 1980s (Robberecht et al. 1980; Caldwell et al. 1982; Barnes et al. 1987), the only "arctic results" to our knowledge come from UV-B enhancement experiments established in a subarctic heathland in Abisko, Northern Sweden (68 °N). Some of these experiments started as early as 1991, and have resulted in many publications (reviewed by Björn (2002)). Generally, these studies simulated an 11 % depletion of the ozone layer (according to the function by Caldwell (1971)) in squarewave lamp systems. Now, a "second generation" of results after 7-8 years of UV-enhancement are being published (Phoenix et al. 2001; Semerdjieva et al. 2003 a,b), allowing comparisons with the first results from the middle 1990s (Johanson et al. 1995; Gwynn-Jones and Johanson 1996, Gwynn-Jones et al. 1997, Phoenix et al. 2000).

The research group, which started the experiments in Abisko, also established UV-B enhancement installations in the arctic Adventdalen, Svalbard (78 °N) in 1996. The first results from these studies are published, (Solheim et al. 2002; Bjerke et al. 2003; Nybakken et al. 2004 a, b).

Most studies of UV-effects on alpine plants are investigations of epidermal screening in plants growing along natural UV-B gradients (see section 1.4 below). The few enhancement studies that exist will be reported on in section 1.3.

12.3 Effects of UV radiation on higher plants

DNA is damaged by absorption of UV-B quanta in aromatic residues, resulting in structural alterations such as formation of cyclobutane dimers (CPDs), and, to a lesser extent, pyrimidine (6-4) pyrimidone photoproducts (Britt 1996). Such photoproducts can inhibit replication or even cause

Fig. 12.1 UV-B enhancement experiment in the Adventdalen valley, Svalbard. The UV-B fluorescent tubes installed on frames above the natural vegetation simulate an 11% depletion of the ozone layer.

mutations, thereby affecting gene expression. DNA modifications may also disrupt cellular metabolism (Britt 1996; Mitchell & Karentz 1993). Normally, damage due to UV-B radiation is efficiently repaired (Britt 2002), but the enzymatic repair may be retarded in cold regions due to low temperature (Li et al. 2002). Therefore, plants in arctic-alpine environments may possibly be more sensitive to an increase in UV-B radiation than plants in warmer areas.

Early UV-B research mainly dealt with adverse effects on the photosynthetic apparatus, and growth chamber studies confirmed negative effects of UV-B (Tevini & Teramura 1989). Caldwell et al. (1982) found that photosynthetic inhibition by UV radiation was consistently more pronounced in arctic populations of the genus *Oxyria*, *Taraxacum* and *Lupinus* than in their closely related counterparts from alpine environments at middle and low latitudes. Their plants were grown from seeds in a growth chamber, and UV-B treated indoors. Allen et al. (1998) concluded in a review that effects on photosynthetic competence and light-interception measured at high UV-B irradiance relative to PAR and UV-A, disappear under field conditions with realistic spectral ratios between the different wavebands. This conclusion was confirmed by a recent field study of the alpine plants *Saussurea superba* and *Gentiana straminea* (at 3200 m a.s.l.) under UV-B enhancement simulating nearly 14% ozone reduction. Shi et al. (2004) found that neither CO_2 uptake nor photosynthetic O_2 evolution rate were decreased after 5 month's treatment.

Increased production of free radicals is another possible threat for plants under high UV-irradiation (Hideg et al. 2002, Barta et al. 2004). As a result, peroxidation of unsaturated lipids occurs as a response to UV-B

(Panagopoulos et al. 1990). Since organisms living in cold environments need a high proportion of unsaturated fatty acids to maintain membrane fluidity (Nishida and Murata 1996), they are presumably vulnerable to radiation-induced peroxidation (Björn 2002). However, Taulavuori et al. (1998), did not find any evidence of oxidative stress in subarctic *Vaccinium vitis-idaea* exposed to enhanced UV-B radiation in the field.

Many plants exhibit morphological changes due to increased UV-B radiation. Examples of such changes are decreased plant height and leaf area (Tevini & Teramura 1989; Johanson et al. 1995; Mepsted et al. 1996), thicker leaves (Rozema et al. 1997) and reduction in dry weight (Searles et al. 2001). Changes in plant morphology may, however, be beneficial, by e.g. decreasing the amount of tissue exposed to UV-B radiation (Cen and Bornman 1993). After 7 years of UV-B treatment of dwarf shrub vegetation in a subarctic heathland (Abisko, Sweden), Semerdjieva et al. (2003a) found no effects of 15 % enhanced UV-B on leaf or cuticle thickness in the four species studied. Gwynn-Jones and Johanson (1996) studied two subarctic grass species, and found that growth (measured as total plant dry weight) was significantly reduced by 15 % enhancement of UV-B in a greenhouse (reduced tillering). Surprisingly, no reduction of growth occurred at 25 % UV-B enhancement (tillering increased compared with ambient UV-B). A negative correlation between growth and amount of UV-B absorbing compounds in one of the investigated grasses (*Calamagrostis lapponica*) indicate a possible cost for compound synthesis, and, hence, reduced growth.

Reproductive yield may also be affected by UV-B (Searles et al. 2001), by altering both the timing of flowering (Mark et al. 1996; Caldwell 1968; Ziska et al. 1992; Staxén & Bornman 1994) as well as the number of flowers (Saile-Mark & Tevini 1997; Musil 1995; Klaper et al. 1996). However, arctic-alpine plants often rely on clonal growth, and seldom reproduce sexually. Phoenix et al. (2000) investigated regeneration by clonal growth of subarctic dwarf shrubs under enhanced UV-B during four years. All aboveground plant parts were removed from the plots. Species responded differently, but the final biomass for each species was significantly affected by UV-B only for *Calamagrostis lapponica* (reduced to 50 %). U. Johanson and his collegues (pers. comm.) measured growth and reproduction responses in a number of species in the UV-B enhancement experiment on Svalbard. In *Bistorta vivipara* most growth parameters measured were significantly reduced by enhanced UV-B, as were numbers of flowers and reproductive units. Other studied species were not significantly or only slightly affected.

As species differ in susceptibility to UV-B, increased UV-B radiation may have an effect on competition between species, and, hence species composition (Barnes et al 1988, 1995). In a two years study of a subarctic heathland in Northern Sweden, Johanson et al (1995) found that growth of evergreen dwarf shrubs (*Vaccinium vitis-idaea* and *Empetrum hermaphroditum*) was repressed more than that of co-occurring deciduous dwarf shrubs *(Vaccinium myrtillus* and *V. uligonosum*). After five more years under the UV-B enhancement these effects had disappeared again. Phoenix et al (2001) reported that now *V. myrtillus* was the only species significantly affected by UV-B, and that the evergreen species showed few responses. They speculate that even if the leaves of the evergreens are exposed to UV-B for longer time during the growing season, and there is a possibility for damage to accumulate over years (Sullivan and Teramura 1990; Johanson et al. 1995), their bud burst is later and they thus avoid much of the highest radiation period in spring. After 8 years, there were still no observable changes in species frequency in these experimental plots (Björn 2002, citing U. Johanson).

In addition to direct effects on single plants, there is increasing evidence that UV radiation affects many trophic interactions and influences a variety of ecosystem functions (Paul & Gwynn-Jones 2003). Field studies have shown that UV-B may reduce insect herbivory, abundance and performance (Ballaré et al. 1996; Rousseaux et al. 1998; Salt et al. 1998; Mazza et al. 1999). The reasons for these changes are not always clear, but they may be mediated through changes in plant secondary chemistry or alterations in plant nitrogen or sugar content (Hatcher & Paul 1994; Yazawa et al. 1992; McCloud & Berenbaum 1994). Plant fungal and viral diseases respond in a number of different ways to UV-B radiation as shown in several experiments, conducted primarily in laboratory and greenhouse conditions. UV-B radiation has significant effects on leaf litter decomposition by stimulating photochemical breakdown of compounds such as lignin and/or by altering the community of decomposer organisms and so inhibiting microbial decomposition (Duguay & Klironomos 2000; Moody 1999; Paul et al. 1999). In general, enhanced UV-B in field studies in the (sub)-arctic caused so far only modest effects in the plants.

12.4 Studies along natural UV-B gradients

Arctic and alpine environments offer the plants some similar climatic conditions, like low temperatures and short growing seasons, and many plant species in the arctic extend their distribution to alpine areas further south. UV radiation, however, increases with altitude and decreases with latitude and represents a probably important difference between arctic and alpine habitats. Furthermore, this allows us to study adaptation to UV-B radiation in ecotypes of the same species (or at least closely related species) along natural UV-B gradients. Most studies of UV-B effects on plants from alpine areas have been performed along such natural UV-gradients, either from lowland to alpine (elevational) and/or from arctic to alpine (latitudinal).

Larson et al. (1990) compared responses of an alpine (*Aquilegia caerulea*) and a non-alpine plant (*A. canadensis*) to enhanced UV-B in a greenhouse. Both species became significantly shorter (greatest effect on the non-alpine species) and developed a higher number of leaves (most in alpine species); there was, however, no altitudinal difference in the effect of UV-B on parameters like biomass, net photosynthesis and flavonoid content. Sullivan et al. (1992) collected seeds of different species along a 3000 m elevation gradient in Hawaii and grew them in a greenhouse with either no or two different irradiances of UV-B. There was a great variation among species, but in general, sensitivity to UV-B, measured as reduced height and biomass, was reduced as the elevation of plant origin increased. In a subsequent study of plants from the same seed collection, Ziska et al. (1992) noted increases in pigments only in plants from low elevation exposed to UV-B, but the plants from high elevations produced a consistently higher amount of these compounds also in the absence of UV-B radiation. Plants from low elevations were also negatively affected by UV-B, showing reduced plant and floral dry biomass, apparent quantum efficiency, relative water content and light-saturated photosynthetic capacity. Species from high elevation did not display any such negative effects. Rau and Hoffmann (1996) investigated species pairs of five genera from different altitudes in the Alps grown in a greenhouse. They observed a slightly higher UV-B resistance of the alpine species, but only when the plants were preadapted to UV-B radiation. Differences in UV-B sensitivity between different genera were much more pronounced than the differences between lowland and alpine species within the same genus. Hübner and Ziegler (1998) measured photosynthetic CO_2 fixation rate under enhanced UV-B (indoors) in three species of Rumex with natural distribution at different altitudes. The lowland (*R. maritimus*) and the montane (*R. crispus*) species revealed a high UV-B tolerance, even without a preceding adaptation to UV-B irradiance, while the alpine species (*R. alpinus*) required an adaptation to a threshold UV-B irradiance to prevent UV-B caused inhibition of photosynthetic CO_2 fixation. The authors suggest that the unexpected reduced performance of non-adapted *R. alpinus* is due to the UV-B caused reduction in stomatal conductance.

There are very few investigations of UV-B effects on plants along latitudinal gradients. Since they are mainly concerned with epidermal screening and/or accumulation of UV-absorbing compounds, they will be dealt with in the next sections.

In summary, the studies along elevational gradients show evidence for a trend to higher, genetically fixed, UV-B resistance in plants from higher altitudes, although this trend is less conspicuous than one might have expected. Since in all reports, UV-B sensitivity was assessed using greenhouse grown plants, the possible contribution of a genetically fixed potential for phenotypic acclimation of resistance could not fully be evaluated.

12.5 Epidermal UV-screening in arctic and alpine plants

Resistance against UV-B radiation can be achieved in plants by three different means: epidermal screening, high antioxidative status and efficient repair of damage. Since there exist very scarce data from arctic plants in the litterature on the two latter topics, we will mainly concentrate on epidermal screening.

Screening is accomplished by UV-absorbing compounds. Hence, the content of these compounds in leaves has frequently been taken as a measure for UV-B screening. This may be misleading, as discussed extensively by Day (2001) for mainly two reasons. The compounds in whole leaf extracts may not exclusively be located in the epidermis, although only there they can serve as screens. Furthermore, it may not be possible to extract all screening compounds since part of them may be bound to the (epidermal) cell walls. Nevertheless, absorbing compound concentrations may indicate trends.

Semerdjieva et al. (2003b) found contrasting strategies for UV-B screening in three sub-arctic dwarf shrubs. In *Vaccinium myrtillus* (deciduous) the methanol-soluble UV-absorbing compounds were distributed throughout the leaf, but particularly concentrated in chlorophyll containing cells, while in *V. vitis-ideae* (perennial leaves) most phenolic compounds were cell wall bound in epidermal cells. For both species the content of phenolic compounds increased when plants were exposed to enhanced UV-B. The response of *V. uliginosum* (deciduous, but with leaf structure similar to *V.vitis-ideae*) was intermediate between the other two, with high concentrations of cell wall-bound phenolics in the epidermis but with this pool decreasing and the methanol-soluble pool tending to increase after exposure to enhanced UV-B.

In their comparison of species pairs of five genera from different altitudes in the Alps, Rau and Hoffmann (1996) found higher content of absorbing compounds in species from high altitudes for three genera, for one genus there was no difference, while the last (*Saxifraga*) had higher absorbance in the lowland species. Acclimation to low UV-B radiation exaggerated the existing trends in these greenhouse grown plants. In a field study, Caldwell (1968) investigated epidermal UV transmittance (300 – 400 nm) in 33 species along a transect in the Rocky Mountains, USA, from 1680 m to 3750 m altitude. He did not observe any elevational trend in UV transmittances nor in UV absorbing compounds in methanolic extracts. In contrast, Rozema et al. (1997a) found increasing content of UV-B absorbing compounds and increasing leaf thickness with elevation (800 m to 1600 m a.s.l.) in four species collected in the Blue Mountains, Jamaica. Comparing four different pairs of related species in a latitudinal gradient from arctic and alpine environments, all grown under field conditions in Utah, Barnes et al. (1987) observed about equal contents of UV-B absorbing compounds in the arctic and alpine species. However, upon exposure to additional UV-B radiation, only the ecotypes of arctic origin increased their content of UV-B absorbing compounds.

By measuring transmittance of epidermal peels directly, Robberecht et al. (1980) compared 11 species growing at Barrow, Alaska, with 32 different species growing at three different high altitude locations at equatorial or tropical latitudes. Mean epidermal UV-B transmittance of alpine plants from low latitudes was < 2 %, while plants from Alaska had a mean value of > 5 %. An arctic provenance of *Oxyria digyna* had a 20 % lower content of UV-B absorbing compounds and 50 % higher epidermal transmittance than an alpine ecotype under identical artificial irradiance (Caldwell et al. 1982).

The paucity of in-situ studies of epidermal transmittance at alpine and arctic locations is probably related to the their remote location in conjunction with the need for relatively elaborate equipment for proper measurement of epidermal transmittance. A recently introduced fluorescence technique employing a portable apparatus (Bilger et al. 1997, 2001) solved some of the inherent problems and allowed us to determine epidermal transmittance in a large number of arctic and alpine species at their growing sites with a sufficient number of samples (Nybakken et al. 2004 a, b)

In an experiment started 1996 in Adventdalen, Svalbard (Norwegian Arctic, 78°N, 0-50 m a.s.l.) the natural tundra vegetation was irradiated with UV-B corresponding to an 11% depletion of the ozone layer. UV transmittance of seven higher plant species growing under the UV-B lamps were measured during 1999, 2000 and 2001, but the transmittance in irradiated plants were just as high as in controls growing under natural UV-B (Nybakken et al. 2004 a). The epidermis of both control and irradiated plants transmitted from 3.8 to 11.6% of the UV-B radiation (290-320 nm), which is not more than what found for *Vicia faba* growing in the lowland in southern Germany (9.2%, Markstädter et al. 2001). Measurements of five other species not included in the enhancement experiment also showed values in the same order of magnitude. Presumably, the epidermal protection may already be sufficient in these arctic plants, implying that there is no need to invest in extra absorbing compounds under the UV-B regime applied.

Six of the studied arctic species were also measured at two high-altitude locations; Finse, Norway (60°N, 1200-1600 m a.s.l) and the area around Col du Lautaret, French Alps (45°N, 2000-2700 m a.s.l) (Nybakken et al. 2004b). Only one species, *Dryas octopetala*, consistently showed decreasing UV-B transmittance with decreasing latitude/increasing altitude all three years (Fig. 12.2). Three other species, *Bistorta vivipara*, *Saxifraga oppositifolia* and *Silene acaulis*, showed a similar trend some years. The epidermal transmittance of *Oxyria digyna* did not vary between years for any of the ecotypes (Fig. 12.2). This species also displayed low transmittance when grown from seeds at a PFD of 150 μmol m^{-2} s^{-1} in a growthroom in the absence of UV-B (Fig. 12.2). It seems that *O. digyna* exhibits a high constitutive protection, even in the ecotype from Svalbard with low ambient UV-B radiation.

The high epidermal UV-B protection in the studied arctic species and the lack of clear differences in epidermal transmittance between arctic and alpine ecotypes of several species challenges established ideas about what is the driving force behind the build-up of epidermal UV-B defense in arctic and alpine plants. Possibly, today's UV+PAR levels approach levels

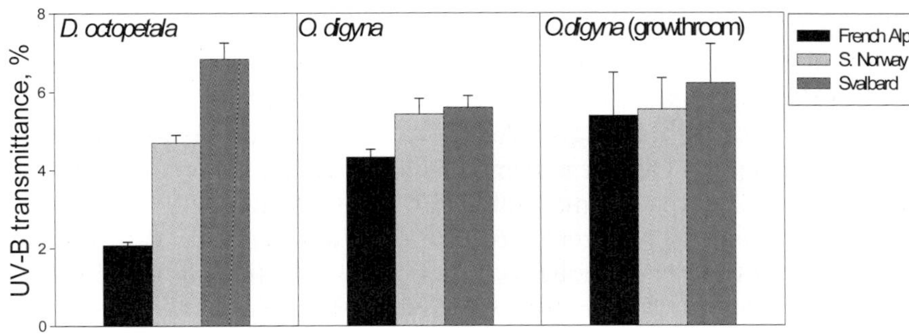

Fig. 12.2 Epidermal UV-B transmittance (%) + S.E of *D. octopetala* and *O. digyna* from three sites along a natural UV-B gradient (mean over three field seasons, 1999-2001) and *O.digyna* of all three ecotypes grown in a growthroom (Nybakken et al. 2004b).

which saturate synthesis of UV-B absorbing compounds in arctic plants. In addition, other environmental factors than light may also influence the epidermal UV-B transmittance. UV-B absorbing metabolites may be induced also by low temperature, herbivory, wounding, drought and by deficiency of certain mineral nutrients (Balakumar et al. 1993; Chaves et al. 1997; de la Rosa et al. 2001; Dixon and Paiva 1995; Hunt and McNeil 1998; Keinänen et al. 1999; Lavola and Julkunen-Tiitto 1994; Levizou et al. 2001; Solecka and Kacperska 1995; Stewart et al. 2001). One or more of these factors may possibly have saturated flavonoid synthesis, so that induction by UV-B radiation became insignificant (also discussed by Liakoura et al. 2003). Differences in temperature and precipitation along the latitudinal gradient may have counterbalanced the gradient in UV-B.

12.6 Conclusions

There are few studies on UV-B effects in arctic-alpine plants. These studies include few species, and often display contradictory results, suggesting species and habitat specific responses. Reduced growth and increased thickness of leaves are common results of UV-B enhancement, but there are no reports of any severe effects at the plant or ecosystem level. Molecular studies, including DNA-damage or other parameters do not exist. However, if there were an increased incidence of damage induced by enhanced UV-B at the cellular level, it doesn't become apparent, possibly due to repair activities. Many results are based on relatively short-term UV-B enhancement, so that it is very difficult to predict long-term effects. But also in those studies where vegetation was monitored for several years, few effects of UV-B were found.

One reason for the missing severe UV-B effects could lie in an efficient epidermal screening. Our measurements indicated that the epidermis of the studied species absorbs a considerable part of the incident UV-B radiation. Other mechanisms such as repair capacity and antioxidative systems may further contribute to plant resistance. However, very little is known about photoreactivation *in situ* or antioxidative capacity in arctic and alpine plants.

The studies presented raises a lot of new questions concerning the mechanisms and driving forces behind the synthesis of UV-B absorbing compounds in arctic and alpine plants. Future research should investigate the importance of other enviromental factors for synthesis of UV-B absorbing compounds, and provide knowledge about the metabolic cost of UV-B protection.

References

Allen DJ, Nogues S, Baker NR (1998) Ozone depletion and increased UV-B radiation: is there a real threat to photosynthesis? J Exp Bot 49:1775-1788

Balakumar T, Hani Babu Vincent V, Paliwal K (1993) On the interaction of UV-B radiation (280-315 nm) with water stress in crop plants. Physiol Plant 87:217-222

Ballaré CL, Scopel AL, Stapleton AE, Yanovsky M (1996) Solar ultraviolet-B radiation affects seedling emergence, DNA integrity, plant morphology, growth rate, and attractiveness to herbovore insects in *Datura ferox*. Plant Physiol 112:161-170

Barnes PW, Flint SD, Caldwell MM (1987) Photosynthesis damage and protective pigments in plants from a latitudinal arctic/alpine gradient exposed to supplemental UV-B radiation in the field. Arct Alp Res 19:21-27

Barnes PW, Flint SD, Caldwell MM (1995) Early-season effects of supplemented solar UV-B radiation on seedling emergence, canopy structure, simulated stand photosynthesis and competition for light. Glob Change Biol 1:43-53

Barnes PW, Jordan PW, Gold WG, Flint SD, Caldwell MM (1988) Competition, morphology and canopy structure in wheat (*Triticum aestivum* L.) and wild oat (*Avena fatua* L.) exposed to enhanced ultraviolet-B radiation. Funct Ecol 2:319-330

Barta C, Kálai T, Hideg K, Vass I, Hideg E (2004) Differences in the ROS-generating efficacy of various ultraviolet wavelengths in detached spinach leaves. Funct Plant Biol 31:23-28

Bilger W, Veit M, Schreiber L, Schreiber U (1997) Measurement of leaf epidermal transmittance of UV radiation by chlorophyll fluorescence. Physiol Plant 101:754-763

Bilger W, Johnsen T, Schreiber U (2001) UV-excited chlorophyll fluorescence as a tool for the assessment of UV-protection by the epidermis of plants. J Exp Bot 52: 2007-2014

Bjerke JW, Zielke M, Solheim B (2003) Long-term impacts of simulated climatic change on secondary metabolism, thallus structure and nitrogen fixation activity in two cyanolichens from the Arctic. New Phytol 159:361-367

Björn LO (2002) Effects of Ultraviolet-B on Terrestrial Organisms and Ecosystems. In: Hessen (ed) UV radiation and Arctic Ecosystems. Ecological Studies 153. Springer.Verlag, Berlin Heidelberg, pp 93-121

Britt AB (1996) DNA damage and repair in plants. Ann Rev Plant Physiol Plant Mol Biol 47:75-100

Britt AB (2002) Repair of damaged bases. In: Somerville CR and Meyerowitz EM (eds) The Arabidopsis Book. American Society of Plant Biologists, Rockville, MD doi/10.1199/tab.0005 http://www.aspb.org/publications/arabidopsis/

Caldwell M (1971) Solar UV irradiation and the growth and development of higher plants. In: Giese AC (ed) Photophysiology. Current Topics in Photobiology and Photochemistry. Academic Press, New York, pp 131-174

Caldwell MM (1968) Solar ultraviolet radiation as an ecological factor for alpine plants. Ecol Monogr 38:243-268

Caldwell MM, Björn LO, Bornman JF, Flint SD, Kulandaivelu G, Teramura AH, Tevini M (1998) Effects of increased solar ultraviolet radiation on terrestrial ecosystems. Photochem Photobiol 46:40-52

Caldwell MM, Robberecht R, Nowak RS, Billings WD (1982) Differential photosynthetic inhibition by ultraviolet radiation in species from the arctic-alpine lifezone. Arct Alp Res 14:195-202

Cen YP, Bornman JF (1993) The effect of exposure to enhanced UV-B radiation on the penetration of monochromatic and polychromatic UV-B radiation in leaves of *Brassica napus*. Physiol Plant 87:249-255

Chaves N, Escudero JC, GutierrezMerino C (1997) Role of ecological variables in the seasonal variation of flavonoid content of *Cistus ladanifer* exudate. J Chem Ecol 23:579-603

de la Rosa TM, Julkunen-Tiitto R, Lehto T, Aphalo PJ (2001) Secondary metabolites and nutrient concentrations in silver birch seedlings under five levels of daily UV-B exposure and two relative nutrient addition rates. New Phytol 150:121-131

Day TA (2001) Ultraviolet Radiation and Plant Ecosystems. In: Cockell CS and Blaustein AR (eds) Ecosystems, Evolution, and Ultraviolet Radiation, pp 80-117. Springer-Verlag, New York

Dixon RA, Paiva NL (1995) Stress-induced phenylpropanoid metabolism. Plant Cell 7:1085-1097

Duguay K, Klironomos J (2000) Direct and indirect effects of enhanced UV-B radiation on the decomposing and competitive abilities of saprobic fungi. Appl Soil Ecol 14:157-164

Gwynn-Jones D, Johanson U (1996) Growth and pigment production in two subarctic grass species under four different UV-B irradiation levels. Physiol Plant 97:701-707

Gwynn-Jones D, Lee JA, Callaghan TV (1997) Effects of enhanced UV-B radiation and elevated carbon dioxide concentrations on a sub-arctic forest heath ecosystem. Plant Ecol 128: 242-249

Hatcher P, Paul ND (1994) The effect of elevated UV-B radiation on herbivory of pea by *Autographa gamma*. Entomolog Exp Appl 71:227-233

Hideg E, Vass I (1996) UV-B induced free radical production in plant leaves and isolated thylakoid membranes. Plant Sci 115:251-260

Hideg E, Barta C, Kalai T, Vass I, Hideg K, Asada K (2002) Detection of singlet oxygen and superoxide with fluorescent sensors in leaves under stress by photoinhibition or UV radiation. Plant Cell Physiol 43: 1154-1164

Hunt JE, McNeil DL (1998) Nitrogen status affects UV-B sensitivity of cucumber. Austral J Plant Physiol 25:79-86

Hübner C, Ziegler H (1998) UV-B effects on physiological parameters of three species of *Rumex* with natural distribution at different altitudes. J Plant Physiol 152: 336-344

Johanson U, Gehrke C, Björn LO, Callaghan TV (1995) The effects of enhanced UV-B radiation on the growth of dwarf shrubs in a subarctic heathland. Funct Ecol 9:713-719

Keinänen M, Julkunen-Tiitto R, Mutikainen P, Walls M, Ovaska J, Vapaavuori E (1999) Trade-offs in phenolic metabolism of silver birch: effects of fertilization, defoliation, and genotype. Ecology 80:1970-1986

Klaper R, Frankel S, Berenbaum MR (1996) Anthocyanin content and UVB sensitivity in *Brassica rapa*. Photochem Photobiol 63:811-813

Landry LG, Chapple CCS, Last RL (1995) Arabidopsis mutants lacking phenolic sunscreens exhibit enhanced ultraviolet-B injury and oxidative damage. Plant Physiol 109:1159-1166

Larson RA, Garrison WJ, Carlson RW (1990) Differential responses of alpine and non-alpine Aquilegia species to increased ultraviolet-B radiation. Plant Cell Environ 13:983-987

Lavola A, Julkunen-Tiitto R (1994) The effect of elevated CO_2 and fertilization on primary and secondary metabolites in birch (*Betula pendula* Roth)? Oecologia 99:315-321

Levizou E, Manetas Y (2001) Enhanced UV-B radiation, artificial wounding and leaf chemical defensive potential in *Phlomis fruticosa* L. Plant Ecol 154:213-217

Li S, Paulsson M, Björn LO (2002) Temperature-dependent formation and photorepair of DNA damage induced by UV-B radiation in suspension-cultured tobacco cells. J Photochem Photobiol 66:67-72

Liakoura V, Bornman JF, Karabourniotis G (2003) The ability of abaxial and adaxial epidermis of sun and shade leaves to attenuate UV-A and UV-B radiation in relation to the UV absorbing capacity of the whole leaf methanolic extracts. Physiol Plant 117:33-43

Lois R, Buchanan BB (1994) Severe sensitivity to ultraviolet radiation in an Arabidopsis mutant deficient in flavonoid accumulation. II. Mechanisms of UV-resistance in *Arabidopsis*. Planta 194:504-509

Mark U, Saile-Mark M, Tevini M (1996) Effects of solar UVB radiation on growth, flowering and yield of central and southern European maize cultivars (*Zea mays* L.). Photochem Photobiol 64:457-463

Markstädter C, Queck I, Baumeister J, Riederer M, Schreiber U (2001) Epidermal transmittances of leaves of *Vicia faba* for UV radiation as determined by two different methods. Photosynth Res 67:17-25

Mazza CA, Zavala J, Scopel AL, Ballaré CL (1999) Perception of solar UVB radiation by phytophagous insects: behavioural responses and ecosystem implications. Proc Natl Acad Sci USA. 96:980-985

McCloud E, Berenbaum MR (1994) Stratospheric ozone depletion and plant-insect interactions: effects of UVB radiation on foliage quality of *Citrus jambhiri* for *Trichoplusia ni*. J Chem Ecol 20:525-539

Mepsted R, Paul ND, Stephen J, Corlett JE, Nogués S, Baker NR, Jones HG, Ayres PG (1996) Effects of enhanced UV-B radiation on pea (*Pisum sativum* L) grown under field conditions in the UK. Glob Change Biol 2:325-334

Musil CF (1995) Differential effects of elevated ultraviolet-B radiation on the photochemical and reproductive performances of dicotyledonous and monocotyledonous arid-environment ephemerals. Plant Cell Environ 19:1017-1027

Nishida I, Murata N (1996) Chilling sensiticity in plants and cyanobacteria: the crucial contribution of membrane lipids. Annu Rev Plant Physiol Plant Mol Biol 47:541-568

Nybakken L, Bilger W, Johanson U, Björn LO, Zielke M, Solheim B (2004a) Epidermal UV-screening in vascular plants from Svalbard (Norwegian Arctic). Polar Biol 27:383-390

Nybakken L, Aubert S, Bilger W (2004b) Epidermal UV-screening of Arctic and Alpine plants along a latitudinal gradient in Europe. Polar Biol 27:391-398

Panagopoulos I, Bornman JF, Björn LO (1990) Effects of ultraviolet radiation and visible light on growth, fluorescence induction, ultraweak luminescence and peroxidase activity in sugar beet plants. J Photochem Photobiol 8:73-87

Paul ND, Callaghan TV, Moody S, Gwynn-Jones D, Johanson U, Gehrke C (1999) UV-B impacts on decomposition and biogeochemical cycling. In: Rozema J (ed) Stratospheric ozone depletion: the effects of enhanced UV-B radiation on terrestrial ecosystems Backhuys Publishers, Leiden, pp 117-133

Paul ND, Gwynn-Jones D (2003) Ecological roles of solar UV radiation: towards an integrated approach. Trends Ecol Evol 18:48-55

Phoenix GK, Gwynn-Jones D, Lee JA, Callaghan TV (2000) The impacts of UV-B radiation on the regeneration of a sub-arctic heath community. Plant Ecol 146: 67-75

Phoenix GK, Gwynn-Jones D, Callaghan TV, Sleep D, Lee JA (2001) Effects of global change on a sub-Arctic heath: effects of enhanced UV-B radiation and increased summer precipitation. J Ecol 89:256-267

Rau W, Hofmann H (1996) Sensitivity to UV-B of plants growing in different altitudes in the Alps. J Plant Physiol 148:21-25

Robberecht R, Caldwell MM, Billings WD (1980) Leaf ultraviolet optical properties along a latitudinal gradient in the arctic-alpine lifezone. Ecology 61:612-619

Rousseaux, M.C., Ballaré CL, Scopel AL, Searles PS, Caldwell MM (1998) Solar ultraviolet-B radiation affects plant-insect interactions in a natural ecosystem of Tierra del Fuego. Oecologia 116:528-535

Rozema J, Chardonnens A, Tosserams M, Hafkenscheid R, Bruijnzeel S (1997a) Leaf thickness and UV-B absorbing pigments of plants in relation to an elevational gradient along the Blue Mountains, Jamaica. Plant Ecol 128:150-159

Rozema J, van de Staaij J, Björn LO, Caldwell M (1997b) UV-B as an environmental factor in plant life: stress and regulation. Trends Ecol Evol 12:22-28

Saile-Mark M, Tevini M (1997) Effects of solar UV-B radiation on growth, flowering and yield of central and southern European bush bean cultivars (*Phaseolus vulgaris* L.). Plant Ecol 128:115-125

Salt D, Moody J, Whittaker J, Paul ND (1998) Effects of enhanced UV-B on populations of the phloem feeding insect *Strophingia ericae* (Homoptera: Psylloidea) on heather (*Calluna vulgaris)*. Glob Change Biol 4:91-96

Searles PS, Flint SD, Caldwell MM (2001) A meta-analysis of plant field studies simulating stratospheric ozone depletion. Oecologia 127:1-10
Semerdjieva SI, Phoenix GK, Hares D, Gwynn-Jones D, Callaghan TV, Sheffield E (2003a) Surface morphology, leaf and cuticle thickness of four dwarf shrubs from a sub-Arctic heath following long-term exposure to enhanced levels of UV-B. Physiol Plant 117:289-294
Semerdjieva SI, Sheffield E, Phoenix GK, Gwynn-Jones D, Callaghan TV, (2003b) Contrasting strategies for UV-B screening in sub-Arctic dwarf shrubs. Plant Cell Environ 26: 957-964
Shi S-B, Zhu W-Y, Li H-M, Zhou D-W, Han F, Zhao X-Q, Tang Y-H (2004) Photosynthesis of *Sassurea superba* and *Gentiana straminea* is not reduced after long-term enhancement of UV-B radiation. Environ Exp Bot 51:75-83
Solecka D, Kacperska A (1995) Phenylalanine ammonia-lyase activity in leaves of winter oilseed rape plants as affected by acclimation of plants to low temperature. Plant Physiol Biochem 33:585-591
Solheim B, Johanson U, Callaghan TV, Lee JA, Gwynn-Jones D, Björn LO (2002) The nitrogen fixation potential of arctic cryptogram species is influenced by enhanced UV-B radiation. Oecologia 133:90-93
Staxén I, Bornman JF (1994) A morphological and cytological study of *Petunia hybrida* exposed to UV-B radiation. Physiol Plant 91:735-740
Stewart AJ, Chapman W, Jenkins GI, Graham I, Martin T, Crozier A (2001) The effects of nitrogen and phosphorous defieciency on flavonol accumulation in plant tissues. Plant Cell Environ 24:1189-1197
Sullivan JH, Teramura AH (1990) Field study of the interaction between solar ultraviolet-B radiation and drought on photosynthesis and growth in soybean. Plant Physiol 92:141-146
Sullivan JH, Teramura AH, Ziska LH (1992) Variation in UV-B sensitivity in plants from a 3,000-m elevational gradient in Hawaii. Am J Bot 79:737-743
Taulavuori E, Bäckman M, Taulavuori K, Gwynn-Jones D, Johanson U, Laine K, Callaghan T, Sonesson M, Björn LO (1998) Long-term exposure to enhanced ultraviolet-B radiation in the sub-arctic does not cause oxidative stress in the dwarf shrub *Vaccinium myrtillus*. New Phytol 140:691-697
Tevini M, Teramura AH (1989) UV-B effects on terrestrial plants. Photochem Photobiol 50:479-487
Yazawa M, Shimizu T, Hirao T (1992) Feeding response of the silkworm *Bombyx mori,* to UV radiation of mulberry leaves. J. Chem Ecol 18:561-569
Ziska LH, Teramura AH, Sullivan JH (1992) Physiological sensitivity of plants along an elevational gradient to UV-B radiation. Am J Bot 79:863-871

13 Effects of UV radiation in arctic and alpine freshwater ecosystems

Dag O. Hessen

University of Oslo, Dept of Biology, P.O.Box 1066, Blindern, 0316 Oslo, e-mail: dag.hessen@bio.uio.no

13.1 Introduction

Alpine and arctic waterbodies range from large, deep and ultraoligotophic lakes to small ponds that typically dominate over large areas in the Arctic. One may question whether there are some truly unique properties that are shared amongst Arctic and alpine freshwaters. Are there some distinct features with regard to physicochemical properties or biology? The numerous small and shallow tundra ponds that are widespread in the circumpolar Arctic represent a quite distinct type of ecosystems that also possess a unique biota. These ponds are situated on permafrost, making them highly susceptible also to global warming, changed precipitation and permafrost thawing. Due to their general shallowness (often < 0.5 m) and fairly high water transparency, the major proportion of solar energy reach the sediment surface. Thus, these high Arctic localities may support a substantial benthic production despite a scarcity or absence of benthic macrophytes. Dense layers of cyanobacterial mats, epibenthic diatoms, heterotrophic bacteria and fungus may cover the sediment surface, and may be major players for nutrient dynamics and carbon fluxes in these wind-exposed and well mixed localities (Vincent 2000). In general, these shallow tundra ponds are devoid of fish, and this allows for establishment of dense populations of large-sized species of crustaceans. The high Arctic does also include a few deeper and larger lakes which resembles the typically deep and oligotrophic alpine lakes. These somewhat deeper lakes (> 3 m) that do not freeze to the bottom may also house populations of fish, where Arctic charr (*Salvelinus alpinus)* are the sole species in the truly high Arctic. While there apparently are very few *species* that are endemic to the Arctic and alpine freshwaters, there is nevertheless a wide span in clonal and genetic variability among some of these groups.

There are some fairly obvious climatic properties of the high arctic that need not to be reiterated at length here. This chapter will focus on observed and potential effects of ultraviolet radiation (UVR) for the inhabitants of these ecosystems, yet touching upon various abiotic challenges that may be superimposed on the UV-stress. During summer, photon flux of short wave radiation may also negatively affect both primary and secondary production. Short wave PAR (blue light, near 400 nm) may cause a suite of cellular damages, notably DNA-breaks and membrane lipid peroxidation. In general UVR is considered the most harmful part of the spectrum, even though the total photon flux is rapidly decreasing towards shorter wavelengths due to ozone absorption of UV in the stratosphere. There are a number of reviews that cover various aspects of UVR effects on aquatic organisms, including alpine (Sommaruga 2001) and arctic (Hessen 2002) freshwaters. Here, focus is here aspects of water properties with specific relevance to the biota, as well as life history traits and evolutionary perspectives that are particularly relevant in the context of UVR.

13.2 Physico-chemical properties of arctic lakes and ponds

The key determinants of alpine and arctic freshwater ecosystems (as in fact for most systems) are light, temperature and nutrients, although immigration history and immigration barriers can be superimposed on these abiotic factors for several areas. Deglaciation and land rise also cause a chronosequence of water bodies that has major bearings for their biota and biodiversity. While probably a few lakes survived the last deglaciation in ice-free refuges, most northern alpine and arctic localities are < 10 000 years, and some are quite recently formed by ice retreat or land rise.

Few arctic lakes are perennially frozen, but a number of high arctic localities may experience open water for very brief periods only. Consequently, epilimnetic temperatures in larger water bodies rarely exceed 5 °C (commonly less), and for major parts of the year the water temperature may be < 1 °C . This is partly caused by the fact that a substantial part of the annual heat budget is involved in the melting of ice. Even surface layer of larger water bodies may never reach maximum density (4 °C), meaning that thermal stratification will never occur. Similarly alpine likes are characterized by low temperatures and short growing season. Needless to say, these low temperatures throughout the water column pose constraints on metabolic processes and hence productivity and challenge life cycle adaptations for the biota of these systems, and it is of relevance for temperature dependent damage from UVR. On the other hand, the shallowness of smaller water bodies like tundra ponds allow for a strong benthic heat absorption, and temperature may rise substantially in summer during periods with 24 h solar irradiation. For tundra ponds at Barrow, Miller (1980) typically recorded temperatures from 8 – 12 °C during July and early August, yet with a temperature range from 2 to 16 °C within 2-3 days at the extreme.

While the temperature and nutrient regimes are not basically different in arctic and alpine localities of similar size and morphometry, the light regimes may differ. Truly alpine and arctic lakes and ponds have a brief time window allowing for primary production during summer, even though sufficient light also may penetrate through the ice (provided absence of snow) to allow for a high primary productivity also before ice-off. In spite of a 24 h sun exposure during summer in the Arctic, light *intensities* are substantially lower compared with southern, alpine habitats. By using a radiation transfer model (Dahlback and Stamnes 1991), it was estimated that even during the Arctic summer, maximum biologically harmful radiation would be some 20-30 % lower at near 80 °N relative to those at 60 °N (Hessen 1996).

Furthermore, Arctic lakes are frequently ice-covered until late July, where maximum doses are reduced by another 35 % relative to midsummer.

Nevertheless, the dose received by the biota may commonly be higher in Arctic ponds compared with lakes at lower latitudes, due to their shallowness and low attenuation of UV. In contrast, alpine lakes receives higher surface doses of UV than lowland lakes at the same altitude, and typically integrated UV-dose increase by roughly 10% per 1000 m elevation, while this altitudinal effect for longer wavelengths (PAR) are lower (Hessen 1993, Blumthaler et al. 1997). Hence there will commonly be a higher UV/PAR ratio in alpine lakes compared with lowland localities. There is an almost paradoxical relationship between light and life in the Arctic, however; levels of PAR, 380-700 nm) probably is the major limiting factor for Arctic life in general, UVR(< 380 nm) is a potential constraint on ecosystem productivity. This ratio may be highly relevant for repair capacity of UV-damage. Siebeck and Böhm (1994) found that peak recovery radiation was in the blue region (420-440 nm), and found that recovery (light repair) strongly increased when radiation was provided after UV-exposure. This positive effect was further increased when recovery radiation was provided also under UV-exposure. The recovery was rapidly decreasing with increased post-exposure darkness. The importance of photorecovery in alpine copepods was also convincingly demonstrated by Zagarese et al. (1997). These authors also noted a pronounced species-specific variability in photorecovery.

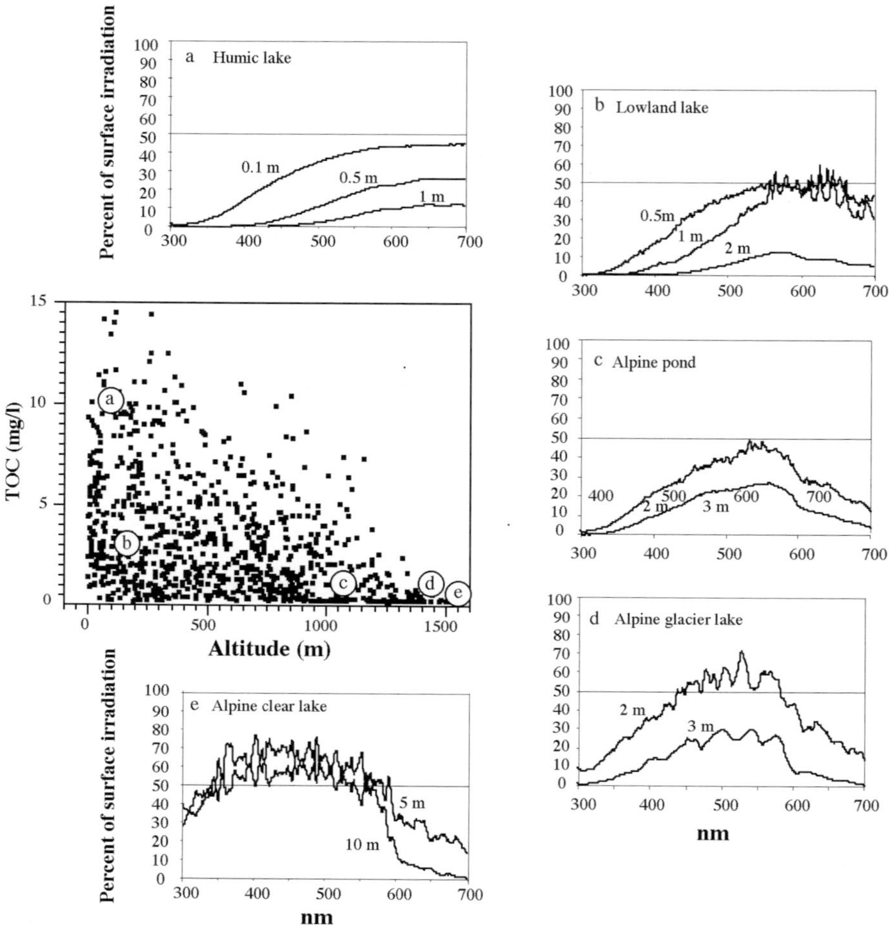

Fig. 13.1. Scatterplot of concentration of total organic carbon (TOC) versus height above sea in 997 Norwegian lakes. Spectral profiles as percent of surface (air) from selected depth in 5 types of lakes; a: Humic lake Skjervatjern, b: Lowland Lake Trollvann, c: Alpine pond at Finse, d: Alpine glacier fed Lake Gjende, e: Alpine clear Lake Bessvatn.

The key determinant of UV regimes in most lakes is the concentrations of coloured dissolved organic matter (CDOM), notably terrestrially derived humus. Most alpine and arctic localities have low concentrations of dissolved organic carbon (DOC), and typically the DOC-content decrease strongly with altitude, yielding tremendous differences on spectral properties and light attenuation (Fig. 13.1). For comparison, in the deep, clear alpine

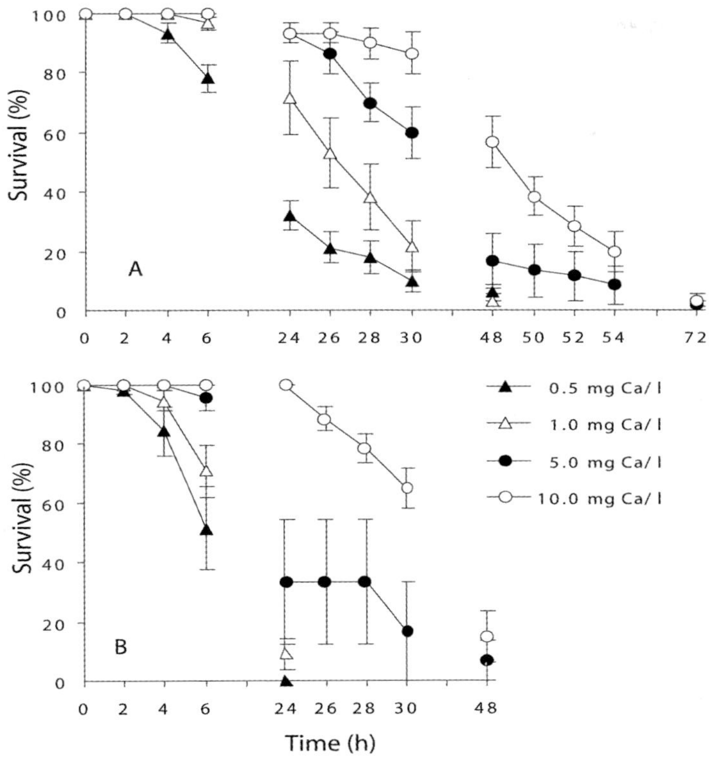

Fig. 13.2. Mean survival (±SE) of (A) *D. magna* and (B) *D. tenebrosa* reared under different Ca concentrations after different times of UV-exposure. Each point represents the mean of 6 replicates with 10 animals. Repeated 6 h exposure periods, with 35.95 W m^{-2} intensity over 300 - 400 nm (after Hessen and Rukke 2000).

Lake Bessvatn (1373 m above sea, < 1 mg DOC l^{-1}), a higher proportion of surface UVR remained at 10 m depth than at 0.1 m depth in humic lake Skjervatjern (12 mg DOC l^{-1}). The alpine lake Gjende had fairly high attenuation of UVR in spite of very low levels of DOC. This is caused by high inputs of particles from glacial runoff.

Most alpine lakes are characterized by high transparency of UVR owing to low levels of DOC and low productivity. The same holds true for most high arctic freshwaters, although for tundra-ponds in more productive areas with developed soils or in wetland areas, DOC concentrations may exceed 5 mg DOC l^{-1} (Hobbie et al. 1996). For Svalbard lakes, concentrations of < 1 – 4 mg C l^{-1} is recorded (Ørbæk et al. 2002), but even localities with 4 mg C l^{-1} have low UVR attenuation due to low influx of allochthonous humic matter, and a dominance of less colored matter of autocthonous origin, caused by a frequent mixing in these shallow water bodies.

Species richness of planktonic species may also be governed by the UV regimes related to concentrations of DOC (Williamsson et al. (2001) provided evidence that macrozooplankton community structure in a set of lakes along a deglaciation chronosequence in Glacier Bay (Alaska) could be attributed to the UV attenuation in these lakes.

Most alpine and arctic lakes and ponds have low levels of nutrients. There is, however, a tremendous scatter both in nutrients and concentrations of various ions. Coastal ponds may be strongly influenced by bird droppings, and may also possess high levels of calcium (Ca). Also local geology may faciliate high ionic content that could be an important indirect mediator of UVR susceptibility. Ca concentrations may in fact be of vial importance for a number of Ca demanding crustaceans, and it has been demonstrated that *Daphnia* is more susceptible to UVR under low ambient Ca-concentrations (Fig. 13.2).

13.3 UVR-effects and adaptations

13.3.1 Evidence for UVR-effects?

There is scattered knowledge on the UVR effects on various aquatic taxa and even less on species or various developmental stages within species, yet there is several studies that demonstrate strong effects of UVR (both UV-A and UV-B) of alpine and arctic phytoplankton and bacteria (Villafane et al. 1999; Sommaruga and Garcia-Pichel 1999; Calleri et al. 2001, Van Donk et al. 2001; Carrillo et al. 2002), zooplankton (Hebert and Emery 1990; Hessen 1993; Zagarese et al. 1997; Zellmer 1998; Tartarotti et al. 1999) and benthic communities (Vinebrooke and Leavitt 1999; Tank et al. 2003). Palelimnological studies like those of Leavitt et al. 1997; 2003) indeed suggest the balance between DOC and UVR as a major determinant of phytoplankton communities and biomass in alpine and subalpine lakes. Heterotrophic bacteria in general are assumed to be highly susceptible to UV-B (Sommaruga et al. 1997), while cyanobacteria and some phytoplankton species that rely on PAR for their biosynthesis have evolved various strategies to cope also with the short-wave part of the spectrum (Vincent 2000; Vincent and Belzile 2002). Notably the benthic bluegreens that are key producers in shallow tundra ponds display very efficient UV-defence strategies (Quesada et al. 1999; Vincent 2000). Amongst phytoplankton, the major effect may be community shifts rather that reduced primary production. There are contrasting reports as to which taxa and species that are most vulnerable to UVR. While Laurion and Vincent (1998) found arctic picocyanobacteria to be well protected against UV, Van Donk et al. (2001) found that these, and other small species, was comparatively more susceptible than larger species. In a series of experiments with the natural phytoplankton communities from three high Arctic lakes on Spitsbergen (79 °N), in situ effects at the community level was assessed under natural UV radiation. These community shifts resulted in a decreased fraction of grazable algal species in UV-transparent bottles, while no effects were seen for PAR-light only. In the more productive lake, there appeared to be negligible effects of UV on the community composition. This suggests that UV could pose strong indirect “bottom up” effects on grazers in systems with high UVR transparency. Carrillo et al. (2002) also found that UVR in an alpine lake indirectly stimulated bacterial production by inducing increased release of DOC from phytoplankton.

13.3.2 UVR and oxidation of fatty acid

For obvious reasons, indirect effects of UVR may be harder to test and detect compared with direct effects. At the food web level, one might list a number of potential impacts of UVR, i.e. like the shift to less edible phytoplankton species mentioned above. One effect that may be of specific relevance for alpine and arctic ecosystems, is the effect of UVR on fatty acid composition. Here, UVR could play a major role in the food web in shallow and clear waters, in particular by reducing the long-chained polyunsaturated fatty acids (PUFA) that play a key role in arctic ecosystems and low-temperature systems. Fatty acid profiles, and notably a few specific PUFA´s like eicosapentaenoic acid (EPA, 20:5□3) and docosahexaenoic acid (DHA, 22:6□3) have been considered as a major determinant to herbivore success and animal nutrition in general. There is some evidence that PUFAs are susceptible to oxidations by UVR (Wang and Chai 1994; Hessen et al. 1997), yet contrasting results are reported (Tank et al. 2003). Consumers cannot synthesize PUFAs de novo, and thus depend on dietary inputs of these, and thus UVR effects on algae will most likely affect also the higher trophic levels.

13.3.3 Susceptible periods and growth rate

UVR effects may probably be most severe when circumstances are especially unfavourable, like when periods with low ozone levels, clear sky and specific mixing regimes coincide with susceptible life cycle stages of key species. Based on studies from Svalbard, the general pattern for *Daphnia* is two generations per year (Hessen et al. 2004). Typically, *Daphnia* hatch from resting eggs (ephippia) during ice off around mid July, and then grow to maturity (around 2 mm) and release the first clutch of eggs after 2-3 weeks. Hence a rather synchronous cohort may develop. The second generation thus produce resting eggs normally by late August. Maximum UV-B exposure under synchronous hatching of unpigmented juvenile *Daphnia* during ice off could typically be such a "window of susceptibility".

The main target for damage is DNA, but the site of eventually breaks will critically depend on the function of the particular part of the genome. Especially, it will make a big difference if coding or non-coding regions are subject to damage. Organisms with miniaturized genomes, such as *Daphnia*, the key players in many arctic and alpine localities, could thus be more susceptible than species with more introns. In general, small organism could be more susceptible than larger ones since all body components including reproductive organs can be reached by UVR. Also rapid cell division or growth, implying a very active gene expression, could be a major determinant of susceptibility (Naganuma et al. 1997). The selective pressure for fast growth in arctic species could thus paradoxically render these organisms highly vulnerable to UVR in spite of low ambient temperature.

13.4 Protection and adaptations

13.4.1 UVR-screening

Organisms may cope with the direct UV radiation effects in different ways, and several lines of defence may be identified that play a different role in different organisms. There may either be seasonal or spatial means of UVR avoidance by timing of reproductive season, depth distribution or diurnal migration. A second line of defence is UV screening compounds (e.g. pigments, mycosporine-lime amino acids). Third, there are several enzymes and other macromolecules serving as radical scavengers that in various ways cope with ambient or intracellular harmful photo-products. A last major defence would be the various means of enzymatic photo-repair that is a common property of all organisms. Vertical migration in zooplankton as a direct response of UVR have been demonstrated both in arctic (Hebert and Emery 1990) and alpine (Hessen 1993) localities.

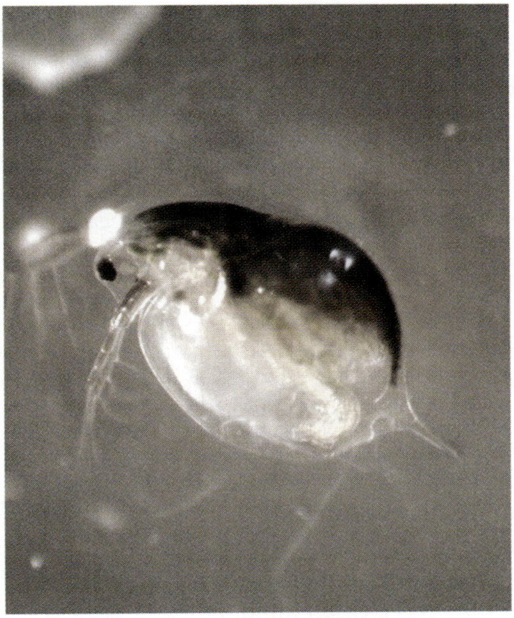

Fig. 13.3. Example of dorsal melanin pigmentation in an arctic *Daphnia tenebrosa*, Svalbard

Many alpine and arctic autotrophs produce UV-screening compounds including scytonemin and mycosporine-like amino acids (MAA) (Vincent 2000). Several phytoplankton species of alpine lakes were found to produce MAAs, with the highest concentrations of MAAs and photoprotective carotenoids recorded in the clearest lakes (Tartarotti et al. 2001; Laurion et al. 2002). The surface of the benthic mats exposed to bright light is typically rich in carotenoids and the cellular concentration of carotenoids increases with increasing PAR and UV and also with decreasing temperature, and is probably crucial for the survival of these organisms under high UV exposure (Roos and Vincent 1998, Vincent and Belzile 2002). The extent of these various UV protection mechanisms varies greatly among taxa and species of autotrophs. Both carotenoids and MAAs are synthesized in autotrophs, but are also extensively used by heterotrophs for photoprotection. Grazer thus typically depends on their food for obtaining these substances, but the ability to use these for photoprotection among heterotrophs differ among taxa and species (Sommaruga and Garcia-Pichel 1999). Tartarotti et al. (2001) reported frequently high concentrations of MAAs in alpine calanoid copepods, moderate levels in sympatric rotifers, while virtually no MAAs were detected in cladocera. This point to some important taxonomic differences in pigmentation strategies at least between copepods and cladocera.

The extensive need for photoprotection in alpine zooplankton species was recognized in early works like those of Brehm (1938) and Merker (1940) who accredited the conspicuous red coloration of alpine and highly light exposed zooplankton high levels of tissue carotenoids. While the role of carotenoid photoprotection seems well justified in copepods (cf. Ringelberg et al. 1984), it is more obscure in the cladocera (Hebert and Emery 1990; Hessen and Sørensen 1990). Sub-Arctic alpine copepods (*Heterocope*) were found to have ten times more carotenoids than sympatric populations of cladocerans, and even low-land transparent copepods have higher carotenoid levels than highly light exposed *Daphnia* (Hessen and Sørensen 1990). A major difference between these groups is that while carotenoids in copepods may be distributed not only in lipid droplets, but also body fluids and tissue, they are most often (though not always) restricted to ovaries or pre-ovarian lipids in *Daphnia*. Pigmentation may also undego ontogonetic shifts. Typically, larval, pelagic stages of the Notostracan *Lepidurus arcticus* are bright red from carotenoids, while the benthic adults have lower levels (Hessen unpubl).

Most of the work on UV in alpine and arctic freshwaters have been devoted to zooplankton however, and particularly on the cladoceran *Daphnia* that possess a truly remarkable adaptation to UV-stress by a conspicuous

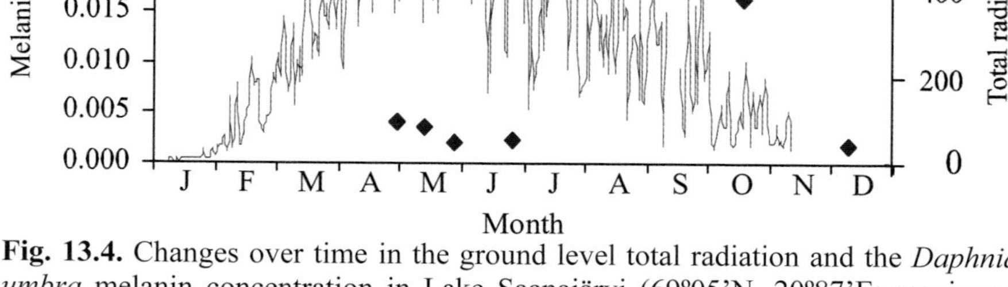

Fig. 13.4. Changes over time in the ground level total radiation and the *Daphnia umbra* melanin concentration in Lake Saanajärvi (69°05'N, 20°87'E; maximum depth = 24 m, altitude = 679 m a.s.l., DOC = 1.6 mg L^{-1}), showing the effects of increased underwater light intensity on pigment synthesis. The thickness of the ice exceeds 1 m in late spring. (From Rautio and Korhola 2002).

carapace melanization (Fig. 13.3). This melanization in zooplankton is apparently a unique property of Arctic and alpine cladocera, and in some clones it is also inducible by presence of UVR (Fig. 13.4, and Rautio and Korhola 2002). Studies on absorbance of the carapace demonstrate a very low transmission of short-wave radiation in the very dark dorsal part of melanic individuals, and a fairly low transmission also in the dorsal, less pigmented parts of the carapace relative to that of non-pigmented animals (Hessen 2003). For arctic ponds, Hebert and Emery (1990) and Hessen et al. (1999) found that melanic clones ranged freely through the water column, while unpigmented clones were restricted to the pond bottom under high light intensities. By comparing UVR tolerance of melanic and non-melanic morphs/clones both *in situ* (Hebert and Emery 1990; Zellmer 1998) and in the laboratory (Hessen 1996; Hessen et al. 1999), it is evident that melanism offer a quite efficient photo-protection. Since the melanin has to be resynthesized after each moult and this synthesis apparently is an energy demanding process. For other invertebrates, and aquatic vertebrates, there is a general lack of knowledge, yet several studies have demonstrated that fish and amphibia may suffer skin lesions, catharact-like symptoms and immunosuppression under high ambient UVR (cf. Hessen 2003).

Finally, a variety of anti-oxidants may together with repair enzymes serve as a final line of defense. However screening of anti-oxidants an a number of arctic, alpine and lowland population of *Daphnia*, including both melanized and hyaline populations did however not reveal any clear-cut differences between the populations (Borgeraas and Hessen 2002).

13.4.2 Evolutionary adaptations and “the ghost of UVR in the past”.

In nature there are rarely dramatic effects of UVR-exposure, but organisms in exposed habitats display several sign of adaptation. Adaptations such as vertical migration or the presence of UVR protective pigments suggest a major need for UV-protection in arctic and alpine localities. Evolutionary adaptations may also include a higher expression of anti-oxidants (under low temperature), more efficient repair system for DNA-breaks or membrane damage etc. Siebeck and Böhm (1994) found higher UVR susceptibility in lowland *Daphnia* compared with an alpine populations, and this probably illustrate a common phenomenon: highly light exposed organisms may tolerate very high doses of UVR. Typically, copepods from lakes at very high latitudes may tolerate truly extreme levels of UVR (Vinebrooke and Leavitt 1999; Sommaruga 2001). In general, such adaptations involve some costs, and the metabolic costs associated with photoprotection and active means of UVR avoidance could be more important effects of UV than lethal or sublethal cellular damage. Such overall fitness costs are clearly seen for melanized *Daphnia* as reduced growth rates compared with their hyaline counterparts (Weider 1987; Hessen 1996). The costs are also indicated by the down-regulation of melanin synthesis in the absence of UVR (Hessen 1996), or the increased risk of fish predation due to increased visibility (Sægrov et al. 1996, Hansson 2000). Thus present tolerance to UVR reflects a series of evolutionary adaptations and alpine and arctic freshwaters possess species and communities that must be regarded as "the ghost of UVR in the past".

13.5 Conclusions

We have seen that there are some aspects of alpine and arctic freshwater localities, in spite of low intensities of UVR, that could make organisms in these ecosystems particularly vulnerable to UVR;

1.Their frequent shallowness, offering no depth refugium for the biota

2. Their low levels of DOC and other UV-absorbing compounds.

3. The low levels of nutrients of thus low primary production

4. The frequent low temperatures that could slow down the anti-oxidant expression and cellular repair mechanisms that are coping with UV-induced damage.

There are nevertheless rather scattered and somewhat contrasting information on effects of UVR in Arctic and alpine habitats. Part of this can be accredited experimental condition and stage or condition of the tested organisms, but it does probably also reflect a real difference in UVR susceptibility among taxa and species. Adaptations such as vertical migration or the presence of UVR protective pigments suggest a major need for UV-protection in arctic and alpine localities. Thus alpine and arctic freshwaters communities are evolutionary shaped towards tolerance to a variety of challenges including high exposure to UVR. These evolutionary adaptations of behaviour, pigmentation and enzymes reflects the organisms´ past experience with UV. Such adaptations clearly have their costs, yet such costs may be quite subtle and hard to detect.

References

Blumthaler M, Ambach W, Ellinger R (1997) Increase in solar UV radiation with altitude. J Photochem Photobiol B 39: 130-134.

Borgeraas J, Hessen DO (2002) Species-specific and clone-specific expression of anti-oxidants and pigmentation in Daphnia spp. Hydrobiologia 477: 15-30.

Brehm V (1938) Die Rotfarbung von Hochgebirgsorganismen. Biol Rev 13: 307-318.

Calleri C, Morabito G, Huot Y, Neale PJ, Litchman E (2001) Photosynthetic response of pico- and nanoplanktonic algae to UVB, UVA and PAR in a high mountain lake. Aquat Sci 63: 286-293.

Carrillo P, Medina-Sanchez JM, Villar-Argaiz M (2002) The interaction of phytoplankton and bacteria in a high mountain lake: importance of the spectral composition of solar radiation. Limnol Oceanogr 47: 1294-1306.

Dahlback A, Stamnes K (1991) A new spherical model for computing the radiation field available for photolysis and heating at twilight. Planet Space Sci 39: 671-683.

Hansson LA (2000) Induced pigmentation in zooplankton: a trade-off between threats from predation and ultraviolet radiation. Proc Royal Soc Lond Ser B 267: 2327-2331.

Hebert PDN, Emery CJ (1990) The adaptive significance of cuticular pigmentation in Daphnia. Funct Ecol 4: 703-710.

Hessen DO, (1993) DNA-damage and pigmentation in alpine and arctic zooplankton as bioindicators of UVRadiation. Verh Internat Verein Limnol 25: 482-486.

Hessen DO (1996) Competitive trade-off strategies in Arctic Daphnia linked to melanism and UV-B stress. Polar Biol 16: 573-579.

Hessen DO (2002) (ed), UV Radiation and Arctic Ecosystems, Ecological Series 153, Springer Verlag, Berlin, Heidelberg.

Hessen DO (2003) UV-radiation and pelagic metazoans. In: Helbling, E.W. and Zagarese, H. (Eds.) UV-effects in Aquatic Organisms and Ecosystems. Comprehensive Series in Photochemical and Photobiological Sciences. pp. 399-430.

Hessen DO, Borgeraas J, Kessler K and Refseth U H (1999) UV-B susceptibility and photoprotection of Arctic Daphnia morphotypes. Polar Res 18: 345-352.

Hessen DO, Blomqvist P, Dahl-Hansen G, Drakare S, Lindström ES (2004) Production and food web interactions of Arctic freshwater plankton and responses on increased DOC. Arch Hydrobiol 159: 289-307.

Hessen, D.O., van Donk, E. and de Lange, M. (1997) UV-induced changes in phytoplankton cells and its effect on grazers. Freshwater Biology 38: 513-524

Hessen DO, Rukke, NEA (2000) Increased UV-susceptibility in Daphnia at low calcium concentrations. Limnol Oceanogr 45: 1834-1838.

Hessen DO, Sørensen K (1990) Photoprotective pigmentation in alpine zooplankton populations. Aqua Fennica 20:165-70

Hobbie JE (1996) Polar limnology. In Taub FB (ed), Lakes and Reservoirs. Elsevier, Amsterdam, pp. 63-86.

Laurion I, Lami A, Sommaruga R (2002) Distribution of mycosporine-like amino acids and photoprotective carotenoids among freshwater phytoplankton assemblages. Aquatic Microbial Ecol 26: 283-294.

Laurion I, Vincent WF (1998) Cell size versus taxonomic composition as determinants of UV sensitivity in natural phytoplankton communities. Limnol Oceanogr 43: 1774-1779.

Leavitt PR, Vinebrooke RD, Donald DB, Smol JP, Schinder DW (1997) Past ultraviolet radiation environments in lakes derived from fossil pigments. Nature 388:457-459

Merker, E (1940) Der Lichtschutz in belebtem Wasser. Int Rev ges Hydrobiol 40: 174-196.

Miller MC, Prentki RT, Barsdate RC (1980) Physics. In: Hobbie JE. (Ed). Limnology of tundra ponds. Barrow, Alaska. Dowden, Hutchinson and Ross, Stroudsburg, Penn.

Naganuma T, Inoue T, Uye S (1997) Photoreactivation of UV-induced damage to embryos of a planktonic copepod. J Plankton Res 19: 783-787

Ørbæk JB, Svenøe T Hessen DO (2002) Spectral properties and UV-attenuation in Arctic freshwater systems. p 57-72 in D. O. Hessen (ed): UV radiation and Arctic Ecosystems, Ecological Series 153, Springer-Verlag.

Quesada A, Vincent WF, Lean DRS (1999) Community and pigment structure of Arctic cyanobacterial assemblages: the occurrence and distribution of UV-absorbing compounds. FEMS Micobiol Ecol 28: 315-323

Rautio M, Korhola A (2002) UV-induced pigmentation in subarctic Daphnia. Limnol Oceanogr 47:295-299

Ringelberg J, Keyser AL, Flik BJG (1984) The mortality effect of ultraviolet radiation in a red morph of Acanthodiaptomus denticornis (Crustacea: Copepoda) and its possible ecological relevance. Hydrobiologia 112: 217-22

Roos JC, Vincent WF (1998) Temperature dependence of UV radiation effects on Antarctic cyanobacteria. J Phycol 34: 118-125.

Siebeck O, Böhm U (1994). Challenges for an appraisal of UV-B effects upon planktonic crustaceans under natural conditions with a non-migrating (Daphnia pulex obtusa) and a migrating cladocran (Daphnia galeata). Ergebnisse der Limnologie Archiv für Hydrobiol Beih 43:197-206

Sommaruga R (2001) The role of solar UV radiation in the ecology of alpine lakes. J Photochem Photobiol B 62: 35-42.

Sommaruga R, Garcia-Pichel F (1999) UV-absorbing mycosporine-like compounds in planktonic and benthic organisms from a high-mountain lake. Arch Hydrobiol 144: 255-269.

Sommaruga R, Obernosterer I, Herndl GJ, et al. (1997) Inhibitory effect of solar radiation on thymidine and leucine incorporation by freshwater and marine bacterioplankton Appl Environ Microbiol 63: 4178-4184

Sægrov H, Hobaek, A, L'Abee Lund, JH (1996) Vulnerability of melanic Daphnia to brown trout predation. J Plankton Res 18: 2113—2118

Tank SE, Schindler DW, Arts MT (2003) Direct and indirect effects of UV radiation on benthic communities: epilithic food quality and invertebrate growth in four montane lakes. Oikos 103: 651-667.

Tartarotti B, Cabrera S, Psenner R, Sommaruga R (1999) Survivorship of Cyclops abyssorum tatricus (Cyclopoida, Copepoda) and Boeckella gracilipes (Calanoida, Copepoda) under ambient levels of solar UVB radiation in two high-mountain lakes. J Plankton Res 21: 549-560

Tartarotti B, Laurion I, Sommaruga R (2001) Large variability in the concentration of mycosporine-like amino acids among zooplankton from lakes located across an altitude gradient. Limnol Oceanogr 46: 1546-1552

Van Donk E, Faafeng B, De Lange HJ, Hessen DO (2001) Differential sensitivity to natural ultraviolet radiation among phytoplankton species in Arctic lakes (Svalbard, Norway). Plant Ecol 154: 213-223

Villafane VE, Andrade M, Lairana AV, Zaretti F, Helbling EW (1999) Inhibition of phytoplankton photosynthesis by solar ultraviolet radiation: studies in Lake Titicaca, Bolivia. Freshw Biol 42: 215-224.

Vincent WF (2000) Cyanobacterial dominance in the polar regions. In Whitton B and Potts M (eds), Ecology of the Cyanobacteria: Their diversity in Space and Time. Kluwers Academic Press, pp 321-340.

Vincent WF, Belzie C (2002) UV effects on aquatic microbial food webs in northern lakes and rivers In. Hessen DO (ed), UV Radiation and Arctic Ecosystems, Ecological Series, Springer Verlag, Berlin, Heidelberg.

Vinebrooke RD, Leavitt PR (1999) Differential responses of littoral communities to ultraviolet radiation in an alpine lake. Ecology 80: 223-237.

Wang KS, Chai T (1994) Reduction of omega-3 fatty acids by UV-B irradiation in microalgae. J Applied Phycol 6: 415-421.

Weider L J (1987) Life history variation among low-arctic clones of obligately parthenogenetic Daphnia pulex; a diploid-polyploid complex. Oecologia 73: 251-256.

Williamson CE, Olson OG, Lott SE, Walker ND, Engstrom DR, Hargreaves BR (2001) Ultraviolet radiation and zooplankton community structure following deglaciation in Glacier Bay, Alaska. Ecology 82:1748-1760

Zagarese HE, Feldman M, Williamson CE (1997) UV-B-induced damage and photoreactivation in three species of Boeckella (Copepoda, Calanoida). J Plankton Res19 : 357-367

Zellmer ID (1998) The effect of solar UVA and UVB on subarctic Daphnia pulicaria in its natural habitat. Hydrobiologia 379: 55-62.

14 Climate control of biological UV exposure in polar and alpine aquatic ecosystems

Warwick F Vincent, Milla Rautio and Reinhard Pienitz

Centre d'Études Nordiques, Université Laval, Québec, Canada G1K 7P4, e-mail: warwick.vincent@bio.ulaval.ca

14.1 Introduction

The severe depletion of stratospheric ozone in Antarctica over the last two decades has generated much concern about the effects of rising UV-B radiation on marine and freshwater ecosystems (de Mora et al. 2000; Cockell and Blaustein 2001; Sommaruga 2001; Hessen 2002; Perin and Lean 2004) and has led to a broad range of research on UV optics, photobiology and photochemistry in the aquatic environment. The studies to date imply that certain ecosystem types may be especially prone to major changes in their spectral UV regime: clear lakes and oceans in the polar regions where ozone depletion is occurring and where UV wavelengths penetrate deeply into the water column (Vincent and Belzile 2003), and oligotrophic alpine waters which have little UV-screening protection because of their low concentrations of colored dissolved organic matter (CDOM; Morris et al 1995; Williamson et al. 1996; Laurion et al. 2000; Markager and Vincent 2000). The biological communities in these ecosystems may also be more vulnerable to UV toxicity because of the inhibiting effects of cold temperatures on cellular repair of UV damage (Rae et al. 2000; Hoffman et al. 2003; MacFadyen et al. 2004).

The UV waveband is a highly reactive component of all environments exposed to the sun, and is subject to large fluctuations at multiple timescales. It is increasingly apparent that climate may exert a strong control on these fluctuations, and that the amplitude of such effects can be much greater than those caused by moderate stratospheric ozone depletion. Major shifts in underwater UV are likely to accompany future climate change, with implications for many important photobiological and photochemical processes in the aquatic environment.

The aim of this chapter is to review the mechanisms, models and observations that link underwater UV exposure and climate change. We first describe a general model that has been applied with success and variously modified to address questions about how changes in the environment translate into UV exposure and responses. We then examine each of the terms in this model and their sensitivity to climate, and how they combine to result in climate control of biological UV exposure in the underwater environment. Finally, we illustrate by way of several case studies how paleo-ecological analysis can provide a valuable approach towards assessing the long-term role of climate in controlling biological UV exposure in alpine and polar aquatic ecosystems.

14.2 Model description

One approach towards assessing the effects of environmental change on spectral UV exposure is by way of the index 'weighted transparency', T^*, where the * indicates that the radiation at each wavelength is multiplied by its biological effectiveness. This index has been applied to cold lakes and oceans (Vincent et al. 1998; Pienitz and Vincent 2000; Gibson et al. 2000; Vincent and Belzile 2002) and has been modified and extended for more general use (Neale 2001; Lehmann et al. 2004). The parameter T^* allows the effects of stratospheric ozone depletion and changes in water column attenuation of UV to be assessed on a common, biologically relevant scale, and it is calculated by integrating the transparency of the water column to biologically weighted irradiance at each wavelength λ over the UV waveband 280-400 nm:

$$T^* = \int T^*(\lambda)\, d\lambda \qquad (1)$$

For an ice-covered lake or ocean, $T^*(\lambda)$ may be partitioned as follows:

$$T^*(\lambda) = E_d(0^+)\, F\, (1\text{-}r)\, (1\text{-}f)\, (1/K_{dUV})\, \varepsilon \qquad (2)$$

where the parameters at each wavelength λ are:
$E_d(0^+)$ = the incident solar irradiance in relative energy units;
F = factor modifying that flux as a function of ozone depletion in the stratosphere;
r = the fraction of UV that is removed by surface reflection (albedo) from the snow, ice or water;
f = the fraction of UV that is removed by attenuation by overlying snow and ice;

K_{dUV} = the diffuse attenuation coefficient for downwelling UV irradiance in the water column;

ε = the biological weighting factor (BWF) which expresses the relative damage incurred by UV (ε = 1.0 at 300 nm). In the following sections we examine the implications of climate change for each of the terms in Eq. (2).

14.2.1 Incident UV irradiance ($E_d(0^+)$ F)

Climate has the potential to substantially modify incident spectral irradiance through at least two mechanisms: changes in heat balance and transport processes in the stratosphere, and through changes in tropospheric cloud cover. The greenhouse energy-trapping effect in the troposphere is accompanied by less outgoing long-wave radiation into the lower stratosphere. In the longer term this results in stratospheric cooling, a strengthening of the polar vortex, and a prolongation of conditions that favor ozone loss (Staehelin et al. 2001). The Arctic stratosphere is thought to be on the brink of major ozone loss and would appear to be especially sensitive to a small amount of additional cooling (Dahlback 2002). There may also be larger scale transport effects. Greenhouse warming could lead to increased zonal flow in mid-latitudes causing the polar vortex to be more stable, again favoring ozone loss and a delay in the eventual recovery from CFC emissions (Shindell et al. 1998).

Cloud cover and precipitation regimes are likely to change as a result of shifts in global climate, and these effects may be especially pronounced in the polar regions (Houghton et al. 2001). Increased open sea conditions will be accompanied by increased evaporation into the overlying air masses that at warmer temperatures can hold more moisture for subsequent cloud formation and precipitation. Global warming is also likely to accelerate the hydrological cycle at lower latitudes, resulting in increased atmospheric transport of water to the polar regions. At Spitsbergen, there has been a significant rise in precipitation from the 1960s to 1990s (Hanssen-Bauer and Forland 1998), however other sectors of the Arctic have shown little change or decreases (Vincent et al. 2001).

Cloud cover can have a strong effect not only on total incident UV radiation, but also on UV spectral composition. Gautier et al. (1994) found that incident UV-B irradiance in Antarctica was more closely correlated with cloud cover than stratospheric ozone. There was a positive correlation between the ratio of DNA-damage weighted UV-B irradiance to UV-A irradiance when the surface was covered by reflective snow and ice, and a negative correlation during open ocean conditions. These authors concluded that cloud cover affects UV spectral ratios but that the effect is modulated by multiple back scattering between the surface and clouds. UV measurements over snow-covered ground have shown that incident erythemal (sunburn)-weighted UV on overcast days can be 80% higher than predicted from cloud attenuation because of these multiple reflections (Renaud et al. 2000).

14.2.2 Albedo effects (*1-r*)

Albedo, the ratio of upwelling (E_u) to downwelling (E_d) irradiance, is high in snow-covered alpine and polar environments and exerts a major control on the underwater UV radiation field. UV albedo for snow and white ice is generally > 80%, but decreases to 50-60% for bare melting ice and to as little as 30% for ponding meltwater over ice. The latter is a common feature over Arctic sea ice during late spring and summer (Perovich et al. 1998; Belzile et al. 2000), while in the drier Antarctic environment a scattering surface layer with high albedo can form during summer (Trodahl and Buckley 1990). Any climate change altering snow cover, duration of ice cover or the surface reflective properties of the ice, for example through meltwater ponds or rainwater over the ice surface, will drastically affect this albedo term and thus T^*. A snow-clearing experiment in Hudson Bay at the edge of the Arctic Ocean underscored the large albedo effect on UV transmittance through sea ice (details in Vincent and Belzile 2002). The removal of surface snow caused a 3 to 16 fold increase in the under-ice UV, thus a similar increase in T^*. Even a layer of snow only 2-cm thick reduced UV by a factor of 3, with slightly greater effects at the shorter wavelengths. Similarly, the removal of snow over a high Arctic ice-covered lake decreased albedo by 20% in the UV range and 13% in the PAR range (Belzile et al. 2001). Lake white ice, that is ice with a high air content, has a high albedo for PAR and UV, and lake ice whitening can occur with gradual warming over spring and summer (Howard-Williams et al. 1998).

14.2.3 Attenuation by snow and ice (*1-f*)

There are few published UV attenuation coefficients for snow and ice because of the difficulties of making measurements in these solid media. In early spring, about 1% of incident UV-B is transmitted through ~1.6 m of sea ice (Arctic: Perovich et al. 1998; Antarctic: Trodahl and Buckley 1990), however the development of ice algae can decrease UV transmittance by an order of magnitude (Perovich et al. 1998). Autochthonous

CDOM derived from ice algae has a significant impact on UV attenuation in sea ice, and the melting of the snow cover, ponding of the ice surface and flushing of the ice algae all tend to increase UV-B transmittance (Belzile et al. 2000). In Antarctica, a highly scattering sea-ice surface forms as the melt progresses, resulting in decreased UV-B transmission relative to early spring values (Trodahl and Buckley 1990). Freshwater ice is typically more transparent to UV irradiance than sea ice because of the absence of brine pockets and minimal ice-biota. Lake ice may even be more UV-transparent than the underlying lake water because of the exclusion of high molecular weight UV-absorbing CDOM during the freeze-up process (Belzile et al. 2002). Even Antarctic lake ice several meters thick can allow measurable transmission of UV to the waters beneath (Vincent et al. 1998), although if such ice contains high concentrations of light-scattering inclusions such as glacial sediments ('dirty ice') or bubbles ('white ice') the UV-transmittance is minimal. For example, the 5-m thick summer ice over Lake Hoare in the McMurdo Dry Valleys was highly scattering, and no UV radiation less than 310 nm was detected in the water column beneath the ice (Kepner et al. 2000).

The dynamics and timing of ice cover melting in spring have substantial effects on the UV exposure of biota in the underlying water column. In alpine lakes for example, UV irradiances at the top of the water column can be an order of magnitude higher after melt-out than under 1.8 m of snow and ice cover at similar solar zenith angles (Sommaruga 2001). Rautio and Korhola (2002) showed that the *Daphnia* population of a subarctic lake responded to ice cover break-up by immediate synthesis of UV-absorbing melanin pigment (Fig. 14.1). Towards fall, the melanin concentration decreased suggesting the pigmentation was a direct response to the increased UV exposure during the peak radiation period. Shorter ice cover duration and exposure to higher UV during the spring ozone depletion could potentially alter the aquatic communities in lakes and require them to invest more in such UV protection mechanisms. This melanin content of cladocera also offers potential to estimate ice cover regimes in the past by analysis of lake sediments, analogous to the cyanobacterial UV-pigment reconstructions that have been applied to date (see below).

Substantial melting of seasonal snow and ice cover caused by rising temperatures has already been recorded in many parts of the world for example over cold lakes and rivers (Magnuson et al. 2000), arctic sea ice (Serreze et al. 2003) and lakes in the Antarctic Peninsula region (Quayle et al. 2002). Such changes are likely to have been accompanied by drastic shifts in underwater UV exposure. T^* calculations for an Arctic marine site

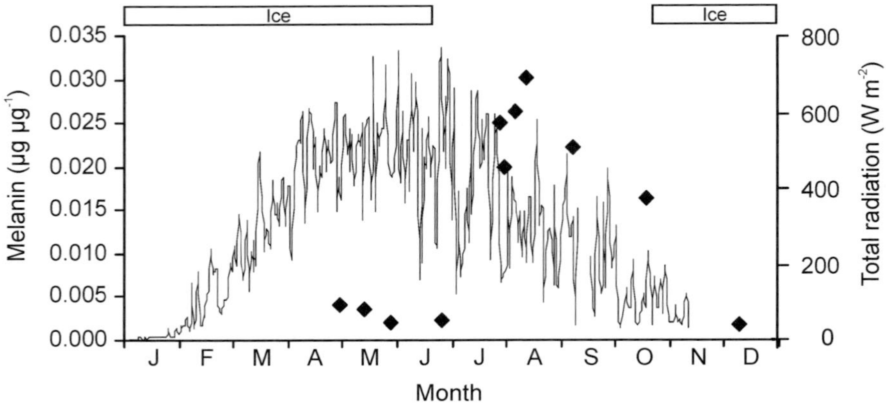

Fig. 14.1. Changes over time in the ground level total radiation and the *Daphnia umbra* melanin concentration in Lake Saanajärvi (69°05'N, 20°87'E; maximum depth = 24 m, altitude = 679 m a.s.l., DOC = 1.6 mg L^{-1}), showing the effects of increased underwater light intensity on pigment synthesis. The thickness of the ice exceeds 1 m in late spring. From Rautio and Korhola (2002)

showed that the removal of ice increased T^*_{PI} (T^* weighted for photosynthetic inhibition) by a factor of 12 and T^*_{DNA} by a factor of 22. The Arctic is likely to be more prone to this effect than the Antarctic in the immediate future. General circulation models (GCMs) predict that global warming will occur first and most intensely at high northern latitudes (Houghton et al. 2001) and there is evidence of accelerated impacts at some sites (Mueller et al. 2003). The north polar ice cap has experienced thinning and contraction over the last three decades and GCMs predict a complete loss of sea ice across the Arctic Ocean basin during summer by the end of this century (Vincent et al. 2001 and refs therein).

14.2.4 Water column transparency ($1/K_{dUV}$)

Colored dissolved organic matter (CDOM) is the primary attenuator of UV radiation in most aquatic environments. This still poorly defined mixture contains many substances, but the UV absorption is mostly due to aromatic polymers, notably humic and fulvic acids derived from terrestrial soils and vegetation. Alpine and polar lakes generally have very low concentrations of CDOM because of their sparse catchment vegetation, and the absorption coefficients for UV radiation per unit DOC may also be low because of the low relative importance of allochthonous versus aromatic, less UV-absorbing autochthonous carbon in these waters (Laurion et al. 2000). There is a non-linear relationship between UV-transparency and CDOM concentration, and between UV spectral ratios such as UVB/UVA (Laurion et al. 1997), and many high latitude and alpine waters are in a range that is responsive to small changes in CDOM. However the large arctic rivers and the coastal waters that they discharge into contain much higher CDOM concentrations because they are fed by catchments that extend far to the south, below the treeline. Antarctica has no major rivers and no well-developed terrestrial vegetation, and as a consequence there are striking differences in the attenuation of UV radiation in coastal Antarctic versus Arctic waters (Fig. 14.2). The relative lack of CDOM in Antarctic waters means that these ecosystems may be more sensitive to other factors controlling biological UV exposure such as changes in depth, ice and snow cover and atmospheric properties.

Climate change has the potential to exert a major influence on CDOM loading, with consequent impacts on underwater UV exposure. The mechanisms of effect include changes in catchment vegetation type and productivity, changes in precipitation and hydrology, and melting of permafrost soils and subsequent release of dissolved organic matter into the downstream receiving waters. For each of these processes, increasing atmospheric CO_2 and temperature are likely to result in increasing CDOM (e.g., Pastor et al. 2003; Freeman et al. 2004). This in turn will substantially release the biota from UV stress, although this positive effect may be

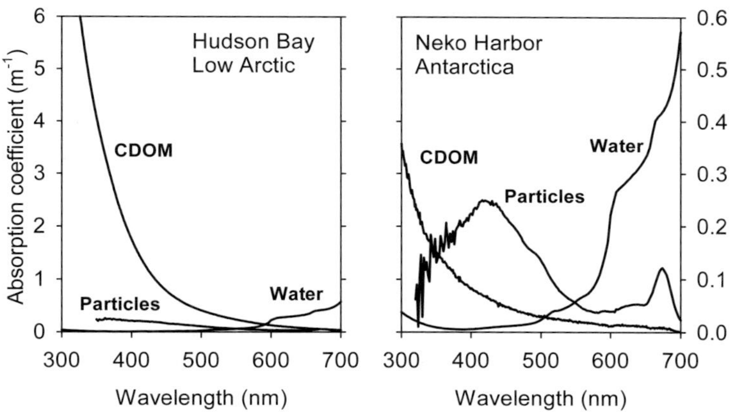

Fig. 14.2. CDOM influence in north versus south high latitude seas on underwater spectral radiation: Hudson Bay and Antarctica. Spectral absorption of water, particles and CDOM for Hudson Bay, 14 August 1999, and for Neko Harbor, Antarctic Peninsula, 3 January 2001.. Note the x10 difference in absorption scales (from Vincent and Belzile 2003).

offset by a reduced availability of PAR for photosynthesis (Arrigo and Brown 1996; Pienitz and Vincent 2000; Neale 2001; Lehmann et al. 2004).

There are steep spatial gradients in CDOM concentration across the Arctic Ocean. Coastal waters in some regions are strongly influenced by the CDOM inputs from large rivers. For example, the black waters of the Lena River have been observed up to 400 km offshore (>5 mg L^{-1} dissolved organic carbon at 4 m; Cauwet and Sidorov 1996) while in other regions, such as the North Water Polynya (NOW) between Greenland and Baffin Island, CDOM concentrations are extremely low and UV transparency is within the range reported for the Southern Ocean (Vincent and Belzile 2002). Major shifts in current patterns across the Arctic Ocean are likely to accompany any substantial change in climate. This in turn will lead to a redistribution of CDOM and large regional increases or decreases in underwater UV exposure. Large-scale shifts in the distribution of water masses have already been observed in the Arctic Ocean (Dickson 1999).

To evaluate the magnitude of spatial variability in north polar waters, T^* was calculated for an inshore water column at Hudson Bay (Canada) and for an offshore water column in the Arctic NOW polynya, both under ice-free conditions (Vincent and Belzile 2002). Moderate ozone depletion caused only a 5% increase in biological exposure weighted for UV-inhibition of photosynthesis (T^*_{PI}), whereas the difference in CDOM accounted for a 29-fold difference in underwater UV exposure between sites. For T^* weighted for DNA damage (T^*_{DNA}), moderate ozone depletion increased UV exposure by a factor of 2.3, but this was small relative to the 27-fold increase caused by differences in CDOM between sites.

In the lowest CDOM environments such as some alpine lakes (Laurion et al. 2000; Sommaruga 2001) and the Southern Ocean (Vincent and Belzile 2003), particles may also play a significant role in UV attenuation. Any climate-related effects that increase particle loading or the *in situ* growth of particles (bacteria and protists) therefore have the potential to affect underwater UV exposure. Some phytoplankton in alpine lakes contain UV-screening mycosporine-like amino acids (Sommaruga and Garcia-Pichel 1999), and stimulation of their growth by the nutrient enrichment accompanying even small rises in temperature (Sommaruga-Wögrath et al. 1997) would substantially reduce the UV transparency of the water column. Turbid glacial inputs are also known to decrease UV transparency, for example at Arctic Kongsfjord at Spitsbergen (Hanelt et al. 2001) and in the coastal Southern Ocean (Stambler et al. 1997). Climatic warming may therefore extend the magnitude and duration of these UV-attenuating effects through increased glacial melt.

14.2.5 Mixing and stratification (integral $1/K_{dUV}$)

Climate influences the depth, frequency and intensity of mixing of the surface layer of lakes and oceans, and this will directly affect UV exposure of planktonic organisms. Biota trapped at the surface during diurnal stratification will experience severe UV exposures, while deeper mixing communities will experience only intermittent, short-term UV stress. Studies on sub-arctic (Milot-Roy and Vincent 1994) and alpine lakes (Vincent et al. 1984) have shown the importance of diurnal thermoclines for UV-and PAR-dependent photoinhibition of algal photosynthesis. Diurnal stratification is likely to become more pronounced in a warmer climate, and may even select for more UV-tolerant genotypes. Deep mixing can eventually expose all of the water column plankton to surface UV, and if their rate of repair of the resultant damage is slower than the timescale of mixing then this deep circulation could potentially enhance the net water column UV damage (Neale et al. 1998b). Such effects may be especially pronounced in cold polar and alpine waters where repair is slowed by the ambient low temperatures (Rae et al. 2000). Most studies, however, show decreased UV-inhibition of total water column photosynthesis associated with deep mixing (Neale et al. 1998b; Barbieri et al. 2002). A modeling study by Huot et al. (2000) on bacterial DNA damage by UV, with repair by photoreactivation and excision, showed that variations in mixed layer depth affected the net damage in the mixed layer (and to a much greater extent than changes in CDOM concentration), and that the speed of mixing affected the vertical distribution of damage. They noted that the latter effect could be significant if biological processes such as survival, productivity and viral lysis respond non-linearly to net DNA damage.

Any future shift in high latitude and alpine lakes from cold monomixis (continuous mixing during summer) to dimixis (stratification during summer) will also result in communities exposed to much greater UV as well as PAR. There is evidence from a Finnish subarctic lake (Korhola et al. 2002) of this type of thermal shift during the 20th century. This was accompanied by an apparent increase in productivity and the appearance of cladocerans in the zooplankton community. These Finnish lake results imply that the positive effects of increased PAR on ecosystem productivity were much greater than the negative effects of increased UV exposure.

The thermal effects of global climate change are also likely to be felt more strongly in clear lakes in general. In an analysis of interannual variations in climate and lake thermal regimes, Snucins and Gunn (2000) found that the volume of cold water was significantly reduced in clear lakes during warm summers, while colored lakes (DOC > 4 mg L^{-1}) were less sensitive to the warming effects of climate change. They also suggested that clear lakes will experience large thermal regime shifts in response to small variations in DOC loading.

The Arctic Ocean has a much shallower mixed layer than the open waters of the Southern Ocean because of the strong freshwater influence, and any change in precipitation and runoff into the Arctic Basin could modify mixed layer depth and thus UV exposure. Similarly, coastal Antarctic seas are often stratified by meltwater input (e.g., Antarctic Peninsula: Mitchell and Holm-Hansen 1991; Terra Nova Bay: Arrigo et al. 2000) and changes in glacial runoff and sea ice melt would affect mixing and UV. The loss of sea ice in a warmer polar climate would drastically increase surface UV exposure, but this would be at least partially offset by the wind-induced mixing of biota throughout a deepened surface layer.

14.2.6 Biological weighting factors (ε)

Biological weighting factors (BWF) express the quantitative relationship between biological damage and UV wavelength, and generally increase sharply with decreasing wavelength. The exact form of this relationship differs greatly among different types of biological effects as well as with duration of exposure, the pre-acclimation characteristics of the biological community, species composition and a variety of other factors that are currently subject to discussion and ongoing research. For example, the modeling study by Gibson et al. (2000) for the Arctic Ocean indicated the combined effect of ozone depletion and changes in CDOM in controlling the extent of DNA damage, and the much greater influence of CDOM in controlling the variations in UV-inhibition of photosynthesis. This was because the BWF for UV-photoinhibition is strongly influenced by UV-A as well as UV-B (both controlled by CDOM) whereas DNA-damage rises much more steeply with decreasing UV wavelength and UV-A.

Climate change is likely to result in species shifts, that in turn affect the capacity of the phytoplankton to acclimate. For example, Neale et al. (1998a) found that the UV-tolerance of phytoplankton in the Weddell-Scotia Confluence was highest in assemblages from shallow mixed layers, suggesting acclimation and-or selection for tolerant genotypes. Warmer temperatures in the present changing climate may allow increased efficiency of repair, and thereby shift the BWF curves of all species.

14.3 Paleo-ecological evidence of climate-UV effects

Paleolimnological or paleoceanographic reconstructions of past irradiance regimes ('paleo-optics', Pienitz and Vincent 2000) can substantially contribute to a better understanding of the magnitude, causes and consequences of temporal variability in UV radiation. Using microfossil remains (siliceous diatoms), photo-protective pigments (e.g., scytonemin) or bulk sedimentary characteristics (% organic matter) preserved in long sedimentary sequences (cores), such retrospective analyses permit reconstructions of past concentrations of UV-absorbing dissolved organic carbon compounds (DOC) that can be used to infer water column transparency, term ($1/K_{dUV}$) in Eq. (2), and other ecological important optical properties such as UV/PAR ratios. The application of paleolimnological approaches to lakes at treeline show both that climatically-induced changes in the export of terrestrial (allochthonous) DOC to lakes are up to 100-fold more effective than modern stratospheric ozone depletion at altering biological exposure to UV (Pienitz and Vincent 2000), and that naturally occurring droughts can increase UV exposure in lakes by up to 10-fold (Yan et al. 1996). Analyses of fossil remains in arctic/boreal and alpine lakes suggest that historical variability in UV exposure has been high (e.g., Leavitt et al. 1997, 2003a,b; Pienitz and Vincent 2000; Saulnier-Talbot et al. 2003), and that lakes may have received twice as much UV prior to ca. 3000-4000 years ago than they do at present. Finally, when used in combination with long-term environmental monitoring (Schindler et al. 1996, 1997), historical reconstructions have proven valuable at identifying the importance of UV relative to other stressors in regulating lake structure and function. Thus, although further research is required to validate fossil interpretations, paleoecological analyses of lake sedimentary records can provide valuable insights into the history of UV exposure and its potential impacts on aquatic ecosystems.

Multivariate analysis of modern diatom assemblages preserved in surface lake sediments has been used to develop statistical models (transfer functions) based on the relationships between the present-day species composition and environmental conditions (e.g., Pienitz and Smol 1993; Fallu and Pienitz 1999; Fallu et al. 2005). Application of these models to fossil diatom assemblages in long sedimentary sequences (cores) allows quantitative reconstructions of past lake conditions over 100s to 1000s of years, including DOC content, water colour and optical regime (e.g., Pienitz and Vincent 2000; Dixit et al. 2001; Ponader et al. 2002; Saulnier-Talbot et al. 2003; Fallu et al. 2004). Further details on the calibration data set approach and standard procedures involved in the development of diatom-based inference models for DOC and other variables are given in Pienitz and Smol (1993) and Fallu and Pienitz (1999).

14.3.1 Lakes at the subarctic treeline

To address the potential impact of long-term climate change relative to that of ozone depletion, Pienitz and Vincent (2000) combined paleolimnological analyses with bio-optical models based on present-day conditions in lakes of northern Canada. This new paleo-optical approach allowed them to estimate past underwater light conditions from DOC concentrations that

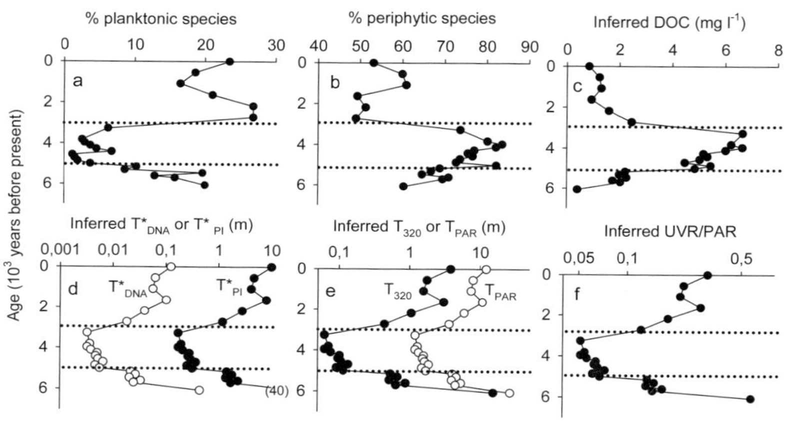

Fig. 14.3. Changes in fossil diatom community structure and inferred optical conditions in Queen's Lake, Northwest Territories (Canada). Diatom data are expressed as a percentage of the total number of valves in each sample associated with planktonic or benthic taxa. Diatom species data were used to infer DOC concentration (mg DOC L^{-1}), biologically weighted UV exposure (T^*_{PI} or T^*_{DNA}) and underwater spectral balance (water column transparency for 320 nm UVR [T_{320}], PAR [T_{PAR}], and the ratio between the two [UVR/PAR]) over the last 6000 years. The dotted lines delimit the period of mid-Holocene maximum forest cover. This analysis demonstrates that biotic exposure to UVR varies substantially due to changes in catchment vegetation and DOC supply, and that during the mid-Holocene climatic warm period, UVR exposure declined by two orders of magnitude. Climatic cooling beginning at ca. 3000 yr BP reduced DOC inputs by reducing soil development and DOC supply, and led to increases in UVR penetration that were up to 4000-fold more significant than those expected to arise from a moderate (30%) ozone depletion. From Pienitz and Vincent (2000)

were inferred from fossil diatom assemblages preserved in Holocene sedimentary deposits from a lake near the northern treeline (Queen's Lake; lat. 64°07'N, long. 110°34'W) in the central Northwest Territories, Canada (Fig. 14.3). Analysis of fossil pollen records indicates that regional vegetation cover was sparse and tundra-like following deglaciation ca. 8000 yr BP and persisted until trees colonized the catchment ca. 5000 years ago (MacDonald et al. 1993).

Diatom community structure and inferred DOC levels showed three distinct and abrupt changes during the history of Queen's Lake. Both diatom biomass and inferred DOC concentrations were low (< 2 mg DOC L^{-1}) during the initial lake phase, with particularly few fossils recovered from sediments dating prior to 5000 yr BP. This initial period of lake existence was followed by major and rapid shifts in fossil species composition and inferred chemical conditions ca. 5000 yr BP, with increased ratios of periphytic:planktonic taxa to >70% of the total diatom assemblage.

The period of forest advance corresponded to a major increase in algal production, recorded as the sediment mass-specific concentration of diatom valves, as well as a three-fold increase in inferred DOC levels. Based on fossil pollen analyses, Pienitz and Vincent (2000) argued that changes in lake chemistry and production resulted from climatic warming that stimulated treeline advance and increased forest density for about 2000 years. Finally, diatom-based reconstructions indicated that DOC concentrations declined >85% after 3000 yr BP, concomitant with the onset of Neoglacial climatic cooling and a southward retreat of treeline (MacDonald et al. 1993).

The large and rapid changes in DOC imply that Queen's Lake experienced major shifts in the underwater optical environment over the last 6000 years as a consequence of climate-induced variation in forest development (Pienitz et al. 1999). Consistent with this hypothesis, application of T^* calculations for UV-inhibition of photosynthesis and UV-damage of DNA showed that the inferred DOC shifts were equivalent to a two order-of-magnitude decrease in exposure to biologically-effective UV during the mid-Holocene vegetation maximum between ca. 5000 and 3000 yr BP. In contrast, the most recent 3000 years were characterised by a >50-fold increase in levels of damaging UV. Overall, changes in DOC concentrations arising from climatic variability increased exposure to photosynthetically damaging UV to an order-of-magnitude greater extent than a moderate (30%) decline in stratospheric ozone levels. Pienitz and Vincent (2000) noted, however, that the large photoinhibiting effect of UV irradiance after treeline retreat would be at least partially offset by the major increase in PAR for photosynthesis under these conditions of much reduced CDOM.

More recently, Lehmann et al. (2004) have modified and refined the biologically weighted transparency model by extending it out into the PAR waveband and by including separate terms for light-dependent photosynthesis and light-dependent inhibition of photosynthesis. This approach allows a quantitative estimate of water column photosynthesis ($\int P$) under different UV regimes via the relationship:

$$\int P = a\,(T_{PUR})^{b} ./\,(T_{PIR})^{c} \qquad (3)$$

where T_{PUR} is transparency weighted for photosynthetically utilisable radiation calculated over the waveband 400-700 nm; T_{PIR} is transparency weighted for photosynthesis-inhibiting radiation over the full waveband 280 to 700 nm; and a, b and c are statistically fitted constants. Lehmann et al. (2004) applied this model to the Queen's Lake paleolimnological data set and their results confirmed that decreasing CDOM concentrations during the retreat of the treeline would result in a significant increase in UV-inhibition of photosynthesis. However, this effect (as % reduction in $\int P$) was a factor of ten less than the T^* calculations predicted, in particular because the increasing water clarity also allowed greater exposure to PAR. This in turn resulted in more total water column photosynthesis that substantially offset the negative effects of increased biological UV exposure. These results have important implications for climate impact studies because they draw attention to the full spectral consequences of changing light regimes, and the need to examine the net effects of both positive and negative responses to environmental change.

14.3.2 Climate change effects on UV exposure in Rocky Mountain lakes

Climatic control of past UV exposure has also been identified as a key factor regulating lake production and algal community composition in lakes at the alpine treeline. Using fossil pigments and organic matter preserved in lake sediments, Leavitt et al. (2003b) quantified past UV exposure in the 12,000 year history of subalpine Crowfoot Lake, located near the alpine treeline in the Rocky Mountains (Alberta, Canada). Their study revealed highest concentrations of UV-absorbing algal pigments and declining algal biomass during periods of early lake existence following deglaciation as well as during the last ca. 4,000 years, when external DOM inputs are believed to have been lowest and UV exposure highest. Based on fossil pigment and bulk organic matter-derived estimates of irradiance penetration, they calculated that algal abundance was reduced 10 to 25-fold during these periods of high UV exposure. However, at all other times in the lake's history (ca. 10,050-4,000 yr BP), photoprotective pigments associated with high UV were absent from sediments and the total algal abundance was high.

14.3.3 Past UV exposure in Antarctica

The lakes at high latitudes are especially suitable for UV studies as they have limited or no terrestrial sources of UV-absorbing DOM within their catchments (Vincent and Pienitz 1996). Therefore changes in UV would be expected to arise solely from changes in solar production, atmospheric transmission, ice and snow cover or lake depth. Using a similar pigment approach as above, Leavitt el al. (2003a) showed that shallow lakes in Antarctica have been exposed to variable levels of UV of the Holocene, and higher levels in the preceding interglacial (Hodgson et al. 2005). Their analyses of lake sediments indicated that photo-protective pigments (scytonemin and its derivatives) were relatively more abundant in sediments older than 4000 years. The reconstruction further indicated that receipt of UV irradiance by benthic cyanobacteria in antarctic lakes has varied by at least 400% during the past 13 000 years. At present, photoprotective pigments, including scytonemin, are common in algal communities in antarctic lakes that are shallow (< 4-6 m depth) and transparent, and their abundance has been shown to increase as a function of algal exposure to UV (Hodgson et al. 2005).

14.3.4 Deglaciation responses in a coastal subarctic lake, Hudson Bay

Saulnier-Talbot et al. (2003) used a diatom-based paleo-optical approach to estimate past depths of UV penetration in coastal Lake Kachishayoot (northwestern Québec, Canada; lat. 55°20.0' N, long. 77°37.4'W; 102 m a.s.l.) following its isolation from the marine waters due to glacio-isostatic rebound of Hudson Bay lowlands. Prior comparisons of optical environments in coastal systems have revealed that shifts from marine to freshwater conditions are accompanied by increased DOC, changes in C-specific UV attenuation and declines in UV penetration (Conde et al. 2000). Consistent with these modern observations, Saulnier-Talbot et al's (2003) multi-proxy investigation revealed abrupt increases in diatom-inferred DOC concentrations and water color that coincided with the retreat of postglacial marine waters and the arrival of spruce trees within the landscape and catchment of the study site. Their investigation also revealed large changes in the underwater irradiance environment over the course of the postglacial period, from extremely high UV exposure following the initial formation of the lake and its isolation from the sea, to an order-of-magnitude lower exposure associated with the development of spruce forests in the catchment. Furthermore, the use of additional macrofossil markers revealed that UV penetration remained low even following forest retreat due to the development of alternate DOC sources in the catchment such as *Sphagnum* wetlands.

14.3.5 UV-exposure in lakes during glacial retreat at Glacier Bay, Alaska

A final paleo-ecological example is based on a set of modern day ecosystems, but environments that span a gradient of different ages and degrees of maturity in successional development. Major variations in CDOM and in underwater UV radiation have been observed in the chronosequence of lakes associated with the retreat of ice over the last 12,000 years in the Glacier Bay area, Alaska. Lakes near the glacier face and up to a few decades old have high pH and low DOC. With the gradual development of alder, coniferous forest and peatlands, the lakes have become more acidic, more dilute and have accumulated DOC associated with the slow but continuous accretion of soil organic matter. This soil development has been accompanied by a shift in hydrology from groundwater flow to increased near-surface runoff resulting in increased flow through the peat and weathered soil horizons, increasing the efficiency of transfer of organic matter from soil to water (Engstrom et al. 2000). Measurements of UV penetration in some of these lakes showed that 1% of surface UV at 320 nm was correlated with DOC content, and ranged from 0.6 m in a 90-year old coloured lake to more than 14 m in a 10-year old clear lake. These disparate optical conditions also corresponded with major differences in zooplankton community structure, and transplant experiments confirmed that UV exposure had a strong effect on zooplankton survival, but to an extent that varied according to species (Williamson et al. 2001). These results provide strong evidence that climate effects on vegetation and hydrology are likely to have a major influence on underwater UV exposure and associated impacts.

14.4 Conclusions

There are many mechanisms linking UV exposure to climate in the aquatic environment. Some are caused by climate-induced changes in the stratosphere, while many are associated with temperature-related effects on the main UV-attenuating components of natural waters, notably changes in snow and ice cover, and in CDOM associated with shifts in terrestrial vegetation. There are also a variety of hydrological effects of climate with major implications for underwater UV such as changes in snow and rainfall (with runoff, vegetation and albedo effects), evaporation and cloud cover, pathways of runoff and degree of interaction between the soil and water. Finally, there are more subtle effects such as changes in wind-induced mixing and climate-induced shifts in species composition towards more or less UV-tolerant species. Many of these effects are likely to be especially pronounced in arctic, antarctic and alpine aquatic ecosystems where there is a precarious balance between freezing and melting throughout summer, and in which CDOM concentrations are often in a range where small changes have a large effect on UV transparency. As shown with the many examples and models summarized here, several of these climate change mechanisms can have much greater impacts on biological UV exposure than the recently observed human-induced changes in stratospheric ozone. The paleo-ecological examples provide several lines of evidence for the overriding control exerted by external catchment processes, in turn mediated by climate, on the underwater optical conditions in lakes. All arctic investigations to date converge on the conclusion that biotic exposure to UV should have been greatest early in lake history immediately after deglaciation or retreat of postglacial seas, prior to the development of terrestrial sources of UV-absorbing DOM. Because past episodes of UV exposure appear to have been greater than those arising from many modern processes, these analyses offer insights (potentially at annual resolution) into the unique impacts of UV on ecosystem processes, including those occurring immediately after lake formation when DOM inputs are lowest. In the Antarctic, the near absence of UV-absorbing CDOM means that the absorption and transmission characteristics of ice, snow and the atmosphere are likely to be more important. Finally, to what extent will the future warming of polar and alpine regions result in UV-mediated changes in biological and biogeochemical processes in the aquatic environment? Higher temperatures will mean longer open water conditions (more UV) but also more vegetation and CDOM input. The mobilisation of DOC within alpine and northern permafrost soils has the potential to cause a massive influx of CDOM, and substantial increases in UV-shading. However, the beneficial effects of these changes may be offset by the concomitant decrease in availability of light in the visible range for photosynthetic production. The future assessment of climate impacts in polar and alpine regions will require close attention to the net result of these opposing effects.

Acknowledgements

This review was written while WFV was a visiting professor at the Universidad Autónoma de Madrid with support from Université Laval, Canada, and the Ministerio de Educación, Cultura y Deporte, Spain. Our research on UV and climate is funded by the Natural Sciences and Engineering Research Council of Canada; the Canada Research Chair program; Fonds de recherche sur la nature et les technologies, Québec; Centre d'Études Nordiques, Québec; and Indian and Northern Affairs Canada; with logistic support from the Polar Continental Shelf Project for field work in the Arctic. We thank Drs Antonio Quesada, Dominic Hodgson and Jon Børre Ørbæk for their review comments on the manuscript.

References

Arrigo KR, Brown CW (1996) Impact of chromophoric dissolved organic matter on UV inhibition of primary productivity in the sea. Mar Ecol Prog Ser 140:207-216

Arrigo KR, DiTullio GR, Dunbar RB, Robinson DH, Van Woert M, Worthen DL, Lizotte MP (2000) Phytoplankton taxonomic variability in nutrient utilization and primary production in the Ross Sea. J Geophys Res 105:8827-8846

Barbieri ES, Villafañe VE, Helbling EW (2002) Experimental assessment of UV effects on temperate marine phytoplankton when exposed to variable radiation regimes. Limnol Oceanogr 47:1648-1655

Belzile C, Gibson JAE, Vincent WF (2002) Colored dissolved organic matter and dissolved organic carbon exclusion from lake ice: implications for irradiance transmission and carbon cycling. Limnol Oceanogr 47:1283-1293

Belzile C, Vincent WF, Gibson JAE, Van Hove, P (2001) Bio-optical characteristics of the snow, ice and water column of a perennially ice-covered lake in the high Arctic. Can J Fish Aquat Sci 58:2405-2418

Belzile C, Johannessen SC, Gosselin M, Demers S, Miller WL (2000) Ultraviolet attenuation by dissolved and particulate constituents of first-year ice during late spring in an Arctic polynya. Limnol Oceanogr 45:1265-1273

Cauwet G, Sidorov I (1996) The biogeochemistry of Lena River: organic carbon and nutrients distribution. Marine Chemistry 53:211-227

Cockell C, Blaustein AR (eds.) (2001) Ecosystems, Evolution, and Ultraviolet Radiation. Springer, New York

Conde D, Aubriot L, Sommaruga R (2000) Changes in UV penetration associated with marine intrusions and freshwater discharge in a shallow coastal lagoon of the southern Atlantic Ocean. Mar Ecol Prog Ser 207:19-31

Dahlback A (2002) Recent changes in surface UV radiation and stratospheric ozone at a high arctic site In: Hessen DO (ed) UV Radiation and Arctic Ecosystems. Springer-Verlag, Berlin, pp 3-22

de Mora SJ, Demers S, Vernet M (2000) The Effects of UV Radiation in The Marine Environment. Cambridge University Press, United Kingdom
Dickson RR (1999) All change in the Arctic. Nature 397:389-91
Dixit SS, Keller W, Dixit, AS, Smol JP (2001) Diatom-inferred dissolved organic carbon reconstructions provide assessments of past UV-B penetration in Canadian Shield lakes. Can J Fish Aquat Sci 58:543-550
Engstrom DR, Fritz SC, Almendinger JE, Juggins S (2000) Chemical and biological trends during lake evolution in recently deglaciated terrain. Nature 408: 161-166
Fallu M-A, Pienitz R (1999) Diatomées lacustres de Jamésie-Hudsonie (Québec) et modèle de reconstitution des concentration de carbone organique dissous. Ecoscience 6:603-620
Fallu M-A, Allaire N, Pienitz R (2002) Distribution of freshwater diatoms in 64 Labrador (Canada) lakes: species-environment relationships along latitudinal gradients and reconstruction models for water colour and alkalinity. Can J Fish Aquat Sci 59:329-349
Fallu M-A, Pienitz R, Walker IR, Lavoie M (2005) Paleolimnology of a shrub-tundra lake and response of aquatic and terrestrial indicators to climatic change in arctic Québec, Canada. Palaeogeography, Palaeoclimatology, Palaeoecology 215:183-203.
Freeman C, Fenner N, Ostle NJ, Kang H, Dowrick DJ, Reynolds B, Lock MA, Sleep D, Hughes S, Hudson J (2004) Export of dissolved organic carbon from peatlands under elevated carbon dioxide levels. Nature 430:195-198
Gautier C, He G, Yang, S, Lubin D (1994) Role of clouds and ozone on spectral ultra violet-B radiation and biologically active UV dose over Antarctica. In: C.S. Weiler and P.A. Penhale (eds) Ultraviolet radiation in Antarctica: measurements and biological effects, American Geophysical Union. Antarctic Res Ser 62:83-91.
Gibson JAE, Vincent WF, Nieke B, Pienitz R (2000) Biological exposure to UV radiation in the Arctic Ocean: the relative importance of ozone versus riverine dissolved organic carbon. Arctic 53:372-382
Hanelt D, Tug H, Bischof K, Gross C, Lippert H, Sawall T, Wiencke C (2001) Light regime in an Arctic fiord: a study related to stratospheric ozone depletion as a basis for determination of UV effects on algal growth. Mar Biol 138:649-58
Hanssenbauer I, Forland EJ (1998) Long term trends in precipitation in the Norwegian Arctic: can they be explained by changes in atmospheric circulations patterns? Climate Res 10:143-153
Hessen DO (2002) UV Radiation and Arctic Ecosystems. Springer-Verlag, Berlin
Hodgson, DA, Vyverman W, Verleyen E, Sabbe K, Leavitt PR, Taton A, Squier AH, Keely BJK (2005) Environmental factors influencing the pigment composition of *in situ* benthic microbial communities in east Antarctic lakes. Aquat Microb Ecol 37:247-263
Hoffman JR, Hansen LJ, Klinger T (2004) Interactions between UV radiation and temperature limit inferences from single factor experiments. J Phycol 39:268-72

Houghton JT, Ding Y, Griggs DJ, Noguer M, van der Linden PJ, Dai X, Maskell K, Johnson CA (2001) Climate Change 2001: The Scientific Basis. (Contribution of Working Group I to the Third Assessment Report of the Intergovernmental Panel on Climate Change (IPCC)

Howard-Williams C, Schwarz A-M, Hawes I, Priscu JC (1998) Optical properties of McMurdo Dry Valleys lakes. Ant Res Ser 72:189-203

Huot Y, Jeffrey WH, Davis RF, Cullen JJ (2000) Damage to DNA in bacterioplankton: a model of damage by ultraviolet radiation and its repair as influenced by vertical mixing. Photochem Photobiol. 72:62-74

Kepner RL, Wharton RA, Collier RD, Cockell CS, Jeffrey WH (2000) UV radiation and potential biological effects beneath the perennial ice cover of an antarctic lake. Hydrobiologia 427:155-165

Korhola A, Sorvari S, Rautio M, Appleby PG, Dearing JA, Hu Y, Rose N, Lami A, Cameron NG (2002) A multiproxy analysis of climate impacts on the recent development of subarctic Lake Saanajärvi in Finnish Lapland. J Paleolimnol 28:59-77

Laurion I, Vincent WF, Lean DRS (1997) Underwater ultraviolet radiation: development of spectral models for northern high latitude lakes. Photochem Photobiol 65:107-114

Laurion I, Ventura M, Catalan J, Psenner R, Sommaruga R (2000) Attenuation of ultraviolet radiation in mountain lakes: Factors controlling the among- and within-lake variability. Limnol Oceanogr 45:1274-1288

Leavitt PR, Hodgson DA, Pienitz R (2003a) Past UVR environments and impacts on lakes. In: Helbling EW, Zagasese H (eds) UV effects in aquatic organisms and ecosystems. Comprehensive studies in photochemical and photobiological sciences. Royal Society of Chemistry Publishers, Cambridge, 509-545

Leavitt PR, Vinebrooke RD, Donald DB, Smol JP, Schinder DW (1997) Past ultraviolet radiation environments in lakes derived from fossil pigments. Nature 388:457-459

Leavitt PR, Cumming BF, Smol JP, Reasoner M, Pienitz R, Hodgson D (2003b) Climatic control of UV radiation impacts on lakes. Limnol Oceanogr 48:2062-2069.

Lehmann MK, Davis RF, Huot Y, Cullen JJ (2004) Spectrally weighted transparency in models of water-column photosynthesis and photoinhibition by ultraviolet radiation. Mar Ecol Prog Ser 269:101-110

MacDonald GM, Edwards TWD, Moser KA, Pienitz R, Smol JP (1993) Rapid response of treeline vegetation and lakes to past climate warming. Nature 361:243-246

MacFadyen EJ, Williamson CE, Grad G, Lowery M, Jeffrey WH, Mitchell DL (2004) Molecular response to climate change: temperature dependence of UV-induced DNA damage and repair in the freshwater crustacean *Daphnia pulicaria*. Global Change Biol. 10: 408-416.

Magnuson JJ, Robertson DM, Benson BJ, Wynne RH, Livingstone DM, Arai T, Assel RA, Barry RG, Card V, Kuusisto E, Granin NG, Prowse TD, Stewart KM, Vuglinski VS (2000) Historical trends in lake and river ice cover in the Northern Hemisphere. Science 289:1743-1746

Markager, S. and W.F. Vincent 2000. Spectral light attenuation and the absorption of UV and blue light in natural waters. Limnol. Oceanogr. 45: 642-650.

Milot-Roy V, Vincent WF (1994) Ultraviolet radiation effects on photosynthesis: the importance of near-surface thermoclines in a subarctic lake. Arch. Hydrobiol. 43:171-184

Mitchell BG, Holm-Hansen O (1991) Bio-optical properties of Antarctic Peninsula waters: differentiation from temperate ocean models. Deep-Sea Res 38:1009-1028

Morris DP, Zagarese H, Williamson CE, Balseiro EG, Hargreaves BR, Modenutti B, Moeller R, Queimalinos C (1995) The attenuation of solar UV radiation in lakes and the role of dissolved organic carbon. Limnol Oceanogr 40:1381-1391

Mueller DR, Vincent WF, Jeffries MO (2003) Break-up of the largest Arctic ice shelf and associated loss of an epishelf lake. Geophys Res Lett 30:10.1029/2003GL017931

Neale PJ (2001) Modeling the effects of ultraviolet radiation on estuarine phytoplankton production: impact of variations in exposure and sensitivity to inhibition. J Photochem Photobiol 62:1-8

Neale PJ, Cullen JJ, Davis RF (1998a) Inhibition of marine photosynthesis by ultraviolet radiation: Variable sensitivity of phytoplankton in the Weddell-Scotia Confluence during the austral spring. Limnol Oceanogr 43:433-448

Neale PJ, Cullen JJ, Davis RF (1998b) Interactive effects of ozone depletion and vertical mixing on photosynthesis of Antarctic phytoplankton. Nature 392:585-589

Pastor J, Solin J, Bridgham SD, Updegraff K, Harth C, Weishampel P, Dewey B (2003) Global warming and the export of dissolved organic carbon from boreal peatlands. Oikos 100:380-386

Perin S, Lean DRS (2004) The effects of ultraviolet-B radiation on freshwater ecosystems of the Arctic: Influence from stratospheric ozone depletion and climate change. Environ Rev 12:1-70

Perovich DK, Roesler CS, Pegau WS (1998) Variability in Arctic sea ice optical properties. J Geophys Res 103:1193-1208

Pienitz R, Smol JP (1993) Diatom assemblages and their relationship to environmental variables in lakes from the boreal forest-tundra ecotone near Yellowknife Northwest Territories Canada. Hydrobiologia 269/270:391-404

Pienitz R, Vincent WF (2000) Effect of climate change relative to ozone depletion on UV exposure in subarctic lakes. Nature 404:484-487

Pienitz R, Smol JP, MacDonald GM (1999) Paleolimnological reconstruction of Holocene climatic trends from two boreal treeline lakes, Northwest Territories, Canada. Arc. Antarct Alp Res 31:82-93

Ponader K, Pienitz R, Vincent WF, Gajewski K (2002) Limnological conditions in a subarctic lake (northern Québec, Canada) during the late Holocene: analyses based on fossil diatoms. J. Paleolimnol 27:353-366

Quayle, WC, Peck LS, Peat H, Ellis-Evans JC, Harrigan PR (2002) Extreme responses to climate change in Antarctic lakes. Science 295:645

Rae R, Howard-Williams C, Hawes I, Vincent WF (2000) Temperature-dependence of photosynthetic recovery from solar damage in Antarctic phytoplankton. In: Davison W, Howard-Williams C, Broady PA (eds) Antarctic Ecosystems: Models for Wider Ecological Understanding, SCAR VII Proceedings, The Caxton Press, Christchurch, New Zealand, p 183-189

Rautio M, Korhola A (2002) UV-induced pigmentation in subarctic *Daphnia*. Limnol Oceanogr 47:295-299

Renaud A, Staehelin J, Fröhlich C, Philipona R, Heimo A (2000) Influence of snow and clouds on erythemal UV radiation: Analysis of Swiss measurements and comparison with models. J Geophys Res 105:4961-4969

Saulnier-Talbot É, Pienitz R, Vincent WF (2003) Holocene lake succession and palaeo-optics of a subartic lake northern Québec (Canada). Holocene 13:517-526

Schindler DW, Curtis PJ, Parker BR, Stainton MP (1996) Consequences of climatic warming and lake acidification for UV-B penetration in North American boreal lakes. Nature 379:705-708

Schindler DW, Curtis PJ, Bailey SE, Parker BR, Beaty KG, Stainton MP (1997) Climate-induced changes in the dissolved organic carbon budget of boreal lakes. Biogeochemistry 36:9-28

Serreze MC, Maslanik JA, Scambos TA, Fetterer F, Stroeve J, Knowles K, Fowler C, Drobot S, Barry RG, Haran TM (2003) A record minimum arctic sea ice extent and area in 2002. Geophys Res Lett 30:10.1029/2002GL016406

Shindell D, Rind D, Lonergan P (1998) Increased stratospheric ozone losses and delayed eventual recovery owing to increasing greenhouse-gas concentrations. Nature 392:589-592

Snucins E, Gunn J (2000) Internannual variation in the thermal structure of clear and colored lakes. Limnol Oceanogr 1639-1646

Sommaruga R (2001) The role of solar UV radiation in the ecology of alpine lakes. J Photochem Photobiol B:Biol 62:35-42

Sommaruga R, Garcia-Pichel F (1999) UV-absorbing mycosporine-like compounds in planktonic and benthic organisms from a high-mountain lake. Arch Hydrobiol 144:255-269

Sommaruga-Wögrath S, Koinig K, Schmidt R, Sommaruga R, Tessadri R, Psenner R (1997) Temperature effects on the acidity of remote alpine lakes. Nature 387:64–67

Staehelin J, Harris NRP, Appenzeller C, Eberhard J (2001) Ozone trends: a review. Rev Geophysics 39:231-90

Stambler N, Lovengreen C, Tilzer MM (1997) The underwater light field in the Bellingshausen and Amundsen Seas (Antarctica). Hydrobiologia 344:41-56

Trodahl HJ, Buckley RG (1990) Enhanced ultraviolet transmission of Antarctic sea ice during the austral spring. Geophys Res Lett 17:2177-2179

Vincent WF, Pienitz R (1996). Sensitivity of high latitude freshwater ecosystems to global change: temperature and solar ultraviolet radiation. Geosci Can 23:231-236

Vincent WF, Belzile C (2002) UV effects on aquatic microbial communities in northern lakes and rivers. In: Hessen DO (ed) UV Radiation and Effects in the Arctic. Springer-Verlag, Berlin, pp137-155

Vincent WF, Belzile C (2003) Biological UV exposure in the polar oceans: Arctic-Antarctic comparisons. In: Huiskes AHL, Gieskes WWC, Rozema J, Schorno RML van der Vies SM, Wolff WJ (eds) Antarctic Biology in a Global Context. Backhuys Publishers, Leiden, The Netherlands, pp176-181.

Vincent WF, Neale PJ, Richardson PJ (1984) Photoinhibition: Algal responses to bright light during diel stratification and mixing in a tropical alpine lake. J Phycol 20:201-211

Vincent WF, Pienitz R, Laurion I (1998) Arctic and Antarctic lakes as optical indicators of global change. In: Budd WF (ed) Antarctica and Global Change. Annals Glaciol 27:691-696

Vincent WF, Gibson JAE, Jeffries MO (2001) Climate change, ice shelf collapse and habitat loss in the Canadian High Arctic. Polar Res 37:133-42

Williamson CE, Stemberger RS, Morris DP, Frost TM, Paulsen SG (1996) Ultraviolet radiation in North American lakes: Attenuation estimates from DOC measurements and implications for plankton communities. Limnol Oceanogr 41:1024-1034

Williamson CE, Olson OG, Lott SE, Walker ND, Engstrom DR, Hargreaves BR (2001) Ultraviolet radiation and zooplankton community structure following deglaciation in Glacier Bay Alaska. Ecology 82:1748-1760

Yan ND, Keller W, Scully NM, Lean DRS, Dillon PJ (1996) Increased UV-B penetration in a lake owing to drought-induced acidification. Nature 381:141-143

15 Effects of UV radiation on seaweeds

Dieter Hanelt[1], C. Wiencke[2] and K. Bischof[3]

[1] Biocenter Klein Flottbek, University of Hamburg, Ohnhorststr. 18, D-22609 Hamburg, Germany. [2] Foundation Alfred Wegener Institute for Polar and Marine Research, Am Handelshafen 12, D-27570 Bremerhaven, Germany. [3] Institute for Polar Ecology, University of Kiel, Wischhofstr. 1-3, D-24148 Kiel, Germany, e-mail: dieter.hanelt@botanik.uni-hamburg.de

15.1 Introduction

Depletion of stratospheric ozone over the Arctic region (Müller et al. 1997; Rex et al. 1997) causes an increase of UV-B radiation (280-315 nm) and could affect algal distribution patterns in the unique marine ecosystem (Wiencke et al. 2000). To date, most of the recent investigations on biological effects of UV radiation on marine ecosystems have concentrated on the Southern Ocean due to the huge "ozone hole" over Antarctica (Holm-Hansen et al. 1993a,b). Increasing ozone depletion over the northern Hemisphere (Blumthaler and Ambach 1990; Wängberg et al. 1996) may also have the potential to change growth conditions for macroalgae and other organisms, especially in Arctic waters. Therefore, in recent years several studies on the distribution and the physiology of several algal species growing in the Arctic environment have been undertaken to investigate the effects of ultraviolet radiation on seaweeds (e.g. Hanelt et al. 1997a; Bischof et al. 1998; Karsten et al. 1998, 1999). In coastal sea waters, UV radiation (UV) and blue light are strongly attenuated (Björn 1993), and this depends largely on the input of dissolved organic matter (DOM) from terrestrial ecosystems. This occurs often during the warmer seasons due to rainfall or melt water from snow layers and glaciers. Moreover, the absolute UV irradiation depends also on the clouds, solar altitude or sun angle which is much lower in the Arctic than at temperate zones. That means that already double of UV-B radiation is impinging on the earth surface in a temperate latitude 20° lower, which needs to be considered for discussing the UV problem in polar regions. However, in comparison to temperate species polar algae are more sensitive to UV.

In spring, coinciding with low temperatures and clear water conditions, the harmful UV wavelengths penetrate deeply into the water column in Kongsfjorden (Svalbard; Hanelt et al. 2001). The threshold irradiance with the potential to affect primary plant productivity negatively was still found at about 5-6 m depth. With increasing temperature in summer, snow layers and glacier ice melt, resulting in a high discharge of turbid fresh water into the fjord overlaying the more dense sea water. This causes a stratification in the optical features, salinity and temperature of the water body, strongly attenuating solar radiation in the first meter of the water column. Consequently, organisms in deeper water are more protected against harmful UV-B radiation. Nevertheless, organisms growing in the intertidal are still exposed to high UV-B during low tide and high sun position also in summer. The possible impact of global climate change on the radiation conditions underwater and the effects on primary production of seaweeds can be higher during the spring season, as organisms in the eulittoral and upper sublittoral zone are affected by UV radiation throughout the polar day. In addition, melt water input is only a phenomenon during the warmer summer season. This applies to Arctic shorelines in half-open fjord systems where the water exchange with the clearer oceanic water is retarded. At open coastlines the melt water will be exchanged much faster with oceanic water which will diminish the observed turbidity effects on light penetration.

15.2 Effects on the molecular level

Due to the high energy of UV-B and the special absorption characteristics of aromatic and sulphur containing aminoacids (e.g. tyrosine, tryptophan, sulfhydril group of cysteine and nucleic-acids, especially pyrimidines), or by production of free radicals within the chloroplast and their reaction with nucleophilic amino acid side chains, UV can induce severe damage on the cellular level (Bornman and Teramura 1993; Jordan 1996). On the molecular level, proteins (e.g. destruction of C-C bonds, or the tertiary protein structure due to S-S bond cleavage) and lipids (e.g. peroxidations at C=C bonds), as well as nucleic acids (especially cyclobutyl pyrimidine dimers and pyrimidine (6-4) pyrimidone photoproducts) are particularly affected. Conformation changes of certain proteins involved in photosynthesis, such as the reaction center protein (D_1) of photosystem II or the CO_2 fixing enzyme in the Calvin cycle (ribulose-1,5-bisphosphate carboxylase/oxygenase, RubisCO) lead to an inhibition of photosynthesis, and consequently to a decrease in the productivity of the plants which will be demonstrated below (for review on the effects see Holm-Hansen et al. 1993b; Bornman and Teramura 1993; Jordan 1996; Franklin and Forster 1997; Häder and Figueroa 1997). Moreover, different life stages in the life cycle of an individual species may differ in its sensitivity to UV radiation (Wiencke et al. 2000). The small zoospores can be easily penetrated by damaging UV radiation. However, several macroalgal species may be able to protect themselves against UV radiation by synthesis and accumulation of photoprotective substances, e.g. of mycosporine-like amino acids (MAAs; Karsten et al. 1998, 2003). The capacity for formation of MAAs induced by UV radiation has been shown in particular for shallow water red macroalgae (Karsten et al. 1999; Hoyer et al. 2002, 2003).

Animals feeding on MAA-rich algae might be protected against UV radiation (Dunlap and Shick 1998; Cockell and Knowland 1999). Adams and Shick (1996, 2001) demonstrated in a feeding experiment with the red alga *Mastocarpus stellatus* and the sea urchin *Strongylocentrotus droebachiensis* that the sea urchin eggs became less sensitive to UV if sea urchins were previously fed with MAA-rich algae. The division of the eggs, normally delayed by UV, occurs faster, when the eggs were loaded with MAAs. A possible protective effect of the MAAs as UV filter in macroalgae was demonstrated by Bischof et al. (2000b) by a comparison of the UV sensitivity of the two space competitive red algae *Mastocarpus stellatus* and *Chondrus crispus* from Helgoland. Quantum yield of photochemistry as well as maximal electron transport rate in *C. crispus* were more strongly affected by UV-B radiation than in *M. stellatus* (Fig.15.1). But no negative effects of the respective radiation conditions were found on total activity of RubisCO. Total MAA content in *M. stellatus* was up to six fold higher than in *C. crispus* and the composition of MAAs in the two species was also different. The results indicate that, among others, UV-B sensitivity may be a factor restricting *C. crispus* to the lower intertidal and upper sublittoral, whereas *M. stellatus* is better adapted to UV radiation and is therefore more competitive in the upper intertidal.

Besides the MAAs, which mainly were found in red and a few green algal species (Karsten et al. 1998) other molecular groups may prevent damage by the establishment of a physical barrier to shield critical cellular components (e.g. the photosynthetic apparatus, DNA) against harmful radiation (Beggs et al. 1986; Karentz 1994). Previous research on marine macroalgae has stressed the induction of different UV screening components in various algal groups.

In the Mediterranean alga *Dasycladus vermicularis* an unknown UV absorbing compound is excreted into the surrounding water under UV stress and protects photosynthesis and enzyme activity against harmful UV-B (Gómez et al. 1998). Generally, cellular concentrations of algal UV

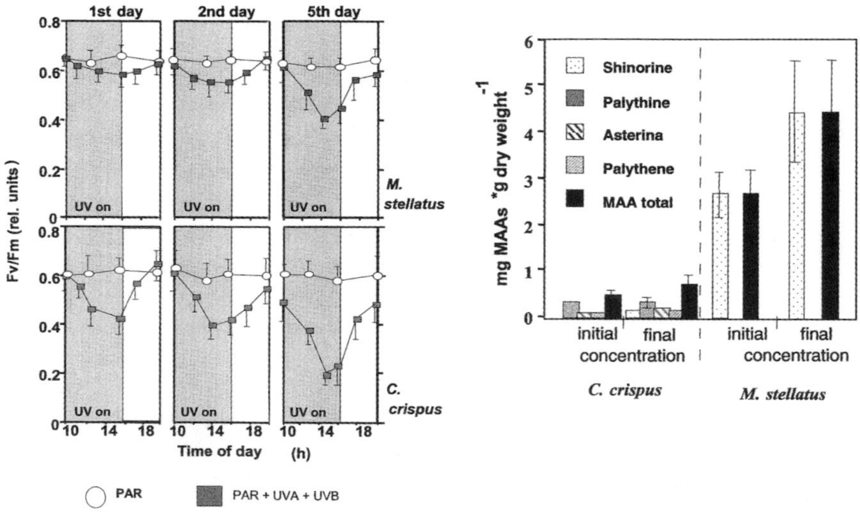

Fig. 15.1. Comparison of the photosynthetic activity and MAA content of two space competitive red algae *Mastocarpus stellatus* and *Chondrus crispus*. Left side: photosynthetic activity under UV stress was measured on five consecutive days by in vivo fluorescence (Fv/Fm). Additional to PAR, UV was irradiated always from 10 to 16 o'clock. The photosynthetic activity of *C. crispus* decreased much stronger under additional UV than in *M. stellatus*. One reason might be the different content of UV absorbing compounds (MAAs) within each species. The right side of the figure shows the content of different MAAs before and after the irradiation period. *C. crispus* possesses a higher diversity in these compounds but the concentration is much lower than in *M. stellatus*, which also means a lower UV absorption by these sunscreens (modified after Bischof et al. 2000b).

absorbing compounds seem to depend on the degree of exposure to UV, thus determined by the season, the algal position in the canopy and the respective growth depth. Wood (1987) described the UV induced production of unspecified UV absorbing compounds in the kelp *Ecklonia radiata* from Western Australia. More recent research indicate that especially in brown algae, phenolic compounds play a major role as UV sunscreens (Schoenwaelder et al. 2003). In *Ascophyllum nodosum* from the west coast of Sweden, Pavia et al. (1997) demonstrated that UV is an important factor related to intraspecific variation in phlorotannin content. In higher plants it is known that UV-B penetration through epidermal layers is much lower in conifer than in herbaceous species. It was concluded that the lower UV-B transmittance through the anticlinal cell walls is caused by the presence of more phenolic compounds (Jordan 1996). As brown algae contain a huge amount of phenolic substances, this may act as a possible UV-B screen in this algal class, comparable to the findings in higher plants.

Under UV stress there is an increased formation of reactive oxygen species such as oxygen radicals, which can induce an oxidation of many essential biomolecules susceptible to oxidative damage (Halliwell and Gutteridge 1989; Franklin and Forster 1997; Franklin et al. 2003). At a high reduction state of ferredoxin, due to high light exposure, electrons transferred from PS I to oxygen can generate superoxide radicals (similar to the Mehler reaction). Reactive oxygen species (ROS) would oxidize chlorophylls and proteins, for instance, the D_1 protein in the reaction center of PS II (Andersson et al. 1992). To prevent those high light damages, many marine organisms possess specific scavenging enzymes for detoxifying the damaging oxygen species (especially superoxide-dismutase, catalase, gluthation reductase, ascorbate peroxidase; Aguilera et al. 2002b; Dummermuth et al. 2003). However, the antioxidative enzyme system may also be inhibited under UV stress and the inhibition of the photoprotective xanthophyll cycle by UV-B may lead to an increase of the ROS production (Bischof et al. 2003). Beside these enzymological systems, various hydrophilic and lipophilic scavenging antioxidants are also present (especially vitamin E and C).

DNA damage can be repaired photoenzymatically in the presence of UV-A or blue light. This repair mechanism is known as photoreactivation or photoenzymatic repair, and it reverses the photodimer products (cyclobutane-pyrimidine dimers; CPDs) to its normal monomeric form. The dimers are cytotoxic because they block DNA and RNA polymerase and consequently inhibit genome replication and expression (Jordan 1996). Specific photolyases have been identified for thymine and 6-4 photoproducts. There is also another repair mechanism, which is independent of light, and it is called nucleotide excision repair or "dark repair". This mechanism, in which several enzymes are involved, consists of excising the damaged oligonucleotide and replacing it by a newly synthesized fragment. Whereas UV induced CPD formation can be rapidly repaired in light, the excision repair works in darkness but more slowly (Britt et al. 1993). CPD induction can exceed repair rates as sun radiation mediated CPD accumulation has been detected in phyto- and bacterioplankton and plants from a wide range of latitudes (Ballaré et al. 1996; Jeffrey et al. 1996a,b; Rousseaux et al. 1999; Boelen et al. 2000; Buma et al. 2000, 2001a, b; van de Poll et al. 2001). The accumulation of CPDs is accompanied by reduced growth rates in several phytoplankton and macroalgal species (Buma et al. 2000; van de Poll et al. 2001, 2002). The DNA damage burden in Arctic macrophytes mediated by ozone depletion may elevate and consequently depress productivity and alter species composition.

However, the ecological relevance of DNA damage as a stress factor for macrophytes is uncertain at high latitudes because relatively low irradiances of the DNA harming short wavelengths of UV-B reach the ground in the Arctic (Hanelt et al. 2001). Exclusion of UV radiation from the natural solar spectrum resulted in an elevated overall activity of RubisCO in *Ulva lactuca* (Bischof et al. 2002b), related to an increase in its cellular concentration. That means UV-B may cause an additional stress on the activity of the Calvin cycle, decreasing the CO_2 fixation rate. Among the photosynthetic pigments, lutein concentration was substantially elevated under UV exclusion and also the ratio of zeaxanthin concentration to the total xanthophyll content increased (Bischof et al. 2002b). Zeaxanthin is one of the important pigments which dissipate excess energy harmless by heat which will be further explained below. Therefore, these results indicate adverse effects of UV-B on the efficiency of photoprotection under high irradiances of PAR. The results confirm a marked impact already of present UV-B levels on macroalgal physiology under natural field conditions.

Marine animals living in the coastal ecosystem are often protected against UV radiation by UV impermeable shells or scales. Moreover, many animals are able to move vertically and thereby to withdraw from the influence of surface radiation (Obermüller et al. 2003). On the other hand, various animals are exposed to UV induced formation of toxic oxygen radicals and hydrogen peroxide, especially in places where - due to a dense algal vegetation or phytoplankton concentration - high oxygen concentrations are present. Hydrogen peroxide penetrates relatively easily through the body surface of soft-skinned animals and is therefore able to damage even organisms which are protected against UV radiation by their shells or the sediment tubes they inhabit (Abele-Oeschger and Oeschger 1995; Viarengo et al. 1998). As in algae, the relevant protecting mechanisms are antioxidants and antioxidative enzymes protecting the functional proteins and membrane lipids against photooxidative damage. By this way, the function of the cellular metabolism is generally maintained as long as the oxygen radical scavenging system is not overstressed.

15.2.1 Inhibition of the photosynthetic performance

Seaweed productivity depends strongly on the daily impinging photon fluences and this determines generally the lower distribution limit in the sublittoral (Gómez et al. 1997; Hanelt et al. 2003). For planktonic algae, generally a lower depth limit of 1% of the surface light is regarded as euphotic zone (Steemann Nielsen 1975). Brown kelps reach their lower growth limit at about 0.6 - 1.2 % of surface light, whereas for the deep growing red macroalgae a minimum of 0.001 - 0.05% was determined (Lüning 1985). Deepest crustaceous macroalgae seem to survive at an absolute light minimum of about 0.01 µmol $m^{-2}s^{-1}$ (Littler et al. 1986). In the intertidal and close below the water surface the survival of the organism depends, however, also on the capacity to cope with temporary high radiation stress. Sunny days can depress photosynthetic activity of marine macroalgae, especially in the intertidal and upper subtidal (Huppertz et al. 1990; Henley et al. 1991, 1992; Hanelt et al. 1993). First results of Larkum and Wood (1993) indicate that increasing UV levels of solar radiation also cause similar effects as observed under strong PAR. UV radiation depresses photosynthetic performance by impairing several target sites which absorb UV-B. It can be envisaged that an increase in the UV component of the light impinging on cells will increase their susceptibility to photodamage, and, hence, affect the kinetics of the recovery processes after high-light induced photoinhibition in the daily course (Hanelt et al. 1997a). Moreover, first data indicate also an effect of UV radiation on N-uptake in macroalgae (Döhler et al. 1995), a result which makes more intensive investigations necessary.

In the field, high irradiances of PAR are generally accompanied by harmful UV-B irradiation. Although the measurable effects (such as reduced photosynthetic efficiency) caused by both wavebands, are similar, the mechanisms behind PAR and UV induced inhibition of photosynthesis are very different. Therefore, both events should be carefully distinguished (Neale et al. 1993). To protect photosynthesis from high irradiances of PAR, plants activate different mechanisms. The increase of thermal energy dissipation removes excessively absorbed energy before generation of reactive oxygen species (ROS) can occur. The mechanism is proposed to include an increase in heat dissipation within the antennae, by the interconversion of violaxanthin to zeaxanthin within the xanthophyll cycle, which is also described for green and brown algae (Demmig-Adams 1990; Uhrmacher et al. 1995; Schofield et al. 1998). Zeaxanthin may act as a direct quencher of excited triplet chlorophyll, thus avoiding energy transfer to triplet oxygen (Frank et al. 1994). ROS may be scavenged by carotenoids and other antioxidants (as e.g. glutathione), or they may be enzymatically degraded by superoxide dismutase (SOD), catalase or peroxidase (Asada and Takahashi 1987; Barber and Andersson 1992). However, if PAR stress exceeds the protective capacity of the radical quenching enzymes and carotenoids, damage to the photosynthetic apparatus occurs, and plants bleach due to the photooxidation of their pigments (Björkman 1981; Krause 1988). This may be enhanced by the concomitant high UV irradiances as the down regulation of the photosynthetic activity due to dynamic photoinhibition seems to be negatively affected by UV-B. The recovery process from photoinhibition is delayed (Hanelt et al. 1997a) and the protective zeaxanthin content is diminished (Bischof et al. 2002b). Moreover, RubisCO activity is inhibited by UV-B exposure (Strid et al. 1990; Bischof et al. 2000a) and this is likely to be caused by protein degradation as well as inactivation of the enzyme.

Due to numerous effects of UV-B radiation to the respective biomolecules involved in photosynthesis, the effects of UV exposure are manifold (see Vass 1997 for review). The common consequences on photosynthetic function are decreased CO_2-fixation and oxygen evolution (Renger et al. 1986; Allen et al. 1997). This could be caused by several molecular events. While most studies have found that PS I is only minimally affected by UV-B (by inhibiting PS I-mediated cyclic photophosphorylation; Iwanzik et al. 1983; Renger et al. 1986), PS II seems to be a more important target (Bornman 1989; Jordan 1996). It is likely that UV-B causes an inhibition of energy transfer within the PS II reaction center by blocking the electron flow. This may be due to interactions with the electron transfer from phaeophytin to plastoquinone or by directly affecting the plastoquinones Q_A and Q_B, both showing a strong absorption within the UV-B region (Iwanzik et al. 1983). UV radiation may also cause structural modification of the Q_A and Q_B apoproteins. Furthermore, the function of the D_1 protein may be impaired by the UV-B induced fragmentation of the protein (Renger et al. 1989; Vass 1997). On the oxidizing side of PS II, the oxygen evolving system (water splitting complex) is another sensitive target of UV-B (Renger et al. 1989). Furthermore, it has been suggested that UV-B may affect the light-harvesting complex (LHC) by its functional disconnection from the photosystem, resulting in an impairment of energy transfer to the reaction center (Renger et al. 1986; Lorenz et al. 1997). In addition to the direct damage to PS II, structural disturbance of the membranes is likely to result in a reduced photosynthetic activity, e.g. due to dilation of the thylakoid membranes and rupture of the chloroplast double membrane (Iwanzik et al. 1983; Strid et al. 1994; Poppe et al. 2002, 2003). A decrease in photosynthetic activity may also be due to the photodestruction of pigments. Within the pigment group of chlorophylls, Chl *a* has been observed to be more affected than Chl *b* (Teramura 1983; Strid et al. 1990) and in brown algae Chl *c* seems to be more sensitive than Chl *a* (Roleda et al. 2004a). An additional effect of UV-B on reactions related to photosynthesis represents the inactivation of chloroplast ATPase (Strid et al. 1990). Impairment of any of the above-mentioned components can contribute to lower the photosynthetic activity during and following UV exposure.

Recently, the gene regulation of the CO_2 fixing enzyme RubisCO has been shown to be another critical component in UV induced inhibition of photosynthesis. The UV-induced decline in its activity is related to the decreasing amount of both subunits as well as the corresponding mRNA levels (Strid et al. 1990; Jordan et al. 1992; Bischof et al. 2000a). Transcripts encoding several other photosynthesis-related proteins were significantly reduced by UV-B radiation (Strid et al. 1994; Jordan 1996; Mackerness et al. 1999). These authors demonstrated that reactive oxygen species are a component in the signal transduction pathway leading to the down-regulation of gene expression in response to UV-B radiation. In contrast, cyanobacteria react to moderate doses of UV-B radiation with increased levels of psbA transcripts encoding the D_1 reaction center subunit of photosystem II (Campbell et al. 1998; Máté et al. 1998). We investigated firstly the effects of UV radiation on the regulation of photosynthetic proteins in marine macroalgae. The chloroplast-encoded psbA and rbcL RNA transcript levels, encoding for D_1 protein of the PS II reaction center and the large subunit of RubisCO, have been determined in Northern blot experiments, in samples exposed to higher UV-B irradiance (0.97 W m^{-2}) over the course of 11 days. psbA showed a rise of up to 400 % of the control after 11 days of UV exposure (Fig. 15.2). rbcL transcripts followed by an increase to 180 % of the control level. These results suggest that UV radiation affects a transcriptional regulation of proteins involved in photosynthesis. A stronger destruction of the D_1 protein due to UV-B induced chronic photoinhibition and a decrease of the RubisCO content may induce a higher transcription rate of the encoding genes for higher synthesis of both proteins to counteract the negative effects of UV-B.

In contrast to PAR, UV can not be regarded as being an "excessive energy input" in a proper sense. The maximal irradiance is much smaller than of PAR (1.8 Wm^{-2} UV-B compared to 368 Wm^{-2} PAR at 70°S in Antarctica or 1.2 Wm^{-2} UV-B compared to 274 Wm^{-2} PAR at 79° N on Spitsbergen; Hoyer et al. 2001; Svendsen et al. 2002). Thus, the UV wavebands do not contribute specifically to energy supply for photosynthetic chemistry.

As emphasized in this chapter, UV exhibits adverse effects on photosynthesis in a more direct way, such as its absorption by many aromatic and sulfhydril containing biomolecules leading to a direct molecular damage or by indirect evolution of reactive oxygen species. The high potential of UV to inhibit photosynthesis was firstly demonstrated by Jones and Kok (1966). Whilst within the PAR range, the action spectrum of photoinhibition runs in parallel with the action spectrum of photosynthesis (Nultsch et al. 1987; Hanelt et al. 1995), and therefore is directly related to photosynthetic pigment absorption, UV-B induced inhibition correlates much

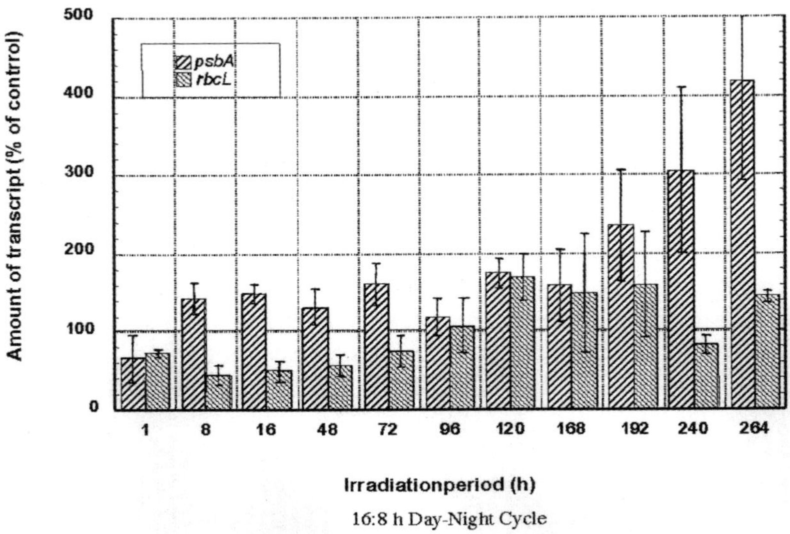

Fig. 15.2. *Palmaria decipiens*: Effect of UV-B on the amount of the mRNA encoding for the D_1 protein (psbA) and the large subunit of RubisCO (rbcL). Samples (n=5 ± SD) for RNA analysis were taken at the end of the respective exposure period (PAR + UV-A,B).

more to the absorption by DNA and proteins, conversely (Jones and Kok 1966; Setlow 1974).

15.3 Effects at the cellular level

UV-B radiation causes clear effects in cellular ultrastructure, especially in chloroplasts and mitochondria of marine macroalgae. In red algae from the Antarctic (*Palmaria decipiens*, *Phycodrys austrogeorgica)* and the Arctic (*Palmaria palmata*, *Bangia atropurpurea)* chloroplast thylakoids formed abnormal vesicles upon exposure to artificial UV radiation (Poppe et al. 2002, 2003). In *P. decipiens*, most vesicles were developed after 8 hours, in the less UV sensitive *P. palmata* after 48 hours of UV exposure. In *B. atropurpurea*, a strong vesiculation of thylakoids was observed after 72 hours of UV irradiation. In the highly UV sensitive red alga *Ph. austrogeorgica*, chloroplast envelope and thylakoid membranes were already strongly damaged and the phycobilisomes became detached from the thylakoids after 12 hours of UV exposure. UV induced changes in the membrane structure of mitochondria were observed in *P. decipiens* and *P. palmata*. However, in *P. decipiens* changes in membrane structure were reversible as was the damage in chloroplast fine structure after 12 hours of UV treatment. Protein crystals occurring in the cytoplasma of *Ph. austrogeorgica* showed a degradation after UV radiation. These findings gave first insight into the fine structural changes during and after UV exposure and confirmed the former results on inhibition of the photosynthetic performance by UV.

15.4 Effects on the early settlement stages

Early settlement stages are normally shade-adapted, as they live sheltered under adult canopies, crevices or even below stones. There is a high mortality in early post-settlement stages, and the specimens which can survive represent a gene pool for the subsequent generation (Coelho et al. 2000). The environment causes a conflict between maximum photosynthetic efficiency under low light conditions and a minimum of inhibition or even photodamage during possible exposures in high light periods. A large number of propagules are normally produced by macroalgae which disperse away from the parent population and colonize new sites. The settlement stages of subtidal macroalgae show, however, rather low light adaptations (e.g. low saturation point for photosynthesis and growth) which enable their survival in a low light environment, than under high light conditions (Lüning and Neushul 1978; Kain and Norton 1990; Franklin and Forster 1997). In many polar species growth rates and propagation are highest in spring, when the ozone destruction also occurs (Wiencke 1990a, b) and within a period when early settlement stages or young thallus parts are more sensitive to UV (Wood 1987; Dring et al. 1996). Spores seem to be the most light-sensitive life history stage found in the studied brown algae and are strongly affected by increased UV-B radiation, both with respect to their photosynthetic performance and their susceptibility to DNA damage (Wiencke et al. 2000). As already mentioned above, Schoenwaelder et al. (2003) demonstrated that zygotes and embryos of *Fucus spiralis* were less sensitive to UV-B than those of *F. serratus* and *F. distichus*. *F. spiralis* grows on the rocky shore at high light exposed positions in the upper intertidal whereas *F. serratus* is a species growing in the middle and lower intertidal, thus, is more protected against UV radiation. This is similar to *F. distichus* which grows only at higher latitudes under lower UV-B irradiances. In a recent study, Wiencke (2000) found that zoospores of several species of brown algae from Spitsbergen and southern Spain were damaged to a different degree. Their vulnerability was directly re-were damaged to a different degree. Their vulnerability was directly related to the depth distribution of their parental sporophytes, e.g. spores mortality of Laminariales is significantly caused in Spain when exposed to the radiation conditions in depths < 5m, restricting the growth to the middle and deep sublittoral. These results are supported in a subsequent study on all five kelp species from Arctic Spitsbergen (Wiencke et al. 2004). One prominent cytological feature of UV exposed spores from these species was the enlargement of phenolic vesicles (physodes), particularly seen in *Saccorhiza dermatodea* and *Alaria esculenta*, which may have a protective function against UV.

Carpospore photosynthesis and germination of the intertidal red algae *Mastocarpus stellatus* and *Chondrus crispus* are low light adapted. Roleda et al. (2004b) demonstrated that UV-A has low impact on the reproduction units of both species while a combined effect of UV-A plus UV-B with increasing exposure time is more pronounced in *C. crispus*. Photosynthetic recovery of *M. stellatus* progresses faster in carpospores when the algae were previously exposed to UV, suggesting a positive effect of UV on recovery processes. Dose-response relationship (BED_{50}) showed that *C. crispus* is more sensitive to UV as higher fluences were necessary to achieve 50% germination inhibition in *M. stellatus*. This corresponds to an efficient DNA damage repair capacity in this species. However, photosynthesis of young gametophytes was completely acclimated to UV after only three days and mature thalli of both species showed no DNA damage. The susceptibility of carpospores to UV was therefore regarded as an important feature which determines recruitment capacity in the upper intertidal apart from other factors already described above (Bischof et al. 2000b).

UV also affects the motility of unicellular algae or swarmers, e.g. for phytoplankton Ekelund (1990, 1994) reported effects on growth and speed of some dinoflagellates and diatom species. Motility of the flagellate *Euglena gracilis* was shown to be very sensitive to exposure to UV (Häder and Häder 1988). Photomovement of macroalgal zoospores is also affected. Flores-Moya et al. (2002) found that the response of photomovement of swarmers in the brown algae *Scytosiphon lomentaria* and *Petalonia fascia* declined linearly with the logarithm of the applied UV fluence. The zoospores of *Laminaria saccharina* ceased their movement after treatment with UV-B, although the spores retained their viability (Makarov and Voskoboinikov 2001). In addition to motility, this study showed that UV-B also affects the spore release and germination of *L. sacharina* negatively. A consequence could be, if the swarmers are the sexual gametes, fertilization is affected under UV-B and the life history may be blocked to a certain extent. But if they are non-sexual gametes at least recruitment of new thalli by asexual reproduction can be diminished as was demonstrated by Wiencke et al. (2000, 2004). According to Dring et al. (1996) the sensitivity of *Laminaria* species to UV is higher in the young sporophytes than in the older ones, and gametophytes are most sensitive. Already moderate photosynthetic active radiation (PAR) can reduce growth of the eulittoral germlings of red algae (Leukart and Lüning 1994) not considering that red algae are often already quite sensitive to UV.

Most studies on the adverse effects of UV, however, have been conducted on mature sporophytes, neglecting that during growth and development macroalgae undergo unicellular life history stages. It was previously stated by Lüning (1980) that depth distribution patterns of large kelps (Laminariales) reflect light requirements of their establishment stages (zoospores, germlings), but hitherto the question whether early developmental stages of macroalgae are more susceptible to UV than large sporophytes has received little attention (Dring et al. 1996; Hanelt et al. 1997b; Yakovleva et al. 1998). Houvinen et al. (2000) demonstrated the high susceptibility to UV in zoospores and gametophytes of *Macrocystis pyrifera*. They found that UV exposure is most dramatically affecting recruitment when interfering with DNA synthesis. Wiencke et al. (2000) showed a direct relationship between DNA damage and mortality of brown alga zoospores. Apart from that it is likely that UV has the ability to impair the motility of brown algal zoospores during their planktonic phase when they swim actively due to the action of two flagella as mentioned above. Altamirano et al. (2003) investigated the relationship between temperature and UV radiation with germlings of different *Fucus* species. High temperature in combination with high fluences of UV-B caused generally the death of the germlings, whereas at low temperature the rate of survival increased. All these manifold results showed that it is absolutely necessary to measure the effects of increased UV levels on the sensitivity of the early settlement stages when considering the future consequences for the marine ecosystem.

15.5 UV effects on growth

Growth-related studies have mostly been performed in the laboratory (Dring et al. 1996; Grobe and Murphy 1997; van de Poll et al. 2001, 2002) and natural sunlight was used only in a few long-term experiments (Grobe and Murphy 1998; Aguilera et al. 1999; Altamirano et al. 2000a, b; Pang et al. 2001; Michler et al. 2002). With a few exceptions, only a low number of species has been included in the studies so far, reflecting the lack of knowledge about individual responses of a broad variety of species. Generally, reduced growth activity may be caused by an inhibition of energy generating processes, such as inactivation or degradation of any component of the photosynthetic process or respiration. Also, an alteration of energy consuming processes, such as pigment synthesis, which are involved in protection mechanisms of the plant, could operate at the expense of growth (Altamirano et al. 2000b). Furthermore, UV radiation is assumed to influence cell division and cell elongation (Grobe and Murphy 1997). UV related damage on a molecular level could also directly affect genes responsible for growth. In a recent laboratory study on different marine red macrophytes, a reduction in growth was mostly attributed to the accumulation of DNA damage (van de Poll et al. 2001). However, growth is considered to be an integrative parameter (Altamirano et al. 2000b) since it integrates stress at multiple levels. All negative effects on physiological processes such as nutrient uptake, protein synthesis, photosynthesis, DNA repair or transcription would decrease the ability of a plant to use light energy and nutrients for primary production and growth. Therefore, growth is a useful indicator to assess UV effects in long-term experiments and should be used more often in combination with other parameters. Despite the high significance of growth data, only few results on UV effects on macroalgal growth of Arctic species are hitherto published. This applies especially to *in situ* experiments. Changes of *in vivo* fluorescence of photosystem II of macroalgae were not always consistent with the measurable changes in growth rate, indicating that physiological processes leading to an inhibition of growth may act independently of changes in photosynthetic activity (Dring et al. 1996; Michler et al. 2002). In numerous studies on higher plants and phytoplankton it was demonstrated that UV-B exposure has the potential to seriously affect growth, and, thus, reduce primary production rates (Holm-Hansen et al. 1989; Björn et al. 1999; McMinn et al. 1999). Macroalgae grown under laboratory conditions and exposed to artificial UV are strongly impaired in terms of gaining biomass. This was shown in a variety of species from the cold temperate to the tropical zone (Dring et al. 1996; Han 1996 a,b). In studies on the cold temperate red alga *Chondrus crispus* it was confirmed that UV induced reduction in growth rate was accompanied by a relatively small reduction in maximal quantum yield of photosynthesis and detectable levels of thymine dimer formation of the DNA (van de Poll et al. 2003). This shows, that reduction in growth might be due to the impairment of several physiological processes and not only by lowered photosynthetic activity. Grobe and Murphy (1994) found strong UV-B induced reduction in growth of the chlorophyte *Ulva expansa* from the Californian coast line and it was concluded that it is cell division which is inhibited by UV-B rather than cell expansion. On Spitsbergen, Aguilera et al. (1999) transplanted algae from deep to shallow waters. The samples exposed to the full solar spectrum mostly exhibit growth rate reductions related to its respective depth distribution in the field. Compared to specimens exposed to PAR only, UV-A also showed an adverse effect on algal growth rates, but was not as severe as UV-B. The only one of the investigated species without significant UV effects on growth was the intertidal brown alga *Fucus distichus*. Apart from these experiments, Aguilera et al. (1999) performed *in situ* growth measurements in summer under turbid water conditions. However, at their natural growth site, the investigated species of the *Laminariaceae* did not show any difference in growth rates due to natural UV. These results are in line with data from the same location obtained by Brouwer et al. (2000) who incubated older sporophytes of *Laminaria saccharina* for 3 weeks at 1 m water depth. As result, specimens did not show significant differences in growth rates when receiving the whole solar spectrum compared to those shielded from UV. Under the present radiation levels, UV impacts on growth of adult Arctic macroalgae might occur only under conditions of shallow water or during periods of extreme water transparency e.g. in Arctic spring after the break up of sea ice. Therefore, it remains open whether increased UV-B will seriously affect growth of the investigated older sporophytes in the field. However, reduced growth rates were also observed at ambient UV-B levels in a variety of other macrophytes (Grobe and Murphy 1994; Aguilera et al. 1999; Franklin et al. 1999; Makarov 1999; Altamirano et al. 2000a,b; Kuhlenkamp et al. 2001; Pang et al. 2001; Michler et al. 2002).

In conclusion, polar species show inhibition of growth, at least the young sporophytes, which is related to the natural vertical zonation pattern. However, polar species are exposed to solar radiation in their natural environment only for a limited time during the year (Kirst and Wiencke 1995; Hanelt et al. 2001), and adaptation time to higher UV irradiation due to thinning of the ozone layer could be a crucial aspect concerning tolerance towards UV. In contrast to polar species, species from mid-latitudes such as from the rocky shore of Helgoland, showed no clear relationship between vulnerability towards UV and depth distribution (Michler at al. 2002). Due to very low transmission of UV through waters around Helgoland (Lüning and Dring 1979; Franklin et al. 1999), the zonation of sublittoral red macroalgae may be also not affected by higher UV-B irradiation. However, observed sensitivities in species from the lower eulittoral such as *Codium fragile* or *Chaetomorpha* sp. indicate that changing UV radiation conditions have the potential to influence the structure and productivity of macroalgal communities in the mid latitudes. Apparently, long term UV effects are more evident on growth of sporophytes than on their photosynthetic processes. In order to elucidate ecological consequences of ozone depletion, future studies should concentrate more on the integrative parameters such as growth and reproduction.

15.6 Acclimation to seasonal increase of UV-B irradiances

Changes in UV-B irradiances are not only caused by anthropogenic effects or meteorological reasons but also occur in the normal course of the season, e.g. breakup of sea ice generate a strong increase of irradiance including the harmful UV-B waveband (Hanelt et al. 2001). Photosynthesis of certain macroalgae species are able to adapt to changes in the radiation environment within a few days (Bischof et al. 1999; Roleda et al. 2004b). However as discussed above, that does not necessarily induce concomitantly also an adaptation in their growth rate (Michler et al. 2002; Roleda et al. 2004b). Ability to acclimate and adjust photosynthesis and growth to the rapid changes in light climate must be one prerequisite for macroalgal life under strong seasonal changes in the Arctic. Increasing light availability after the break up of sea ice is linked to the increase of maximum photosynthetic rates in several Arctic species (Bischof et al. 2002a). This acclimation is reversible as soon as underwater radiation conditions are diminished due to decreasing solar irradiance and increasing water turbidity. A high ability to adjust photosynthesis to the respective radiation conditions may be due to a multitude of specific regulatory events including e.g. a de- and acceleration of dark reactions (Scheibe 1990; Schmid and Dring 1996), an adjustment of the absorption cross section (Hanelt and Nultsch 1991), antennae size and number of active reaction centers (Dau 1994) as well as pigment composition (Gómez and Wiencke 1997). Moreover, seasonal changes in pigment composition, a substantial increase in the concentration of UV absorbing mycosporine-like amino acids (MAAs), the antioxidative activities of the enzymes SOD and catalase and other biochemical defense systems are related to seasonal changes in irradiances as well as variation in nutrient levels (Aguilera et al. 1999, 2002a,b). This enables a flexible response of the algae to a highly variable marine environment.

15.7 Conclusions

Organisms living in the littoral ecosystem have to a certain extent developed the capability to adapt to an increasing stress factor, e.g. the increasing UV radiation. From the studies conducted so far, it can be concluded that the presently measured UV levels are not likely to seriously affect the macroalgal communities in Arctic coastal ecosystems. Although photosynthesis and growth were shown to be sensitive to UV exposure when algae were sampled or transplanted from their original location, all species studied exhibited sufficient acclimation potential to the respective variations in radiation conditions at their specific growth site. However, in combination with other stress factors, e.g. global warming, increased storm frequencies or pollution the impact on the plant will increase or may even become exponential so that the organism is strongly harmed and the species might become endangered. As a result, this will affect not only the plant community but also the grazers and carnivores which adopt a higher position in the respective biotic pyramid. In future, more attention has to be directed to the small developmental stages of the macroalgae. Zonation pattern depends mostly on this life history stage with the highest susceptibility to UV. Due to the tiny size or the delicate structure of the early developmental stages, UV penetrates easily and can produce strong damage to the young organisms. Due to the central role of macroalgae as primary producers at the Arctic coast line these investigations will provide important information on the function of this unique ecosystem.

References

Abele-Oeschger D, Oeschger R (1995) Enzymatic antioxidant protection in spawn, larvae and adult worms of *Phyllodoce mucosa* (Polychaeta). Ophelia 43: 101-110

Adams NL, Shick JM (1996) Mycosporine-like amino acids provide protection against ultraviolet radiation in eggs of the green sea urchin *Strongylocentrotus droebachiensis*. Photochem Photobiol 64: 149-158

Adams NL, Shick JM (2001) Mycosporine-like amino acids prevent UVB-induced abnormalities during early development of the green sea urchin *Strongylocentrotus droebachiensis*. Mar Biol 138: 267-280

Aguilera J, Karsten U, Lippert H, Vögele B, Philipp E, Hanelt D, Wiencke C (1999) Effects of solar radiation on growth, photosynthesis and respiration of marine macroalgae from the Arctic. Mar Ecol Prog Ser 191: 109-119

Aguilera J, Bischof K, Karsten U, Hanelt D, Wiencke C (2002a) Seasonal variation in ecophysiological patterns in macroalgae from an Arctic fjord: II. Pigment accumulation and biochemical defence systems. Mar Biol 140: 1087-1095

Aguilera J, Dummermuth A, Karsten U, Schriek R, Wiencke C (2002b) Enzymatic defences against photooxidative stress induced by ultraviolet radiation in Arctic marine macroalgae. Polar Biol 25: 432-441

Allen DJ, McKee IF, Farage PK, Baker NR (1997) Analysis of limitations to CO_2 assimilation on exposure of leaves of *Brassica napus* cultivars to UV-B. Plant Cell Environ 20: 633-640

Altamirano M, Flores-Moya A, Figueroa FL (2000a) Long-term effects of natural sunlight under various ultraviolet radiation conditions on growth and photosynthesis of intertidal *Ulva rigida* (Chlorophyceae) cultivated *in situ*. Bot Mar 43: 19-126

Altamirano M, Flores-Moya A, Figueroa FL (2000b) Growth seasonality, photosynthetic pigments, and C and N content in relation to environmental factors: a field study on *Ulva olivascens* (Ulvales, Chlorophyta). Phycologia 39: 50-58

Altamirano M, Flores-Moya A, Figueroa FL (2003) Effects of UV radiation and temperature on growth of germlings of three species of *Fucus* (Phaeophyceae). Aquatic Bot 75: 9-20

Andersson B, Salter AH, Virgin I, Vass I, Styring S (1992) Photodamage to photosystem II - primary and secondary events. J Photochem Photobiol B: Biol 15: 15-31

Asada K, Takahashi M (1987) Production and scavenging of active oxygen in photosynthesis. In: Kyle DJ, Osmond CB, Arntzen CJ (eds) Photoinhibition. Topics in Photosynthesis 9, Elsevier Science Publishers, Amsterdam, pp 89-109

Ballaré CL, Scopel AL, Stapleton AE, Yanovsky MJ (1996) Solar ultraviolet-B radiation affects seedling emergence, DNA integrity, plant morphology, growth rate, and attractiveness to herbivore insects in *Datura ferox*. Plant Physiol 112: 161-170

Barber J, Andersson B (1992) Too much of a good thing: light can be bad for photosynthesis. Trends Biochem Science 17: 61-66

Beggs CJ, Schneider-Ziebert U, Wellmann E (1986) UV-B radiation and adaptive mechanisms in plants. In: Worrest RC, Caldwell MM (eds) Stratospheric ozone reduction, solar ultraviolet radiation and plant life. NATO ASI Series, vol G8, Springer, New York, pp 235-250

Bischof K, Hanelt D, Tüg H, Karsten U, Brouwer PEM, Wiencke C (1998) Acclimation of brown algal photosynthesis to ultraviolet radiation in Arctic coastal waters (Spitsbergen, Norway). Polar Biol 20: 388-395

Bischof K, Hanelt D, Wiencke C (1999) Acclimation of maximal quantum yield of photosynthesis in the brown alga *Alaria esculenta* under high light and UV radiation. Plant Biol 1: 435-444

Bischof K, Hanelt D, Wiencke C (2000a) Effects of ultraviolet radiation on photosynthesis and related enzyme reactions of marine macroalgae. Planta 211: 555-562

Bischof K, Kräbs G, Hanelt D, Wiencke C (2000b) Photosynthetic characteristics and mycosporine-like amino acids under UV-radiation: a competitive advantage of *Mastocarpus stellatus* over *Chondrus crispus* at the Helgoland shore line? Helgol Mar Res 54: 47-52

Bischof K, Hanelt D, Aguilera J, Karsten U, Vögele B, Sawall T, Wiencke C (2002a) Seasonal variation in ecophysiological patterns in macroalgae from an Arctic fjord: I. Sensitivity of photosynthesis to ultraviolet radiation. Mar Biol 140: 1097-1106

Bischof K, Kräbs G, Wiencke C, Hanelt D (2002b) Solar ultraviolet radiation affects the activity of ribulose-1,5-bisphosphate carboxylase-oxygenase and the composition of photosynthetic and xanthophyll cycle pigments in the intertidal green alga *Ulva lactuca* L. Planta 215: 502-509

Bischof K, Janknegt PJ, Buma AGJ, Rijstenbil JW, Peralta G, Breeman AM (2003) Oxidative stress and enzymatic scavenging of superoxide radicals induced by solar UV-B radiation in *Ulva* canopies from southern Spain. Sci Mar 67: 353-359

Björkman O (1981) Responses to different quantum flux densities. In: Encyclopaedia of plant physiology, N. S. 12 A, pp 57-107

Björn LO (1993) Light. Technical report, Dept. of Plant. Physiol. University of Lund

Björn LO, Callaghan TV, Gehrke C, Johanson U, Sonesson M (1999) Ozone depletion, ultraviolet radiation and plant life. Chemosphere: Global Change Sci 1: 449-454

Blumthaler M, Ambach W (1990) Indication of increasing solar ultraviolet radiation flux in alpine regions. Science 248: 206-208

Boelen P, Boer MK de, Kraay GW, Veldhuis MJW, Buma AGJ (2000) UVBR-induced DNA damage in natural marine picoplankton assemblages in the tropical Atlantic ocean. Mar Ecol Prog Ser 193: 1-9

Bornman JF (1989) Target sites of UV-radiation in photosynthesis of higher plants. J Photochem Photobiol B: Biol 4: 145-158

Bornman JF, Teramura AH (1993) Effects of ultraviolet-B radiation on terrestrial plants. In: Young AR, Björn LO, Moan J, Nultsch W (eds) Environmental UV photobiology, Plenum Press, New York, pp 427-471

Britt AB, Chen JJ, Wykoff D, Mitchell D (1993) A UV-sensitive mutant of *Arabidopsis* defective in the repair of pyrimidine-pyrimidinone (6-4) dimers. Science 261: 1571-1574

Brouwer PEM, Bischof K, Hanelt D, Kromkamp J (2000) Photosynthesis of two Arctic macroalgae under different ambient radiation levels and their sensitivity to enhanced UV radiation. Polar Biol 23: 257-264

Buma AGJ, Oijen T van, Poll WH van de, Veldhuis MJW, Gieskes WWC (2000) The sensitivity of *Emiliania huxleyi* (Prymnesiophyceae) to ultraviolet-B radiation. J Phycol 36: 296-303

Buma AGJ, Boer MK de, Boelen P (2001a) Depth distributions of DNA damage in Antarctic marine phyto- and bacterioplankton exposed to summertime UV radiation. J Phycol 37: 200-208

Buma AGJ, Helbling EW, Boer MK de, Villafane VE (2001b) Patterns of DNA damage and photoinhibition in temperate South-Atlantic picophytoplankton exposed to solar ultraviolet radiation. J Photochem Photobiol B: Biol 62: 9-18

Campbell D, Eriksson MJ, Öquist G, Gustafsson P, Clarke AK (1998) The cyanobacterium *Synechococcus* resists UV-B by exchanging photosystem II reaction-center D_1 proteins. Proc Nat Acad Sci 95: 364-369

Cockell CS, Knowland J (1999) Ultraviolet radiation screening compounds. Biol Rev 74: 311-345

Coelho SM, Rijstenbil JW, Brown MT (2000) Impacts of anthropogenic stresses on the early development stages of seaweeds. J Aquat Ecosyst Stress Recovery 7: 317-333

Dau H (1994) Short-term adaptation of plants to changing light intensities and its relation to photosystem II photochemistry and fluorescence emission. J Photochem Photobiol B: Biol 26: 3-27

Demmig-Adams B (1990) Carotenoids and photoprotection in plants: a role for the xanthophyll zeaxanthin. Biochim Biophys Acta 1020: 1-24

Döhler G, Hagmeier E, David C (1995) Effects of solar and artificial UV irradiation on pigments and assimilation of ^{15}N ammonium and ^{15}N nitrate by macroalgae. J Photochem Photobiol B: Biol 30: 179-187

Dring MJ, Makarov V, Schoschina E, Lorenz M, Lüning K (1996) Influence of ultraviolet-radiation on chlorophyll fluorescence and growth in different life-history stages of three species of *Laminaria* (phaeophyta). Mar Biol 126: 183-191

Dummermuth AL, Karsten U, Fisch KM., König M, Wiencke C (2003) Responses of marine macroalgae to hydrogen-peroxide-stress. J Exp Mar Biol Ecol 289: 103-121

Dunlap WC, Shick JM (1998) Ultraviolet radiation-absorbing mycosporine-like amino acids in coral reef organisms: a biochemical and environmental perspective. J Phycol l31: 418-430

Ekelund NGA (1990) Effects of UV-B radiation on growth and motility of four phytoplankton species. Physiol Plant 78: 590-594

Ekelund NGA (1994) Influence of UV-B radiation on photosynthetic light-response curves, absorption spectra and motility of four phytoplankton species. Physiol Plant 91: 696-702

Flores-Moya A, Posudin YI, Fernández JA, Figueroa FL, Kawai H (2002) Photomovement of the swarmers of the brown algae *Scytosiphon lomentaria* and *Petalonia fascia*: Effect of photon irradiance, spectral composition and UV dose. J Photochem Photobiol B 66: 134-140

Frank HA, Cua A, Chynwat V, Young A, Gosztola D, Wasielewski MR (1994) Photophysics of the carotenoids associated with the xanthophyll cycle in photosynthesis. Photosynth Res 41: 389-395

Franklin LA, Forster RM (1997) The changing irradiance environment: consequences for marine macrophyte physiology, productivity and ecology. Eur J Phycol 32: 207-232

Franklin LA, Yakovleva I, Karsten U, Lüning K (1999) Synthesis of mycosporine-like amino acids in *Chondrus crispus* (Florideophyceae) and the consequences for sensitivity to ultraviolet B radiation. J Phycol 35: 682-693

Franklin LA, Osmond BC, Larkum AWD (2003) Photoinhibition, UV-B and algal photosynthesis. In: Larkum AW, Douglas SE, Raven JA (eds) Photosynthesis in Algae, Kluwer Academic Publisher, Dordrecht, pp 351-384

Gómez I, Wiencke C (1997) Seasonal growth and photosynthetic perfomance of the Antarctic macroalga *Desmarestia menziesii* (Phaeophyceae) cultivated under fluctuating Antarctic daylengths. Bot Acta 110: 25-31

Gómez I, Weykam G, Klöser H, Wiencke C (1997) Photosynthetic light requirements, daily carbon balance and zonation of sublittoral macroalgae from King George Island (Antarctica) Mar Ecol Prog Ser 148: 281-293

Gómez I, Perez-Rodriguez E, Viñegla B, Figueroa FL, Karsten U (1998) Effects of solar radiation on photosynthesis, UV-absorbing compounds and enzyme activities of the green alga *Dasycladus vermicularis* from Southern Spain. J Photochem Photobiol B: Biol 47: 46-57

Grobe CW, Murphy TM (1994) Inhibition of growth of *Ulva expansa* (Chlorophyta) by ultraviolet-B radiation. J Phycol 30: 783-790

Grobe CW, Murphy TM (1997) Artificial ultraviolet-B radiation and cell expansion in the intertidal alga *Ulva expansa* (Setch.) S. and G. (Chlorophyta). J Exp Mar Biol Ecol 217: 209-223

Grobe CW, Murphy TM (1998) Solar ultraviolet-B radiation effects on growth and pigment composition of the intertidal alga *Ulva expansa* (Setch.) S. and G. (Chlorophyta). J Exp Mar Biol Ecol 225: 39-51

Häder DP, Figueroa FL (1997) Photoecophysiology of marine macroalgae. Photochem Photobiol 66: 1-14

Häder DP, Häder M (1988) Inhibition of motility and phototaxis in the green flagellate, *Euglena gracilis*, by UV-B radiation. Arch Microbiol 32: 903-906

Halliwell B, Gutteridge JMC (1989) Free radicals in biology and medicine. Clarendon Press, Oxford

Han T (1996a) Effect of ultraviolet-B radiation on *Ulva pertusa* Kjellman (Chlorophyta) I. Growth and pigment content. Algae (Kor J Phycol) 11: 155-159

Han T (1996b) Far-UV action on growth, pigmentation and photosynthesis of *Ulva lactuca* (Chlorophyta). Algae (Kor J Phycol) 11: 203-206

Hanelt D, Nultsch W (1991) The role of chromatophore arrangement in protecting the chromatophores of the brown alga *Dictyota dichotoma* against photodamage. J Plant Physiol 138: 470-475

Hanelt D, Huppertz K, Nultsch W (1993) Daily course of photosynthesis and photoinhibition in marine macroalgae investigated in the laboratory and field. Mar Ecol Prog Ser 97: 31-37

Hanelt D, Uhrmacher S, Nultsch W (1995) The effect of photoinhibition on photosynthetic oxygen production in the brown alga *Dictyota dichotoma*. Bot Acta 108: 99-105

Hanelt D, Wiencke C, Nultsch W (1997a) Influence of UV-radiation on the photosynthesis of Arctic macroalgae in the field. J Photochem Photobiol B: Biol 38: 40-47

Hanelt D, Wiencke C, Karsten U, Nultsch W (1997b) Photoinhibition and recovery after high light stress in different developmental and life-history stages of *Laminaria saccharina* (Phaeophyta). J Phycol 33: 387-395

Hanelt D, Tüg H, Bischof K, Gross C, Lippert H, Sawall T, Karsten U, Wiencke C (2001) Light regime in an Arctic fjord: a study related to stratospheric ozone depletion as a basis for determination of UV effects on algal growth. Mar Biol 138: 649-658

Hanelt D, Wiencke C, Bischof K (2003) Photosynthesis in marine macroalgae. In: Larkum AW, Douglas SE, Raven JA (eds) Photosynthesis in Algae, Kluwer Academic Publisher, Dordrecht, pp 413-435

Henley WJ, Levavasseur G, Franklin LA, Lindley ST, Ramus J, Osmond CB (1991) Diurnal responses of photosynthesis and fluorescence in *Ulva rotundata* acclimated to sun and shade outdoor culture. Mar Ecol Prog Ser 75: 19-28

Henley WJ, Lindley ST, Levavasseur G, Osmond CB, Ramus J (1992) Photosynthetic response of *Ulva rotundata* to light and temperature during emersion on an intertidal sand flat. Oecologia 89: 516-523

Holm-Hansen O, Mitchell BG, Vernet M (1989) Ultraviolet radiation in Antarctic waters: Effect on rates of primary production. Antarct J US 24: 177-178

Holm-Hansen O, Helbling EW, Lubin D (1993a) Ultraviolet radiation in Antarctica: inhibition of primary production. Photochem Photobiol 58: 567-570

Holm-Hansen O, Lubin D, Helbling EW (1993b) Ultraviolet radiation and its effects on organisms in aquatic environments. In: Young AR, Björn LO, Moan J, Nultsch W (eds) Environmental UV Photobiology, Plenum Press, New York, pp 379-425

Houvinen PS, Oikari AOJ, Soimasuo MR, Cherr GN (2000) Impact of UV radiation on the early development of the giant kelp (*Macrocystis pyrifera*) gametophytes. Photochem Photobiol 72: 303-313

Hoyer K, Karsten U, Sawall T, Wiencke C (2001) Photoprotective substances in Antarctic macroalgae and their variation with respect to depth distribution, different tissues, and developmental stages. Mar Ecol Prog Ser 211: 117-129

Hoyer K, Karsten U, Wiencke C (2002) Induction of sunscreen compounds in Antarctic macroalgae by different radiation conditions. Mar. Biol. 41: 619-627

Hoyer K, Karsten U, Wiencke C (2003) Inventory and factors controlling the content of UV-absorbing mycosporine-like amino acids in polar macroalgae. In: Huiskes AHL, Gieskes WWC, Rozema J, Schorno RML, van der Vies SM, Wolff WJ (eds) Antarctic Biology in a Global Context, Backhuys Publishers, Leiden, pp 56-62

Huppertz K, Hanelt D, Nultsch W (1990) Photoinhibition of photosynthesis in the marine brown alga *Fucus serratus* as studied in field experiments. Mar Ecol Prog Ser 66: 175-182

Iwanzik W, Tevini M, Dohnt G, Voss M, Weiss W, Gräber O, Renger G (1983) Action of UV-B radiation on photosynthetic primary reactions in spinach chloroplasts. Physiol Plant 58: 401-407

Jeffrey WH, Aas P, Lyons MM, Coffin RB Pledger RJ, Mitchell DL (1996a) Ambient solar radiation-induced photodamage in marine bacterioplankton. Photochem Photobiol 64: 419-427

Jeffrey WH, Pledger RJ, Aas P, Hager S, Coffin RB, von Haven R, Mitchell DL (1996b) Diel and depth profiles of DNA photodamage in bacterioplankton exposed to ambient solar ultraviolet radiation. Mar Ecol Progr Ser 137: 283-291

Jones LW, Kok B (1966) Photoinhibition of chloroplast reactions. I. Kinetics and action spectra. Plant Physiol 41: 1037-1043

Jordan BR (1996) The effects of UV-B radiation on plants: a molecular prospective. In: Callo JA (ed) Advances in botanical research, Academic Press, New York, pp 97-162

Jordan BR, He J, Chow WS, Anderson JM (1992) Changes in mRNA levels and polypeptide subunits of ribulose-1,5-bisphosphate carboxylase in response to supplementary ultraviolet-B radiation. Plant Cell Envir 15: 91-98

Kain J, Norton T (1990) Marine ecology. In: Cole K, Sheath R (eds) Biology of the red algae. Cambridge University Press, pp 377-422

Karentz D (1994) Ultraviolet tolerance mechanisms in Antarctic marine organisms. In: Ultraviolet radiation in Antarctica: measurements and biological effects. Antarctic Res Ser 62: 93-110

Karsten U, Sawall T, Hanelt D, Bischof K, Figueroa FL, Flores- Moya A, Wiencke C (1998) An inventory of UV-absorbing mycosporine-Like amino acids in macroalgae from polar to warm-temperate regions. Bot Mar 41: 443-453

Karsten U, Bischof K, Hanelt D, Tüg H, Wiencke C (1999) The effect of UV-radiation on photosynthesis and UV-absorbing substances in the endemic Arctic macroalga *Devaleraea ramentacea* (Rhodophyta). Physiol Plant 105: 58-66

Karsten U, Dummermuth A, Hoyer K, Wiencke C (2003) Interactive effects of ultraviolet radiation and salinity on the ecophysiology of two Arctic red algae from shallow waters. Polar Biol 26: 249-258

Kirst GO, Wiencke C (1995) Ecophysiology of polar algae. J Phycol 31: 181-199

Krause GH (1988) Photoinhibition of photosynthesis. An evaluation of damaging and protective mechanisms. Physiol Plant 74: 566-574

Krause GH, Weis E (1991) Chlorophyll fluorescence and photosynthesis: the basics. Annu Rev Plant Physiol Plant Mol Biol 42: 313-349

Kuhlenkamp R, Franklin LA, Lüning K (2001) Effect of solar ultraviolet radiation on growth in the marine macroalga *Dictyota dichotoma* (Phaeophyceae) at Helgoland and its ecological consequences. Helgol Mar Res 55: 77-86

Larkum AWD, Wood WF (1993) The effect of UV-B radiation on photosynthesis and respiration of phytoplankton, benthic macroalgae and seagrasses. Photosyn Res 36: 17-23

Leukart P, Lüning K (1994) Minimum spectral light requirements and maximum light levels for long-term germling growth of several red algae from different water depths and a green alga. Eur J Phycol 29: 103-112

Littler MM, Littler DS, Blair SM, Norris JM (1986) Deep water plant communities from an uncharted seamount off San Salvador Island, Bahamas: distribution, abundance and primary productivity. Deep Sea Res 33: 881-892

Lorenz M, Schubert H, Forster RM (1997) *In vitro-* and *in vivo*-effects of ultraviolet-B radiation on the energy transfer in phycobilisomes. Photosynthetica 33: 517-527

Lüning K, Neushul M (1978) Light and temperature demands for growth and reproduction of laminarian gametophytes in southern and central California. Mar Biol 45: 297-309

Lüning K, Dring MJ (1979) Continuous underwater light measurements near Helgoland (North Sea) and its significance for characteristic light limits in the sublittoral region. Helgol Wiss Meeresunters 32: 403-424

Lüning K (1980) Critical levels of light and temperature regulating the gametogenesis of three *Laminaria* species (Phaeophyceae). J Phycol 16: 1-15

Lüning K (1985) Meeresbotanik. Georg Thieme Verlag, Stuttgart, Germany

Makarov MV (1999) Influence of ultraviolet radiation on the growth of the dominant macroalgae of the Barents Sea. Chemosphere: Global Change Sci 1: 461-467

Makarov MV, Voskoboinikov GM (2001) The influence of ultraviolet-B radiation on spore release and growth of the kelp *Laminaria saccharina*. Bot Mar 44: 89-94

Mackerness S A-H, Jordan BR, Thomas B (1999) Reactive oxygen species in the regulation of photosynthetic genes by ultraviolet-B radiation (UV-B: 280-320 nm) in green and etiolated buds of pea (*Pisum sativum* L.) J Photochem Photobiol B: Biol 48: 180-188

Máté Z, Sass L, Szekeres M, Vass I, Nagy F (1998) UV-B induced differential transcription of psbA genes encoding the D1 protein of photosystem II in the cyanobacterium *Synechocystis* 6803. J Biol Chem 273: 17439-17444

McMinn A, Ashworth C, Ryan K (1999) Growth and productivity of Antarctic sea ice algae under PAR and UV irradiances. Bot Mar 42: 401-407

Michler T, Aguilera J, Hanelt D, Bischof K, Wiencke C (2002) Long-term effects of ultraviolet radiation on growth and photosynthetic performance of polar and cold-temperate macroalgae. Mar Biol 140: 1117-1127

Müller R, Crutzen PJ, Grooß JU, Brühl C, Russel JM, Gernandt H, Mc Kenna DS, Tuck AF (1997) Severe ozone loss in the Arctic during the winter of 1995-96. Nature 389: 709-712

Neale PJ, Cullen JJ, Lesser MP, Melis A (1993) Physiological bases for detecting and predicting photoinhibition of aquatic photosynthesis by PAR and UV radiation. In: Yamamoto HY and Smith CM (eds) Photosynthetic responses to the environment, American Society of Plant Physiologists, Rockville, Maryland, pp 61-77

Nultsch W, Pfau J, Materna-Weide M (1987) Fluence and wavelength dependence of photoinhibition in the brown alga *Dictyota dichotoma*. Mar Ecol Prog Ser 41: 93-97

Obermüller B, Karsten U, Pörtner HO, Abele D (2003) Effects of UV-radiation on oxidative stress parameters and uptake of mycosporine-like amino acids (MAAs) in polar marine amphipods. In: Huiskes AHL, Gieskes WWC, Rozema RML, Schorno SM, Vies SM, Wolff WJ (eds) Antarctic Biology in a Global Context, Backhuys Publishers, Leiden, The Netherlands, pp 63-68

Pang S, Gómez I, Lüning K (2001) The red macroalga *Delesseria sanguinea* as a UVB-sensitive model organism: selective growth reduction by UVB in outdoor experiments and rapid recording of growth rate during and after UV pulses. Eur J Phycol 36: 207-216

Pavia H, Cervin G, Lindgren A, Aberg P (1997) Effects of UV-B radiation and simulated herbivory on phlorotannins in the brown alga *Ascophyllum nodosum*. Mar Ecol Prog Ser 157: 139-146

Poppe F, Hanelt D, Wiencke C (2002) Changes in ultrastructure, photosynthetic activity and pigments in the Antarctic red alga *Palmaria decipiens* during acclimation to UV radiation. Bot Mar 45: 253-261

Poppe F, Schmidt RAM, Hanelt D, Wiencke C (2003) Effects of UV radiation on the ultrastructure of several red algae. Phycol Res 51: 11-19

Renger G, Voss M, Gräber P, Schulze A (1986) Effect of UV irradiation on differential partial reactions of the primary processes of photosynthesis. In: Worrest RC, Caldwell MM (eds) Stratospheric ozone reduction, solar ultraviolet radiation and plant life. NATO ASI Series, Vol. G8; Springer, Heidelberg, pp 171-184

Renger G, Völker HJ, Eckert J, Fromme R, Hohm-Veit S, Gräber P (1989) On the mechanisms of photosystem II deterioration by UV-B irradiation. Photochem Photobiol 49: 97-105

Rex M, Harris NRP, von der Gathen P, Lehmann R, Braathen GO, Reimer E, Beck A, Chipperfield MP, Alfier R, Allaart M, o´Connor F, Dier H, Dorokhov V, Fast H, Gil M, Kyrö E, Litynska Z, Mikkelsen IS, Molyneux MG, Nakane H, Notholt J, Rummukainen M, Viatte P, Wenger J (1997) Prolonged stratospheric ozone loss in the 1995-96 Arctic winter. Nature 389: 835-838

Roleda MY, Hanelt D, Kräbs G, Wiencke C (2004a) Tissue damage, growth and photosynthetic parameters in *Laminaria ochroleuca* (Laminariales, Phaeophyta) under UV radiation, Phycologia (in press).

Roleda MY, van de Poll W, Hanelt D, Wiencke C (2004b) Impact of UVB-induced DNA damage on photosynthesis, spore viability and growth of *Mastocarpus stellatus* and *Chondrus crispus*: Implication on recruitment and zonation pattern. Mar Ecol Prog Ser (in press)

Rousseaux MC, Ballaré CL, Giordano CV, Scopel AL, Zima AM, Szwarchberg-Bracchitta M, Searles PS, Caldwell MM, Diaz SB (1999) Ozone depletion and UVB radiation: impact on plant DNA damage in southern South America. Proc Nat Acad Sci USA 96: 15310-15315

Scheibe R (1990) Light/dark modulation: regulation of chloroplast metabolism in a new light. Bot Acta 103: 327-334

Schmid R, Dring MJ (1996) Blue light and carbon acquisition in brown algae: an overview and recent developments. Sci Mar 60: 115-124

Schoenwaelder MEA, Wiencke C, Clayton MN, Glombitza KW (2003) The effect of elevated UV radiation on *Fucus* spp. (Fucales, Phaeophyta) zygote and embryo development. Plant Biol 5: 366-377

Schofield O, Evens TJ, Millie DF (1998) Photosystem II quantum yields and xanthophyll-cycle pigments of the macroalga *Sargassum natans* (Phaeophyceae): responses under natural sunlight. J Phycol 34: 104-112

Setlow RB (1974) The wavelengths in solar radiation effective in producing skin cancer: a theoretical analysis. Proc Nat Acad Sci USA 71: 3363-3366

Steemann Nielsen E (1975) Marine photosynthesis. Elsevier, Amsterdam, The Netherlands

Strid A, Chow WS, Anderson JM (1990) Effects of supplementary ultraviolet-B radiation on photosynthesis in *Pisum sativum*. Biochim Biophys Acta 1020: 260-268

Strid A, Chow WS, Anderson JM (1994) UV-B damage and protection at the molecular level in plants. Photosynth Res 39: 475-489

Svendsen H, Beszczynska-Møller A, Hagen JO, Lefauconnier B, Tverberg V, Gerland S, Ørbæk JB, Bischof K, Papucci C, Zajaczkowski M, Azzolini R, Bruland O, Wiencke C, Winther JG, Dallmann W (2002) The physical environment of Kongsfjorden-Krossfjorden, an Arctic fjord system in Svalbard. Polar Res 21: 133-166

Teramura AH (1983) Effects of ultraviolet-B radiation on the growth and yield of crop plants. Physiol Plant 58: 415-427

Uhrmacher S, Hanelt D, Nultsch W (1995) Zeaxanthin content and the photoinhibitory degree of photosynthesis are linearly correlated in the brown alga *Dictyota dichotoma*. Mar Biol 123: 159-165

van de Poll WH, Eggert A, Buma AGJ, Breeman AM (2001) Effects of UV-B induced DNA damage and photoinhibition on growth of six temperate marine red macrophytes: distinct habitat related differences in ultraviolet-B tolerance. J Phycol 37: 30-37

van de Poll WH, Eggert E, Buma AGJ, Breeman AM (2002) Temperature dependence of UV radiation effects in Arctic and temperate isolates of three red macrophytes. Eur J Phycol 37: 59-68

van de Poll WH, Bischof K, Buma AGJ, Breeman AM (2003) Habitat related variation in UV tolerance of tropical marine red macrophytes is not temperature dependent. Physiol Plant 118: 74-83

Vass I (1997) Adverse effects of UV-B light on the structure and function of the photosynthetic apparatus. In: Pessarakli M (ed) Handbook of photosynthesis, Marcel Dekker Inc, New York, pp 931-949

Viarengo A, Abele-Oeschger D, Burlando B (1998) Effects of low temperature on prooxidants and antioxidant defence systems in marine organisms. In: Pörtner HO, Playle R (eds) Cold Ocean Physiology, Cambridge University Press, Cambridge, pp 213-235

Wängberg SA, Selmer JS, Ekelund NGA, Gustavson K (1996) UV-B effects on Nordic marine ecosystems. Tema Nord. Nordic Council of Ministers, Copenhagen

Wiencke C (1990a) Seasonality of brown macroalgae from Antarctica - a long-term culture study under fluctuating Antarctic daylengths. Polar Biol 10: 589-600

Wiencke C (1990b) Seasonality of red and green macroalgae from Antarctica - a long-term culture study under fluctuating Antarctic daylengths. Polar Biol 10: 601-607

Wiencke C, Gómez I, Pakker H, Flores-Moya A, Altamirano M, Hanelt D, Bischof K, Figueroa FL (2000) Impact of UV radiation on viability, photosynthetic characteristics and DNA of brown algal zoospores: Implications for depth zonation. Mar Ecol Prog Ser 197: 217-229

Wiencke C, Clayton MN, Schoenwaelder M (2004) Sensitivity and acclimation to UV radiation of zoospores from five species of Laminariales from the Arctic. Mar Biol 145: 31-39

Wood WF (1987) Effect of solar ultra-violet radiation on the kelp *Ecklonia radiata*. Mar Biol 96: 143-150

Yakovleva IM, Dring MJ, Titlyanov EA (1998) Tolerance of North Sea algae to UV and visible radiation. Russ J Plant Physiol 45: 45-54

16 Climate and ozone change effects on UV-radiation and health risks

Harry Slaper, Peter den Outer and Gert Kelfkens

Laboratory for Radiation Research, the National Institute for Public Health and the Environment (RIVM), po box 1, 3720 BA Bilthoven, the Netherlands, e-mail: harry.slaper@rivm.nl

16.1 Introduction

The stratospheric ozone layer effectively serves as a protective shield that reduces the most harmful part of the solar ultraviolet (UV) radiation reaching the earths surface. As a consequence, the depletion of stratospheric ozone observed over the last two decades has probably led to an increase in ambient UV radiation. An increase in UV at ground-level may induce a wide range of harmful effects like an increase in skin cancer, cataracts, a decrease in bio-mass production and crop yields, and suppression of the human immune system (UNEP 1999, 2003). Ozone depletion is probably related to the large scale emissions of halocarbons, among which the chlorofluorocarbons (CFC's), in the seventies and eighties of the previous century. The Vienna Convention in 1985 was the starting point for international policy agreements to protect the ozone layer by aiming at reducing the production and emission of ozone depleting substances, and by providing the framework for the restrictive protocols that were agreed upon later.

The Montreal protocol in 1987 provided the first restrictive countermeasures and in view of the compelling evidence that ozone depletion occurred, this protocol was subsequently strengthened in several more restrictive Amendments. In the latest Amendments the production of the most potent ozone depleting substances is completely phased out in 1996. A longer phase out period is allowed for the developing countries. These countermeasures are expected to lead to a slow recovery of the ozone layer over the next century.

To evaluate the consequence of ozone depletion in terms of health risks such as skin cancer incidence, the full source-risk chain has to be taken into account. The Assessment Model for UV Radiation and Risks

(AMOUR) developed at RIVM, in collaboration with NOAA, Utrecht/Leiden University and KNMI, incorporates the various aspects of the source risk chain, including: the link between production and emission of halocarbons, and the resulting depletion of the ozone layer, the consequences of ozone changes for ambient effective UV doses and the corresponding skin cancer risks. The effects of the agreed countermeasures in terms of a recovery of the ozone layer, the (future) UV-radiation levels and their effect on the excess skin cancer risks associated with ozone depletion were previously analysed for the USA and Europe (Slaper et al., 1996).

In line with other studies (UNEP, 1999) it was expected that a slow recovery of the ozone layer will occur with a return to 'normal' (1980) levels around 2050. Excess skin cancer risks, caused by ozone depletion, are in those scenarios expected to rise until 2050-2070. Slaper et al (1996) clearly showed the potential success of the countermeasures in reducing the future excess skin cancer risks associated with the chemical depletion of ozone.

A limitation of the previous analyses is the basic assumption that ozone depletion is purely chemically driven, and that no interaction with climate change occurs. Thus, it is assumed that the ozone layer will return to the "normal" 1980-levels as soon as the active chlorine levels in the stratosphere reach the 1980 levels. However, in view of the ongoing climate change this assumption might well be too simple, and especially polar regions might play a crucial role in the future developments.

This paper focuses on the possible effects of past, present and future developments in ozone depletion on UV-radiation levels and the skin cancer risks associated with these changes, using an updated version of the AMOUR-model with a first attempt to include possible effects of climate ozone interactions (Kelfkens et al. 2002). We provide an update of the previous analysis of skin cancer risks for various scenarios for ozone depletion in relation to countermeasures taken and possible effects of ozone-climate interactions. Special focus is given to the role of climate and ozone changes in the arctic region on the future risks at mid-latitudes in densely populated areas in Europe.

16.2 Effects of UV on human health

UV-exposure of skin and eye can lead to a wide variety of adverse health effects, among which the development of erythema (sunburn), photokeratitis (snowblindness), an impairment of the immune system, with possible suppressive effects on the resistance to infections and the effectivity of vaccination programs, skin ageing, cataracts and skin cancer. A positive effect of UV-exposure of the skin is the production of vitamin-D, which is important for bone-formation in humans. It should be noted however that only a relatively small amount of UVB exposure in every day life is adequate for the required vitamin D production. Vitamin D requirements can also be met by a proper diet with sufficient consumption of fatty fish (or food that is artificially supplemented with vitamin D).

In response to the cellular damages in the skin invoked by UV exposure a natural adaptive response occurs through thickening of the skin and an increase of skin pigmentation (tanning). Especially the skin thickening offers some protection against sunburn effects in new exposure situations simply by shielding the viable cells in the epidermis. Tanning is often acknowledged as a positive effect, but it is in fact a response to the damage and offers limited protection in caucasian skin.

Many of the adverse health effects associated with UV-exposure, like skin cancer, skin ageing and possibly cataracts, are related to doses received by individuals over prolonged time periods, i.e. exposure over many years up to a lifetime. Other effects like an impairment of the immune system can be related to doses received over a period of weeks to months. Some short term effects, like skin erythema (sunburn) and snow-blindness, are primarily related to doses received over a day, but are influenced by the exposure history of the previous days to weeks. Adverse effects to plants and terrestrial and aquatic ecosystems are also primarily related to doses received over at least days to months, or in case of ecological changes years. Effects of UV on materials, especially polymers, and paints are also related to prolonged exposure periods.

In order to evaluate risks associated with ozone depletion and changes in UV-exposure it is essential to acquire information on the dose-time-effect relationships for the effects under study and on the relative effectiveness of UV-exposure at different wavelengths, i.e. the determination of the biological weighting functions or action spectra. For many of the aforementioned adverse effects the relative effectiveness of radiation at different wavelengths varies considerably in the range of the solar UV-spectrum: in order to induce the same effect an exposure at wavelengths around 340 nm requires 1000-10000 times higher physical doses than exposure at wavelengths around 300 nm.

16.2.1 Skin-cancer and it's relationship to UV-exposure

Skin cancer is the most common cancer type in light-skinned populations and the yearly incidence in mid-latitudinal regions like Europe and North

America is 1000-2000 new cases per year per million inhabitants. The incidence increases with lower latitudes and thus with higher solar radiation levels. Three major skin cancer types can be distinguished: basal cell carcinoma (BCC), squamous cell carcinoma (SCC) and malignant melanoma. Of all skin cancers around 70-80 % are BCC. This most common type is not very aggressive and the lethality rate is lower than 1%. The second most common skin cancer type is SCC (10-20%) with a lethality of around 2-3%. Melanoma are least common (around 10%) but they are the most aggressive type with a lethality of around 20-25%. BCC and SCC originate from the malignant transformation of keratinocytes, the major structural cells in the epidermis of the skin. BCC and SCC are together also referred to as non-melanoma skin cancers (NMSC). Melanoma results from malignant transformation of the skin's pigment producing cells in the epidermis, ie the melanocytes.

Evidence that UV exposure induces skin cancer comes from three main sources of information: epidemiology, animal experiments, and studies on the biological mechanisms of skin carcinogenesis.

16.2.2 Epidemiological evidence

Skin cancer occurs with higher probability in humans with a sun-sensitive skin, and especially light skinned people from Celtic origin like Irish or English are at highest risks. Risks for Asian and black populations are much lower. Risks are especially high in Xeroderma Pigmentosum (XP) patients, who lack the ability to repair solar induced cellular DNA-damage and who develop multiple skin cancers very early in life. Furthermore, among populations with a similar skin complexity incidences are higher in areas with higher ambient solar radiation levels, i.e. at lower latitudes, and skin cancer primarily occurs on sun-exposed areas. Non-melanoma skin cancers are primarily found on head, neck and hands, and melanoma also occur frequently on occasionally exposed areas like the trunk.

Within a population, those with highest exposure are at highest risks, i.e. groups with outdoor jobs. This holds for the NMSC, but not for the melanoma. For melanoma intermittent exposure is indicated as a risk factor, and highest risks are seen in indoor workers with relatively low overall exposure but excessive exposure in holiday-periods.

The evidence further suggests that SCC are primarily related to the cumulative UV-dose received by the skin, but for BCC and melanoma exposures early in life appear to be most relevant (de Gruijl et al. 2003). This could be an indication that the UV-related processes are early events in the carcinogenic process for the BCC and melanoma, whereas for SCC the specific order of events appears to be less relevant.

16.2.3 Animal experiments and action spectra

UV irradiation of the skin has been shown to lead to the incidence of SCC skin cancer in laboratory animals. Extensive dose-time-response studies have demonstrated the relationship between exposure and the incidence of skin cancers (de Gruijl et al. 1983). These studies provided dose-time-effect relationships which could be used in the analysis of epidemiological data and are the basis for quantitative risk models (see next section). The most common type produced by UV in laboratory animals, i.e. mice and rats, are SCC. The accumulated UV-dose over a life time appears to be the primary risk factor. Recently also melanoma mammalian models in transgenic mice and the opossum have been found for the induction of melanoma through UV-exposure. However, in contrast with the induction of SCC for melanoma the exposures early in life appear to be the dominant risk-factor (see de Gruijl et al. 2003).

In addition to the dose-time-effect information animal studies have provided essential information on the relative effectiveness of UV-radiation from different wavelength regions in the induction of skin cancers. Such information cannot be obtained from epidemiological data. The observed latitude dependency of the incidence does not give any insight in the relative importance of different wavelength regions, because the exposure is lower at higher latitudes for all UV and visible wavelengths in the solar spectrum. Most extensive information on action spectra from animal studies is available for SCC. De Gruijl et al. (1993) extracted an action spectrum for skin cancer development in hairless mice, SCUP-m. This murine action spectrum was transferred to humans by accounting for the differences in skin transmission between mice and human skin, and thus de Gruijl and van der Leun (1994) constructed the SCUP-h action spectrum, which is plotted in fig. 16.1. The SCUP-h action spectrum closely matches an action spectrum for pyrimidine dimer formation in cellular DNA following a correction for transmission of the epidermis.

16.2.4 Mechanistic evidence from molecular biology

UV radiation, and in particular the UVB region of the spectrum (280-315 nm) is capable of causing damage and mutations to cellular DNA. UVB is particularly effective in the induction of pyrimidine dimers. For a large majority of SCC and half of the BCC skin cancers specific mutations in the

P53 tumour suppressor gene are found, which have characteristics of UV-induced point mutations. The molecular evidence that UV is directly involved in the development of melanoma is not as apparent since the mutations regularly found in melanoma were not identified as specific UV-induced mutations. However, UVB induced DNA damage is not restricted to point mutations and can involve larger sections of DNA. Such defects can also be caused by other agents, and thus cannot be identified as specific UV-related defects (de Gruijl et al. 2003).

16.2.5 A quantitative risk assessment model for skin cancer in relation to UV-exposure

In order to provide risk assessments it is essential to use proper dose-time-response relationships and a proper action spectrum. We used the quantitative model that was used and described in Slaper et al (1996). The basic quantitative relationships were derived from detailed dose-time response studies in hairless mice (de Gruijl et al. 1983; Slaper et al. 1996). In case of continued exposure the cumulated number of skin cancer cases in a population is described by:

$$Y(a) = \gamma (Da)^{c} a^{d-c} \qquad (1)$$

where D stands for the daily (or yearly) effective UV-dose received, a for the time since the start of the exposure (or age), and γ, c and d are the basic model-parameters reflecting respectively the sensitivity of the population, and the dose and age dependency of the carcinogenic process. The parameter c is sometimes referred to as the biological amplification factor indicating the percent change in incidence related to a 1% increase in UV-doses received.

Epidemiological data relating ambient UV-radiation levels to the skin cancer incidences were used to fit the model-parameters for each of the three skin cancer types. Equation (1) requires additional interpretation in case of changing exposures over a life-time. Epidemiological data and animal experiments indicate that for squamous cell carcinoma the life time cumulative dose is the dominant factor. This implies that equation (1) could be used interpreting the factor Da as the cumulative lifetime dose ($\Phi(a)$). In terms of processes this implies that UV and time related changes can occur independently of each other. For melanoma and basal cell carcinoma epidemiological data suggest that UV-exposure early in life is more important than the cumulative dose, which is an indication that at least a part of the UV-related processes have to occur prior to the time related processes. In that case equation (1) is modified to (Slaper et al. 1996):

$$Y(a) = \gamma \, \Sigma \, D(x) \, (\Phi(x))^{c-1} \, (a-x)^{d-c} \tag{2}$$

where the summation Σ is over all ages x from birth to age a.

In the presented analysis the SCUP-h action spectrum (fig. 16.1) is used as the biological weighting function. It should be noted that the SCUP-h action spectrum also roughly resembles the action spectrum for erythema, which is often used characterizing the ambient UV-radiation levels (for instance in the solar UV-index).

In addition to the SCUP-h action spectrum fig. 16.1 also provides an example of a solar spectrum measured at ground level (high overhead sun at mid-latitudes), and the two multiplied with each other which indicates the effective spectral irradiance, which peaks in the UVB range of the solar spectrum. The effective spectral irradiance integrated over the full wavelength range is referred to as the effective dose-rate (marked area in fig. 16.1) and integrated over time the effective dose is obtained. Also indicated in fig. 16.1 is the extraterrestrial spectrum which lacks the strong ozone induced cut off in the UVB range of the spectrum (below 315 nm) that occurs in the solar spectrum at ground level.

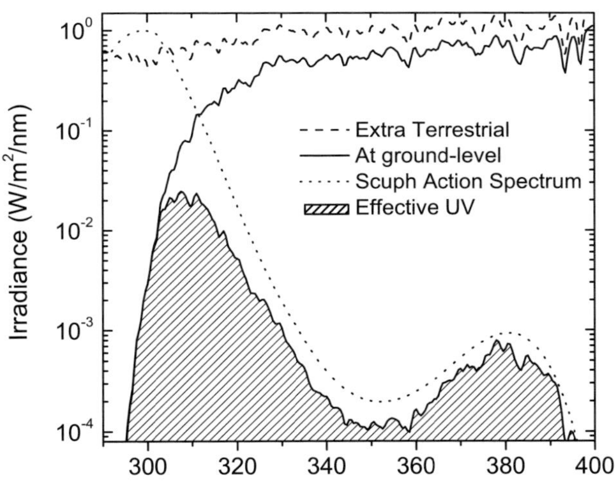

Fig. 16.1. Solar spectrum (extra terrestrial and at the ground (RIVM)), action spectrum for photo-carcinogenesis (SCUP-h), and weighted effective UV irradiance for skin cancer

For melanoma, previous studies in a fish model seemed to indicate that the UVA part of the solar spectrum could be more effective than UVB, however, recent evidence gathered in two mammalian models suggest that similar to the action spectra discussed above, UVB is indeed much more effective in producing melanoma. For the risk assessment, provided in this paper we use the SCUP-h action spectrum for all three skin cancer types (de Gruijl and van der Leun, 1994).

16.3 UV-exposure

UV-exposure of the skin is not restricted to solar exposure, since many artificial sources emit or can emit UV-radiation. Although detailed knowledge on the overall UV-exposure to the sun and artificial sources is lacking, it is highly likely that the sun dominates the effective UV-exposure of the overall population and estimated for the Netherlands amount to over 90% of the mean exposures. It should be noted however that for groups with excessive exposure to artificial tanning booths the yearly effective dose received using this equipment could well be equal or higher than the solar exposure. The same holds for psoriasis patients receiving UVB phototherapy.

The skin receives only a fraction of the solar UV that is measured on a horizontal plane, since not all time is spent outside, and the skin usually is not fully oriented towards the sun and open sky. In addition clothes and the use of sunscreens can influence the exposure received by the viable layers of the skin. Furthermore, the local environment like mountains, buildings, snow and trees can shade and reflect part of the UV and influence the radiation levels actually received.

It should be noted that in solar exposure roughly half of the effective UV comes directly from the sun and half from the sky. In situations where the ground is snow covered the reflectance from the ground can also considerably contribute to the exposure.

Indications of the exposure received in every day life are obtained using personal UV-dosimeters. Indoor workers receive on average around 2-3% of the available ambient effective dose on the most exposed skin parts (face, neck and hands). Holiday periods in the summer could double the year round exposure of indoor workers to 5%. For outdoor workers it is estimated that around 9% of the ambient dose is received. Large variations occur among different individuals depending strongly on the leisure activities.

It should be noted that changes in behaviour influencing the exposure would also alter the (excess) risks estimated in relation to environmental change. Excess risks are altered in two ways: first the excess dose received due to environmental change is changed in relation to a different exposure fraction, and second the non-linearity of the dose-time response relationships implies that excess risks due to a specific additional dose are depending on the "normal" exposure. In that respect it should be noted that skin cancer incidence has been reported to be increased over the past decades (de Gruijl et al. 2003). This increase is probably related to behavioural factors rather than the environmental change, simply because it is too soon to identify the influence of environmental change.

In the scenario-analysis we assume that exposure habits do not change over time, and assuming that the exposure received is a certain be it not exactly known fraction of the ambient dose, we can use equations 1 and 2 and use the ambient doses in stead of the actual dose received (the fraction received is accounted for in the parameter γ). Assessing the long term risks related to UV-exposure requires life-time analysis of the UV-exposure doses received.

16.4 Determination of the UV-climate combining measurements and modelling

The effective UV radiation received at the ground primarily depends on the solar height, the total ozone column, clouds, aerosols and snow cover. Skin cancer risk assessments require the life-time UV-dose received, and thus a long term assessment of the UV-dose is necessary.

The most direct way to assess changes in the UV-climate would be to use long-term data from highly accurate ground based UV-monitoring stations, however such long-term data are largely lacking at present (WMO, 1999). An assessment of changes over prolonged periods of time is therefore based on modelled UV-transfer, using ozone and other atmospheric data from either ground based or satellite based sources. Such model approaches can be at least partly validated by a comparison with ground based measurements for shorter time periods (den Outer et al. 2005, Slaper et al. 2001, den Outer et al. 2000). In the next sections we will first compare modelled and measured effective UV in relation to ozone and clouds, and then provide an assessment of the historical changes in UV-radiation levels combining models and measurements.

16.4.1 Ozone and cloud dependence of the effective UV radiation

The effect of an ozone change on the effective UV-dose received at the ground is illustrated in fig. 16.2a, which shows the clear anti-correlation between skin cancer weighted UV and ozone. The measured results, which are corrected for cloud effects, are obtained using daily integrated UV-measurements from the RIVM UV-monitoring system in Bilthoven for the period 1996-2000, and are compared with modelled results. On the vertical axis the relative deviation of the measured ratio of daily SCUP-h doses and daily UVA doses from the modelled ratio for the same day using an ozone value of 350 DU. This is plotted versus the measured ozone value, thus obtaining a view of the relative increase in effective UV related to a change in ozone. Measured data-points for 3 days with similar ozone values are averaged to obtain a smooth result, and in addition a modelled curve is provided for the same days, showing good agreement between modelled and measured ozone dependence of the effective UV-dose. From this plot a Radiation Amplification Factor (RAF), indicating the increase in effective UV-dose related to a one percent change in ozone, could be derived, which amounts to 1.3 for the SCUP-h weighted UV. This RAF for SCUP-h is slightly higher than the RAF for the erythemal action spectrum (1.1-1.2). Similar approaches for other locations are summarised in WMO (1999). In addition it has been shown that an upward trend in clear-sky UV-dose is correlated to a downward trend in ozone (Zerefos et al.1998; Mckenzie, 1999).

The effects of clouds on the UV-irradiance are highly variable and related to a broad range of cloud parameters like: cloud cover, cloud shapes, cloud optical thickness, water content etc. Previous studies have shown empirical relationships between the cloud induced reduction of global solar radiation and cloud effects on biologically weighted effective UV (Bordewijk et al. 1995, Bodeker and McKenzie, 1996). Fig. 16.2b illustrates the non-linear empirical relationship between the reduction of effective UV and global solar radiation. It should be noted that this relationship is depending on the solar zenith angle and becomes slightly more linear at high sun and more curved at low sun conditions. An extensive analysis of cloud induced reduction on UV and global solar radiation is given in den Outer et al. (2005).

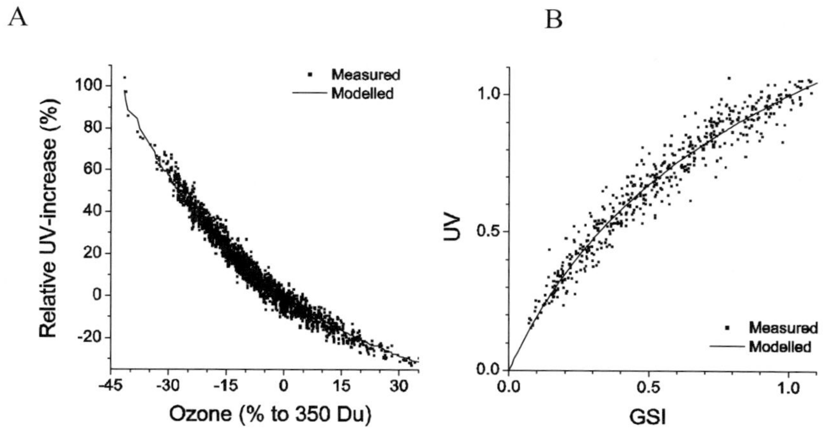

Fig. 16.2. Relation between changes in ozone and changes in effective UV-radiation levels (left hand panel 2A), relation between the cloud-induced reduction of global solar irradiation (GSI) and the reduction of UV-irradiance (shown for 45-60 degrees Solar Zenith Angle)(right hand panel 2B). Measurements from RIVM UV-monitoring spectroradiometer (den Outer et al. 2005).

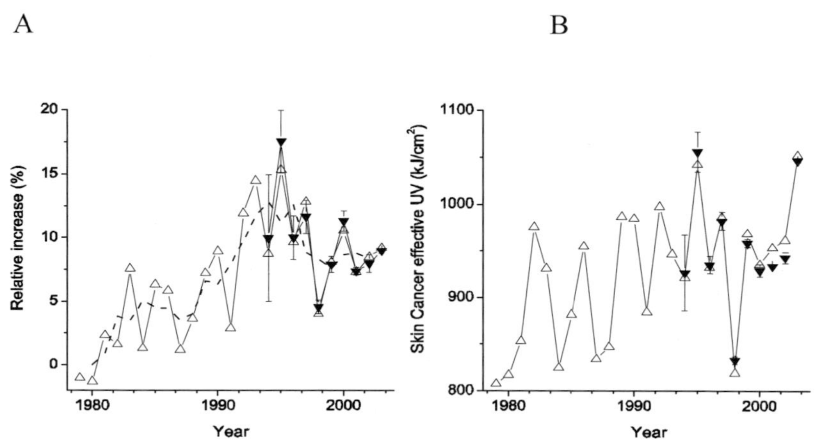

Fig. 16.3. Change in yearly effective UV-dose (SCUP-h weighted) in Bilthoven, the Netherlands (52 N, 5.2 E) related to ozone changes (left panel, 3A), and related to ozone and cloud changes (right panel, 3B). Filled triangles derived from RIVM ground based UV-monitoring, open triangles modeled UV-doses (method described in den Outer et al 2005).

16.4.2 Changes in ambient UV-radiation in the past decades

Using the above models and available data from ground based and satellite based measurements on ozone and global solar radiation it is possible to analyse the past UV-climate. Fig. 16.3 shows an analysis of yearly ambient effective UV-doses for the past 25 years for Bilthoven in the Netherlands (derived from den Outer et al. 2005). This modelled analysis is compared with ambient UV-measurements available from RIVM's UV-monitoring system since 1994. The left hand panel (3a) shows the ozone related changes in the UV relative to the average 1979-1981 yearly UV-dose. Here the measurements are corrected for cloud variations from year to year using the above illustrated empirical cloud model backwards. The results in fig. 16.3a show that highest UV-radiation levels are calculated for the period 1992-1995. These high levels are probably related to enhanced ozone depletion caused by heterogeneous chemical interaction of halocarbons with stratospheric dust from the volcanic eruption of the Pinatubo (WMO, 1999, 2003). The right hand panel (3b) shows the skin cancer effective dose actually received in the same period including the cloud variability. An upward trend is apparent in both figures, and using a linear least squares fit an upward trend of 3.4±0.7 % per decade for the ozone related changes and 5.7±2.0 % per decade for the overall trend including clouds and ozone changes (den Outer et al. 2005). It should be noted that the difference in these trends is not significant due to the high variability of the cloud effects from year to year. Highest UV-doses, over the past 25 years were received in the Netherlands in 1995 and 2003. Both years show relatively low ozone values and a low reduction by clouds, but in 1995 the effect of ozone depletion dominates, whereas in 2003 the low reduction by clouds dominates the deviation from other years.

Satellite data on ozone and (cloud) reflectance can be used to analyse the UV-climate at the ground, and previous studies comparing satellite and ground based modelling approaches have validated this (Matthijsen et al 2000, Williams et al. 2004, Kaurola et al 2000, Slaper et al. 2004). Using TOMS ozone data (version 7) we analysed the ozone related UV-trends for the European continent as shown in fig. 16.4. The figure shows that the largest increasing UV-trends related to ozone depletion are observed over large parts of central and western Europe and amount to a 3-4 % increase per decade. The area most affected includes all most densely populated areas on the European continent, including Germany, the Netherlands, Belgium, France and northern Italy, and ranging from the southeast of England to Poland, and from the southwest of France to former Yugoslavia.

The effects of clouds are also important with respect to climate change. Although no decisive evidence is available yet, climate change may well

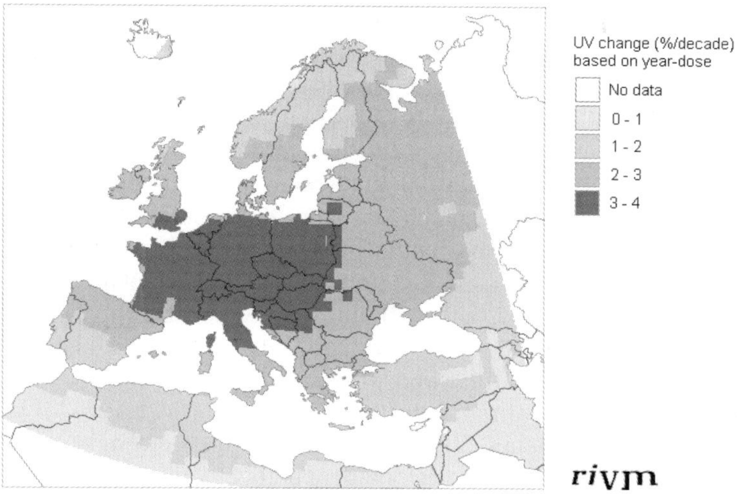

Fig. 16.4. Ultraviolet trends over Europe derived from satellite based changes in ozone in the period 1979-2000 (using TOMS ozone data; methodology described in Slaper et al. 2001)

affect cloud cover and cloud optical thickness in a way dependent on season and location. A change in cloudiness and temperature will, in turn, influence (recreational) exposure with potential impacts on the overall risk of the population. At the moment, effects of structural change in cloudiness or a change in human behaviour due to climate change are not incorporated in our scenario analysis for (future) human risks.

Although there is a lack of direct observations of UV trends in relation to ozone changes, it is highly likely that the yearly UV doses received in large parts of Europe have increased by 5-10% over the past two decades. Further monitoring is required to substantiate the analysis and to establish how the climate-ozone changes influence the future UV radiation levels at the ground. In view of the high year to year variability of clouds and vortex effects, one might argue that a prolonged observation period will be required to fully establish UV trends.

16.5 Ozone-depletion and Ozone-climate interactions

Ozone is found as a trace gas throughout the atmosphere, with highest concentrations in the stratosphere at heights around 25 km above the earths surface. At standard atmospheric pressure and temperature the total atmospheric amount of ozone consists of a layer of no more than 3 mm. Ozone molecules are constantly produced and destroyed in a series of (photo)chemical reactions in the stratosphere. Molina and Rowland (1974) were the first to predict that large scale emissions of halocarbons might lead to a disturbance of the natural photochemical ozone-balance leading to a decrease in ozone. CFC's were widely and increasingly used during the sixties and seventies, and without countermeasures a further increase could have been expected. CFC's are non-toxic and inert gases, which are not readily destroyed in the troposphere. They slowly diffuse to the stratosphere and are photo-dissociated there under the influence of the shortwave solar UV releasing effective chlorine which strongly enhances the destruction of ozone. The first evidence that ozone depletion occurred came from the discovery of, what is usually referred to as the Antarctic ozone hole: a strong decrease in stratospheric ozone during the early spring in the Antarctic region which was observed since the beginning of the eighties (Farman et al. 1985). The discovery of the Antarctic ozone hole came as a surprise because it had not been predicted that the Antarctic region would be exceptionally vulnerable.

It was found that the photochemical breakdown of ozone by active chlorine is strongly catalysed in the presence of Polar Stratospheric Clouds (PSC) and/or stratospheric aerosols, such as resulting from volcanic eruptions. The early Antarctic spring provides a stable polar vortex with very low stratospheric temperatures, and thus optimal conditions for the formation of PSC, which contribute to the catalytic breakdown of ozone by active chlorine. In the arctic region the polar vortex conditions are less stable and the temperatures in the stratosphere are higher. Ozone depletion, however was not restricted to the Antarctic and a decrease in ozone has now also been well established over mid-latitudes at the northern and southern hemisphere (WMO 2003).

The observations seem to indicate that ozone depletion started around 1980, which is surprisingly late in view of the fact that the stratospheric chlorine load at that time was al ready elevated compared to pre-industrial times (from around 0.8 ppbv to nearly 2 ppbv). Taking into account the information available on the emission of halocarbons and a scenario of future releases many efforts were made to analyse the past and future trends using predictive 2D and 3D modelling of stratospheric ozone. Fig. 16.5

provides a summary plot, adapted from Kelfkens et al. 2002, showing the range of calculated developments for mid-latitudes in the northern hemisphere, obtained with 2D and 3 D modelling efforts (grey band indicates the results). Two lines indicating results obtained with the AMOUR-model include an empirical fit of ozone depletion observed in the 1980-2000 period in NW-Europe and a scenario analysis for future developments taking into account the countermeasures to prevent further ozone depletion. It can be seen that the historical trend observed (represented by the AMOUR-model line) is underestimated by the 2D and 3D modelling approaches. Two different scenario-lines indicate the future developments with the AMOUR-model. The dashed line represents the situation assuming that ozone depletion is purely chemically driven and that no climate ozone interactions are involved in future developments. This evaluation closely resembles previous modelling attempts provided by Slaper et al. (1996), the difference being that the scenario for halocarbon releases has been updated to include the Montreal Amendments of the Montreal Protocol. The second scenario analysis, represented by the full line, attempts to include possible effects of climate-ozone interactions as discussed below.

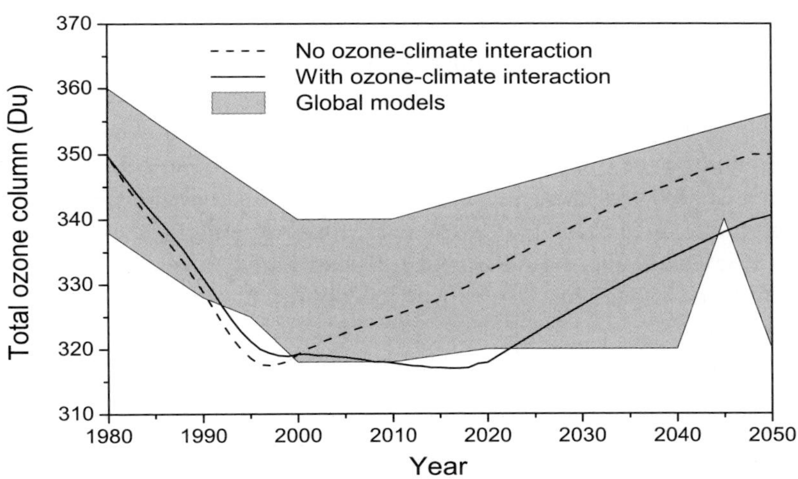

Fig. 16.5. Range in prognosis for the future development of the stratospheric ozone layer (grey modified from Kelfkens et al. 2002, including results from Shindell et al. 1998, Austin et al. 2001, 2000, and Dvortsov and Salomon, 2001). The curves are obtained using the RIVM AMOUR model for the NW-European region.

Most studies assuming ozone depletion to be purely chemically driven, estimate that the ozone layer will return to the normal 1980 levels as soon as the active chlorine levels in the stratosphere reach the 1980 levels. However, other relevant atmospheric parameters such as stratospheric temperatures and troposphere heights might change in relation to ongoing climate change and this on its term can have important implications for ozone. Polar regions might well play a crucial role in the future developments of the ozone layer and the effects of climate ozone interactions. As a consequence of global warming it is expected that temperatures in the stratosphere will decrease, and thus conditions in the arctic region might become more favourable for PSC formation and an enhanced ozone depletion. The implications however are not restricted to the polar regions, since the low ozone concentrations can be transported to mid-latitudes and thus affect the heavily populated areas in Europe and North America. We use results reported by Kelfkens et al (2002) to simulate the effects of such climate ozone interactions on ozone levels at mid-latitudes.

16.6 Ozone-climate change scenario-analysis

The calculated development of the ozone layer over Europe during the period 1980-2100 is based on a scenario for chlorine in the stratosphere resulting from the countermeasures taken in the Montreal Amendments of the Montreal Protocol (Madronich et al., 1999) and on the ozone trend over 1980-2000 observed by the NASA-TOMS satellite. Monthly ozone values enable us to calculate - with a UV-transfer model – the future effective UV-levels at the earth's surface. These future UV-levels yield via a dose-response model for skin cancer, eventually the increased skin cancer incidence.

In this chapter the baseline risk, determined by the situation with no ozone-climate interaction, is compared with a scenario including ozone-climate interaction. The magnitude of the interaction is based on a literature survey, 2D model calculations for the Arctic vortex and on expert judgement. For the ozone-climate interaction we formulate the following preconditions:

- Chlorine scenario A1 (Madronich et al. 1999)
- Stratospheric temperature according to IPCC, IS92a (Velders, 1997)
- Stratospheric cooling results – via a shift in the chemical equilibrium for ozone production/destruction - in 0.9% extra ozone per one Kelvin drop in temperature.

- Splitting up of the arctic vortex induces a month and latitude dependent ozone depletion over the Northern hemisphere. The magnitude of the ozone loss is calculated by 2 D simulation (Velders, 1997).
- Of the observed ozone trend 25% can be attributed to non-vortex dynamical changes.

For details on the AMOUR model, the calculations and the interaction between the ozone layer and climate change the reader is referred to Kelfkens et al. (2002). The result of the AMOUR assessment can be given at a fixed location in terms of the increase in effective UV and skin cancer incidence with and without ozone-climate interaction. Finally AMOUR produces maps of the difference in effective UV and skin cancer incidence over Europe. Here, it should be noted that the excess risks due to ozone depletion are calculated for a population with the genetically determined skin sensitivity of the Northwest European population.

Fig. 16.6 shows the increase in effective UV for Bilthoven, the Netherlands. The triangles are three year running mean yearly doses derived from UV-monitoring at RIVM in the period 1994-2003, and the dash-dotted line represents the three year running mean for the modelled yearly UV-dose (from fig. 16.3a, den Outer et al 2005). The dotted curve shows the AMOUR model result if no countermeasures would have been taken to protect the ozone layer (scenario according to Slaper et al 1996). This curve shows a rapid upward trend, and the observed UV-radiation levels in the past six years do not follow this upward curve, indicating the success of the countermeasures in slowing down the ozone and UV trends. The dashed curve shows the AMOUR model result of the Montreal Amendments of the Montreal Protocol assuming that no climate ozone interaction occurs, and in that scenario the UV-radiation levels are estimated to slowly decrease from the year 2000 onward. The full line indicates AMOUR model results obtained using the same emission scenario, but now incorporating ozone-climate interactions. As can be seen in that scenario the UV-radiation levels will slowly rise until around 2020 and thereafter start a slow decrease.

The consequences of the latter two scenarios in terms of future excess skin cancer risks related to ozone depletion are shown in fig. 16.7, where it can be seen that without ozone-climate interaction future excess skin cancer risks might increase to around 2055 with 60 extra cases per million inhabitants per year, whereas including climate ozone interactions the risks will increase until 2065 and amount to 95 extra cases per million inhabitants per year. Thus the difference in terms of excess risks is considerable between the scenarios with and without climate-ozone interactions.

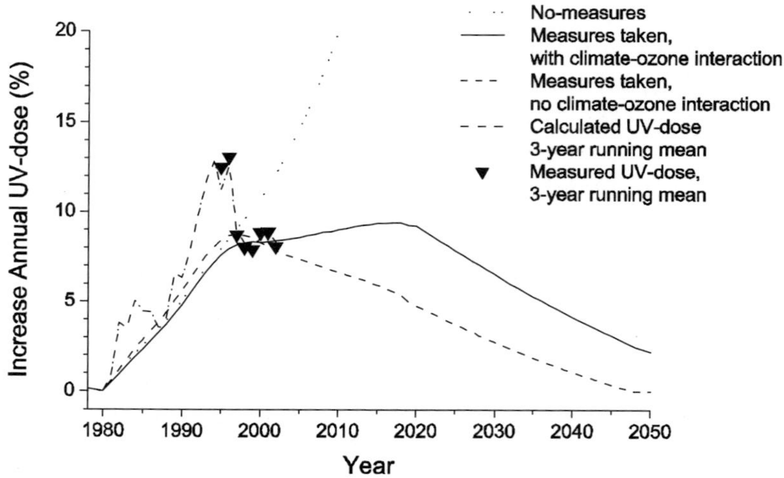

Fig. 16.6. Relative changes of yearly UV-doses for the Netherlands due to ozone depletion

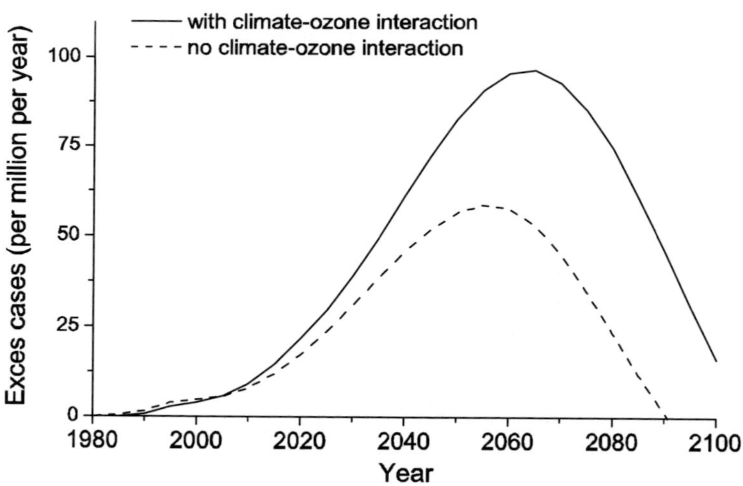

Fig. 16.7. Excess skin cancer incidence caused by ozone depletion calculated in relation to past and future ozone changes for the Netherlands (representative for large part of NW-European population).

Fig. 16.8 provides a map of excess skin cancer risks related to ozone changes in the ozone-climate interaction scenario. The map shows that the calculated excess risks for 2070 amount to 75-125 additional cases per year per million inhabitants for large parts in central Europe. The map is calculated for a population with the sensitivity of the Dutch population, and could be regarded as representative for NW-European populations.

The results given have a large uncertainty because the stratospheric ozone concentration at mid-latitudes is a subtle balance of chemical destruction of ozone, dilution by ozone poor air from the Arctic vortex, destruction/dilution by non-vortex dynamical effects and production of ozone due to chemical equilibrium shift as a consequence of temperature changes. Scientific understanding of these processes, especially of the non-vortex dynamics and of the Arctic vortex itself, is far from complete. The observed changes do presently not allow a distinction between the presented scenarios. In that respect it should also be noted that climate ozone interactions through changes in polar vortex might introduce a strong year to year variability which is not indicated in the scenario analysis. Thus, the analysis of future developments requires further knowledge on the climate ozone interactions, and further monitoring of changes and trends using satellite and ground based observations for ozone and UV.

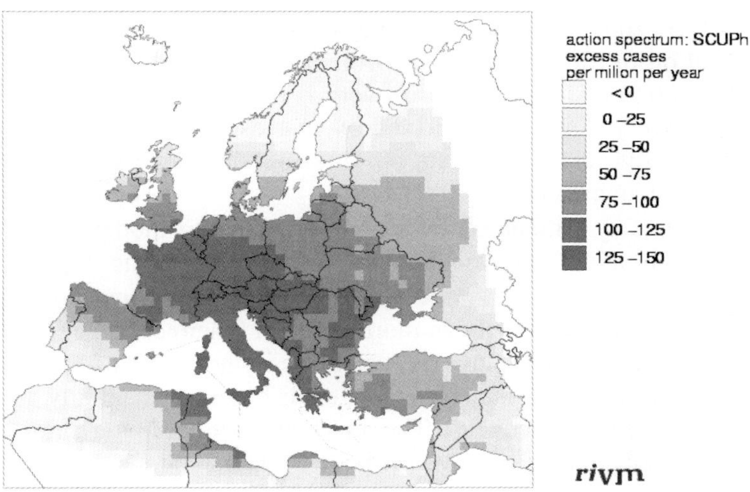

Fig. 16.8. Excess skin cancer risks due to ozone depletion calculated for 2070 with the AMOUR-model including climate and ozone interactions

16.7 Conclusions

Ozone depletion has occurred over large parts of the globe, with implications for the effective UV-radiation received on the ground. Over large parts of Europe the UV-radiation levels have probably increased by 5-10 % since 1980. As a consequence future skin cancer risks will probably increase.

A slow recovery of the ozone layer might be expected as a consequence of the countermeasures taken to restrict the emissions of ozone depleting substances. However, a substantial delay in ozone recovery, might occur due to ongoing climate change and the interaction between climate and ozone change. The arctic region might well play a crucial role in the future developments regarding ozone depletion and it's influence has been shown to involve large parts of the northern hemisphere including North-America and Europe.

Future developments however are regarded highly uncertain. This uncertainty arises from the fact that the stratospheric ozone concentration at mid-latitudes is a subtle balance of chemical destruction, dilution and production of ozone, and atmospheric dynamics. Scientific understanding of the underlying processes, especially of the non-vortex dynamics and of the Arctic vortex itself, is far from perfect. With this in mind we conclude:

- The ozone layer is expected to recover as a result of the implementation of the Montreal protocol. Based on the expected decrease in chlorine and bromine a recovery is expected around 2050. The full recovery is possibly delayed by increased concentrations of greenhouse gases with effects on atmospheric temperatures and atmospheric dynamics.
- The delay calculated by Chemistry-Climate models show a large variation, ranging from about 10 years to more than 20 years.
- A stronger winter Arctic vortex - triggered by lower stratospheric temperatures – results in more efficient (chemical) ozone loss. Thus, the Arctic vortex plays a crucial role in the future ozone levels, also at mid-latitudes, and possibly contributes substantially to the delayed ozone recovery. The mechanism underlying this enhanced ozone depletion, however, is poorly understood.
- Dynamic changes in the climate system, not coupled to the Arctic vortex, may also contribute substantially to a delay in ozone recovery. But, at present there is little consensus on the magnitude and relative importance of this contribution. Also dynamical changes could lead to substantial regional differences in ozone recovery.

- A delay in recovery of the ozone layer, induced by ozone layer - climate change interactions, leads to a higher and more persistent increase in effective UV-levels over Europe. In 2020 UV-levels over mid-Europe, accounting for ozone-climate interaction, are estimated to be 4-6% higher compared to the UV-levels without interaction.
- Elevated ambient UV-levels result in a higher and more persistent increase in skin cancer incidence. Under the assumption that other conditions, in particular cloudiness, are not affected and that climate change does not influence behaviour of the population, the excess risk due to ozone changes in the Netherlands (representative for large parts of the central-western part of Europe) increases from approximately 60 extra cases per million inhabitants per year (no ozone-climate interaction, 2055) to about 95 extra cases per million per year (ozone-climate interaction taken into account 2065).

Acknowledgement

This study was financially supported by NIVR as part of the COEUR-project (contract 5.1 IB-01), and by EU as part of the SCOUT project (contract 505390-GOCE-CT-2004). Bram Bregman (KNMI) is acknowledged for his contribution to fig. 16.5.

References

Austin J, Butchart N, and Knight J (2001) Three-dimensional model simulations of the ozone layer: 2015-55, Q. J. Meteorol. Soc., 127, 959-974

Austin, J., J.R. Knight, and N. Butchart (2000), Three-dimensional model simulations of the ozone layer: 1979-2015, Q. J. Meteorol. Soc., 126, 1533-1556

Bodeker GE, Mckenzie RL (1996) An algorithm for inferring surface uv irradiance including cloud effects. Journal-of-Applied-Meteorology 35(10):1860-77

Bordewijk JA, Slaper H, Reinen HAJM, Schlamann E (1995) Total solar radiation and the influence of clouds and aerosols on the biologically effective UV. Geophysical-Research-Letters 22(16):2151-4

de Gruijl, F.R., Van der Meer, J.B. and Van der Leun, J.C. (1983) Dose-time dependency of tumor formation by chronic UV exposure, *Photochem. Photobiol.* **37**, 53-62

de Gruijl FR, Sterenborg HJCM, Forbes PD Davies R.E.; Cole C.; Kelfkens G.; van Weelden H.; Slaper H.; van der Leun J.C. (1993) Wavelength dependence of skin cancer induction by Ultraviolet Irradiation of Albino hairless mice. Cancer Res 53:53-60

de Gruijl FR, van der Leun JC (1994) Estimate of the wavelength dependency of ultraviolet carcinogenesis in humans and its relevance to the risk assessment of a stratospheric ozone depletion, Health Physics 67, 319-325

de Gruijl FR, Longstreth J, Norval M, Cullen AP, Slaper H, Kripke ML, Takizawa Y, and Van Der Leun JC (2003) Health effects from stratospheric ozone depletion and interactions with climate change. Photochemical-and-photobiological-sciences 2(1), 16-28.

den Outer PN, Slaper H, Matthijsen J, Reinen HAJM, Tax R (2000)Variability of ground-level ultraviolet: model and measurement. Radiation-Protection-Dosimetry 91:105-10

den Outer PN, Slaper H, Tax R (2005) UV radiation in the Netherlands: Assessing long-term variability and trends in relation to ozone and clouds. Journal-of-Geophysical-Research-Atmospheres 110 (doi: 10.1029/2004 jd 004824) : D02203

Dvortsov VL and Solomon S (2001), Response of the stratospheric temperatures and ozone to past and future increases in stratospheric humidity, J. Geophys. Res. 106, 7505-7514

Farman JC, Gardiner BG, and Shanklin JD (1985), Large losses of total ozone in Antarctica reveal seasonal ClOx/NOx interaction, Nature 315, 207

Kaurola J, Taalas P, Koskela T, Borkowski J, Josefsson W (2000) Long-term variations of uv-b doses at three stations in northernEurope. Journal-of-Geophysical-Research-Atmospheres 105(D16):20813-20

Kelfkens G, Bregman A, de Gruijl FR, van der Leun JC, Piquet A, van Oijen T, Gieskes WWC, van Loveren, H., Velders, GJM., Martens P, Slaper H (2002) Ozone layer - climate change interactions Influence on UV levels and UV related effects. Dutch National Research Programme on Global Air Pollution and Climate Change (NRP/NOP), 410 200 112

Madronich S, Velders GJM, Daniel JS, Murari Lal M, McCullogh A and Slaper H (1999), chapter 11 in WMO, Scientific Assessment of Ozone depletion: 1998, Geneva, Switzerland, WMO report 44

Matthijsen J, Slaper H, Reinen HAJM, Velders GJM (2000) Reduction of solar uv by clouds: a comparison between satellite-derived cloud effects and ground-based radiation measurements. Journal-of-Geophysical-Research-Atmospheres 105(D4):5069-80

McKenzie RL, Conner B, Bodeker GE (1999) Increased summertime uv radiation in new zealand in response to ozone loss. Science 285(5434):1709-11

Molina MJ and Rowland FS (1974), Stratospheric sink for chlorofluoromethanes: chlorine atom-catalyzed destruction of ozone, Nature 249, 810, 1974.

Shindell DT, Rind D, and Lonergan P (1998), Increased polar stratospheric ozone loss and delayed eventual recovery owing to increasing green-house gas concentrations, Nature, 392, 589-592

Slaper H, Velders GJM, Daniel JS, Degruijl FR, Vanderleun JC (1996) Estimates of ozone depletion and skin cancer incidence to examine the vienna convention achievements. Nature 384(6606):256-8

Slaper H, Matthijsen J, den Outer PN, Velders GJM (2001) Climatology of Ultraviolet Budgets using Earth Observation (CUBEO): mapping UV from the perspective of risk assessments. BCRS, USP-2 report 00-17

Slaper H, den Outer PN, Williams J, Bais A, Kaurola J, Koskela T, Lakkala K, Seckmeyer G, Weihs P. (2004) Towards a European UV-climatology utilising ground-based spectral data and satellite based analysis. Oral presentation at the Quadrennial Ozone Symposium, 1- 8 June, Kos, Greece, 2004. In proceedings *Quadrennial Ozone Symposium, 1-8 June 2004*, p. 1063-1064Kos Greece. Edt. C.S. Zerefos, Univ. of Athens, Greece, May 2004

United Nations Environmental Program (UNEP) (1999) Environmental effects of ozone depletions: 1998 assessment. Photochem. Photobiol. B 46:1-108.

United Nations Environmental Program (UNEP) (2003) Environmental effects of ozone depletion and its interactions with climate change: 2002 assessment, ISBN92-807-2312-X

Velders GJM (1997) Effect of greenhouse gas emissions on stratospheric ozone depletion, the National Institute for Public Health and the Environment, RIVM-report nr 72201011

Williams JE, den Outer PN, Slaper H, J.Matthijsen J, Kelfkens G. (2004) Cloud induced reduction of solar UV-radiation: a comparison of ground-based and satellite based approaches. Geophysical-Research-Letters 31(3):L03104.

World Meteorological Organization. (1999) Scientific assessment of ozone depletion 1998. Global Ozone Reaserach and Monitoring Project . Geneva, Switserland: World Meteorological Organization, 44

World Meteorological Organization. (2003) Scientific assessment of ozone depletion 2002. Global Ozone Reaserach and Monitoring Project . Geneva, Switserland: World Meteorological Organization, 44

Zerefos CS, Meleti C, Balis D, Tourpali K, Bais AF. (1998) Quasi-biennial and longer-term changes in clear sky uv-b solar irradiance. Geophysical-Research-Letters 25(23):4345-8.

Long range pollutants transport and ecological impacts

17 Contaminants, global change and cold regions

R.W. Macdonald

Dep. of Fisheries and Oceans, Inst.of Ocean Sciences, P.O. Box 6000, Sidney BC, Canada, V8L 4B2, e-mail: macdonaldrob@pac.dfo-mpo.gc.ca

17.1 Introduction

Perhaps the first example of climate change and contaminants interacting with fatal consequences occurred during the Franklin expedition of 1845-1848. Franklin's group, which entered the Canadian Archipelago hoping to discover a Northwest Passage during a cool period of climate (the end of the so-called 'little ice age' (Bradley and Jones 1993)), was doomed to failure simply by poor ice conditions. They brought with them food that was contaminated by lead from soldering in early experiments with canning (Keenleyside, et al. 1997). The harsh environmental conditions, which stalled progress and prevented retreat, also made game hard to find leading to a strong reliance on the food they brought. Although there remains controversy over what role lead poisoning might have played in Franklin's demise, this example well illustrates how the environment and contaminants can operate together; toxicity is often a consequence of two or more factors, any one of which might not cause irreparable harm by itself, and sometimes there is a shortcut pathway, or vector, that provides a surprise.

How global change might impact semi-volatile organic contaminants was considered almost a decade ago (Harner 1997), but the focus was mainly on the effect of temperature rise. Intuitively, warmer temperatures enhance volatility, force chemicals back out of water, and perhaps increase degradation rates. This sounds like good news and McKone et al. (1996) concluded from model results that exposure of aquatic foodwebs to HCB (hexachlorobenzene) would be reduced should warming occur. However, in Alpine and Arctic regions, the important change is not likely to be temperature rise per se, but rather phase change (ice-water, snow-rain) and change in systems (foodwebs, organic carbon cycle, hydrological cycle, ocean/ice interaction). These system changes alter transfers between media and contaminant magnifying processes within the environment. Recent detailed reviews of change in the Arctic and inferences of what these changes might mean for contaminants reveal a surprising number of ways that climate variation and climate change can affect pathways, and how these pathway changes then alter risks to ecosystems (Macdonald, et al. 2003; 2005). Here, it is not my intention to repeat these lengthy reviews but rather to provide a synopsis of how global change may alter the risks to humans and ecosystems from semi-volatile contaminants in cold regions.

17.2 Transient Emissions of Global Contaminants

Semi-volatile organic contaminants have been released in large quantities predominantly by agriculture and industry in Northern Hemispheric temperate countries (Fig. 17.1). For example, the estimated cumulative pesticide usage up to the year 2000 was 10 Mt for HCH (hexachlorocyclohexane),

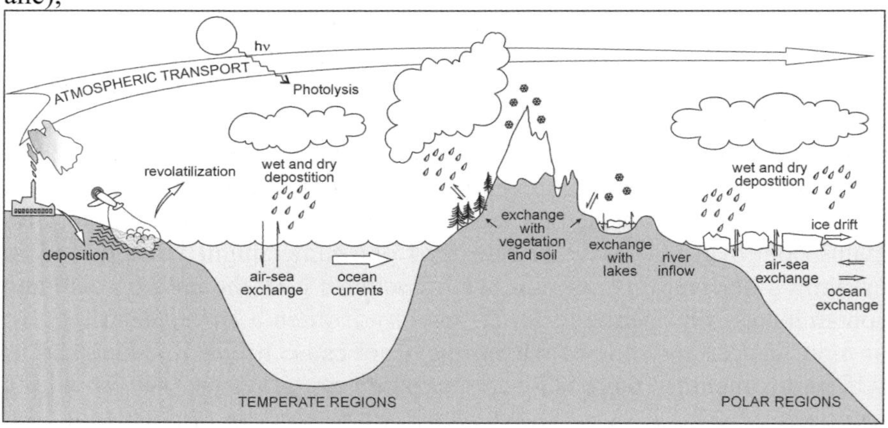

Fig. 17.1. The transport of contaminants from emission to receptor regions in the environment. On the left hand side, climate factors can operate on the transporting mechanisms including wind and water pathways and on the coupling between these two (snow and rain, air-particle partitioning, air-water exchange). On the right hand side, climate operates on the capturing mechanisms within the receptor region. Climate variation and change alters the balance between gaseous and particulate contaminant, the amount of precipitation and its type (snow/rain) or the amount of open water for air-water exchange.

4.5 Mt for DDT (dichlorodiphenyltrichloroethane), 1.33 Mt for toxaphene and 0.34 Mt for endosulfan (Li and Macdonald 2005). To these may be added industrial products such as polychlorinated biphenyls (PCBs; 1.2 Mt) or by-products such as PCDD/Fs (Macdonald, et al. 2000) and, more recently, the poly-brominated diphenyl ether (PBDE) flame retardants (estimated demand in 1999 on the order of 0.2 Mt (de Wit 2002)).

With respect to global pathways and transport, it is important to note that semi-volatile contaminants have generally entered the environment in a transient pulse, which for the classic organochlorine contaminants started in about 1940-1950. An ensuing rapid rise in emission is eventually followed by a period of decline (after about 1970) when various countries banned or restricted the use of many of these substances, or found ways to eliminate them as by-products. Eventually, we have seen complete banning of some of these chemicals, at least in the industrialized world.

Initially, the atmosphere provides the major means for rapid, widespread dispersal as discovered by atmospheric measurements in remote regions (Ottar 1981). But as environmental reservoirs become loaded (e.g., surface water, soils, vegetation, snow and ice), these media then begin to exert a greater influence on distribution patterns and processes. When industrial or agricultural emissions are stopped, the environmental reservoirs then account for all of the circulating chemical. What this means, specifically, is that the transport of any given chemical and its loading into foodwebs evolves with time and is likely to be very different between early and late periods in a chemical's history. Furthermore, variations in physical and chemical properties (volatility, partitioning, degradation) between contaminants means that, even if the emission of two chemicals has been very similar (e.g., the CFCs (chlorofluorocarbons) and HCHs (hexachlorocyclohexane), they may behave very differently once they begin cycling in the environment. This is especially true late in a contaminant's history when phase partitioning and environmental cycles (temperature, precipitation) become dominant sources of transport and transfer.

17.3 Environmental Concentrating Processes

As soon as a contaminant is released to air or water, dispersion begins. If advection and diffusion were the only environmental processes at play, contaminants would become more dilute as they drifted away from their sources such that remote regions would exhibit far lower concentrations than close to emissions. Contradicting this, the distribution of HCH in the Northern Hemisphere ocean surface water in the early 1990s, which had higher concentrations in the Arctic Ocean than in North Pacific waters near Asian points of release (Li, et al. 2002; Wania, et al. 1999), clearly demonstrates that powerful processes in the environment may reverse the effects of dispersion. The environment actually contains several surprisingly effective concentrating processes, aquatic foodwebs providing perhaps the best-known example (Muir and Norstrom 1994). But, before discussing how global change can alter transfers and transports of chemicals, it is useful to consider the concentrating mechanisms, which come in two fundamentally different forms (Macdonald, et al. 2002). The first mechanism involves transfers between phases – often described and modeled with reference to partition coefficients (e.g., air-water, air-particle, water-particle, water-oil). These processes involve the spontaneous switching of a contaminant from one medium to another, often with an increase in concentration in one of the media. For example, the counter-intuitive distribution of HCH in surface ocean waters can be explained entirely by Henry's Law gas partitioning; HCH tends to partition strongly into water and especially strongly into cold water (Li, et al. 2002; 2004). The second mechanism involves the reduction of the volume of the phase in which the contaminant is contained. Unlike the switching mechanism, this process requires an input of energy and can, therefore, push a contaminant's chemical potential above thermodynamic equilibrium. Indeed, an observation of a contaminant in, for example, a marine mammal, at concentrations above thermodynamic equilibrium with the surrounding water points strongly to a solvent reduction process.

17.4 Recent Change in the Arctic

Perhaps one of the greatest surprises of the 1990s was how rapidly change could occur in the Arctic and how many forms that change could take (Macdonald 1996; Macdonald, et al. 2003; ACIA 2005). Although many of the observed changes have been linked to the Arctic Oscillation (AO), which is a variable, or cyclic, distribution of northern hemispheric atmospheric sea-level pressure (Wallace and Thompson 2002), there is evidence that long-term change associated with greenhouse gases and global warming may in fact produce changes similar to those we have witnessed under the strong positive phase of the AO during the early 1990s. A leading question is whether any of these changes will reverse themselves as part of a cycle or whether they portend a change toward a new Arctic Ocean that is seasonally clear of ice.

What were these changes? Following an atmospheric pressure drop over the North Pole during the late 1980s, the Arctic's atmospheric circulation became more cyclonic. Winds connecting the eastern seaboard of North America and the Nordic Seas became more intense, and the storm track to the east of Greenland penetrated more deeply into the Arctic delivering larger amounts of precipitation. At the same time, there were observations of loss of ice cover over the Arctic's shelves and a general shift from multiyear ice to seasonal (first-year ice). Freeze-up became delayed by weeks in some places, and spring advanced. Within the Arctic Ocean, the ice-drift pathways changed such that the Beaufort Gyre became smaller and disconnected from the rest of the Arctic Ocean and the Transpolar Drift diverted from its normal direct path over the Lomonosov Ridge to the Greenland Sea, to a path that placed sea ice from the Laptev Sea into the Canada Basin. Water masses changed as well. The domain in the upper ocean traditionally occupied by waters of Pacific origin became smaller as Atlantic water pushed itself across the Lomonosov Ridge to realign the Pacific/Atlantic front at the Alpha Mendeleev Ridge – this change entailed a displacement of perhaps 20% of the Arctic Ocean's surface water. The Arctic's hydrological cycle appears also to have changed in several ways possibly including more precipitation, a change in the relative amounts of snow and rain, higher river inflow from Russian rivers entering the Kara and Laptev Seas (Peterson, et al. 2003), a diversion eastward of the pathways of river water entering the Russian Shelves and an alteration of the storage of freshwater within the Arctic Ocean. Over land other changes have been reported; for example, snow cover has disappeared earlier in some locations, ice has melted earlier on lakes, permafrost is melting especially in discontinuous permafrost zones, and terrestrial vegetation has responded to these changes (Vörösmarty, et al. 2001). Other intriguing changes, which are likely to be highly significant but difficult to quantify, include extreme events (e.g., storms, rain, coastal erosion; (see Stirling and Smith 2004; ACIA 2005)).

These primary physical changes then produce change in biological systems and organic carbon cycling (Macdonald, et al. 2003; Tynan and DeMaster 1997). Clearly, changes altering the distribution of, or access to, biological resources will have enormous significance for humans living in the Arctic and depending on country foods.

17.5 Contaminant Pathways and Change

17.5.1 Transport from Emission to Receptor

The first opportunity for global change to affect contaminant pathways occurs immediately upon the release of a chemical from a source (Fig. 17.1, left side). It is now well recognized that large scale pressure changes in the atmosphere (e.g., the North Atlantic Oscillation (NAO), Arctic Oscillation (AO), El Niño-Southern Oscillation (ENSO), Pacific North American (PNA) pattern) insert decadal or longer signals into wind pathways and the distribution of precipitation (Dickson, et al. 2000; Serreze, et al. 2000). On the large scale, general eastward atmospheric transport determines who is downwind from whom (e.g., see Wilkening 2001). Climate cycles then change these wind tracks to alter the length of time of transport and the connection between source and sink. Changes in precipitation along the atmospheric path alter the transfer to surfaces especially for contaminants that partition strongly into water (HCH) or strongly onto particles (highly chlorinated polychlorinated biphenyls (PCBs). The effect of climate cycles on contaminant transport is a neglected topic, largely due to the paucity of appropriate long-term observations; nevertheless, recent papers have begun to seek – and find – such connections (Ma et al. 2003; 2004). To disentangle the effects of the NAO or AO, for example, from other sources of variance requires the collection of coherent data over a climate cycle or two – that is, for a decade or longer. But there are examples where such effects may have been recorded. The decline of aerosol metal concentrations at Alert after about 1991, ascribed by Sirois and Barrie (1999) to the collapse of industry following the fragmenting of the former Soviet Union, could alternatively be explained by changes in wind and precipitation patterns at the end of the 1980s due to the strong positive phase of the AO/NAO at that time (Macdonald, et al. 2003), or to a combination of the two. More recently, Ma et al. (2004) have found correlations between climate variables (NAO, ENSO, PNA) and air concentrations of HCHs, HCB (hexachlorobenzene) and several PCBs over the Great Lakes and the Arctic. As these authors point out, the correlation may be a consequence of changed wind pathways or altered temperatures which then change upwind re-emission from soils. Finally, altered wind fields together with increased numbers of extreme events may produce a stochastic process of contaminant delivery – perhaps resulting in larger contaminant depositions at certain locations (Stirling and Smith 2004; Welch, et al. 1991) under climate change scenarios.

The ocean also provides a potent pathway of transport. In contrast to the atmosphere, which is quick to load and quick to transport contaminants (a few days for emissions to reach remote regions), the ocean is slow to load and takes years to decades to transport properties long distances. However, once the upper ocean becomes loaded with a contaminant, it can become the major reservoir and transporting mechanism (Gobeil, et al. 2001; Li, et al. 2002; Li, et al. 2004; Macdonald, et al. 2000) and, indeed, may delay atmospheric turndown subsequent to emission control by evading gas back out of the ocean (Jantunen and Bidleman 1995; Shen, et al. 2004). Accordingly, it is important to note that surface ocean currents and temperatures are important components of climate change as are connections between oceans – for example the passages at Bering and Fram Straits which allow the Arctic Ocean to exchange water with the Pacific and Atlantic Oceans respectively (Macdonald, et al. 2003). Changes in ocean currents, and the locations of fronts and surface temperature anomalies, while apparently important to budgets (Li, et al. 2004), have been little studied in the context of contaminant transport. One exception to this ironically involves artificial radionuclides: Smith et al. (1998) document the rapid transport of European coastal water into the Arctic Ocean and then around the basin margins using radiotracers like ^{137}Cs and ^{129}I which were disposed under license by the nuclear reprocessing industry.

17.5.2 Capture and Re-emission at the Receptor

The second opportunity for global change to affect contaminant pathways to cold regions occurs in the mechanisms by which contaminants are deposited (Fig. 17.1, right side). Capture may occur through precipitation (rain/snow), which removes contaminants through gas exchange (into rain or onto snow surfaces) or by removing particles onto which contaminants have exchanged. Contaminants can also deposit to surfaces through dry deposition (particles) or air-water exchange at the surface. There are three primary means by which climate change may directly alter this deposition. Firstly, there is temperature change, which alters partitioning onto particles, particularly important for chemicals with log K_{OA} values in the range of 11 to 14 (Macdonald, et al. 2003), and air-water exchange (McKone, et al. 1996; Wania 2003). Both of these affect the efficiency of removal of contaminants from the atmosphere by precipitation. Secondly, there is the amount of precipitation; the Arctic is practically a desert (20-30 cm/yr) and the efficiency of deposition for many contaminants may be rather low (e.g., see Akeredolu, et al. 1994) so that increases in precipitation as projected by change scenarios will be accompanied by increased capture of contaminants. Thirdly, the forms of precipitation (rain versus snow) have very different contaminant exchange characteristics (Macdonald, et al. 2000; Macdonald, et al. 2002; Wania and Halsall 2003; Daly and Wania 2005). For cold regions the change between snow and rain offers an enormous potential for altered pathways and for strong thermodynamic forcing when snow loses its surface area upon sintering or melting.

In addition to these primary changes, there are other climate-related variables that alter net capture. For example, the percent ice cover on lakes, rivers and oceans, which is affected by temperature and winds, controls air-water exchange. The cycle of soil temperature affects the re-emission of many contaminants (multi-hopping), and the timing and rate of melting in the Arctic affect the coupling of deposited contaminants into the hydrological cycle (see, for example Diamond, et al. 2003; Shepson, et al. 2003). Alteration in ice cover may affect primary production which then affects how the lake's sediments capture persistent organic pollutants (POPs) (Stern, et al. 2005). Change from tundra to vascular plants such as willows can also alter exchange between air and surfaces (Wania and McLachlan 2001) either upwind from cold regions or within these regions themselves. Other climate-related impacts on contaminant cycling at the site may affect the distribution after deposition; for example, deposition during climatically cold periods may store contaminant in permanent snow fields only to be released during a subsequent climatically warm period (Blais, et al. 2001).

17.5.3 Concentration Processes at the Receptor

Concentrating processes – often involving the reduction of solvent (Fig. 17.2) – offer subtle means to change the exposure, and hence risks, of contaminants to biota. Several of these processes are unique to the cryosphere and, thus, may make such regions especially vulnerable to long-range contaminant transport. In particular, snow provides an important exchange surface for many contaminants. Once snow deposits, it weathers, which may involve a slow loss of surface during sintering or a complete loss of the surface during melting. These losses of surface provide an opportunity for very large thermodynamic forcing (Macdonald, et al. 2002; Wania 1998); upon melting the sorbed contaminant must go somewhere else, either back into the atmosphere or into other media such as soil or biota.

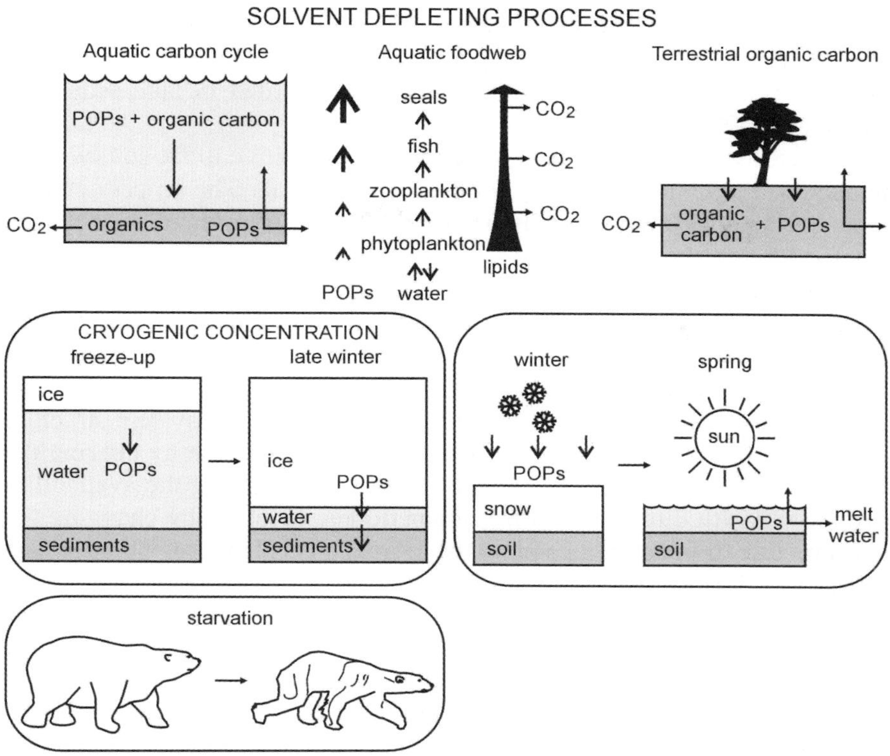

Fig. 17.2. Examples of processes in the environment that lead to high concentration of organochlorine contaminants through the removal of solvent. These include the metabolism of organic carbon in aquatic sediments and terrestrial soils, the aquatic food chain (inefficient lipid transfer up food webs), the exclusion of contaminants into a dwindling volume of water under ice as it forms, the loss of snow surfaces during melting and the loss of lipids in animals that undergo starvation cycles.

If snow is an important contributing factor to the cryosphere's attraction for contaminants, then climate change, which alters the seasonality of snowfall and snow melt, and which may alter the frequency and timing of important random events such as freezing rain, will have a great opportunity to alter that attraction. We do not understand post-depositional processes in snow sufficiently to say how these changes might occur – only to say they are likely to be important (Shepson, et al. 2003). As is the case for scavenging of contaminants by precipitation within the Arctic, it appears that most of the contaminants that enter the hydrological cycle do not get captured by the lakes they enter but, rather, pass under the ice cover in a stratified layer in spring to exit in the lake's outflow (Macdonald, et al.

2000). Inefficient processes of capture like these offer great leverage for change in that efficiency, and such changes delivered by either the hydrological cycle or the biological cycle, can be recorded in lake sediments (Stern, et al. 2005).

Aquatic systems tend to be among the most sensitive to fat-soluble contaminants due to the mechanism of biomagnification. The process of contaminant amplification depends on the inefficient transfer of fat between trophic levels (Macdonald, et al. 2002) and the presence of a large number of trophic steps – often as many as five (Kidd, et al. 1998). Climate change can impact aquatic foodwebs, sometimes radically and often rapidly (Hare and Mantua 2000; Macdonald, et al. 2003). Such changes can lead to an alteration of the number of trophic steps, which then changes the contaminant exposure of apex feeders. Alternatively, ecosystem structure and function can be changed in several ways, for example by altering the balance between pelagic and benthic coupling, by altering the availability of food (growth dilution) to a particular trophic level, or by changing the age structure of populations within the system. This latter would have particular significance, for example, for predatory lake fish where age correlates directly with mercury burden. Change can also come in the form of altered fat cycles within animals that undergo periods of starvation (i.e., solvent depletion) during which fat-soluble contaminants may be released depending on how severe the fasting has been (cf. De Swart, et al. 1996). Fat cycling and starvation cycles are common to most aquatic animals including anadromous fish (Ewald, et al. 1998; Krümmel, et al. 2003), polar bears (Stirling, et al. 1999), gray whales (Le Boeuf, et al. 1999) and gray seals (Reilly 1991) to name a few.

Change in biological systems can occur through bottom-up forcing initiated by oceanographic or limnological properties such as ice cover, nutrient supply, temperature, salinity or stratification. But change can also be effected by changing the distribution or prevalence of dominant species at the top of the foodweb (Parsons 1992; Pauly, et al. 2003). The Arctic is especially sensitive to such top-down changes because many of the dominant species depend in some way on the ice (e.g., bowhead whales, belugas, walrus, polar bears, seals, eiders, arctic cod). These top-down changes then have great potential to alter trophic structure (Tynan and DeMaster 1997) and diet of humans who depend on many of these species for food (ACIA 2005).

Change in foodwebs can also be produced by shifts in the distributions of species in response to climate change (e.g., Gaston, et al. 2003) or because exotic species are introduced accidentally into regions where they can then expand as did the zebra mussel in the Great Lakes (Morrison, et al. 1998). Other changes mediated by human activities can also be envisioned – for example the introduction of new species (e.g. the rainbow smelt into the Hudson Bay drainage (Remnant, et al. 1997)) which is now displacing indigenous species, or heavy predation on selected species such that age or trophic structure changes (Myers and Worm 2003). An ice-reduced climate in the Arctic Ocean enhances all of these possibilities.

17.5.4 Vectors and Surprises

Perhaps the most interesting – and threatening – interactions between climate change and contaminants will occur due to combinations of unforeseen circumstances. In such cases there are likely to be two components to the threat – the setting of the stage and a triggering event. Recently, it has been shown that mass mortality in marine mammals occurs due to several factors including population stress, population size, population immunity (or naïvety) to specific pathogens, condition of animals, contaminant burdens, and disease vectors (cf. Ross 2002; Daszak, et al. 2000). Each of these can be affected by climate change (e.g., Harvell, et al. 1999) and it is the complexity of this interaction that provides fertile ground for surprises. Some disease outbreaks have been observed following migrations associated with large-scale ecological change while others have derived from the introduction of viruses from domestic animals. But the addition of immunotoxic chemicals, like many of the POPs, may further set the stage for an epidemic to occur (Ross, et al. 2000). The widespread distribution of phocine distemper virus (member of Genus *Morbillivirus*) in seals from Greenland, led Dietz, et al. (1989) to speculate that a long migration of a harp seal or seals (*Phoca groenlandica*) from the Norwegian Coast to north-western Europe in 1986-87 might have provided a disease vector. The co-factors of a European seal population naïve to this virus and also manifesting a suppressed immune system due to PCB contamination would then produce an epidemic (Ross, et al. 1996; De Swart, et al. 1996). In northwestern European waters, these factors conspired to kill 20,000 harbour seals (*Phoca vitulina*; up to 60% of some populations), and a further 400 grey seals (*Halichoerus grypus*) (cf Ross 2002; Dietz, et al. 1989).

Within the Arctic, top predators would be at greatest risk to such an epidemic due to their relatively high exposure to contaminants. Marine mammals like polar bears face the added stress of change in ice climate which alters their expenditure of energy and ability to find food (Stirling and Derocher 1993; Stirling, et al. 1999). Therefore, evidence of immunosuppression in polar bears, northern fur seals and glaucous gulls (de Wit, et al. 2003) should be viewed with concern, and it seems that the polar bears of the Kara Sea, Franz Jozef Land, East Greenland and Svalbard are especially vulnerable. First, they exhibit inordinately high contaminant burdens (Norstrom, et al. 1998) and these high burdens may very well derive partly from the enhanced connectivity between this region and Europe/North America under the high AO/NAO indices of the 1990s. Second, change in ice climate and in marine ecosystems may have provided the added stress of malnourishment and a change in pathogen transmission pathway via migration (e.g., Daszak, et al. 2000). Finally, it seems these bears already have sufficient contaminant burdens to exhibit health effects (Skaare, et al. 2001).

Sea ice provides another pathway that is subject to climate change. Although the potential in the Arctic for sea ice to transport contaminants has long been recognized (Pfirman, et al. 1995), detailed studies are still lacking. Material transport by ice involves three main steps including incorporation into the ice, concentration at the ice surface during seasonal cycling and, finally, release from the ice when it melts. Suspension freezing over the middle shelves occurs when cold winds blow over open seawater at its freezing temperature; sediments are resuspended while at the same time frazil ice is produced and this scavenges suspended particles and incorporates them into the newly forming ice cover (Eicken, et al. 2000). This process, which can account for a large off-shelf transport of shelf sediment and organic carbon (Stein and Macdonald 2003), also incorporates contaminants associated with the sediments. As the ice drifts across the ocean in time scales that span two to six years (Rigor, et al. 2002), ice melts at the surface during summer but more ice is formed at the bottom of the ice in winter. This process tends to push particles and contaminants toward the surface of the ice where they may concentrate in mud balls. Additionally, as the ice drifts across the Arctic Ocean, contaminants deposit from the atmosphere depending on processes discussed above such as precipitation. Finally, when the ice travels far enough south (to the Polar Front near Fram Strait, for example), it melts and drops its load back into the water column. Climate can change the intensity of the suspension freezing process, the length of time ice takes to transport across the Arctic Ocean (and accumulate contaminants), the intensity of scavenging by snow or rain, the rate of ice export from the Arctic and the location at which the ice finally melts. Clearly, this pathway offers some 'shortcut' surprises. For example, depending on the ice-drift trajectories, radionuclide contaminants in sediments of the Russian Shelves could be transported to the Greenland Sea or into the Canadian Archipelago, becoming concentrated in a dirty surface layer along the way. Similarly, under high Arctic Oscillation index, oil spilled over the Beaufort shelves would tend to drift around in the ice of the Beaufort Gyre whereas oil spilled during low AO index would be rapidly exported into the East Siberian Sea within the ice drift.

Among the most important vectors for contaminant transport are high-trophic level aquatic biota. These are the carriers of biomagnified fat-soluble contaminants like the organochlorines and, through long migration or very dense migrations, focus these contaminants to particular locations or into human diets. Anadromous fish provide perhaps one of the clearest examples; by returning in millions to natal streams and lakes, they deliver sufficient contaminant to affect the fish native to these lakes (Ewald, et al. 1998). For lakes exhibiting high sockeye returns, this pathway can account for up to eight times the amount of PCB delivered through the atmosphere (Krümmel, et al. 2003) and the focusing of this contaminant within the salmon's life cycle may put eggs and fry at risk. Interestingly, anadromous fish are not the only means by which organochlorine contaminants can 'vector' into lakes. Evenset et al., (2004) have shown that bird guano has, likewise, impacted the nutrient and contaminant cycles within a lake on Bear Island in the Barents Sea, and birds are well-known vectors for contaminants accumulated during long migrations (Braune, et al. 1999). Biological vectors are exceptionally sensitive to climate change which is well known to alter migration pathways (Babaluk, et al. 2000; Moore, et al. 2003) and intensities (Finney, et al. 2002).

17.6 The Special Case of Mercury

Mercury has a natural biogeochemical cycle with a high level of complexity that allows the operation of climate change at many points (see Fig. 17.3). Human activities have mobilized mercury during the past two centuries such that we presently account for as much as one half to two thirds of the cycling mercury (Lamborg, et al. 2002; Pacyna and Pacyna 2001). The long atmospheric residence time (~1-2 years) easily distributes mercury globally, but it is the phenomenon of the mercury depletion event (MDE) that provides a deposition pathway unique to polar regions. MDEs are associated with polar sunrise and are initiated by photochemistry that oxidizes halogens and produces an oxidized, bioavailable form of mercury (see Fig. 17.3), which is then rapidly deposited to the surface (Lindberg, et al. 2002; Lu, et al. 2001). The marginal seas of the Arctic Ocean appear to be the prominent location for this process, which requires ice surfaces, snow, freezing temperatures and sea salt. However, it is not at all clear what conditions of ice, snow and water are optimum for this process and it has been argued that MDEs may actually be a recent phenomenon associated with change in the Arctic (Lindberg, et al. 2002). It is certain that the intensity of the process is related to climate variables including the distribution of open water, the types of ice (multi-year, first year) and the amount of primary production. In this context, the increase in first-year ice over the past decades (Yu, et al. 2004), especially in the marginal seas, may

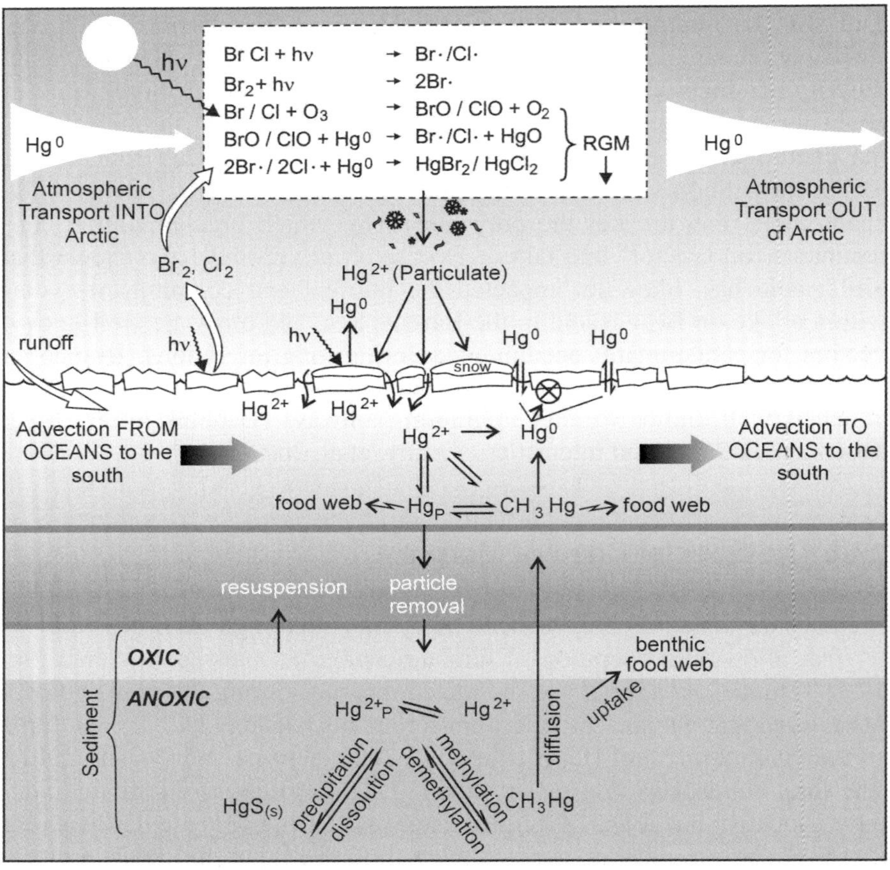

Fig. 17.3. The mercury cycle showing the many processes subject to climate variation and change. In particular, mercury depletion events are affected by ice and snow cover and the presence of sea salts, and the methylation of mercury can be enhanced by flooding land or, possibly, by melting permafrost.

have led to more intense MDEs which might then explain why mercury concentration has been increasing in some biota counter to atmospheric trends and emissions (Lindberg, et al. 2002; Pacyna and Pacyna 2001; Steffen, et al. 2003). Nevertheless, there remains much uncertainty about how much of the deposited mercury actually enters aquatic systems and how much returns to the atmosphere (Macdonald, et al. 2003; Shepson, et al. 2003), a process also subject to climate variation. Recent findings suggest that MDEs play almost no role in Hg(II) loading in the Arctic marine system (Kirk and St. Louis 2004; Stern and Macdonald 2004) and, therefore, do not provide a mechanism that can explain observed increases in Hg in marine mammal tissue (cf. Wagemann et al. 1995; Fisk et al. 2003).

Methyl-mercury is the predominant toxic form that accumulates in fish and other animals. It seems clear, therefore, that processes altering the production, accumulation or destruction of methyl-mercury are of greater concern than MDEs (Fig. 17.3) and yet we know little about what controls methyl mercury concentrations spatially and temporally. As is well known from the release of methyl mercury in response to flooding of reservoirs (Bodaly and Johnston 1992), change in hydrology may lead enhanced mercury fluxes from terrestrial soils. For the Arctic, where permafrost is melting, there is likely to be a patchwork of responses within terrestrial systems including flooding and drying. Exactly how these changes will alter the balance of methyl-mercury fluxes remains to be seen, and there are likely to be other processes that control the production and destruction of methyl mercury in the ocean about which we presently know little.

17.7 Time Series and Climate Variability

Time series provide unique insight into contaminant sources, sinks and the effects of emission controls. There are relatively few long time series available, and those we have often span a decade or less or are sporadic, containing relatively few points. For example, atmospheric measurements of organochlorines and mercury at Alert (Hung, et al. 2001; Steffen, et al. 2003) are rich in detail but span less than a decade. On the other hand, Addison et al. (1998) have measured organochlorine contaminants in seal blubber collected between 1972 and 1991 at Holman Island, but there are only four sample intervals, and Lockhart et al. (2004) have collated Hg date from belugas hunted between 1981 and 2002. Given the Arctic Oscillation's time scale of 5-7 years, it would be difficult to sort out the effects of emission controls and factors of climate change based solely on these records. Proxy measurements, which have been extracted from sediment, ice and snow cores, tooth samples, tree rings, peat bogs and museum collections can do better in terms of producing a longer record but these, too, are subject to climate variation and change. Very little concern has been paid to the potential influence of climate cycles on such records and, where it has, evidence of cycles has been found (Hung, et al. 2001; Ma, et al. 2004; Stern et al., 2005). The lesson here is clear. To understand contaminant time series we need to collect data for periods that span more than one climate cycle (e.g., decades), the interval of sampling should be at least annual, ancillary measurements (e.g., isotopic composition) should be made to help interpret the record, and trends should be collected for a set of media (e.g., air, water, sediment, low-mid-high trophic level). Furthermore, we should be archiving samples in properly funded and maintained repositories or tissue banks such that emerging issues can be studied with new tools.

17.8 Conclusions

Contaminants enter natural cycles which include transfers between media (air, water, soil, biota), magnification due to loss of phases (snow surfaces, organic carbon, fats), and permanent burial or degradation. Climate variables can impact all of these processes. In particular, each contaminant responds differently in the environment depending on its physical and chemical properties, which often act to produce contaminant separation during long-range transport similar to a global chromatography or global distillation. Air and water, the Earth's two dominant moving phases, pass over and interact with soils, vegetation, and sediments, the Earth's dominant stationary phases. Initially, the transport process is dominated by the atmosphere, but as the other media become loaded, these then dominate future transport once the emissions have been turned off. At this point, the transport, magnification and burial of chemicals occurs on time scales set by the sizes of the environmental reservoirs and the natural processes occurring within and between them. This cadence of turndown is then very much subject to change in climate systems – especially those that alter the balance between phases or exchange between reservoirs.

Nowhere in the world are there regions more sensitive to change than the poles and the mountain alpine and it is these regions we need to observe carefully in the coming decades. Recent reviews of climate change have found that temperature rise is but one manifestation of change. Others include alterations to hydrological cycles, potential stalling of thermohaline circulation, more frequent extreme events, alteration to organic carbon cycles, change in global food webs, and adaptation by humans (Houghton, et al. 2001; ACIA 2005). For contaminants distributed by long-range transport, it is the interaction with these elements of change that will produce inevitable surprises.

Acknowledgements

The author is indebted to Roland Kallenborn for encouragement to write this article and to Drs. Peter Ross and Gary Stern for helpful comments on an earlier draft of the manuscript.

References

ACIA (2005). Impacts of a warming Arctic: Arctic Climate Impact Assessment, Cambridge University Press, in press, (http://www.acia.uaf.edu).

Addison RF, Smith TG (1998) Trends in organochlorine residue concentrations in Ringed Seal (*Phoca hispida*) from Holman, Northwest Territories, 1972-1991. Arctic 51:253-261

Akeredolu FA, Barrie LA, Olson MP, Oikawa KK, Pacyna JM, Keeler GJ (1994) The flux of anthropogenic trace metals into the Arctic from the mid-latitudes in 1979/80. Atmos Environ 28:1557-1572

Babaluk JA, Reist JD, Johnson JD, Johnson L (2000) First records of sockeye (*Onchorhynchus nerka*) and pink salmon (*O. gorbuscha*) from Banks Island and other records of Pacfic salmon in Northwest Territories, Canada. Arctic 53:161-164

Blais JM, Schindler DW, Muir DCG, Sharp M, Donald D, Lafreniere M, Braekevelt E, Comba M, Backus S (2001) Glaciers are dominant source of persistent organochlorines to a subalpine lake in Banff National Park, Canada. Ambio 30:410-415

Bodaly RA, Johnston TA (1992) The mercury problem in hydro-electric reservoirs with predictions of mercury burdens in fish in the proposed Grand Baleine complex, Québec. James Bay Publication Series, Hydro-Electric Development: Environmental Impacts 3:15

Bradley RS, Jones PD (1993) 'Little Ice Age' summer temperature variations: their nature and relevance to recent global warming trends. The Holocene 3:367-376

Braune B, Muir D, DeMarch B, Gamberg M, Poole K, Currie R, Dodd M, Duschenko W, Eamer J, et al (1999) Spatial and temporal trends of contaminants in Canadian Arctic freshwater and terrestrial ecosystems: a review. Sci Total Environ 230:145-207

Daly GL, Wania F (2005) Organic contaminants in mountains. Environ Sci Technol in press

Daszak P, Cunningham AA, Hyatt AD (2000) Emerging infectious diseases of wildlife - threats to biodiversity and human health. Science 287:443-449

De Swart, RL, Ross PS, Vos JG, Osterhaus ADME (1996) Impaired immunity in Harbour Seals (*Phoca vitulina*) exposed to bioaccumulated environmental contaminants: Review of a long-term feeding study. Env Health Pers 104:823-828

de Wit CA (2002) An overview of brominated flame retardants in the environment. Chemosphere 46:583-624

de Wit CA, Hobbs K, Fisk AT, Muir DCG (2003)AMAP II Persistent organic pollutants report. In: AMAP Assessment 2002. Arctic Monitoring and Assessment Programme (AMAP), Oslo, pp. in press.

Diamond M, Helm P, Semkin R, Law S(2003)Mass balance and modelling of contaminants in lakes. In: Bidleman TF, Macdonald RW, Stow J, editors. Canadian Arctic Contaminants Assessment Report II: Sources, Occurrence and Pathways. INAC, Ottawa, pp. 187-197.

Dickson RR, Osborn TJ, Hurrell JW, Meincke J, Blindheim J, Adlandsvik B, Vinje T, Alekseev G, Maslowski W (2000) The Arctic Ocean response to the North Atlantic Oscillation. J Clim 13:2671-2696

Dietz R, Ansen CT, Have P (1989) Clue to seal epizootic. Nature 338:627

Eicken H, Kolatschek J, Lindemann F, Dmitrenko I, Freitag J, Kassens H (2000) A key source area and constraints on entrainment for basin scale sediment transport by Arctic sea ice. Geophys Res Lett 27:1919-1922

Evenset A, Christensen GN, Skotvold T, Fjeld E, Schlabach M, Wartena E, Gregor D (2004) A comparison of organic contaminants in two high Arctic lake ecosystems, Bjørnøya (Bear Island), Norway. Sci Total Environ 318:125-141

Ewald G, Larsson P, Linge H, Okla L, Szarzi N (1998) Biotransport of organic pollutants to an inland Alaska lake by migrating Sockeye salmon (*Oncorhyncus nerka*). Arctic 51:40-47

Finney BP, Gregory-Eaves I, Douglas MSV, Smol JP (2002) Fisheries productivity in the northeastern Pacific Ocean over the past 2,200 years. Nature 416:729-733

Fisk AT, Hobbs K, Muir DCG (2003) Contaminant levels, trends and effect in the biological environment, Canadian Arctic Contaminants Assessment Report II 175pp.

Gaston AJ, Woo K, Hipfner JM (2003) Trends in forage fish populations in Northern Hudson Bay since 1981, as determined from the diet of nestling thick-billed Murres *Uria lomvia*. Arctic 56:227-233

Gobeil C, Macdonald RW, Smith JN, Beaudin L (2001) Lead contamination in Arctic basin sediments tracks Atlantic water flow pathways. Science 293:1301-1304

Hare SR, Mantua NJ (2000) Empirical evidence for North Pacific regime shifts in 1977 and 1989. Prog Oceanogr 47:103-145

Harner T (1997) Organochlorine contamination of the Canadian Arctic, and speculation on future trends. Int J Environ Pollut 8:51-73
Harvell CD, Kim K, Burkholder JM, Colwell RR, Epstein PR, Grimes DJ, Hofmann EE, Lipp EK, Osterhaus ADME, et al (1999) Emerging marine diseases - climate links and anthropogenic factors. Science 285:1505-1510
Houghton JT, Ding Y, Griggs DJ, Noguer M, van der Linden PJ, Xiaosu D (2001)Climate Change 2001: The Scientific Basis.IPCC Secretariat, Geneva.
Hung H, Halsall CJ, Blanchard P, Li HH, Fellin P, Stern G, Rosenberg B (2001) Are PCBs in the Canadian Arctic atmosphere declining? Evidence from 5 years of monitoring. Environ Sci Technol 35:1303-1311
Jantunen LM, Bidleman TF (1995) Reversal of the air-water gas exchange direction of hexachlorocyclohexanes in the Bering and Chukchi Seas. Environ Sci Technol 29:1081-1089
Keenleyside A, Bertulli M, Fircke HC (1997) The final days of the Franklin Expedition: New skeletal evidence. Arctic 50:36-46
Kidd KA, Schindler DW, Hesslein RH, Muir DCG (1998) Effects of trophic position and lipid on organochlorine concentrations in fishes from subarctic lakes in Yukon Territory. Can J Fish Aquat Sci 55:869-881
Kirk JL, St. Louis V. (2004) Exchange of mercury between the atmosphere and snowpack in western Hudson Bay: Cycles of oxidation and reduction. Proceedings of the 7th International Conference on Mercury as a Global Pollutant, Ljubljana, June 27th-July 2nd.
Krümmel E, Macdonald RW, Kimpe LE, Gregory-Eaves I, Demers DJ, Smol JP, Finney B, Blais JM (2003) Anadromous PCB transport via Salmon migration. Nature 425:255-256
Lamborg CH, Fitzgerald WF, O'Donnell J, Torgerson T (2002) A non-steady-state compartmental model of global-scale mercury biogeochemistry with interhemispheric gradients. Geochim Cosmochim Acta 66:1105-1118
Le Boeuf BJ, Pérez-Cortés M, Urbán J, Mate BR, Ollervides F (1999) High gray whale mortality and low recruitment in 1999: potential causes and implications. J Cet Res Manag 2:85-99
Li Y-F, Macdonald RW (2005) Sources and pathways of selected organochlorine pesticides to the Arctic and the effect of pathway divergence on HCH trends in biota: A review. Sci Total Environ in press
Li Y-F, Macdonald RW, Jantunen LMM, Harner T, Bidleman TF, Strachan WMJ (2002) The transport of ß-hexachlorocyclohexane to the western Arctic Ocean: a contrast to a-HCH. Sci Total Environ 291:229-246
Li Y-F, Macdonald RW, Ma JM, Hung H, Venkatesh S (2004) Historical a-HCH budget in the Arctic Ocean: The Arctic mass balance box model (AMBBM). Sci Total Environ 324:115-139
Lindberg SE, Brooks S, Lin C-J, Scott KJ, Landis MS, Stevens RK, Goodsite M, Richter A (2002) Dynamic oxidation of gaseous mercury in the arctic troposphere at polar sunrise. Environ Sci Technol 36:1245-1256
Lockhart WL, Stern GA, Wagemann R, Hunt RV, Metner DA, DeLaronde J, Dunn B, Stewart REA, Hyatt CK, Harwood L, Mount K (2004) Concentra-

tions and trends of mercury in tissues of beluga whales (*Delphinapterus* leucas) from the Canadian Arctic from 1981 to 2002. Sci Total Environ submitted.

Lu JY, Schroeder WH, Barrie LA, Steffen A, Welch HE, Martin K, Lockhart L, Hunt RV, Boila G, et al (2001) Magnification of atmospheric mercury deposition to polar regions in springtime: the link to tropospheric ozone depletion chemistry. Geophys Res Lett 28:3219-3222

Ma J, Hung H, Blanchard P (2004) How do climate fluctuations affect persistent organic pollutants distribution in North America? Evidence from a decade of air monitoring. Environ Sci Technol in press

Ma J, Venkatesh S, Jantunen LM (2003) Evidence of the impact of ENSO events on temporal trends of hexachlorobenzene air concentrations over the Great Lakes. Sci Total Environ 313: 177-184.

Macdonald RW (1996) Awakenings in the Arctic. Nature 380:286-287

Macdonald RW, Barrie LA, Bidleman TF, Diamond ML, Gregor DJ, Semkin RG, Strachan WMJ, Li YF, Wania F, et al (2000) Contaminants in the Canadian Arctic: 5 years of progress in understanding sources occurrence and pathways. Sci Total Environ 254:93-234

Macdonald, R.W., T. Harner and J. Fyfe, 2005. Recent climate change in the Canadian Arctic and its impact on contaminant pathways, Sci Total Environ in press.

Macdonald RW, Harner T, Fyfe J, Loeng H, Weingartner T(2000)The Influence of Global Change on Contaminant Pathways to, within and from the Arctic. In: Macdonald RW, ed AMAP Assessment 2002. Arctic Monitoring and Assessment Programme, Oslo, Norway, pp. 65.

Macdonald RW, Mackay D, Hickie B (2002) Contaminant amplification in the environment: Revealing the fundamental mechanisms. Environ Sci Technol 36:A457-A462

Macdonald RW, Mackay D, Li Y-F, Hickie B (2003) How will global climate change affect risks from long-range transport of persistent organic pollutants? Hum Ecol Risk Assess 9:643-660

Macdonald RW, Sakshaug E, Stein R(2003)The Arctic Ocean: Modern status and recent climate change. In: Stein R, Macdonald RW, editors. The Arctic Ocean Organic Carbon Cycle Present and Past, Chapter 1.2. Springer Publishing Company, Berlin - New York, pp. 6-21.

McKone TE, Daniels JI, Goldman M (1996) Uncertainties in the link between global climate change and predicted health risks from pollution: Hexachlorobenzene (HCB) case study using a fugacity model. Risk Analysis 16:377-393

Moore S, Grebmeier JM, Davies JR (2003) Gray whale distribution relative to forage habitat in the northern Bering Sea: current conditions and retrospective summary. Can J Zool 81:734-742

Morrison HA, Gobas FAPC, Lazar R, Whittle DM, Haffner GD (1998) Projected changes to the trophodynamics of PCBs in the western Lake Erie ecosystem attributed to the presence of zebra mussels (*Dreissena polymorpha*). Environ Sci Technol 32:3862-3867

Muir DCG, Norstrom RJ (1994) Persistent organic contaminants in Arctic marine and freshwater ecosystems. Arctic Research of the United States 8:136-146

Myers RA, Worm B (2003) Rapid worldwide depletion of predatory fish communities. Nature 423:280-283

Norstrom RJ, Belikov SE, Born EW, Garner GW, Malone B, Olpinski S, Ramsay MA, Schlieve S, Stirling I, et al (1998) Chlorinated hydrocarbon contaminants in polar bears from eastern Russia, North America, Greenland, and Svalbard: Biomonitoring of arctic pollution. Arch Environ Contam Toxicol 35:354-367

Ottar B (1981) The transfer of airborne pollutants to the Arctic region. Atmos Environ 15:1439-1445

Pacyna J, Pacyna EG (2001) An assessment of global and regional emissions of trace metals to the atmosphere from anthropogenic sources worldwide. Environ Rev 9:269-298

Parsons TR (1992) The removal of marine predators by fisheries and the impact of trophic structure. Mar Poll Bull 25:51-53

Pauly D, Alder J, Bennett E, Christensen V, Tyedmers P, Watson R (2003) The future for fisheries. Science 302:1359-1361

Peterson BJ, Holmes RM, McClelland JW, Vörösmarty CJ, Lammers RB, Shiklomanov AI, Shiklomanov IA, Rahmstorf S (2002) Increasing river discharge to the Arctic Ocean. Science 298:2171-2173

Pfirman SL, Eicken H, Bauch D, Weeks W (1995) The Potential transport of pollutants by Arctic sea ice. Sci Total Environ 159:129-146

Reilly, J. J. (1991). Adaptations to prolonged fasting in free-living weaned grey seal pups. Am J Physiol 260:R267-R272.

Remnant R, Graveline PG, Bretecher RL (1997) Range extension of the rainbow smelt, *Osmerus mordax*, in the Hudson Bay drainage of Manitoba. Can Field-Nat 111:660-662

Rigor IG, Wallace JM, Colony R (2002) Response of sea ice to the Arctic Oscillation. J Clim 15:2648-2663

Ross PS (2002) The role of immunotoxic environmental contaminants in facilitating the emergence of infectious diseases in marine mammals. Hum Ecol Risk Assess 8:277-292

Ross P, De Swart R, Addison R, Van Loveren H, Vos J, Osterhaus A (1996) Contaminant-induced immunotoxicity in harbour seals: wildlife at risk? Toxicology 112:157-169

Ross PS, Vos JG, Birnbaum LS, Osterhaus ADME (2000) PCBs are a health risk for humans and wildlife. Science 289:1878-1879

Serreze MC, Walsh JE, Chapin FSI, Osterkamp T, Dyurgerov M, Romanovsky V, Oechel WC, Morison J, Zhang T, et al (2000) Observational evidence of recent change in the northern high-latitude environment. Climatic Change 46:159-207

Shen L, Wania F, Lei YD, Teixeira C, Muir DCG, Bidleman TF (2004) Hexachlorocyclohexanes in the North American atmosphere. Environ Sci Technol 38:965-975

Shepson P, Matrai P, Barrie L, Bottenheim J (2003) Ocean-atmosphere-sea ice-snowpack interactions in the Arctic, and global change. Eos Trans Am Geophys Union 84:349, 355

Sirois A, Barrie LA (1999) Arctic lower tropospheric aerosol trends and composition at Alert, Canada: 1980-1995. J Geophys Res 104:11599-11618

Skaare JU, Bernhoft A, Wiig Ø, Norum KR, Eide DM, Derocher AE (2001) Relationships between plasma levels of organochlorines, retinol and thyroid hormones from polar bears (*Ursus maritimus*) at Svalbard. J Toxicol Environ Health 62:227-241

Smith JN, K.M. Ellis, Kilius LR (1998) ^{129}I and ^{137}Cs tracer measurements in the Arctic Ocean. Deep-Sea Res I 45:959-984

Steffen A, Schroeder W, Poissant L, Macdonald RW (2003) Mercury in the Arctic atmosphere. Canadian Arctic Contaminants Assessment Report II :120-138

Stein R, Macdonald RW, editors. The Organic Carbon Cycle in the Arctic Ocean. Springer Publishing Company,: Berlin - New York; 2003.

Stern, GA, Braekevelt E, Hel PA, Bidleman TF, Outridge PM, Lockhart WL, McNeeley R, Rosenberg B, Ikonomou MG, Hamilton P, Tomy GT, Wilkinson P (2005) Modern and historical fluxes of halogenated organic contaminants to a lake in the Canadian Arctic as determined from annually laminated sediment cores. Sci Total Environ in press.

Stern GA, Macdonald RW (2004) Biogeographical provinces of Hg in the Beaufort and Chuchi Seas: Results from the SHEBA drift. Environ Sci Technol submitted.

Stirling I, Derocher AE (1993) Possible impacts of climatic warming on polar bears. Arctic 46:240-245

Stirling I, Lunn NJ, Iacozza J (1999) Long-term trends in the population ecology of polar bears in western Hudson Bay in relation to climatic change. Arctic 52:294-306

Stirling I, Smith TG (2004) Implications of warm temperatures and an unusual rain event on the survival of ringed seals on the coast of southeastern Baffin Island. Arctic 57:59-67

Tynan C, DeMaster DP (1997) Observations and predictions of Arctic climate change: Potential effects on marine mammals. Arctic 50:308-322

Vörösmarty CJ, Hinzman LD, Peterson BJ, Bromwich DH, Hamilton LC, Morison J, Romanovsky VE, Sturm M, Webb RS. The Hydrologic Cycle and its Role in the Arctic and Global Environmental Change: A Rationale and Strategy for Synthesis Study. , Fairbanks, Alaska: Arctic Research Consortium of the U.S.; 2001, pp.

Wallace JM, Thompson DWJ (2002) Annular modes and climate prediction. Physics Today February:28-33

Wania F (1998) On the origin of elevated levels of persistent chemicals in the environment. Eviron Sci Pollut Res 6:11-19

Wania F (2003) Assessing the potential of persistent organic chemicals for long-range transport and accumulation in polar regions. Environ Sci Technol 37:1344-1351

Wania F, Halsall C (2003) Air/snow exchange of persistent organic pollutants. Canadian Arctic Contaminants Assessment Report II :185-190

Wania F, Mackay D, Li Y-F, Bidleman TF, Strand A (1999) Global chemical fate of a-hexachlorocyclohexane. 1. Evaluation of a global distribution model. Environ Sci Technol 18:1390-1399

Wania F, McLachlan MS (2001) Estimating the influence of forests on the overall fate of semivolatile organic compounds using a multimedia fate model. Environ Sci Technol 35:582-590

Welch HE, Muir DCG, Billeck BN, Lockhart WL, Brunskill GJ, Kling HJ, Olson MP, Lemoine RM (1991) Brown snow: A long-range transport event in the Canadian Arctic. Environ Sci Technol 25:280-286

Wilkening K (2001) Trans-Pacific air pollution: scientific evidence & political implications. Water Air Soil Pollut 130:1825-1830

Yu Y, Maykut G, Rothrock DA (2004) Changes in the thickness distribution of arctic sea ice between 1958-1970 and 1993-1997. J Geophys Res in press

18 Modeling of long-range transport of contaminants from potential sources in the Arctic Ocean by water and sea ice

Vladimir Pavlov

Norwegian Polar Institute, Polar Environmental Centre, N-9296, Tromsø, Norway, e-mail: pavlov@npolar.no

18.1 Introduction

The geographical position and climatic features of the Arctic seas mean that their ecological balance is sensitive to disturbance by inputs of man-made pollutants. The Arctic seas represent zones where pollutants naturally accumulate and pollutants are transported between regions where there is active exploitation of natural resources and pollution, and the ecologically clean regions of the central Polar Basin. The processes involved in the transport, transformation and accumulation of contaminants from different possible sources are important in assessing whether we are to forecast the fate of potential pollutant releases. These sources and potential sources are described by several reports and papers (Yablokov Commission Report 1993; Aarkrog 1993; Pavlov and Pfirman 1995; Duursma and Carroll 1996; ANWAP 1997; AMAP 1997; AMAP 1998; Champ et al. 1998; Yablokov 2001). In some Arctic seas, such as the Barents and Kara seas, there were earlier local sources of anthropogenic origin, connected with the nuclear trials at the test site of Novaya Zemlya and dumping of radioactive waste over their areas. Extensive dumping of nuclear materials in the Kara and Barents seaы marine environment is listed in the Yablokov Commission Report (1993) and in Yablokov (2001). Low level liquid radioactive wastes are stored in the following regions of Russia in large volumes: by the northern fleet in the Kola Peninsula in the Murmansk and Severodvinsk regions, and by the Pacific fleet in the Russian Far East in the naval ship yards at Vladivostok. The combined capacity of all Northern Fleet containers amounts to 10,000 m^3 and annual production is estimated to be about 20,000 m^3. Approximately 30 to 40,000 m^3+ of solid radioactive waste is stored at different sites in the Russian Far East and Northwest

(ANWAP 1997; AMAP 1998; Champ et al. 1998). The discharges of fresh water from the large Arctic rivers that have huge catchment areas draining water from land areas and industrial zones also contribute to the input of pollutants into the Arctic Ocean.

Ocean currents and drifting ice are among the most important mechanisms of pollutant transport (Nürnberg et al. 1994; Emery et al. (1997); Pfirman et al. 1997; Nilsson 1997; Rigor and Colony 1997; Smith et al. 1998; Smith and Ellis 1999; Zhang et al. 2000; Rigor et al. 2002; Zhang et al. 2003; Pfirman et al. 2004a; Pfirman et al. 2004b). Severe natural conditions and the year-round presence of drifting ice make direct full-scale observations of currents difficult and expensive. Numerical modelling, supported and validated by in situ field observations, is therefore the only practical possibility for gaining an understanding of water circulation in the Arctic Ocean on different spatial and temporal scales. Many models describing transport and transformation of various pollutants in the water environment of the Arctic have appeared in recent years inspired by increasing anthropogenic effects, especially in coastal zones (Preller and Cheng 1995; Pavlov et al. 1995; Harms 1997; Scott et al. 1997; AMAP 1998; Nies et al. 1999; Harms and Karcher 1999; Harms et al. 2000; Karcher et al. 2004 and others). In these papers, modelling results for the spreading of contaminants from individual sources, mostly located in the Nordic seas and Kara Sea, were discussed. For example, Harms (1997) has described the application of 3-D, baroclinic circulation models to study the dispersal of radioactivity in the Barents and Kara seas. Release is expected to occur at underwater dump sites for radioactive waste in the Kara Sea, used by the former Soviet Union. To cover the wide range of a possible radionuclide dispersion, two different spatial scales were considered: i) the regional scale, which covered the shelves of the Barents and Kara seas and ii) the local scale, which is focused on the bay where some of the dumping took place. The regional-scale model results have suggested that, even for a worst case scenario, the radioactive contamination of Siberian coastal waters would be relatively small compared to observations in other marine systems (e.g. the Baltic Sea and the Irish Sea). Realistic gradual release scenarios show very low concentrations in the central and eastern Kara Sea. Significant contamination of shelf seas such as the Laptev Sea, the Arctic Ocean or the Barents Sea by radioactive waste dispersion from the Kara Sea seems unlikely.

Nies et al. (1999) presented a review of results from a joint project carried out in Germany in order to assess the consequences to the marine environment from the dumping of nuclear waste in the Kara and Barents seas. The project consisted of experimental work on measurements of radionuclides (^{137}Cs, ^{90}Sr, $^{239+240}Pu$, ^{238}Pu, ^{241}Am) in samples from the Arctic marine environment, and numerical modelling of the potential pathways and dispersion of contaminants in the Arctic Ocean. The role of transport by sea ice from the Kara Sea into the Arctic Ocean was assessed. This transport process might be considered as a rapid contribution of pollutants due to entrainment of contaminated sediments into sea ice, followed by export from the Kara Sea by the Transpolar Ice Drift, and subsequent release in the Arctic Ocean in the region of the East Greenland Current. Numerical modelling of pollutant dispersion from the Kara and Barents seas was carried out both on a local scale, for the Barents and Kara seas, and for long range dispersion into the Arctic and Atlantic oceans. 3-D baroclinic circulation models were applied to trace the transport of pollutants. Modelling results show no significant pollution even for worst case scenarios from the radioactive waste dumped in the Kara Sea to other seas in the Arctic or North Atlantic (as in Harms 1997). The results from the dispersion models suggest that, even for worst case scenarios, the contamination of Arctic waters and North Atlantic areas is relatively minor compared to pre-contamination from Sellafield, or global fallout from nuclear weapon testing in the 1960s. Long range simulations of Sellafield discharges of ^{137}Cs since the 1960s correlates well with measured levels. Harms et al. (2000) investigated the role of Siberian river runoff for the transport of possible river contaminants in the Arctic Ocean. 3-D coupled ice-ocean models of different horizontal resolution were applied to simulate the dispersion of river water from the Ob, Yenisei and Lena. Circulation model results explain the main pathways and transit times of Siberian river water in the Arctic Ocean. Kara Sea river water clearly dominates in the Siberian branch of the Transpolar Drift, while the water from the Lena dominates in the Canadian Branch. The model confirms that contaminant transport through sediment laden sea ice offers a short and effective pathway for pollutant transport from Siberian rivers to the Barents and Nordic seas. Karcher et al. (2004) have compared the simulated dispersion of ^{99}Tc in surface water from the sources to the Nordic Seas and the Arctic Ocean as calculated by a hydrodynamic model and in assessment box model with field-observations from 1996 to 1999 to study concentrations, pathways and travel times. The observations cover the northern part of the Nordic Seas. The main sources of ^{99}Tc are global fallout from nuclear weapon testing, and discharges from reprocessing plants for spent nuclear fuel in Northwestern Europe. Radioactive wastes have been discharged from the reprocessing plant at Sellafield (UK) into the Irish Sea, and at La Hague (France) into the English Channel since 1952 and 1966 respectively. The model results were consistent with the observations and have shown typical pathways of dissolved radionuclides from the Irish Sea via the North Sea along the Norwegian coast. The results of the hydrodynamic model have indicated a large variability of surface concentrations in the West Spitsbergen Current.

The importance of sea ice for the climate has led to many efforts to develop different models which study sea-ice morphology, dynamics and thermodynamics of sea ice. Many papers, e.g. Zwally and Walsh (1987), Barry et al. (1993), Zhang et al. (2003), give very good overviews of sea ice modelling. Some comparisons between different modelling approaches and results are given in Pavlov et al. (2004). Large-scale sea ice modelling requires effective use of commensurate observational data sets (sea ice concentration, sea ice extent, sea ice motion) on daily to inter-annual timescales for initialisation and verification. Recent sea ice observation data, measured by satellites, combined with sea ice motion derived from buoys in the International Arctic Buoy Programme (IABP), initiated a large number of studies of ice motion and variability of ice conditions in the Arctic Ocean. This also includes the problem of pollutant transport by sea ice, and provides a good opportunity to develop, compare and verify sea ice models (Pfirman et al. 1997).

Here we discuss the results of our simulation of the transport of passive non-conservative tracers by currents from a number of possible sources in the Arctic Ocean and Nordic seas. For simulation vectors of ocean currents we used a 3-D baroclinic ocean model developed at the Arctic and Antarctic Research Institute (AARI, St.Petersburg, Russia). Model is documented in Pavlov (1995) and Pavlov and Pavlov (1999).

We also include the possible transport of contaminants via sea ice. To estimate the transport of passive tracers by sea ice from potential sources of contamination in the Arctic Ocean we used the Ice Statistical Model (ISMO) developed at the Norwegian Polar Institute (Korsnes et al. 2002; Pavlov et al. 2004). The main approach of ISMO is to reveal the statistical relationship between atmospheric forcing and sea ice conditions derived from satellite imagery, in order to reconstruct the ice drift and ice concentrations, estimate the sea ice fluxes through the main straits in the Arctic Ocean and Arctic marginal seas and calculate forward and backward trajectories of the ice drift from any point in the Arctic Ocean.

18.2 Methods and data

18.2.1 Model of the dispersion of contaminants by ocean currents

To model the dispersion of a soluble contaminant we use the 3-D Eulerian transport equations for a non-conservative passive tracer with properties corresponding to anthropogenic radionuclides such as ^{137}Cs and ^{90}Sr.

$$\frac{\partial c}{\partial t}+u\frac{\partial c}{\partial x}+v\frac{\partial c}{\partial y}+w\frac{\partial c}{\partial z}=-\lambda c+\frac{\partial}{\partial z}\eta\frac{\partial c}{\partial z}+\mu\Delta c+Q\delta(r-r_0) \quad (1)$$

where c is the contaminant concentration; u,v,w are the components of the current velocity in the x, y and z directions; λ is the decay constant; η is the vertical diffusion coefficient; μ is the horizontal diffusion coefficient; $r_0=(x_0,y_0,z_0)$ gives the co-ordinates of the source; δ is Dirac delta function; and Q is the source strength.

Initial conditions are c=0 at t=0 and the boundary conditions are:

$$\frac{\partial c}{\partial n}=0 \qquad u_n\geq 0; \qquad\qquad c=0 \qquad u_n<0 \quad (2)$$

where u_n is the projection of the current speed vector to the external normal to the surface S, restricting the calculation area. These conditions allow for the transport of a contaminant out of the model domain. In general Eq. (1) describes two different physical processes. The first is the transfer of a substance, with its conservation along the trajectory (LHS Eq. (1)) and is described by the equation:

$$\frac{\partial c}{\partial t}+u\frac{\partial c}{\partial x}+v\frac{\partial c}{\partial y}+w\frac{\partial c}{\partial z}=0 \quad (3)$$

The second physical process is connected with the diffusion of the substance and its disintegration in the process of spreading from the source (RHS Eq. (1)) and is described by the equation:

$$\frac{\partial c}{\partial t}=-\lambda c+\frac{\partial}{\partial z}\eta\frac{\partial c}{\partial z}+\mu\Delta c+Q\delta(r-r_0) \quad (4)$$

Marchuk (1982) has shown that splitting the initial Eq. (1) into the two Eqs. (3) and (4), by physical processes, gives a form more suitable for numerical application. Eq. (4) is discretized as finite differences on an Arakawa C-grid and Eq. (3) is solved by the "Flux-Corrected Transport" (FCT) method (Boris and Book 1973) which has low viscosity whilst preserving the monotone and conservative character of the tracer field.

Horizontal and vertical velocities of currents for Eq. (1) were simulated using a 3-D baroclinic ocean level-type model (Pavlov and Pavlov 1999). The model is based on the Boussinesq approximation of the non-linear primitive equations of motion. Model equations including the sea ice parameterisation and numerical methods are in detail described in Pavlov and Pavlov (1999). The Arctic Ocean model has 22 layers in the vertical scale. We used a spherical rotated grid with a spatial resolution of 55 km. The model was driven with monthly mean climatologic atmospheric pressure and wind calculated from NCEP/NCAR gridded data (http://dss.ucar.edu/datasets/) and 3-D seasonal mean potential density calculated from data in the US-Russian atlas (US-Russian Joint Atlas 1996). The climatologic annual cycle of the discharge of the major rivers into the seas of the Arctic Ocean (Ivanov 1976; Omstedt et al. 1994) and data on water exchange through the Bering Strait and straits of the Canadian Archipelago were used to assign boundary conditions at open boundaries. Mean water exchange values through Bering Strait are sufficiently well known and estimates by different authors for different times are very close. For this model the water transport through Bering Strait was prescribed to be 1.2 Sv in summer and 0.7 Sv in winter (Pavlov and Pavlova 1999). The water transport from the Arctic Ocean through the Canadian Archipelago straits is taken from Coachman and Aagaard (1988) to be 2.0 Sv. The water exchange at open boundaries in the straits between Greenland and Iceland and between Iceland and the European continent were not prescribed, but the emission condition for the vertical-averaged horizontal velocity $\bar{u}$ was assumed to be $\bar{u} = \xi(g/h)^{1/2}$, where ξ -sea level elevation, g - Earth gravity acceleration and h -depth.

The annual cycle of all forcing and boundary conditions was not changed during the simulations and as a result after about 30 years we obtained a stable annual cycle of the 3-D water circulation in the Arctic Ocean. Harms et al. (2000) using a similar approach achieved a stable seasonal cycle in the ice and the upper ocean circulation after 35-years running their coupled ice-ocean model of the Arctic Ocean. The annual cycle of the water circulation obtained is used in the integration of Eq. (1).

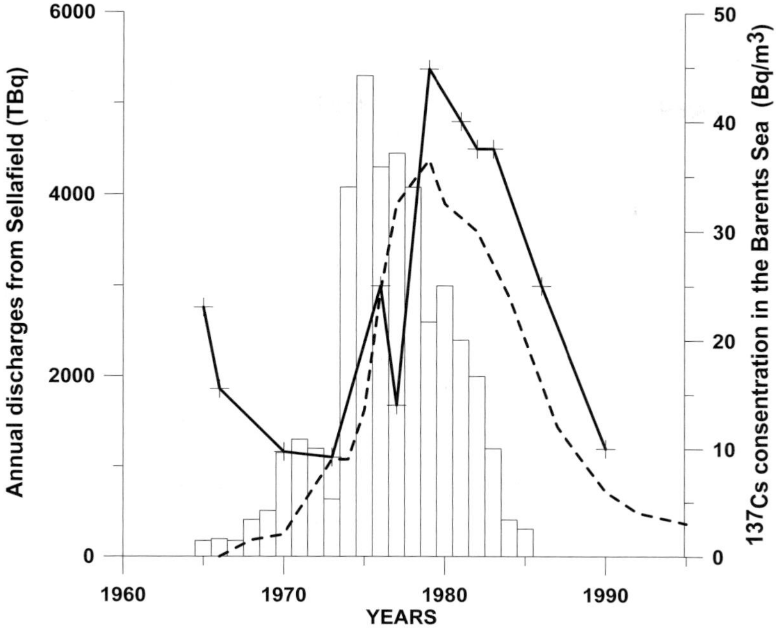

Fig. 18.1. Annual discharge (TBq) of ^{137}Cs from Sellafield (bars); ^{137}Cs concentration (Bq m^{-3}) in the Barents Sea surface waters. Observations from Kershaw and Baxter (1995)-solid line, and modelling results-dashed line (Kulakov and Pavlov 1999)

Kulakov and Pavlov (1999) have verified this model using data on the release of ^{137}Cs from the Sellafield reprocessing plant into the Irish Sea on the Cumbrian coast of England. The discharge from Sellafield started in 1952 (Dahlgaard 1995; Israel et a1. 1993; Matishov et al. 1994; Smith et al. 1990, 1998; Smith and Ellis 1999) and peaked in the mid- to late-1970s (Fig. 18.1). Almost all ^{137}Cs entering the sea with waste from Sellafield was transported from the Irish Sea to the North and Norwegian seas and then via the North-Atlantic Current to the Barents Sea and the Arctic Ocean. This passage has been well described previously; summary is presented in Kershaw and Baxter (1995). Using this model the redistribution and transformation of ^{137}Cs from the source in coastal water of the British Islands was calculated for 30 years from 1965 to 1995 with a time interval of one day. The strength of the source was prescribed in accordance with the real discharge volumes (Fig. 18.1). For the calculations the coefficients of the vertical and horizontal diffusion were assumed to be equal to 5 and $2 \cdot 10^5$ cm^2/s respectively. Kulakov and Pavlov (1999) reported that the calculated ^{137}Cs distribution is both qualitatively and quantitatively close to that in the Nordic seas (Kershaw and Baxter 1995; Vakulovsky et al. 1993).

Fig. 18.1 shows the comparison of the observed and the calculated time variations of the levels of ^{137}Cs concentrations in the surface water of the Barents Sea. Good agreement suggests that the model describes well the processes of transfer and transformation of ^{137}Cs.

18.2.2 Model of the sea ice transport

Sea ice trajectories have been simulated by the ISMO which was well documented in Korsnes et al. (2002) and Pavlov et al. (2004). In the development of ISMO, a multiple linear regression model was used to establish the statistical relationship between: 1) ice motion and the spatial structure of the sea level atmospheric pressure (SLP); and 2) sea ice concentration and both sea surface temperature (SST) and the spatial structure of the sea level pressure (SLP) in the each point of simulating domain.

The following observational data was used for statistical analysis: Monthly mean sea ice concentration and ice drift from the Special Sensor Microwave /Imager data set from the EOS Distributed Active Archive Center (DAAC) at the National Snow and Ice Data Center, University of Colorado, Boulder, CO. Monthly mean gridded sea level atmospheric pressure (SLP) fields and sea surface temperature (SST) from the NCEP/NCAR data set.

This model was verified by comparing simulated and observed ice conditions such as ice drift, ice concentration, ice fluxes through the main straits of the Arctic Ocean and ice tracks. The comparisons showed quite good agreement between the ISMO results and the observational data, and also with results of other models (see Pavlov et al. 2004). Using the ISMO we have computed the vectors of ice velocity for the period 1899-2000.

18.3 Contaminant Transport

18.3.1 Dispersion of passive tracers by water from potential sources of contaminants

Numerical experiments simulating the transport of contaminants from possible sources in different parts of the Arctic Ocean have been performed using calculated 3-D current fields. For these simulations potential pollut-

ant sources have been located in the vicinity of river-mouths of major rivers flowing into the Arctic Ocean, as well as in the Bering Strait, in the bottle-neck of the White Sea, in the Faeroe-Shetland Channel and also in the region of the wreck of the "Komsomolets" nuclear submarine.
What is the reason of such chooses?

The Faeroe-Shetland Channel (not far from the Sellafield reprocessing plant), the nuclear submarine wreck "Komsomolets" as well as the Kara Sea with the Ob and Yenisei river mouths, are chosen due to the public, political and scientific concern, and because there are many papers and reports already published from these sites.

The cumulative, and decay corrected total ^{137}Cs radioactivity released in Sellafield amounts to approximately 40 PBq, 14 PBq of which entered the Arctic regions mainly through the Faeroe-Shetland Channel. Maximum releases from Sellafield occurred in 1975. Releases of ^{137}Cs have been continually reduced since that time (Kershaw and Baxter 1995).

Radioactive sources dumped in the Kara Sea mainly include 17 nuclear ship reactors, seven of them still containing spent fuel. The total radioactive inventory at the time of dumping was 37 PBq (Yablokov Commission Report 1993; NPRA 1996; AMAP 1998). By 1994 this had decayed to approximately 4.7 PBq. The dominant nuclides are ^{137}Cs, ^{90}Sr, ^{63}Ni and ^{241}Pu. The amount of ^{137}Cs is estimated to be approximately 1 PBq for 1994 (IAEA 1997).

The sunken submarine "Komsomolets" contains one nuclear reactor with an inventory of long lived radionuclides comprising of 2800 PBq of ^{90}Sr and 3100 PBq of ^{137}Cs. Two nuclear torpedoes with mixed uranium/plutonium warheads, situated in the forepart of the hull contain about 16 PBq of weapons-grade plutonium (AMAP 1998).

The Kara Sea is distinguished from the other Siberian shelf seas by the strong influence of continental discharge. It receives about 55 % (1290 km^3/year) of the total river runoff discharged to the entire Siberian Arctic. The annual discharge from the Ob River is 400 km^3 and from the Yenisei River is 630 km^3 (Soviet Arctic 1970; Pavlov and Pfirman 1995). The nuclear fuel reprocessing plant Mayak is situated around the headwaters of the river Techa, which ultimately drains into the Kara Sea via the Ob River. The waste management system has been developed on a series of natural and artificial reservoirs and drainage canals. A total of 4000 PBq (decay corrected to 1994) comprising mainly of ^{90}Sr and ^{137}Cs has been released to this restricted system (NRPA 1997). Tomsk reprocessing plant also enters the drainage basin of the Ob. The plant's storage ponds contain an estimated 4800 PBq. The storage ponds at Krasnoyarsk reprocessing plant are believed to contain an inventory of about 2 PBq. As at the other sites, there is a risk of contaminated groundwater migrating into the rivers, in this case the Yenisei river. (AMAP 1998). The observations carried out by Roshydromet, Russia (Vakulovsky et al. 1993) indicate that the amount of ^{90}Sr transported by the Ob during 1961-1989 is about 1 PBq, and 0.1 PBq ^{137}Cs.

In contrast to the possible sources of contamination mentioned above, information about contaminant levels at other possible sources such as the Lena, Kolyma and Mackenzie river mouths, the bottle-neck of the White Sea and the Bering Strait are fragmental or practically absent.

The large rivers of Siberia such as Lena and Kolyma, and Mackenzie in Canada transport large amounts of water over long distances and on their way to the Arctic seas. The annual discharge of the Lena and Kolyma is 525 km^3 and 132 km^3, respectively, and the Mackenzie runoff is 333 km^3/year (AMAP 1998). These areas include agricultural and industrial regions, and also regions of mining, and oil and gas explorations in Siberia and Canada. So, these rivers are expected to be a key source for considerable quantities of several different contaminants

About 1 Sv enter the Arctic Ocean from the Pacific Ocean with the Pacific current through the Bering Strait (Pavlov and Pavlova 1999). These waters wash the industrial regions of the Russian Far East and Northern America where possible sources of contamination are located, such, for example, as the large shipyards for nuclear-powered submarines near Vladivostok, and the regions of drilling activity in Alaska.

The bottle-neck of the White Sea was chosen because the city of Severodvinsk lies on the delta of the Dvina River, close to Archangelsk, and has one of the largest shipyards for nuclear-powered submarines in the former Soviet Union. It is also a disposal site for military nuclear waste (Nilsen and Bøhmer 1994; Champ et al. 1998). Severodvinsk is potentially a major source of radioactive contamination in the White Sea.

However, even if we do not have information about contaminants from these regions, the selected sources could release large amounts of contaminants in the future. For example, release of contaminants through accidents during production, transport, waste disposal and storage, oil and gas exploration and exploitation, and also nuclear submarine waste generated from operations in the Northwest and Far East, or accidents involving nuclear weapons. Potential contaminant release that may occur in the Arctic in the future and source related assessments of potential release are well documented in AMAP (1998).

As we do not have information about a possible release volume, pollutant concentration in all these sources has been set to 100 arbitrary units. The non-conservative parameter has been prescribed to 30 years, this being approximately equal to the averaged half-life of the dangerous anthropogenic long-life radionuclides, such as ^{137}Cs and ^{90}Sr.

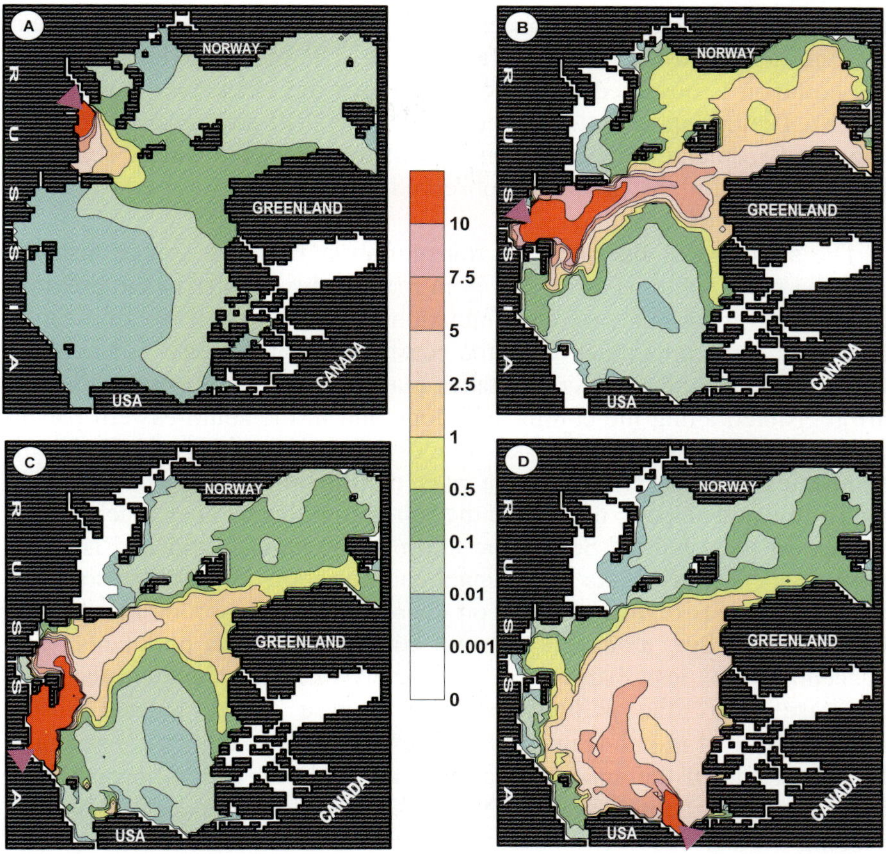

Fig. 18.2. Distributions of pollutant concentration in the surface layer of the Arctic Ocean after 15 years of release from sources in the river-mouths of the major rivers: A - Ob and Yenisei rivers, B - Lena river, C - Kolyma river, and D - Mackenzie river. The scale shows % of pollutant in relation to its concentration at source. ▲ - source location.

Distributions of the pollutant concentration from all the permanently acting sources in the river-mouths of major rivers in the surface layer of the Arctic Ocean after 15 years from the beginning of the release are presented in Fig. 18.2. Depths of the sources near the river-mouths have been set equal to 5 m.

Pollutant spreading from sources in the river mouths of the Siberian shelf is directed predominantly to the north-west. Pollutants from the source in the Kara Sea in the region of the Ob-Yenisei river mouth (Fig. 18.2a) cover a greater part of the Nordic seas, the Laptev Sea and the area near the northern coast of Greenland. However, pollutant concentrations exceeding 5% of the source concentration in the surface layers of the ocean are only observed within the Kara Sea area. The most rapid spreading of the pollutant is observed from the source near the Lena river mouth (Fig. 18.2b). The pollutants entering the Transpolar Drift are transported to the coast of Greenland, spreading later over the region of the Nordic seas. Pollutant concentrations of 5% are observed in the Fram Strait and near the northern coast of Greenland.

The structure of the pollutant transportation from the source near the Kolyma river mouth (Fig. 18.2c) is in many ways similar to the previous one. However, in this case contamination of the Nordic seas significantly decreases, and contamination of the coastal zone of the East Siberian and Chukchi seas increases. As this takes place concentrations exceeding 5% are registered along the continental slope and in the south-eastern part of the Laptev Sea and in the coastal zone of the East Siberian Sea. Pollutants from the source near the Kolyma river mouth are transported also along the continental slope to the east in the sub-bottom layer. They reach the sea surface in the Chukchi Sea to the north of Wrangel Island and near the north-east coast of Alaska. A pollutant concentration of 5% of the source is registered in this region, whereas on the surface of the northern part of the East Siberian Sea, along the transportation route of the pollutants, their concentration is less than 3%. A completely different situation occurs in the case where the pollutant spreads from a source located in the Mackenzie river mouth (Fig. 18.2d). The pollutants in this case essentially fill the region of the anticyclonic gyre, and only an insignificant part of them enters the Laptev Sea and Nordic seas. Pollutant concentrations exceeding 5% are observed in the surface layers in the anticyclonic gyre in the Canadian Basin.

In the second set of numerical experiments, pollutant with 100% concentration are released in the region of the wreck of the "Komsomolets" nuclear submarine in the sub-bottom layer at the depth 1700 m, in the bottle-neck of the White Sea, in the Faeroe-Shetland Channel and in the Bering Strait in the streams of the Atlantic and Pacific currents, respectively, at the depth 5 m. The pollutant-spreading from these sources after 15 years from the beginning of their activity is given in Fig. 18.3.

The pollutants from the "Komsomolets" wreck region crop out at the surface mainly in the area of the central part of the Barents Sea and the north-western part of the Kara Sea (Fig. 18.3a). Regions with concentrations higher than 5% are located near the north-western part of Norway. The pollutants from the source in the surface layer near the bottleneck of the White Sea (Fig. 18.3b) spread to the east along the coasts of the Barents and Kara seas and to the north along the western coast of Severnaya Zemlya.

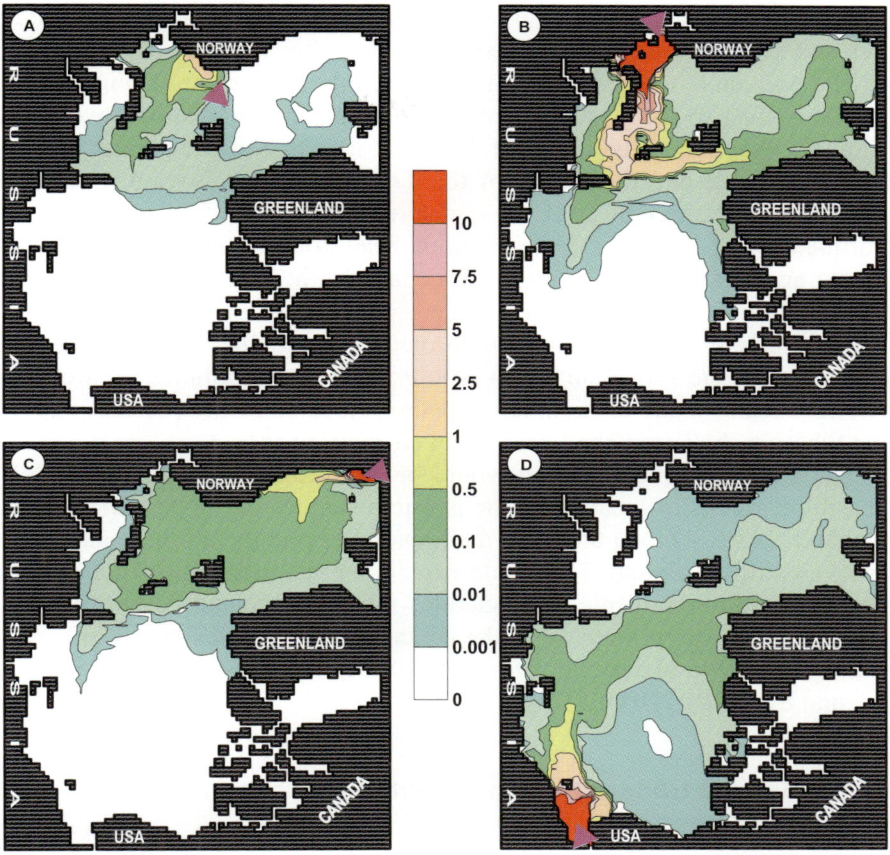

Fig. 18.3. The spread of a pollutant with ocean currents, 15 years after the beginning of a hypothetic pollutant release, from sources located in the following regions: A - the wreck of the "Komsomolets" nuclear submarine, B - the bottle-neck of the White Sea, C - the Faeroe-Shetland Channel in the stream of the Atlantic current and D - the Bering Strait. The scale shows % of pollutant in relation to its concentration at source. ▲ - source location.

Having entered the Transpolar Drift Stream in the northern part of the Kara Sea the pollutants subsequently reach the Greenland Sea. Pollutant concentrations exceeding 5% are registered in the eastern and south-eastern parts of the Barents Sea, in the Kara Sea and in the stream of the Transpolar Drift.

The pollutants from the possible source in the stream of the North Atlantic Current in the Faeroe-Shetland Channel (Fig. 18.3c) spread predominantly in the Norwegian and Barents seas and in the northern part of the Kara Sea. A region with concentrations higher than 5% in the surface layer is observed in the Norwegian Sea, in the western part of the Barents Sea and near the southern and western coasts of Spitsbergen.

The pollutants from the possible source in the Bering Strait (Fig. 18.3d) spread mainly to the west. The zone of possible contamination covers the Chukchi Sea, the northern part of the East Siberian Sea, the greater part of the Laptev Sea and the region of the Transpolar Drift Stream. Pollutant concentrations exceeding 5% are registered in the Chukchi Sea and in the north-eastern part of the East Siberian Sea. Concentrations of 10% and above are located in the Bering Strait and in the Chukchi Sea near the coasts of Alaska and Chukotka.

The numerical experiments performed revealed that the anticyclonic gyre zone in the Canadian Basin would be the least polluted area of the Arctic Ocean for the contamination sources located in the coastal zone of the Siberian shelf seas, the Barents and Norwegian seas.

There are some regions that would become contaminated in nearly all possible variants of location of the possible sources in the coastal zone of the Arctic seas. Among these are the Laptev Sea and the northern and eastern coasts of Greenland. This could explain measured increased concentration of the anthropogenic radionuclides in the Greenland coastal waters, for example a sharp increase in the concentration of ^{137}Cs near Danmarkshavn from 1986 (3.5 Bq m^{-3}) to 1989 (>8 Bq m^{-3}) (Dahlgaard 1995; Pavlov and Stanovoy 2001).

18.3.2 Sea ice transport from potential sources of contaminants

Here we continue the analysis of the simulation results described by Pavlov et al. (2004). In Pavlov et al. (2004), in order to estimate the drift route of the sea ice from the areas containing potential sources of pollution in the Arctic Ocean, trajectories of the ice drift were simulated. The following regions were chosen: in the Kara Sea–the regions of the Ob and Yenisei river mouths; in the Laptev Sea–the region of the Lena river mouth; in the East Siberian Sea–the region of the Kolyma river mouth; in the Chukchi Sea–the region near the Bering Strait; and in the Beaufort Sea–the mouth of the Mackenzie River. For these simulations reconstructed ice drift data for the period 1899-2000 were used. Trajectories were started for each month that the ice concentration at the starting position was more than 10%. Most of the trajectories were launched in the arctic winter time. In this work, based on trajectories simulated in Pavlov et al. (2004) from different potential sources in the Arctic Ocean, we analyse and discuss interannual variability of the travel time (TT) from all selected sources, and

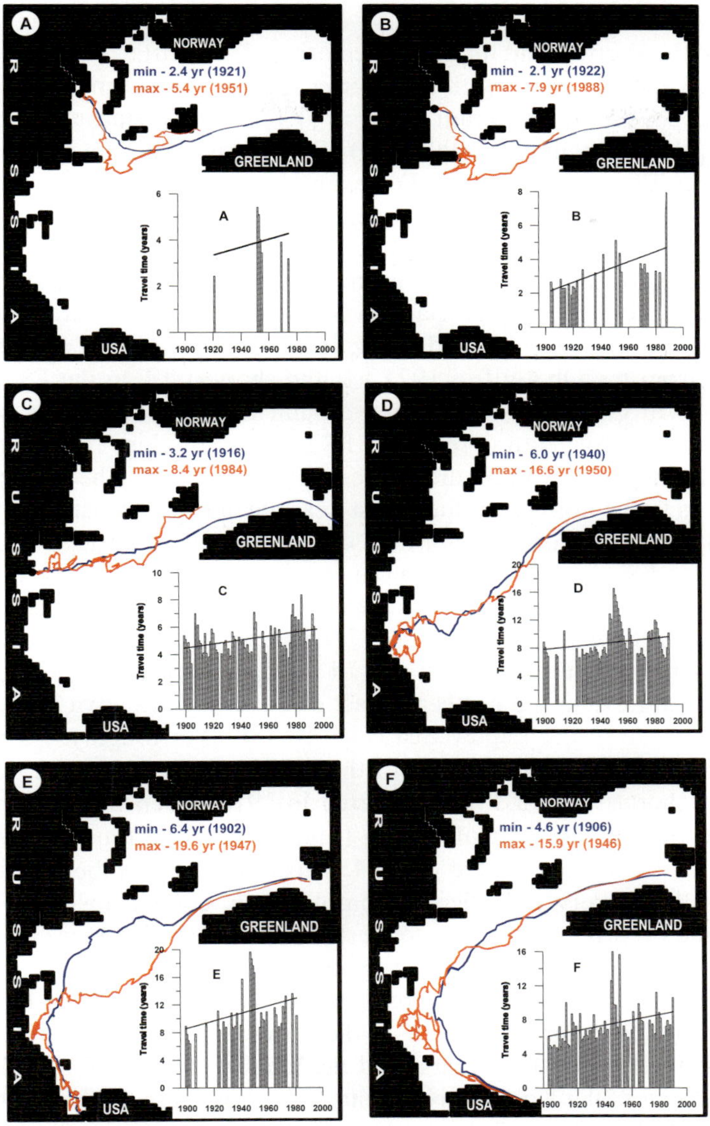

Fig. 18.4. Trajectories for sea ice drift with the minimum (blue) and maximum (red) travel time from selected potential sources to the Fram Strait region (80°N). A-Ob River, B-Yenisei River, C-Lena River, D-Kolyma River, E-the Bering Strait and F-Mackenzie river. Bar diagrams show the travel time to the Fram Strait for sea ice starting in different years in the last century (1899-2000).

estimate the shortest and longest trajectories of ice drift.

Fig. 18.4 shows maps with trajectories and the travel time from selected potential sources to the Fram Strait (FS) for sea ice starting in different years in the last century (1899-2000). Trajectories with the minimum and maximum TT are also shown. For example, sea ice from the sites in the Kara Sea reaches the FS after 2-4 years (Fig. 18.4a, b). The probability of ice from the Ob Gulf region reaching the FS is only 0.7%. From the Yenisei Gulf region the probability is increased to 27%. The most probable route from these sites to the Barents Sea is through the strait between the islands of Novaya Zemlya and Franz Josef Land (Pavlov et al. 2004). Ice starting from the Ob Gulf in 1921 has the shortest TT to the FS. It drifts from the Ob Gulf to the north and around the northern coasts of Franz Josef Land and Svalbard to reach the FS in 2.4 years (Fig. 18.4a).

Ice starting in 1954 has the longest TT (5.4 years). The ice from this area drifts to the north-east, then west near the northern coast of Severnaya Zemlya, and then along the northern coast of Svalbard through the eastern part of the FS. A similar configuration of trajectories with the minimum and maximum TT was simulated for ice drifting from the Yenisei Gulf in 1922 (TT 2.1 years) and in 1988 (7.9 years). The trajectory starting in 1988 reached the north-western part of the Laptev Sea and the ice drifted for a long time near the northern coast of Severnaya Zemlya (Fig. 18.4b). The sea ice from the Laptev Sea (Fig. 18.4c) takes roughly 4-8 years to reach the FS and the probability of this happening is 71%. The trajectory with the shortest TT (3.2 years) started in 1916, when the ice drifted directly to the western part of the FS (Fig. 18.4c). The trajectory with the longest TT (8.4 years) started in 1984. This trajectory has the same general direction but a very complicated configuration, ending up in the eastern part of the FS (Fig. 18.4c). The longest TT is from the Kolyma river mouth (Fig. 18.4d) and the Bering Strait region (Fig. 18.4e).

The TT of the sea ice drift from the Kolyma River ranges form 7-17 years with a probability of reaching the FS of 71%. The ice drift trajectories with the shortest (6.0 years) and the longest (16.6 years) TT are both from the area of the Kolyma river delta. They both pass the North Pole and continue to the south and the western part of the FS. However the trajectory with the longest TT, starting in 1950, has a much more complicated configuration (loops at the shelf of the East Siberian Sea) than the trajectory with the shortest TT, starting in 1940. For ice starting from the Bering Strait, TT ranges between 5-19 years (Fig. 18.4e) and the probability of reaching the FS is decreased to 41%. The trajectory with the shortest TT (6.4 years) started in 1902. It runs along the coast of Siberia, through the De Long Strait and to the northern part of the Laptev Sea along the continental slope. From Severnaya Zemlya island the ice drifts west and southwest to the FS. The trajectory with the longest TT (19.6) started in 1947 and it has a more eastern position in the centre of the Polar Basin (Fig. 18.4e). Ice drifts a long time at the shelf zone of the Chukchi and East Siberian seas. From the site near the Mackenzie River (Fig. 18.4d) sea ice can reach the FS in 5-16 years with a probability of 68% of this happening. Trajectories with the shortest (4.6 yr) and longest (15.9 yr) TT have nearly the same position, but the configuration of the trajectory with the longest TT is much more complicated. Based on analysis of the patterns of the atmospheric circulation (not shown) we can conclude that the shortest TT of ice drift to the FS is connected with the extreme development of anticyclonic circulation above the Arctic Ocean. During the longest TT the atmospheric circulation is in an extreme cyclonic regime.

In the light of climate change in the Arctic it is interesting to consider the inter-annual variability of the TT of ice drift through the Arctic Ocean. We have obtained a generally positive trend (Fig. 18.4). The minimum TT was in the first two decades of the last century. In the last four decades the TT has increased. Maximum values of the TT were obtained at the end of the 1940s and mid 1950s, except for ice drifting from the Lena River where the maximum was at the end of the 1970s and beginning of the 1980s.

These calculated trajectories and TT estimations are in good agreement with the results of Pfirman et al. (1997; 2004a).

The studies of Pfirman et al. (1997) have indicated that drifting Arctic sea ice plays an important role in the redistribution of sediments and contaminants. It was also shown that forward and backward trajectories of sea ice can be calculated to identify regions influenced by sea ice from different source areas and reconstruct the drift path of individual ice floes with some confidence over a period of several years. Authors have demonstrated that ice from the Kara Sea has a strong influence on the Laptev Sea, Barents Sea, Svalbard, the southern portion of the Transpolar Drift Stream, and eastern FS. Ice from the Laptev Sea is mostly advected through the FS and to a lesser degree into the Barents Sea. Ice from the East Siberian Sea is either advected through the FS or is caught up in the Beaufort Gyre and is transported along the northern North American coast. Pfirman et al. (2004a) have analysed sea ice drift from 1979 to 1997 based again on fields of ice motion obtained from IABP. The analysis of potential trajectories of sea ice incorporated in the central Arctic pack between 1979 and 1997 showed extensive changes in the fate of sea ice exported from the Arctic shelves. TT of ice within the central Arctic Basin decreased by at least 1 year, at the same time that the fraction of ice with 4 year TT exported through the FS increased. Ice from distant sources that formerly recirculated in the Beaufort Gyre under lower NAO/AO conditions of the 1980s, was exported more directly through the FS in higher NAO/AO conditions of the 1990s. It was recognised that changes in trajectories of ice with different origins are important, because they affect advection and release of any transported material.

However, in contrast to the ISMO approach for simulation of sea ice trajectories, Pfirman et al. (1997; 2004a) have calculated trajectories based on historical drift data from IABP. Buoy positions have high spatial accuracy and temporal resolution. But the spatial distribution is generally rather coarse, so that buoy data provide only a highly restricted view of the temporal and spatial variability of the large-scale sea ice drift patterns. Buoy data provided by the IABP cover mostly the central part of Arctic Ocean, so in significant areas of the Siberian seas, where there are potential sources of pollution, the buoy data are absent.

18.4 Conclusions

The most rapid spreading of the pollutants and largest contaminated areas result from possible sources located near the Lena and Mackenzie river mouths.

There are two regions, the northern and eastern coastal zone of Greenland and the Laptev Sea, that are contaminated in nearly all variants of location of the possible sources in the coastal zone of the Arctic and Nordic seas.

The calculated trajectories of ice drift from areas of potential sources of pollution allow us to evaluate the character of pollutant transport and the areas of redistribution. From that we can conclude that sea ice from most potential sources of contaminant can reach the open Polar Basin and the FS. Contaminated sea ice from potential sources in the Kara and Laptev seas can reach FS within 2-4 years and from the East Siberian, Chukchi and Beaufort seas within 6-11 years.

Analysis of the inter-annual variability of sea ice TT from different sites to FS has shown a significant positive trend in the last century.

The results of the simulation can also give useful information for the selection of the most representative areas for monitoring contaminants in the Arctic Ocean. Based the structure of passive tracer spreading we obtained and our simulated trajectories of ice drift from different potential sources in the Arctic Ocean we conclude that the most important regions for monitoring of contaminants are FS and Laptev Sea.

Acknowledgements

Support for this work was provided by the Transport and Effects Programme, Phase 2 (financed by the Norwegian Ministries of Environment and of Foreign Affairs). Partly the work was supported by the Barents Sea Production Licenses 182, 225 and 228 (Norsk Hydro, Statoil, Agip, Chevron, Fortum, Petero and Enterprise - contract no. 9000000465).

References

Aarkrog A (1993) Radioactivity in polar regions-main sources. In: Strand P, Holm E (Eds) Environmental Radioactivity in the Arctic and Antarctic, Norwegian Radiation Protection Authority, Østerås, Norway, pp 15-34

AMAP (1998) AMAP Assessment Report: Arctic Pollution Issues. Arctic Monitoring Assessment Programme (AMAP), Oslo, Norway

AMAP (1997) Arctic Pollution Issues: A State of the Arctic Environment report., Arctic Monitoring Assessment Programme (AMAP), Oslo, Norway

ANWAP (1997) Radionuclides in the Arctic Seas from the Former Soviet Union: Potential Health and Ecological Risks. Office of Naval Research, Arlington, VA, USA

Barry RG, Serreze MC, Maslanik JA, Preller RH (1993) The Arctic sea ice-climate system: observations and modelling. Reviews of Geophysics 31(4):397-422

Boris JP, Book DL (1973) Flux-Corrected Transport I: SHASTA-A Fluid Transport Algorithm That Works. J Comput Phys 11:38-69

Champ MA, Makeyev VV, Brooks JM, Delaca TE, van der Horst KM, Engle MV (1998) Assessment of the Impact of Nuclear Wastes in the Russian Arctic. Mar Pollut Bull 35(7-12):203-221

Coachman LK, Aagaard K (1988) Transports through Bering Strait: Annual and Interannual Variability. J Geophys Res 93(C12):15535-15541

Dahlgaard H (1995) Transfer of European coastal pollution to the Arctic-radioactive-tracers. Mar Pollut Bull 31(1-3):3-7

Duursma EK, Carroll J (1996) Environmental Compartments, equilibria and assessment of processes between air, water sediments and biota. Spring-Verlag, Berlin, Germany

Emery WJ, Fowler CW, Maslanik JA, Pfirman S (1997) New satellite derived sea ice motion tracks Arctic contamination. Mar Pollut Bull 35(7-12):345-352

Harms I (1997) Modelling the dispersion of ^{137}Cs and ^{239}Pu released from dumped waste in the Kara Sea. J Mar Syst 13:1-19

Harms I, Karcher MJ (1999) Modeling the seasonal variability of hydrography and circulation in the Kara Sea. J Geophys Res 104(C6):13431-13448

Harms I, Karcher MJ, Dethleff D (2000) Modelling Siberian river runoff – implications for contaminant transport in the Arctic Ocean. J Mar Syst 27:95-115

IAEA (1997) Predicted radionuclide release from marine reactors dumped in the Kara Sea-report of the source team working group of the IASAP, IAEA-TECDOC-938, Vienna, Austria

Israel Y, Tsiban A, Vakulovsky S (1993) Radioactivity and Environmental Problems for the Seas and Ocean. Proc Radioactivity and Environmental Security in the Oceans: New Research and Policy Priorities in the Arctic and North Atlantic. Woods Hole Oceanographic Institution, June 7-9, 1993, Woods Hole, Massachusetts, USA:65-79

Ivanov V (1976) Fresh water balance of the Arctic Ocean. Rosgidrometeoizdat, Leningrad, Trudy AANII 323:138-148 (In Russian)

Karcher MJ, Gerland S, Harms IH, Iosjpe M, Heldal HE, Kershaw PJ, Sickel M (2004) The dispersion of ^{99}Tc in the Nordic Seas and the Arctic Ocean: a comparison of model results and observations. J of Environmental Radioactivity 74:185-198

Kershaw P, Baxter A (1995) The transfer of reprocessing wastes from north-west Europe to the Arctic. Deep-Sea Res II 42(6):1413-1448

Korsnes R, Pavlova O, Godtliebsen F (2002) Assessment of potential transport of pollutants into the Barents Sea via sea ice–an observational approach. Mar Pollut Bull 44(9):861-869

Kulakov MYu, Pavlov VV (1999) Modelling of consequences of damp of radioactive waste from Sellafield plant. Rosgidrometeoizdat, St.Petersburg, Trudy AANII 442:159-164 (In Russian)

Marchuk GI (1982) Mathematical modelling in respect of the environmental problem. Nauka, Moscow, Russia (In Russian)

Matishov G, Matishov D, Shipa Ye, Rissanen K (1994) Radionuclides in the ecosystem of the region of the Barents and the Kara Seas. Apatity, Murmansk, Russia (In Russian)

Nies H, Harms I, Karcher MJ, Dethleff D, Bahe C (1999) Anthropogenic radioactivity in the Arctic Ocean - review of the results from the joint German project. The Science of the Total Environment 237/238:181-191

Nilsen T, Bøhmer N (1994) Sources to radioactive contamination in Murmansk and Archangelsk counties. Bellona rapport 1994: 1, Miljøstiftelsen Bellona, Oslo, Norway

Nilsson A (1997) Arctic pollution issues: A state of the Arctic environment report. Arctic Monitoring Assessment Programme (AMAP), Oslo, Norway

NRPA (1996) Dumping of radioactive waste and investigation of radioactive contamination in the Kara Sea. Results from 3 years of investigations (1992-1994) in the Kara Sea. Joint Norwegian-Russian Expert Group for Investigation of Radioactive Contamination in the Northern Areas, Norwegian Radiation Protection Authority, Østerås, Norway

NRPA (1997) Sources contributing to radioactive contamination of the Techa river and areas surrounding the “Mayak” production association, Ural, Russia. Joint Norwegian-Russian Expert Group for Investigation of Radioactive Contamination in the Northern Areas, Norwegian Radiation Protection Authority, Østerås, Norway

Nürnberg D, Wollenburg I, Dethleff D (1994) Sediments in Arctic sea ice-entrainment, transport and release. Mar Geol 119:185-214

Omstedt A, Carmack E, Macdonald RW (1994) Modeling the seasonal cycle of salinity in the Mackenzie shelf estuary. J Geophys Res 99(C5):10011-10021

Pavlov VK (1995) Modelling of the Thermohaline Water Circulation of the Arctic Ocean. Proc of NDRE Workshop, Modeling Requirements for Water Mass Dynamics, Ice and River Transports in the Kara Sea, Workshop Norwegian Defence Research Establishment, Tjome, 26-30 June 1995, Norway: 200–224

Pavlov V, Stanovoy V (2001) The problem of transfer of radionuclide pollution by sea ice. Mar Pollut Bull 4:319-323

Pavlov VK, Pavlov PV (1999) Features of seasonal and interannual variability of sea level and water circulation in the Laptev Sea. In: Land-Ocean systems in the Siberian Arctic. Dynamics and history. Eds Kassens et al., Springer-Verlag, Berlin Heidelberg, pp 3-16

Pavlov VK, Pavlova OA (1999) Features of seasonal and interannual variability of thermohaline structure, currents and water exchange in the Bering Strait. Rosgidrometeoizdat, St.Petersburg, Trudy AANII 442:16-52 (In Russian)

Pavlov V, Pfirman S (1995) Hydrographic structure and variability of the Kara Sea. Deep-Sea Res II 42(6):1369-1390

PavlovV, Pavlova O, Korsnes R (2004) Sea ice fluxes and drift trajectories from potential pollution sources, computed with a statistical sea ice model of the Arctic Ocean. J Mar Syst 48(1-4):133-157

Pavlov VK, Kulakov MYu, Stanovoy VV (1995) Modeling of the transport and transformation of pollutants in the Arctic Ocean. Proc of NDRE Workshop, Modeling Requirements for Water Mass Dynamics, Ice and River Transports in the Kara Sea, Workshop Norwegian Defence Research Establishment, Tjome, 26-30 June 1995, Norway:225-249

Pfirman S, Haxby WF, Colony R, Rigor I (2004a) Variability in Arctic sea ice drift. Geophys Res Letters 31, L16402, doi:10.1029/2004GL020063

Pfirman S, Haxby W, Eicken H, Jeffries M, Bauch D (2004b) Drifting Arctic sea ice archives changes in ocean surface conditions. Geophys Res Letters 31, L19401, doi:10.1029/2004GL020666

Pfirman S, Colony R, Nürnberg D, Eicken H, Rigor I (1997) Reconstructing the origin and trajectory of drifting Arctic sea ice. J Geophys Res 102(C6):12575-12586

Preller RH, Cheng A (1995) Modeling the Dispersion of Radioactive Contaminants in the Arctic Using a Coupled Ice-ocean Model. Proc Arctic Nuclear Waste Assessment Program Workshop, Woods Hole Oceanographic Institution, Woods Hole, Massachusetts, 1-4 May, 1995

Rigor I, Colony R (1997) Sea-ice production and transport of pollutants in the Laptev Sea, 1979-1993. Science Total Environ 202:89-110

Rigor IG, Wallace JM, Colony RL (2002) Response of sea ice to the Arctic oscillation. J Clim 15:2648-2663

Scott EM, Gurbutt P, Harms I, Heling R, Nielsen SP, Osvath I, Preller R, Sazykina T, Wada A, Sjoeblom KL (1997) Benchmarking of numerical models describing the dispersion of radionuclides in the Arctic Seas. The Science of the Total Environment 202:123-134

Smith JN, Ellis KM (1999) Calculation features in the central Arctic Ocean revealed by nuclear fuel reprocessing tracers from Scientific Ice Expeditions 1995 and 1996. J Geophys Res 104(C12):29663-29677

Smith JN, Ellis KM, Jones EP (1990) Cesium 137 Transport Into the Arctic Ocean Through Fram Strait. J Geophys Res 95(C2):1693-1701

Smith JN, Ellis KM, Kilius LR (1998) ^{129}I and ^{137}Cs tracer measurements in the Arctic Ocean. Deep-Sea Res I 45(6):959-984

Soviet Arctic (1970) Seas and Islands of the Arctic Ocean. Nauka, Moscow (In Russian)

Vakulovsky S, Nikitin A, Chumichev V (1993) Radioactive Contamination of the Barents and Kara Seas. Proc Radioactivity and Environmental Security in the Oceans: New Research and Policy Priorities in the Arctic and North Atlantic. Woods Hole Oceanographic Institution, June 7-9, 1993, Woods Hole, Massachusetts, USA:157-169

US-Russian Joint Atlas (1996) Joint US-Russian Atlas of the Arctic Ocean, winter. National Snow and Ice Data Center, Environmental Working Group, Boulder, Colorado (on CD-Rom)

Yablokov AV (2001) Radioactive waste disposal in seas adjacent to the territory of the Russian Federation. Mar Poll Bull 43(1-6):8-18. This article extracts material from Yablokov Commission Report (1993)

Yablokov Commission Report (1993) Yablokov AV, Karasev VK, Rumyantsev VM, Kokeyev MYe, Petrov OL, Lytsov VN, Yemelyankov AF, Rubtsov PM (1993) Facts and Problems related to Radioactive Waste Disposal in Seas Adjacent to the Territory of the Russian Federation. Office of the President of the Russian Federation, Moscow

Zhang JL, Rothrock D, Steel M (2000) Recent changes in Arctic sea ice: The interplay between ice dynamics and thermodynamics. J Clim 13:3099-3114

Zhang JL, Thomas DR, Rothrock D, Lindsay RW, Yu Y (2003) Assimilation of ice motion observations and comparisons with submarine ice thickness data. J Geophys Res 108(C6),3170,doi:10.1029/2001JC0011041

Zwally HJ, Walsh JE (1987) Comparison of observed and modelled ice motion in the Arctic Ocean. Annals of Glaciology 9:136-144

19 Long-term atmospheric contaminant monitoring for the elucidation of airborne transport processes into polar regions

Roland Kallenborn and Torunn Berg

Norwegian Institute for Air Research, P.O. Box 100, NO-2027 Kjeller, Norway, e-mail: Roland.Kallenborn@unis.no

19.1 Introduction

Present levels of anthropogenic contaminants found in the polar environments cannot be explained by known use and/or release from sources within the region. Based on continuous monitoring and surveillance, atmospheric long-range transport from lower latitudes is known to be the most important source for the presence of many persistent organic (POP) and inorganic pollutants in the Arctic environment today. During the past decades, international long-term atmospheric monitoring programs have proven to be versatile tools for the demanding task of identification and evaluation of hazardous chemicals in the environment. Modern tools for atmospheric monitoring especially in background regions like the polar regions, are designed to:

- control and assess international regulatory measures
- determine "natural" background levels as well as the concentration thresholds when counter measures are required.
- collect information on environmental behaviour (spatial and temporal trends) and serve as "early warning tool". Even minimal changes in levels and distributions may be a first signal for larger global changes expected at a later stage.

This concept has proven highly effective and useful. Many examples illustrate the importance of continuous atmospheric monitoring for surveillance of anthropogenic contamination in the cold polar regions.

The principles and pathways of global atmospheric long-range transport of persistent semi-volatile contaminants across several climate zones into the polar regions have been revealed and confirmed by long-term monitoring of persistent organic pollutants from the temperate zones of the globe into the polar atmosphere (Bindler *et al.* 2001, Macdonald *et al.* 2000, Macdonald *et al.* 2003, Wania 2003b).

The atmospheric burden of lead caused by emissions from fuel combustion increased continuously during the first 70 years of the 20th century, affecting even the Arctic regions (Bindler *et al.* 2001, Hopke *et al.* 2001, Riget *et al.* 2000, Rosman *et al.* 2000). International governmental agreements and regulations lead to considerable reduction of lead additives in standard fuels for vehicles worldwide (Riget *et al.* 2000). Subsequent monitoring of the lead distribution in the Arctic background air documented the continuous reduction of atmospheric lead globally (DeWit *et al.* 2004)

Polychlorinated biphenyls (PCB) have been used and applied as industrial chemicals and as oils in transformers as well as in capacitors. About 1 000 000 tons have been produced and applied since the 1940s. In the 1960s evidence for the presence of PCBs in the biological environment was presented. In the 1970s the presence of PCBs in the Arctic atmosphere was reported from Norwegian and Canadian research groups. The highly toxic effects of PCBs in the Arctic environment (e.g., on top predators and indigenous people of the arctic lead consequently to the international ban of these chemicals as industrial products in the 1980s. Modern monitoring programs are tracking this type of atmospheric contaminants and contribute to the evaluation of suitable regulatory measures and have shown that even more than a decade of continuous monitoring reveals no clear general trends are observed. In the European Arctic as well as Antarctica no decreasing tendency/ trend are observed for POPs in air. (Berg *et al.* 2004, Bignert *et al.* 1998, DeWit *et al.* 2004, Macdonald *et al.* 2000), whereas decreasing trends are reported for the Canadian Arctic (Hung *et al.* 2001, 2002).

The occurrence and detection of the stratospheric "ozone hole" in the Antarctic region is an additional example of the effectiveness of continuous pollution monitoring in the Antarctic. Only through long-term monitoring of high volatile chlorofluoro-carbons (CFC) and other related volatile chemicals, the devastating result of their depletion effects on the stratospheric ozone layer was made visible and adequate countermeasures on an international levels were conducted.

As a direct result of the above-mentioned international efforts on atmospheric contaminant monitoring, large monitoring programs covering also the polar background regions, have been established. A comprehensive "list of action" focusing on national and international initiatives directed towards elimination/ reduction of contaminant burdens in the environment is summarized in UNEP's master list of actions (UNEP, 2000).

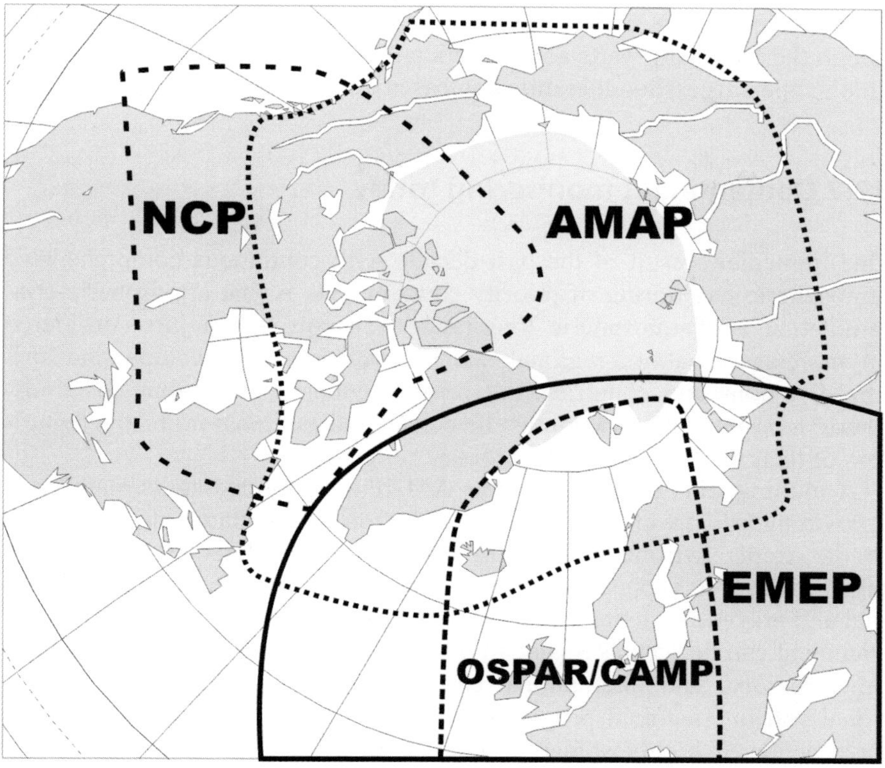

Fig. 19.1. Four major international and national long-term monitoring programs with specific interest in Arctic environments. Abbreviations are explained in Table 19.1. **) NCP is also performing atmospheric monitoring at two stations in the Russian Arctic (Amderma and Dunai)*

A selected number of international atmospheric monitoring programs with focus on Arctic and Alpine environments is shown in fig. 19.2.

All circum-Arctic countries as well as numerous southern nations with interest in the Antarctic, maintain national monitoring and scientific screening initiatives in order to control and survey levels and distribution of priority contaminants. Thus, during the past decades a wealth of scientific information on transport, levels, distribution and fate of priority contaminants was produced and reported to the respective authorities, institutions and stakeholders. However, the availability and transparency of validated monitoring information for scientific purposes represents often a major challenge for environmental sciences. The access to the majority of the collected data is often restricted due to national regulations or simply due to the fact that reports are not officially registered and, thus, not available in open accessible literature databases for environmental studies.

19.2 Contaminant monitoring today

One immediate result of the past decade with continuous comprehensive atmospheric monitoring of priority contaminants is that atmospheric contamination and atmospheric long-range transport is not just considered anymore as a local or a regional challenge restricted to developed industrialized regions and countries. All priority contaminants prone to atmospheric long-range transport must be considered as global problems regardless of their sources or transport routes.

A multinational program such as AMAP, has defined as its main goal: "providing reliable and sufficient information on the status of, and threats to, the Arctic environment, and providing scientific advice on actions to be taken in order to support Arctic governments in their efforts to take remedial and preventive actions relating to contaminants" (www.amap.no). This statement can be seen as a general description for the overall goal of many other regional and international contaminant monitoring programs. National and international programs like NCP, AMAP, CAMP, EMEP etc. have achieved highly valuable results by highlighting and documenting distribution and fate of anthropogenic contamination. Through this type of surveillance, long-range transport of persistent pollutants and, thus, the global transport phenomenon of this type of chemicals has been revealed and brought into the political agenda. Persistent organochlorine pesticides (e.g, chlordane, toxaphene, DDT, HCH), highly important as crop protection agents in North America, Europe or developing countries in Africa and Asia, are transported as atmospheric burden throughout several climate zones into the Arctic regions with significant implications for the health and well being of the indigenous people of the North (Bidleman et al. 1993, Bridgman et al. 1994, Macdonald et al. 2000). Short-term monitoring programs performed in the Antarctic regions, revealed similar patterns for atmospheric long-range transport as already identified for the Northern hemisphere (Kallenborn 2000, Lakaschus et al. 2002).

The main antropogenic sources of heavy metals are various industrial sources, foundries and smelters. Relatively volatile heavy metals and those that become attached to airborne particles can be widely dispersed. Emission data indicate that reductions occurred from 1983 (Nriagu and Pacyna 1988) to the mid-1990s (Pacyna and Pacyna 2002) on a hemispheric scale.

Long term monitoring in the Arctic, however, have revealed no significant increasing or decreasing trends for many metals derived from antropogenic sources (Department of Indian Affairs and Northern Development 2003; Berg et al. 2004). Temporal trends with regard to increasing mercury levels in marine birds and mammals have been observed in some Arctic areas (Bignert *et al.* 2004a, Braune *et al.* 2001, Kim *et al.* 1996, Riget *et al.* 2000). There is also evidence that current mercury exposures pose a health risk to some people and animals in the Arctic. Cadmium levels in some seabirds are high enough to cause kidney damage (Chavez *et al.* 1991, DeWit et al. 2004, Holmes *et al.* 1996).

19.2.1 Restrictions and challenges

19.2.1.1 Data access

Several regional and international long-term monitoring programs are established nowadays focusing on global aspects of contaminant distribution and transport into polar regions (table 19.1). Except for monitoring of aerosols, ozone and CFC, practically no long-term atmospheric monitoring initiative have been established for Antarctica (with a monitoring period over 5 years).

In addition to regional and international initiatives, all circum-Arctic countries as well as a selection of southern nations with interest in the Antarctic, maintain national monitoring initiatives in order to control and survey levels and distribution of priority contaminants. Thus, during the past decades a wealth of scientific information on transport, levels, distribution and fate of priority contaminants was produced and reported to the respective authorities, institutions and stakeholders. However, the availability of validated monitoring information collected during long-term atmospheric monitoring is often restricted due to national regulations or simply due to the fact that reports are not officially registered and, thus, not available in open accessible literature databases. With these restrictions in mind, the Arctic Monitoring and Assessment program (AMAP) as well as the EMEP initiative have established joint data centers collecting and administrating all available data within the monitoring programs.

AMAP's assessments are based to a large extent on information and results from recent (largely unpublished) monitoring and research work (including atmospheric contaminant information). Data from such activities are compiled together with routine monitoring data within AMAP's thematic data centers. Data are made available from the data centers to scientists engaged in AMAP assessments under strict conditions that protect the rights of data originators. These conditions are described in AMAP's Data Policy documentation (see www.amap.no). AMAP Thematic Data Centres have been established to meet the following objectives:

- to provide access to data from recent monitoring and research activities conducted as part of the AMAP National Implemention Plans (NIPs);
- to provide a means to ensure that data are treated in a consistent manner, undergo uniform statistical analysis, etc., including application of objective quality assurance procedures;
- to begin the process of establishing a long-term archive of Arctic-relevant monitoring data, for use in future assessments of, e.g. temporal trends, etc.;
- to meet the terms of reference of Ministerial declarations, charging AMAP with establishing databases of sources, types, and levels of contamination of the atmospheric, aquatic and terrestrial environments of the Arctic and northern areas.

As already mentioned, similar data centers are established within the LRTAP-EMEP monitoring initiative with the potential for comprehensive co-operation and data transfer. However, no official communication/ efforts directed towards the ultimate goal to combine the data centers or allow inter-comparison for common monitoring parameters are established yet. In addition, no long-term atmospheric monitoring of persistent contaminants neither on a national or an international basis is established yet for pristine high-altitude, alpine environments.

19.2.1.2 Quality and comparability

Quality assurance issues are normally an integral component of most monitoring programs and research works. However, national and international monitoring programs (including atmospheric monitoring) have different demands with regard to data quality, priority criteria for contaminant selection, documentation, sampling frequency and methodology (AMAP 2000, Manø *et al.* 2003, Perkins 1995). These simple discrepancies often severely hamper the opportunity to compare monitoring data and perform scientific elucidations of temporal and spatial trends beyond the borders of the assigned region in an effective way. As an additional consequence, effective implementation and source elucidation by comparing the information collected within the arctic monitoring with monitoring information from possible source regions is severely restricted or even impossible due to the above-described differences.

As already stated, atmospheric long-range transport of anthropogenic contaminants is currenlty recognized as a global process with severe implications for ecosystems, wildlife and human populations of the northern regions (Brunstrom *et al.* 2000, Hansen 2000a, Hoekstra *et al.* 2003, Law *et al.* 2003, Muir *et al.* 1999a, Pereg *et al.* 2003, Ross *et al.* 2003, Skaare *et al.* 2002). Since the sources and possible solutions for this challenge are mostly located in the middle latitude temperate zones, the hitherto regionally restricted monitoring efforts of priority contaminants is not sufficient anymore to cope with the challenge of estimating global atmospheric transport and distribution processes. Joint strategies for and co-operation between regional and global atmospheric monitoring programs are necessary to raise the widely scattered scientific efforts on monitoring and surveillance of atmospheric contamination in the Polar Regions onto a global level with strong linkages between the national and international initiatives.

However, not just sampling and method differences restrict the comparability of monitored data. Local characteristic meteorological and climate conditions as well as varying content of particulate materials (including snow formation) might lead to significant differences in the capacity of air masses to adsorb and transport specific contaminant burdens (Daly *et al.* 2004b, Graham 2003, Guo *et al.* 2004, Mandrioli *et al.* 1984, Moore *et al.* 2003, Pacyna 1995, Prospero 1999, Wania 2003a, Zhuang *et al.* 2001). These differences restrict severely the comparability of datasets collected under different meteorological conditions. Based on these conclusions, the Canadian meteorological services have developed a statistical time-series tool (digital filtration = DF) in order to normalize monitoring data based on statistical trend analysis. DF is developed as a statistical trend and seasonal cycle development tool based on time-series development theories (Nakazawa *et al.* 1997). DF will definitively develop into a very useful tool for large-scale data analysis in atmospheric monitoring and consequently allow a better comparison of otherwise different datasets (Hung *et al.* 2002).

Thus, a strong and high priority commitment for co-operation between long-term monitoring programs in order to develop general guidelines in terms of sampling techniques, quality control measures, monitoring parameters (priority contaminants, sampling frequency etc.) is overdue in order to allow investigating and evaluating global aspects of atmospheric contaminant distribution. In order to allow better data comparison across the regional borders, annual laboratory intercomparison excercises should be established.

The official lists of target compounds for the four most important international monitoring initiatives with interest in Arctic environments are

Table 19.1. Main parameters covered by the four major international atmospheric monitoring programs with specific interest in Arctic environments.

Acronym	Name	Sample types	Chemical parameters
LRTAP-EMEP	European Monitoring and Evaluation Program	Precipitation, particles/ aerosol, gaseous phase	Gaseous phase: Particles/aerosols: SO_4^{2-}, Cd, Pb, Cu, Zn, As, Cr, Ni Gaseous + particulate phase: SO_2, NO_2, O_3, VOC (C_2-C_7), Carbonyl-compounds, Hg, HNO_3+ NO_3^-, NH_3+NH_4^+, PAH, PCB, HCB, chlordane, γ–, α-HCH, DDT, DDE Precipitation : SO_4^{2-}, NO_3^-, Cl^-, pH, NH_4^+, Na, Mg, Ca, Hg, Cd, Pb, Cu, Zn, As, Cr, Ni, PAH, PCB, HCB, α–, γ–HCH, DDT, DDE.
AMAP	Arctic Monitoring and Assessment Program	Precipitation, aerosol, gaseous phase	Gaseous + particulate: PAH, PCB, HCB, mirex, toxaphene, aldrin, dieldrin, α–, β–, γ–HCH, DDT, DDD, DDE, chlordanes, TBT, endosulfan, Hg, Pb, Cd, As, Sn, Zn, Cu, Ni, Cr, Precipitation: PAH, PCB, HCB, mirex, toxaphene, aldrin, dieldrin, α–, β– and γ–HCH, DDT, DDD, DDE, chlordanes, TBT, endosulfan, Hg, Pb, Cd, As, Sn, Zn, Cu, Ni, Cr,
OSPAR / CAMP	Comprehensive Atmospheric Monitoring Program	Precipitation, aerosol, gaseous phase	Precipitation: Cd, Hg, NO_3, NH_4^+ optional: As, Cr, Cu, Ni, Pb, Zn, organo-halogens. Aerosol + gaseous phase: Cd, Hg, α-HCH, γ-HCH, HNO_3, NO_3^-, NO_2, NO, NH_3, NH_4^+ Optional: As, Cr, Cu, Ni, Pb, Zn.
WMO-GAW	Global Atmospheric Watch	Gaseous phase	CO_2, O_3, CH_4, N_2O, CFCs, NO_2, H_2O, CH_3CCl_3, $^{14}CO_2$, SO_2

listed in table 19.1. The below presented overview demonstrates clearly the strong potential for joint and harmonised monitoring strategies.

Atmospheric monitoring programmes usually monitor contaminants in ambient air (gaseous and particulate phase) as well as in precipitation (incl. meteorological data). LRTAP-EMEP, OSPAR/CAMP and AMAP report on gaseous and particulate compartments whereas WMO-GAW concentrate on the gaseous phase only. A comprehensive list of target contaminants is presented. Many programmes are concentrating on inorganic pollutants like "greenhouse gases" (incl. NO_x, NH_4^+ etc.) and a variety of trace metals. However, as substantial contaminants chloro-fluoro carbons (CFC) and related compounds are usually included. Only AMAP and LRTAP-EMEP have included a substantial monitoring programme for a suit of priority persistent organic pollutants. However, also OSPAR/CAMP included a few (two) organochlorine pesticide compounds in the monitoring program, whereas WMO/GAW only focuses on greenhouse gases incl. methane, carbon dioxide, CFC etc.

Generally about 60% of the priority target chemicals are included in all major monitoring programmes and, thus, a basis for mutual interest in data exchange and joint quality criteria for analytical procedures and reporting is obvious. In the light of newly ratified international conventions and agreements (e.g., EU- Water frame work directive, Stockholm POP convention, etc.) a close co-operation between modern contaminant monitoring programs is also important from a political point of view.

National and international regulation and control authorities have recognised the here-described urgent needs for harmonisation and cross-linking of international monitoring initiatives as an essential step towards globalisation of contaminant monitoring efforts (AMAP 1998, Prandle 1991) Therefore, the joint initiative of the European Commission and the European Space Agency: "Global Monitoring for Environment and Security (GMES)" has the ambition to develop a joint strategy for global atmospheric observations, which also includes Arctic atmospheric background monitoring. Thus, a cluster of several EU-projects has been established focusing on ozone (O_3) and greenhouse gasses as a first step to co-ordinate monitoring and surveillance initiatives ("Global Atmospheric Observations", GMES-GATO-cluster: http://www.nilu.no/gmes-gato/). The GATO cluster enables the space agencies and atmospheric scientists working in this area to communicate effectively. Coordination of EC and nationally-funded research occurs within GATO through workshops, meetings and working papers.

LRTAP-EMEP has recognised the need for integration and compatibility of local and global initiatives (Tørseth and Hov, 2003, Tørseth, 2004). The program officials acknowledge the need for integration and communication between regional monitoring initiatives aiming at monitoring atmospheric chemical composition. LRTAP-EMEP, therefore, emphasises strong co-operation linkages between the regional monitoring initiatives listed in table 19.2.

The general need is, thus, to establish more formal monitoring requirements for those participating in these co-ordination efforts in order to improve the compliance between the programs. It is those important to ensure data comparability with respect to site location, representativity, network

Table 19.2. List of regional monitoring programmes on atmospheric chemical composition with already established communication linkages.

Regional monitoring Programmes
- European Monitoring and Evaluation Programme (LRTAP-EMEP)
- CLRTAP Working Group on Effects
- World Meteorological Organisation - Global Atmosphere Watch (WMO-GAW)
- Marine Conventions (OSPAR, HELCOM)
- Arctic Monitoring and Assessment Programme (AMAP)
- EC AQFD (EIONET)
- UNEP Global POPs network
- EANET

density, parameters to be requested, temporal resolution, method and future co-operation opportunities.

19.2.2 Interdisciplinary linkage and coordination

The produced monitoring information shows its full potential and value for the evaluation of regional and global distribution processes when combined with other scientific information. Only in combination with various related scientific information such as modelling tools and chemical evaluations, meteorological information and climate data will environmental scientists hopefully be able to unravel the still highly complex challenge of global atmospheric transport and distribution of anthropogenic pollution including pesticide chemicals from tropical regions into the polar climate zones.

19.2.2.1 Modelling

The use and availability of monitoring data for scenario assessments and/or monitoring purposes is one of the most important features of today's international atmospheric monitoring programs. However, advances in modelling have been enhanced by a growing database of physico-chemical properties as a function of temperature. Models of precipitation scavenging of contaminants from air have advanced our understanding of deposition processes (Dethloff *et al.* 2002, Felzer *et al.* 2000, Furevik *et al.* 2003, Groves *et al.* 2002, Holland 2003, Kattsov *et al.* 2000, Proshutinsky *et al.* 1999, Serreze *et al.* 2003, Serreze *et al.* 2000, Walsh *et al.* 2002). Modern multi-compartmental models contribute today significantly to the interdisciplinary efforts to investigate compound-specific atmospheric distribution and transport processes (Beyer *et al.* 2000, Breivik *et al.* 2003, Christensen *et al.* 2004a, Daly *et al.* 2004c, Hansen *et al.* 2004, Karcher *et al.* 2004, Ma *et al.* 2004, Mackay *et al.* 1995, Muir *et al.* 2004, Scheringer *et al.* 2004, Toose *et al.* 2004, Wania *et al.* 1996, Wania 1998, Wania *et al.* 1999b, Wania *et al.* 1999a, Wania *et al.* 2003). However, for calibration and evaluation of new models, empirically derived and quality controlled monitoring data are highly important. Therefore, continuous communication between scientists involved in maintaining atmospheric monitoring, and modellers is essential in order to provide relevant scientifically sound information needed for the development of transport and distribution models.

Already today, a large number of versatile models exist with the potential for joint scenario assessment. Under the LRTAP-EMEP umbrella, six lagrangian and Eulerian models exists already as assessment tools for global transport processes of POPs and trace metals, acid deposition, fate of photo-oxidants and particulate matter (Andersson-Skold *et al.* 1999, Andersson-Skold *et al.* 2001, Bartnicki 1994, Derwent 1993, Jonson *et al.* 1998, Jonson *et al.* 2001, Kahnert *et al.* 2004, Kangas *et al.* 2002, Kruger *et al.* 1997, Olendrzynski *et al.* 2000, Tuovinen *et al.* 2004).

However, as an important measure for optimisation, harmonisation and quality assurance, model inter-comparisons must be performed. In this context currently, model intercomparsions are on the way to elaborate harmonisation and combination of established models. The EMEP Meteorological Synthesising Center East (MSC/E) is evaluating a first comparison of 16 regional and global atmospheric transport model for POPs (Shatalov *et al.* 2003). Many other comparative initiatives are on the way to evaluate and combine model approaches in order to improve the today available tools for scenario and transport evaluations.

19.2.2.2 Compound specific evaluations

One example on interdisciplinary linkage is the Mercury Depletion Phenomenon (MDE's): The Arctic Council encouraged, expanded and accelerated the comprehensive research and recent knowledge on critical aspects of the mercury cycle and budget in the Arctic. Based on the recent state of knowledge, future monitoring initiatives on atmospheric mercury distribution should contribute to investigation of long-range transport, elucidation of mercury deposition mechanisms, as well as to the evaluation of processes leading to biological exposure and effects, and the influence of climate variability and change on these processes (Allengil *et al.* 1995, Bjerregaard *et al.* 2000, Gauthier *et al.* 1998, Landers *et al.* 1995, Macdonald *et al.* 1996, Riget *et al.* 2000). The current finding of photochemical degradation of elementary mercury during the Arctic and Antarctic sunrise gives raise to considerable concern, since degradation products are assumed to be bioavailable and toxic for polar ecosystems (Ariya *et al.* 2004, Asmund *et al.* 2000, Berg *et al.* 2003, Berg *et al.* 2001, Christensen *et al.* 2004b, Ebinghaus *et al.* 2002, Hansen 2000b, Lindberg *et al.* 2002, Lu *et al.* 2001, Poissant *et al.* 2002, Power *et al.* 2002, Schroeder 2002, Schroeder *et al.* 1998, Skov *et al.* 2004, Sommar *et al.* 2004; Sprovieri *et al.* 2002; Temme *et al.* 2003, Wagemann *et al.* 1998; Wängberg *et al.* 2003).

During the atmospheric mercury depletion events (MDE's) gaseous elemental mercury (GEM) disappears almost completely from the atmosphere (Fig. 19.2). During these episodes GEM is transformed to more reactive species of mercury, which have significantly higher deposition rates than GEM.

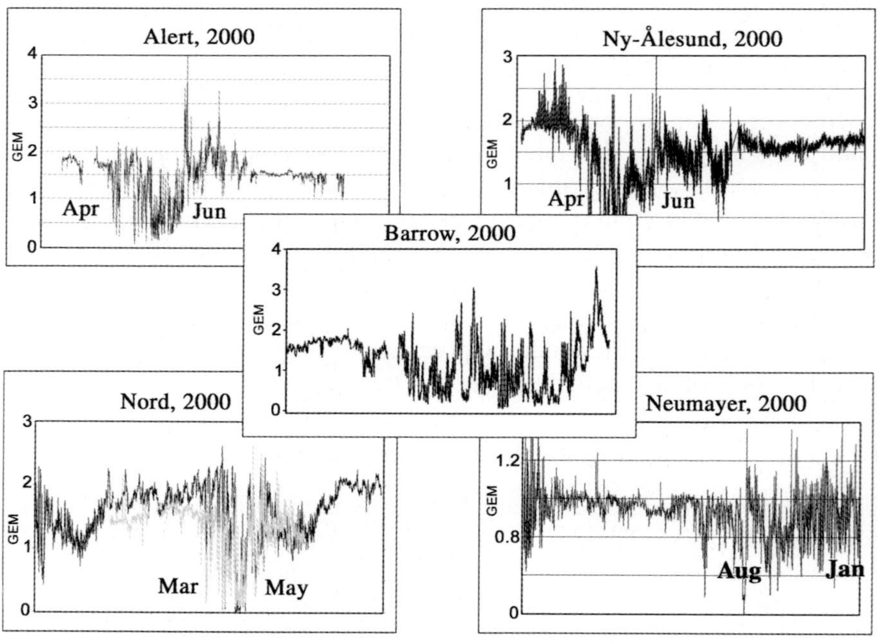

Fig. 19.2. "Polar Spring" mercury depletion events documented during mercury measurements campaigns in the year 2000 Atmospheric monitoring stations presented: Alert (Arctic Canada) (Schroeder *et al.* 1998), Point Barrow (Alaska) (Lindberg *et al.* 2002), Ny-Ålesund (Svalbard) (Berg *et al* 2003), North (Greenland) (Skov *et al.* 2004), Neumayer (German Antarctic station) (Ebinghaus *et al.* 2002) Concentrations are given in [ng/m^3].

19.2.2.3 Meteorology and source elucidation

Atmospheric long-range transport episodes can only be revealed and elucidated by combining unusual contamination patterns in atmospheric samples with meteorological air mass back trajectory calculations (Bailey *et al* 2000, Brankov *et al.* 1998, Eneroth *et al.* 2003, Gebhart *et al.* 2001, Hung *et al.* 2002, Kallenborn *et al.* 1998, Stohl 1996, Subhash *et al.* 1999, Traub *et al.* 2003, Vaughan *et al.* 2002). New scientific studies have also shown that the circulation pattern of air masses in the Arctic influences the long-term distribution trend of atmospheric contaminants (Rinke *et al.* 1999, Stone 1997, Stone *et al.* 2002, Zhao *et al.* 2004). These results implies that even a region with a constant emission could result in an oscillation contribution to the Arctic in a 4-14 years cycle, which may either weaken or amplify contributions of the changes in the anthropogenic emissions to the Arctic (CACAR 2003; Kahl *et al.* 1999).

Seasonally dependent distribution patterns have been found for POPs (Borga *et al.* 2004, Daly *et al.* 2004a, Dietz *et al.* 2004, Fellin *et al.* 1996, Halsall 2004, Hargrave *et al.* 1997, Hung *et al.* 2002, Pavlov *et al.* 2004), volatile organic contaminants (Blake *et al.* 2003, Gautrois *et al.* 2003) as well as for several heavy metals (Berg *et al.* 2004, Bignert *et al.* 2004b, Klungsoyr *et al.* 1995, MacDonald *et al.* 2000, Muir *et al.* 1999b, Muir *et al.* 1992, Riget *et al.* 2004, Thomas *et al.* 1992, Wagemann *et al.* 1996, Zheng *et al.* 2003) and arctic haze compounds (Heidam 1994, Hopke *et al.* 1995, Rasmussen *et al.* 1983, Weber *et al.* 2003) at Arctic monitoring sites. Highest levels of heavy metals were found in winter whereas the lowest concentrations were reported for the summer period. This situation is mainly due to the character of major weather systems: In winter and spring, a high pressure system over Siberia pushes the Arctic front far to the south, so that important polluted areas are within the Arctic air mass.

For some of the POPs, highest levels are found in summer and the lowest concentrations are reported for the winter period. As major reason for this typical distribution pattern, remobilisation from secondary sources (e.g., re-evaporation from open sea surface), which is considerable reduced during wintertime due to low temperature and ice-coverage, is identified (Berg et al. 2004)

19.3 Perspectives and future needs

The major challenge for future atmospheric monitoring of long-range transported contaminants is identified as the need for expansion from regionally restricted atmospheric monitoring towards global monitoring initiatives implementing locally, regionally and continent-related specific parameters. Two possible strategies are identified as promising scientific approaches for effective future global atmospheric monitoring of anthropogenic persistent contaminants with the aim to survey and study atmospheric long-range transport of anthropogenic contaminants into polar regions.

- International and national agreements concerning priority compounds, sample collection, methodology, quality control and data access will lead to global network of regional atmospheric monitoring programs with similar priority compounds and monitoring strategies. This approach will allow an easy comparability of monitored data. All data will be available for multidisciplinary studies and evaluations as well as for regulatory purposes.
- Within the frame of a global intergovernmental environmental organization, a global atmospheric monitoring program will be established or an already ongoing atmospheric monitoring program (e.g., GAW) will be expanded.

Reflecting on today's situation in atmospheric monitoring of relevant persistent contaminants and other pollution, "strategy 1" seems to be the strategy of choice for international monitoring programs. National and international regulatory institutions recognised already a decade ago the importance for harmonisation and standardisation of monitoring initiatives (Oei *et al.* 1994). Thus, already in 1991, the Danish National Environmental Research institute (NERI), department for atmospheric research, launched an initiative on "Harmonisation within Atmospheric Dispersion Modeling for Regulatory Purposes". This initiative still serves as a national and international tool for increased cooperation and standardisation of atmospheric dispersion models for regulatory purposes (http://www.harmo.org/).

19.3.1 Quality control measures

The general need for harmonisation and quality assurance of the collected data within large monitoring and surveillance programmes is recognised by all major international environmental monitoring programmes (e.g., LRTAP-EMEP, AMAP and others). LRTAP-EMEP has recently documented needs and requirements for comprehensive international monitoring programmes (EMEP 2001). In order to meet the requirements of scientific based quality assurance (QA) and harmonisation between the stations, LRTAP-EMEP recommends (EMEP 1988):

- Appointment of a QA Manager in each of the participating units (countries, laboratories). These persons will be responsible for implementing harmonized quality assurance systems within the units, including documentation of standards and reference materials.
- Development of standardized operating procedures based on recognised QA recommendations (e.g., AMAP 1994, EMEP 2001).
- Co-location experiments and instrument comparisons in the various units to document precision and quantify internal network differences.
- Continuation of efforts towards site characterization.

Very comprehensive and up-to-date guidelines on the issues of harmonisation and quality assurance measures are also given in a recently published US-EPA report (US-EPA 2001). On the basis of this documentation, continuous discussions and co-operation on QA and harmonisation are necessary as an integrated part of future integrated atmospheric monitoring co-operations.

In addition, the Arctic Monitoring and Assessment Programme (AMAP) developed already in 1995 a collection of method performance criteria and quality control measures for Arctic environmental samples (AMAP 1994). The quality control criteria developed include recommendations considered having significant influence on the reliable determination of contaminants (particles, pollutants etc.). This includes both sampling and fieldwork. In this context, a tight cooperation between the all laboratories, researchers, modellers and regulatory authorities involved is needed and required.

19.4 Conclusions

The role of long-term atmospheric monitoring in Polar Regions as a versatile tool for the evaluation and assessment of compound specific distribution and transport processes, is today fully recognised by the international scientific community as well as by regulatory authorities. Important phenomena like polar springtime depletion of mercury, POP and trace metal atmospheric long-range transport, as well as "Arctic haze" have been studied in detail through circum Arctic atmospheric monitoring programs.

However, the today ongoing monitoring initiatives (including Arctic and Antarctic monitoring initiatives) are mostly regionally oriented and, thus, face considerable challenges in adapting and harmonising monitoring programs to the aspects of global transport processes.

The need for close co-operation between international atmospheric monitoring programs is recognised and will be a priority task for future monitoring initiatives. Topics like sample and method harmonisation, joint quality control criteria as well as continuous inter-comparison, will be issues to be discussed and implemented in co-operation programs between international and regional atmospheric monitoring initiatives.

Acknowledgement

The help and support of the Editor Dr. Jon Børre Ørbæk (NPI), Dr. Kjetil Tørseth, Dr. Jozef Pacyna (NILU) and Dr. Hayley Hung (Meteorological Services, Canada) during the preparation work is highly appreciated. We thank Finn Bjørklid (NILU) for his substantial help with the preparation of the manuscript.

References

Allengil, S.M. and Martynov, V. G. (1995) Heavy-Metal Burdens in 9 Species of Fresh-Water and Anadromous Fish from the Pechora River, Northern Russia. Sci. Tot. Environ. 161, 653-659

AMAP (1994) Method Performance Criteria and Quality Control Measures for the Determination of Persistent Organic Compounds (POPs) in Ambient air, Precipitation and Water, Working group "Quality Control Programme", Workshop on Techniques of POP measurements in Northern Environments, Waterloo, Ontario, Canada, 16-18.June 1994. Available from Arctic Monitoring and Assessment Programme (AMAP) Secretariat, P.O. Box 8100, Dep., NO-0032 OSLO, Norway.

AMAP (1998) AMAP Strategic Plan: 1998-2003, 1-17.

AMAP (2000) AMAP Trends and Effects Programme, Section E - Quality Assurance., 1-16

Andersson-Skold,Y. and Simpson, D. (1999) Comparison of the chemical schemes of the EMEP MSC-W and IVL photochemical trajectory models. Atmos. Environ. 33, 1111-1129

Andersson-Skold, Y. and Simpson, D. (2001) Secondary organic aerosol formation in northern Europe: A model study. J. of Geophys.l Res.-Atmospheres 106, 7357-7374

Ariya, P.A., Dastoor, A. P., Amyot, M., Schroeder, W. H., Barrie, L., Anlauf, K., Raofie, F., Ryzhkov, A., Davignon, D., Lalonde, J., and Steffen, A. (2004) The Arctic: a sink for mercury. Tellus Series B-Chemical and Physical Meteorology 56, 397-403

Asmund, G. and Nielsen, S. P. (2000) Mercury in dated Greenland marine sediments. Sci. Tot. Environ. 245, 61-72

Bailey, R., Barrie, L.A., Halsall, C.J., Fellin, P., Muir, D.C.G. (2000) Atmospheric organochlorine pesticides in the western Canadian Arctic : Evidence of Transpacific transport. J. Geophys. Res. 105, D9, 11805-11811,\.

Bartnicki, J. (1994) An Eulerian Model for Atmospheric Transport of Heavy-Metals Over Europe - Model Description and Preliminary-Results. Water Air and Soil Poll. 75, 227-263

Berg, T., Bartnicki, J., Munthe, J., Lattila, H., Hrehoruk, J., and Mazur, A. (2001) Atmospheric mercury species in the European Arctic: measurements and modelling. Atmos. Environ. 35, 2569-2582

Berg, T., Kallenborn, R., and Manø, S. (2004) Temporal Trends in Atmospheric Heavy Metal and Organochlorine Concentrations at Zeppelin, Svalbard. Arctic, Antarctic, and Alpine Res. 36/3, 283-290

Berg T., Sekkesæter S, Steinnes E, Valdal A, Wibetoe G (2003) Arctic springtime depletion of mercury in the European Arctic as observed at Svalbard. Sci. Total Environ. 304: 43-51.

Berg, T., Sommar, J., Wangberg, I., Gardfeldt, K., Munthe, J., and Schroeder, B. (2003) Arctic mercury depletion events at two elevations as observed at the Zeppelin Station and Dirigibile Italia, Ny-Alesund, spring 2002. J. de Phys. Iv 107, 151-154

Beyer, A., Mackay, D., Matthies, M., Wania, F., and Webster, E. (2000) Assessing long-range transport potential of persistent organic pollutants. Environ. Sci. & Techno. 34, 699-703

Bidleman, T.F., Walla, M. D., Roura, R., Carr, E., and Schmidt, S. (1993) Organochlorine Pesticides in the Atmosphere of the Southern-Ocean and Antarctica, January March, 1990. Marine Poll. Bull. 26, 258-262

Bignert, A., Olsson, M., Persson, W., Jensen, S., Zakrisson, S., Litzen, K., Eriksson, U., Haggberg, L., and Alsberg, T. (1998) Temporal trends of organochlorines in Northern Europe, 1967-1995. Relation to global fractionation, leakage from sediments and international measures. Environ. Poll. 99, 177-198

Bignert, A., Riget, F., Braune, B., Outridge, P., and Wilson, S. (2004) Recent temporal trend monitoring of mercury in Arctic biota - how powerful are the existing data sets? J. Environ.Monit. 6, 351-355

Bindler, R., Renberg, I., Anderson, N. J., Appleby, P. G., Emteryd, O., and Boyle, J. (2001) Pb isotope ratios of lake sediments in West Greenland: inferences on pollution sources. Atmos. Environ. 35, 4675-4685

Bjerregaard, P. and Hansen, J. C. (2000) Organochlorines and heavy metals in pregnant women from the Disko Bay area in Greenland. Sci. Tot. Enviro. 245, 195-202

Blake, N.J., Blake, D. R., Sive, B. C., Katzenstein, A. S., Meinardi, S., Wingenter, O. W., Atlas, E. L., Flocke, F., Ridley, B. A., and Rowland, F. S. (2003) The seasonal evolution of NMHCs and light alkyl nitrates at middle to high northern latitudes during TOPSE. J. Geophys. Res.-Atmospheres 108

Borga, K., Fisk, A. T., Hoekstra, P. F., and Muir, D. C. G. (2004) Biological and chemical factors of importance in the bioaccumulation and trophic transfer of persistent organochlorine contaminants in arctic marine food webs. Environ. Toxicol. Chem. 23, 2367-2385

Brankov, E., Rao, S. T., and Porter, P. S. (1998) A trajectory-clustering-correlation methodology for examining the long-range transport of air pollutants. Atmosp. Environ. 32, 1525-1534

Braune, B.M., Donaldson, G. M., and Hobson, K. A. (2001) Contaminant residues in seabird eggs from the Canadian Arctic. Part I. Temporal trends 1975-1998. Environ. Poll. 114, 39-54

Breivik, K. and Wania, F. (2003) Expanding the applicability of multimedia fate models to polar organic chemicals. Environ. Sci. & Technol. 37, 4934-4943

Bridgman, H.A. and Bodhaine, B. A. (1994) On the Frequency of Long-Range Transport Events at Point-Barrow, Alaska, 1983-1992. Atmosp. Environ. 28, 3537-3549

Brunstrom, B. and Halldin, K. (2000) Ecotoxicological risk assessment of environmental pollutants in the Arctic. Toxicol. Let. 112, 111-118

Chavez, E., Zazueta, C., Osornio, A., Holguin, J. A., and Miranda, M. E. (1991) Protective Behavior of Captopril on Hg++-Induced Toxicity on Kidney Mitochondria - Invivo and Invitro Experiments. J. of Pharmacol. and Exp. Therap. 256, 385-390

Christensen, J.H., Brandt, J., Frohn, L. M., and Skov, H. (2004b) Modelling of mercury in the Arctic with the Danish Eulerian Hemispheric Model. Atmos. Chem. and Phys. 4, 2251-2257

Christensen, J.H., Brandt, J., Frohn, L. M., and Skov, H. (2004a) Modelling of mercury in the Arctic with the Danish Eulerian Hemispheric Model. Atmos. Chem. and Phys. 4, 2251-2257

Daly, G.L. and Wania, F. (2004c) Simulating the influence of snow on the fate of organic compounds. Environ. Sci. & Technol. 38, 4176-4186

Department of Indian Affairs and Northern Development (2003) Canadian Contaminants Assessment Report II, Sources, Occurrence, Trends and Pathways in the physical environment, Northern Contaminants program, Minister of Indian Affairs and Northern Development, Minister of Public Works and Government Services Canada, p,p. 332.

Derwent, R.G. (1993) Evaluation of the Chemical Mechanism Employed in the Emep Photochemical Oxidant Model. Atmos. Environ. Part A-General Topics 27, 277-279

Dethloff, K,. Schwager, M., Christensen, J. H., Kilsholm, S., Rinke, A., Dorn, W., Jung-Rothenhausler, F., Fischer, H., Kipfstuhl, S., and Miller, H. (2002) Recent Greenland accumulation estimated from regional climate model simulations and ice core analysis. J. of Clim. 15, 2821-2832

DeWit, C.A., Fisk, A. T., Hobbs K.E., Muir, D. C. G., Gabrielsen, G. W., Kallenborn, R., Krahn M.M., Norstrom, R., and Skaare, J. U. (2004) AMAP assessment 2002: Persistent Organic Pollutants in the Arctic., AMAP Assessment 2002: Persistent organic pollutants in the Arctic, pp. 309

Dietz, R., Riget, F. F., Sonne, C., Letcher, R., Born, E. W., and Muir, D. C. G. (2004) Seasonal and temporal trends in polychlorinated biphenyls and organochlorine pesticides in East Greenland polar bears (Ursus maritimus), 1990-2001. Sci. Tot. Environ. 331, 107-124

Ebinghaus, R., Kock, H. H., Temme, C., Einax, J. W., Lowe, A. G., Richter, A., Burrows, J. P., and Schroeder, W. H. (2002) Antarctic springtime depletion of atmospheric mercury. Environ.Sci. Technol. 36, 1238-1244

EMEP (1988) EMEP Quality Assurance Plan. EMEP/CCC-Report 1/88, 1-100

EMEP (2001) EMEP manual for sampling and chemical analysis. Norwegian Institute for Air Research (NILU), NO-2027 Kjeller, Norway, pp. 303.

Eneroth, K., Kjellstrom, E., and Holmen, K. (2003) Interannual and seasonal variations in transport to a measuring site in western Siberia and their impact on the observed atmospheric CO2 mixing ratio. J. of Geophys. Res.-Atmospheres 108

Fellin, P., Barrie, L. A., Dougherty, D., Toom, D., Muir, D., Grift, N., Lockhart, L., and Billeck, B. (1996) Air monitoring in the arctic: Results for selected persistent organic pollutants for 1992. Environm. Toxicol. and Chem. 15, 253-261

Felzer, B., Thompson, S., Pollard, D., and Bergengren, J. (2000) GCM-simulated hydrology in the Arctic during the past 21,000 years. J. of Paleolimn. 24, 15-28

Furevik, T., Bentsen, M., Drange, H., Kindem, I. K. T., Kvamsto, N. G., and Sorteberg, A. (2003) Description and evaluation of the bergen climate model: ARPEGE coupled with MICOM. Clim. Dyn. 21, 27-51

Gauthier, J.M., Dubeau, H., and Rassart, E. (1998) Mercury-induced micronuclei in skin fibroblasts of beluga whales. Environ. Toxicol. Chem. 17, 2487-2493

Gautrois, M., Brauers, T., Koppmann, R., Rohrer, F., Stein, O., and Rudolph, J. (2003) Seasonal variability and trends of volatile organic compounds in the lower polar troposphere. J of Geophys. Res.-Atmospheres 108

Gebhart, K.A., Kreidenweis, S. M., and Malm, W. C. (2001) Back-trajectory analyses of fine particulate matter measured at Big Bend National Park in the historical database and the 1996 scoping study. Sci. of the Tot.l Environ.t 276, 185-204

Graham, B., Guyon, P., Taylor, P. E., Artaxo, P., Maenhaut, W., Glovsky, M. M., Flagan, R. C., and Andreae, M. O. (2003) Organic compounds present in the natural Amazonian aerosol: Characterization by gas chromatography-mass spectrometry. J. of Geophys. Res.-Atmospheres 108

Groves, D.G. and Francis, J. A. (2002) Variability of the Arctic atmospheric moisture budget from TOVS satellite data. J. of Geophys. Res.-Atmospheres 107

Guo, J., Rahn, K. A., and Zhuang, G. S. (2004) A mechanism for the increase of pollution elements in dust storms in Beijing. Atmos.Enviro. 38, 855-862

Halsall, C.J. (2004) Investigating the occurrence of persistent organic pollutants (POPs) in the arctic: their atmospheric behaviour and interaction with the seasonal snow pack. Environ.Poll. 128, 163-175

Hansen, J.C. (2000) Environmental contaminants and human health in the Arctic. Toxicol. Let 112, 119-125

Hansen, K.M., Christensen, J. H., Brandt, J., Frohn, L. M., and Geels, C. (2004) Modelling atmospheric transport of alpha-hexachlorocyclohexane in the

Northern Hemisphere with a 3-D dynamical model: DEHM-POP. Atmos. Chem. and Phys. 4, 1125-1137

Hargrave, B.T., Barrie, L. A., Bidleman, T. F., and Welch, H. E. (1997) Seasonality in exchange of organochlorines between Arctic air and seawater. Environ.Sci. Technol 31, 3258-3266

Heidam, N.Z. (1994) Arctic Aerosol Factor Models - Validation by Marginally Detected Elements. Atmos. Environ. 28, 1549-1555

Hoekstra, P.F., O'Hara, T. M., Fisk, A. T., Borga, K., Solomon, K. R., and Muir, D. C. G. (2003) Trophic transfer of persistent orgranochlorine contaminants (OCs) within an Arctic marine food web from the southern Beaufort-Chukchi Seas. Environ.l Poll. 124, 509-522

Holland, M.M. (2003) The North Atlantic oscillation-arctic oscillation in the CCSM2 and its influence on arctic climate variability. J of Clim. 16, 2767-2781

Holmes, E., Bonner, F. W., and Nicholson, J. K. (1996) Comparative biochemical effects of low doses of mercury II chloride in the F344 rat and the multimammate mouse (Mastomys natalensis). Comp. Biochem. and Physiol. C-Pharmacol. Toxicol. & Endocrinol. 114, 7-15

Hopke, P.K., Barrie, L. A., Li, S. M., Cheng, M. D., Li, C., and Xie, Y. (1995) Possible Sources and Preferred Pathways for Biogenic and Non-Sea-Salt Sulfur for the High Arctic. J. of Geophys. Res.-Atmospheres 100, 16595-16603

Hopke, P.K., Liu, C. H., and Rubin, D. B. (2001) Multiple imputation for multivariate data with missing and below-threshold measurements: Time-series concentrations of pollutants in the Arctic. Biometrics 57, 22-33

Hung, H,, Halsall C.J,, Blanchard, P,, Li, H.H,, Fellin, P,, Stern. G,, Rosenberg. B. (2001) Are PCBs in the Canadian Arctic atmosphere declining? Evidence from 5 years of monitoring. Environ. Sci. Technol. 35/7, 1303-1311.

Hung, H., Blanchard, P., Poole, G., Thibert, B., Chiu, C.H. (2002) Measurement of particle-bound polychlorinated dibenzo-p-dioxins and dibenzofurnas (PCCD/Fs) in Arctic air at Alert, Nunavut, Canada. Atmos. Environ. 36, 1041-1050

Hung, H., Halsall, C. J., Blanchard, P., Li, H. H., Fellin, P., Stern, G., and Rosenberg, B. (2002) Temporal trends of organochlorine pesticides in the Canadian Arctic atmosphere. Environ.Sci. Technol. 36, 862-868

Jonson, J.E., Bartnicki, J., Olendrzynski, K., Jakobsen, H. A., and Berge, E. (1998) EMEP Eulerian model for atmospheric transport and deposition of nitrogen species over Europe. Environ. Poll. 102, 289-298

Jonson, J.E., Sundet, J. K., and Tarrason, L. (2001) Model calculations of present and future levels of ozone and ozone precursors with a global and a regional model. Atmos. Environ. 35, 525-537

Kahl J.D.W., Galbraith JA, Martinez DA (1999) Decadal-scale variability in long range atmospheric transport to the Summit of the Greenland Ice Sheet. Geophys, Res. Let. *26, 481-484.*

Kahnert, M., Lazaridis, M., Tsyro, S., and Torseth, K. (2004) Requirements for developing a regional monitoring capacity for aerosols in Europe within EMEP. J.of Environ. Monit. 6, 646-655

Kallenborn, R, Oehme, M., Wynn-Williams, D. D., Schlabach, M., and Harris, J. (1998) Ambient air levels and atmospheric long-range transport of persistent organochlorines to Signy Island, Antarctica. Sci. of the Tot. Environ. 220, 167-180

Kangas, L. and Syri, S. (2002) Regional nitrogen deposition model for integrated assessment of acidification and eutrophication. Atmos. Environ. 36, 1111-1122

Karcher, M.J., Gerland, S., Harms, I. H., Iosjpe, M., Heldal, H. E., Kershaw, P. J., and Sickel, M. (2004) The dispersion of Tc-99 in the Nordic Seas and the Arctic Ocean: a comparison of model results observations. J of Environ. Radioact. 74, 185-198

Kattsov, V.M. and Walsh, J. E. (2000) Twentieth-century trends of arctic precipitation from observational data and a climate model simulation. J.l of Clim. 13, 1362-1370

Kim, E.Y., Murakami, T., Saeki, K., and Tatsukawa, R. (1996) Mercury levels and its chemical form in tissues and organs of seabirds. Arch. of Environ.l Contam. and Toxicol. 30, 259-266

Klungsoyr, J., Saetre, R., Foyn, L., and Loeng, H. (1995) Mans Impact on the Barents Sea. Arctic 48, 279-296

Kruger, O. and Tuovinen, J. P. (1997) The effect of variable sub-grid deposition factors on the results of the Lagrangian long-range transport model of EMEP. Atmos. Environ. 31, 4199-4209

Lakaschus S,. Weber K, Wania F, Bruhn R, Schrems O (2002). The air-sea equilibrium and time trend of hexachlorocyclohexanes in the Atlantic Ocean between the Arctic and Antarctica. Environ. Sci. Technol. 36: 138-145.

Landers, D.H., Ford, J., Gubala, C., Monetti, M., Lasorsa, B. K., and Martinson, J. (1995) Mercury in Vegetation and Lake-Sediments from the Us Arctic. Water Air and Soil Poll. 80, 591-601

Law, R.J., Alaee, M., Allchin, C. R., Boon, J. P., Lebeuf, M., Lepom, P., and Stern, G. A. (2003) Levels and trends of polybrominated diphenylethers and other brominated flame retardants in wildlife. Environ. Internat. 29, 757-770

Lindberg, S.E., Brooks, S., Lin, C. J., Scott, K. J., Landis, M. S., Stevens, R. K., Goodsite, M., and Richter, A. (2002) Dynamic oxidation of gaseous mercury in the Arctic troposphere at polar sunrise. Environ.Sci. Technol 36, 1245-1256

Lu, J.Y., Schroeder, W. H., Barrie, L. A., Steffen, A., Welch, H. E., Martin, K., Lockhart, L., Hunt, R. V., Boila, G., and Richter, A. (2001) Magnification of atmospheric mercury deposition to polar regions in springtime: the link to tropospheric ozone depletion chemistry. Geophys. Res. Let. 28, 3219-3222

Ma, J.M., Cao, Z. H., and Hung, H. (2004) North Atlantic Oscillation signatures in the atmospheric concentrations of persistent organic pollutants: An analysis using Integrated Atmospheric Deposition Network-Great Lakes monitoring data. J. of Geophys. Res.-Atmospheres 109

Macdonald, R.W., Barrie, L. A., Bidleman, T. F., Diamond, M. L., Gregor, D. J., Semkin, R. G., Strachan, W. M. J., Li, Y. F., Wania, F., Alaee, M., Alexeeva, L. B., Backus, S. M., Bailey, R., Bewers, J. M., Gobeil, C., Halsall, C. J.,

Harner, T., Hoff, J. T., Jantunen, L. M. M., Lockhart, W. L., Mackay, D., Muir, D. C. G., Pudykiewicz, J., Reimer, K. J., Smith, J. N., Stern, G. A., Schroeder, W. H., Wagemann, R., and Yunker, M. B. (2000) Contaminants in the Canadian Arctic: 5 years of progress in understanding sources, occurrence and pathways. Sci. of the Tot.Environ. 254, 93-234

Macdonald, R.W. and Bewers, J. M. (1996) Contaminants in the arctic marine environment: Priorities for protection. Ices J, of Marine Sci. 53, 537-563

Macdonald, R.W., Mackay, D., Li, Y. F., and Hickie, B. (2003) How will global climate change affect risks from long-range transport of persistent organic pollutants? Human and Ecol. Risk Assess. 9, 643-660

Mackay, D. and Wania, F. (1995) Transport of Contaminants to the Arctic - Partitioning, Processes and Models. Sci. of the Tot. Environ. 161, 25-38

Mandrioli, P., Negrini, M. G., Cesari, G., and Morgan, G. (1984) Evidence for Long-Range Transport of Biological and Anthropogenic Aerosol-Particles in the Atmosphere. Grana 23, 43-53

Manø, S. and Schaug, J. (2003) EMEP POP laboratory comparison 2000-2002. CCC-report 10/2003, 1-40

Moore, K.G., Clarke, A. D., Kapustin, V. N., and Howell, S. G. (2003) Long-range transport of continental plumes over the Pacific Basin: Aerosol physiochemistry and optical properties during PEM-Tropics A and B. J.l of Geophys. Res.-Atmospheres 108

Muir, D,. Braune, B., DeMarch, B., Norstrom, R., Wagemann, R., Lockhart, L., Hargrave, B., Bright, D., Addison, R., Payne, J., and Reimer, K. (1999a) Spatial and temporal trends and effects of contaminants in the Canadian Arctic marine ecosystem: a review. Sci. of the Tot. Environ. 230, 83-144

Muir, D.C.G., Teixeira, C., and Wania, F. (2004) Empirical and modeling evidence of regional atmospheric transport of current-use pesticides. Environ. Toxicol. and Chem. 23, 2421-2432

Muir, D.C.G., Wagemann, R., Hargrave, B. T., Thomas, D. J., Peakall, D. B., and Norstrom, R. J. (1992) Arctic Marine Ecosystem Contamination. Sci. of the Tot. Environ. 122, 75-134

Nakazawa, T., Ishizawa, M., Higuchi, K., Trivett, N.B.A. (1997) _Two curve fitting methods applied to CO2 flask data. Environmetircs 8, 197-218

Nriagu J., Pacyna J (1988) Quantitative assessment of worldwide contamination of air, water and soil from trace metals, Nature 333, 134-139.

Oei, H. and Falkenberg, E. (1994) Harmonization of Information-System Modeling and Specification Techniques. Methods and Associated Tools for the Information Systems Life Cycle 55, 151-168

Olendrzynski, K., Berge, E., and Bartnicki, J. (2000) EMEP Eulerian acid deposition model and its applications. Europ. J. of Operat. Res. 122, 426-439

Pacyna, J.M. (1995) The Origin of Arctic Air-Pollutants - Lessons Learned and Future-Research. Science of the Total Environment 161, 39-53

Pacyna E.G., Pacyna J.M. (2002) Global emission of mercury from anthropogenc sources in 1995. Water Air Soil Poll., *137,* 149-165.

Pavlov, V., Pavlova, O., and Korsnes, R. (2004) Sea ice fluxes and drift trajectories from potential pollution sources, computed with a statistical sea ice model of the Arctic Ocean. Journal of Marine Syst. 48, 133-157

Pereg, D., Ayotte, P., Muckle, G., Ryan, J. J., Gingras, S., and Dewailly, E. (2003) Concentrations of persistent organic pollutants in human milk from nunavik (Arctic Quebec, Canada): Changes observed from 1989 to 200. Toxicol. Sci. 72, 185-185

Perkins, P.E. (1995) An Overview of International Institutional Mechanisms for Environmental-Management with Reference to Arctic Pollution. Sci. of the Tot. Environ. 161, 849-857

Poissant, L., Dommergue, A., and Ferrari, C. P. (2002) Mercury as a global pollutant. Journal de Physique Iv 12, 143-160

Power, M., Klein, G. M., Guiguer, K. R. R. A., and Kwan, M. K. H. (2002) Mercury accumulation in the fish community of a sub-Arctic lake in relation to trophic position and carbon sources. J. of Appl, Ecol. 39, 819-830

Prandle, D. (1991) A 5-Year Scientific-Research Program for Managing Coastal Seas. Mar. Poll. Bull. 23, 63-68

Proshutinsky, AY., Polyakov, I. V., and Johnson, M. A. (1999) Climate states and variability of Arctic ice and water dynamics during 1946-1997. Polar Res. 18, 135-142

Prospero, J.M. (1999) Long-range transport of mineral dust in the global atmosphere: Impact of African dust on the environment of the southeastern United States. Proceedings of the National Academy of Sciences of the United States of America 96, 3396-3403

Rasmussen, R.A., Khalil, M. A. K., and Fox, R. J. (1983) Altitudinal and Temporal Variation of Hydrocarbons and Other Gaseous Tracers of Arctic Haze. Geophys. Res. Let. 10, 144-147

Riget, F., Dietz, R., Johansen, P., and Asmund, G. (2000) Lead, cadmium, mercury and selenium in Greenland marine biota and sediments during AMAP phase 1. Sci.of the Tot. Environ.t 245, 3-14

Riget, F., Dietz, R., Vorkamp, K., Johansen, P., and Muir, D. (2004) Levels and spatial and temporal trends of contaminants in Greenland biota: an updated review. Sci. of the Tot. Environ. 331, 29-52

Rinke, A., Dethloff, K., Spekat, A., Enke, W., and Christensen, J. H. (1999) High resolution climate simulations over the Arctic. Polar Res. 18, 143-150

Rosman, K.J.R., Ly, C., Van de Velde, K., and Boutron, C. F. (2000) A two century record of lead isotopes in high altitude Alpine snow and ice. Earth and Planet. Sci. Let. 176, 413-424

Ross, P.S. and Birnbaum, L. S. (2003) Integrated human and ecological risk assessment: A case study of persistent organic pollutants (POPs) in humans and wildlife. Human and Ecological Risk Assessment 9, 303-324

Scheringer, M., Stroebe, M., Wania, F., Wegmann, F., and Hungerbuhler, K. (2004) The effect of export to the deep sea on the long-range transport potential of persistent organic pollutants. Environ. Sci. and Poll. Res. 11, 41-48

Schroeder, W.H. (2002) Emission sources, atmospheric transport, chemical speciation, and fate of mercury entering the arctic environment. Abstracts of Papers of the American Chemical Society 223, U522-U522

Schroeder, W.H., Anlauf, K. G., Barrie, L. A., Lu, J. Y., Steffen, A., Schneeberger, D. R., and Berg, T. (1998) Arctic springtime depletion of mercury. Nature 394, 331-332

Serreze, M.C., Clark, M.P., and Bromwich, D.H. (2003) Monitoring precipitation over the Arctic terrestrial drainage system: Data requirements, shortcomings, and applications of atmospheric reanalysis. J. of Hydrometeorol.4, 387-407

Serreze, M.C. and Hurst, C. M. (2000) Representation of mean Arctic precipitation from NCEP-NCAR and ERA reanalyses. J.. of Clim. 13, 182-201

Shatalov, V., Feldmann, H., Martilli, A., Stull, R.B., and Walker, S.E. (2003) Persistent Organic Pollutants in the Environment. EMEP report, p. 100

Skaare, J.U., Larsen, H. J., Lie, E., Bernhoft, A., Derocher, A. E., Norstrom, R., Ropstad, E., Lunn, N. F., and Wiig, O. (2002) Ecological risk assessment of persistent organic pollutants in the arctic. Toxicol. 181, 193-197

Skov, H., Christensen, J. H., Goodsite, M. E., Heidam, N. Z., Jensen, B., Wahlin, P., and Geernaert, G. (2004) Fate of elemental mercury in the arctic during atmospheric mercury depletion episodes and the load of atmospheric mercury to the arctic. Environ.Sci. Technol 38, 2373-2382

Skov H., Goodsite ME, Christensen J, Geernaert G, Zeuthen Heidam N, Jensen B, Wåhlin P, Feilberg A (2002) Fate of Atmospheric Mercury in Arctic: results from Station Nord. Paper presented at the Arctic Atmospheric Mercury Research Workshop, Toronto, Canada, 2002.

Sommar J., Wängberg I., Berg T.; Gårdfeldt K., Munthe J., Richter A., Urba A., Wittrock F. and Schroeder W.H. (2004) Circumpolar transport and air-surface exchange of atmospheric mercury at Ny-Ålesund (79°N), Svalbard, spring 2002. Atmospheric Chemistry and Physics Discussions 4, 1727-1771.

Sprovieri, F., Pirrone, N., Hedgecock, I.M., Landis, M.S., Stevens, R.K. (2002) Intensive atmospheric mercury measurements a Terra Nova Bay in Antarctica during November and December 2000. Journal of geophysical research, Vol. 107, Art. No. 4722.

Stohl, A. (1996) Trajectory statistics - A new method to establish source-receptor relationships of air pollutants and its application to the transport of particulate sulfate in Europe. Atmos. Environ. 30, 579-587

Stone, R.S. (1997) Variations in western Arctic temperatures in response to cloud radiative and synoptic-scale influences. J. of Geophys. Res.-Atmospheres 102, 21769-21776

Stone, R.S., Dutton, E. G., Harris, J. M., and Longenecker, D. (2002) Earlier spring snowmelt in northern Alaska as an indicator of climate change. J. of Geophys. Res.-Atmospheres 107

Subhash, S. and Honrath, R. E. (1999) Back-trajectory analysis of atmospheric polychlorinated biphenyl concentrations over Lake Superior. Environ.Sci. Technol. 33, 1509-1515

Temme, C., Einax, J. W., Ebinghaus, R., and Schroeder, W. H. (2003) Measurements of atmospheric mercury species at a coastal site in the Antarctic and over the south Atlantic Ocean during polar summer. Environ.Sci. Technol. 37, 22-31

Thomas, D.J., Tracey, B., Marshall, H., and Norstrom, R. J. (1992) Arctic Terrestrial Ecosystem Contamination. Sci. of the Tot. Environ. 122, 135-164

Toose, L., Woodfine, D. G., MacLeod, M., Mackay, D., and Gouin, J. (2004) BETR-World: a geographically explicit model of chemical fate: application to transport of alpha-HCH to the Arctic. Environ. Poll. 128, 223-240

Tørseth K., Hov Ø. (2003) The EMEP monitoring strategy 2004 – 2009. Background document with justification and specification of the EMEP monitoring programme 2004-2009. EMEP/CCC-Report 9/2003. Norwegian Institute for Air Research (NILU), NO-2027 Kjeller, Norway, pp 69.

Tørseth K. (2004) EMEPs contribution to a multi-purpose monitoring capacity for atmospheric composition in Europe being compatible with local and global initiatives: Conventions and Directives, user requirements. Atmospheric Chemistry Applications Workshop, ESTEC, The Netherlands, 20-21. January 2004.

Traub, M., Fischer, H., de Reus, M., Kormann, R., Heland, J., Ziereis, H., Schlager, H., Holzinger, R., Williams, J., Warneke, C., de Gouw, J., and Lelieveld, J. (2003) Chemical characteristics assigned to trajectory clusters during the MINOS campaign. Atmospheric Chemistry and Physics 3, 459-468

Tuovinen, J.P., Ashmore, M. R., Emberson, L. D., and Simpson, D. (2004) Testing and improving the EMEP ozone deposition module. Atmos. Environ. 38, 2373-2385

UNEP (2000) Master list of actions on the reduction and/or elimination of the Releases of Persistent Organic Pollutants (Second Edition), March 2000, UNEP Chemicals, Geneva, Switzerland, pp 217.

US-EPA (2001) Environmental Investigations Standard Operating Procedures and quality Assurance manual. US Environmental Protection Agency, Athens, Georgia, pp 413.

Vaughan, J.M., Maryon, R. H., and Geddes, N. J. (2002) Comparison of atmospheric aerosol backscattering and air mass back trajectories. Meteorol. and Atmos. Phys. 79, 33-46

Wagemann, R., Innes, S., and Richard, P. R. (1996) Overview and regional and temporal differences of heavy metals in Arctic whales and ringed seals in the Canadian Arctic. Sci. of the Tot.Environ. 186, 41-66

Wagemann, R., Trebacz, E., Boila, G., and Lockhart, W. L. (1998) Methylmercury and total mercury in tissues of arctic marine mammals. Sci. of the Tot.l Environ. 218, 19-31

Walsh, J.E., Kattsov, V. M., Chapman, W. L., Govorkova, V., and Pavlova, T. (2002) Comparison of Arctic climate simulations by uncoupled and coupled global models. J.of Clim. 15, 1429-1446

Wängberg I., Sommar J., Berg T., Gårdfeldt K. and Munthe J. W.H. (2003) Interpretation of Mercury Depletion Events Observed at Ny-Ålesund, Svalbard During Spring 2002. J. Phys. IV France 107, 1353-1356.

Wania, F. (2003b) Assessing the potential of persistent organic chemicals for long-range transport and accumulation in polar regions. Environ.Sci. Technol . 37, 1344-1351

Wania, F. (1998) Multi-compartmental models of contaminant fate in the environment. Biotherapy 11, 65-68

Wania, F. (2003a) Assessing the potential of persistent organic chemicals for long-range transport and accumulation in polar regions. Environ.Sci. Technol . 37, 1344-1351

Wania, F. and Dugani, C. B. (2003) Assessing the long-range transport potential of polybrominated diphenyl ethers: A comparison of four multimedia models. Environ. Toxicol. and Chem. 22, 1252-1261

Wania, F. and Mackay, D. (1996) Tracking the distribution of persistent organic pollutants. Environ.Sci. Technol . 30, A390-A396

Wania, F. and Mackay, D. (1999b) Global chemical fate of alpha-hexachlorocyclohexane. 2. Use of a global distribution model for mass balancing, source apportionment, and trend prediction. Environ. Toxicol. and Chem. 18, 1400-1407

Wania, F. and Mackay, D. (1999a) The evolution of mass balance models of persistent organic pollutant fate in the environment. Environ.l Poll. 100, 223-240

Weber, R.J., Orsini, D., Wang, B., Scheuer, E., Talbot, R. W., Dibb, J. E., Seid, G. K., DeBell, L., Mauldin, R. L., Kosciuch, E., Cantrell, C., and Eisele, F. (2003) Investigations into free tropospheric new particle formation in the central Canadian arctic during the winter/spring transition as part of TOPSE. J. of Geophys. Res.-Atmospheres 108

Zhao, P., Zhou, X. J., Jian, Z. M., Sparrow, M., and Han, Y. (2004) Modeling the tropical climate and the impact of the western Pacific sea surface temperature at the Last Glacial Maximum. J. of Geophys. Res.-Atmospheres 109

Zheng, J., Zdanowicz, C., Fisher, D., Hall, G., and Vaive, J. (2003) A new 155-year record of Pb pollution from Devon ice cap, Canada. J. de Phys. Iv 107, 1405-1408

Zhuang, G.S., Guo, J. H., Yuan, H., and Zhao, C. Y. (2001) The compositions, sources, and size distribution of the dust storm from China in spring of 2000 and its impact on the global environment. Chinese Sci. Bull. 46, 895-901

20 Levels and effects of persistent organic pollutants in arctic animals

Geir Wing Gabrielsen

Norwegian Polar Institute, Hjalmar Johansensgt. 14, N-9296 Tromsø, Norway, e-mail: geir@npolar.no

20.1 Introduction

Despite the fact that the Arctic is far from the industrialized world the presence of persistent organic pollutants (POPs) in arctic food webs shows that man-made substances are transported over long distances to these remote areas. Several international reports (AMAP 1998; 2004; EEA 2004) and reviews (Muir et al. 1999; Gabrielsen and Henriksen 2001; Fisk et al. 2003; de Wit et al. 2004) list and document threats to the Arctic ecosystem from long-range transported contaminants such as POPs.

POPs are a diverse group of anthropogenic pollutants that are of industrial and agricultural origin. POPs are rarely used in the Arctic, but they have been documented in arctic wildlife since the beginning of the 1970s. Substances such as polychlorinated biphenyls (PCBs) and chlorinated pesticides (e.g., dichlorodiphenyltrichloroethane (DDT), toxaphenes) were reported in arctic seals, polar bears *(Ursus maritimus)* and glaucous gulls *(Larus hyperboreus)* in the early 1970s (Bogan and Bourne 1972; Bourne and Bogan 1972; Holden 1972; Bowes and Jonkel 1975). Although low levels of POPs have been reported in terrestrial Arctic species, the problems are mainly related to marine organisms. While some POPs (e.g., DDT and PCBs) have decreased in levels during the last 10-20 years, following the introduction of bans and restrictions in use, new persistent pollutants are increasing in the environment due to the fact they are currently being produced in large quantities. Some of these new chemicals include chlorinated naphthalene (PCN), brominated flame retardants (BFRs) and perfluoro-octane sulfonate (PFOS) (AMAP 2004).

Most POPs found in the Arctic are transported from distant industrial and agricultural sources by atmospheric and oceanic currents, as well as river discharges (AMAP 2003). The most important transport route is atmospheric circulation bringing contaminants from the lower latitudes to the Arctic within days (Macdonald et al. 2000). The contaminants are deposited and taken up mainly in the lipid rich food chains of the arctic marine ecosystem (Muir et al. 1992; Borgå et al. 2001; 2004; Fisk et al. 2003). As a result of processes of bioaccumulation and biomagnification, POPs can reach very high concentrations in apex predators such as polar bears, arctic fox (*Alopex lagopus*) and glaucous gulls (Wang-Andersen et al. 1993; Gabrielsen et al. 1995; Bernhoft et al. 1997; de Wit et al. 2004).

The concern about POP levels and possible effects in arctic wildlife increased with the comprehensive surveys performed in the 1980s and 1990s (AMAP 1998; 2004). POPs are known to elicit a range of detrimental effects on biota. For POPs the effects are related to the enzyme-, immune-, hormone- and vitamin systems. Of great concern are contaminants that may have reproductive effects and those that can mimic and disrupt the hormone system (Colborn et al. 1993; Giesy et al. 2003). During the last 5-10 years, a number of field studies have been performed on polar bears and glaucous gulls from Bjørnøya and Svalbard in order to study the relationships between contaminant levels and effects. The levels documented in glaucous gulls and polar bears are high enough to raise concern about the effects on the health of these species.

The present article summarizes some recent studies on the levels and biological effects of POPs in arctic animals. A special emphasis has been put on effect studies on polar bears and glaucous gulls from Svalbard and Bjørnøya.

20.2 Persistent organic pollutants (POPs)

Environmental contaminants include both industrial chemicals, such as PCBs, hexachlorobenzene (HCB), chlorinated pesticides, such as DDT, chlordanes, hexachloro-cyclohexane (HCH), aldrin/dieldrin, polychlorinated boranes (Toxaphenes) and industrial byproducts such as polychlorinated dibenzo-p-dioxins (PCDDs), polychlorinated dibenzofurans (PCDFs), polybrominated biphenyls (PBBs) and polybrominated diphenyl ethers (PBDEs) (Muir et al. 1992). Most of these contaminants were developed and put into production more than 50 years ago. The highest production of most of these contaminants was at the end of the 1960s and the start of 1970s (Blus 1995). For some of the new contaminants (e.g., PBDEs) the use has increased drastically in the past decade. Most of the use is in industrial areas of the northern hemisphere that are potential source regions to the Arctic.

20.2.1 PCBs

PCBs are mixtures of chlorinated hydrocarbons that have been heavily used since 1930 for many industrial purposes such as dielectrics in transformers and large capacitors, heat exchange fluids, paint additives, in carbonless copy paper and plastics (Fisher 1999). At present, the major source of PCB exposure seems to be environmental recycling of PCBs from former usage. There are 209 possible PCBs (congeners), of which about 100 have been found in biological samples (McFarland and Clarke 1989). The properties of PCBs depend on the number of Cl-atoms and their position and include low water solubility, high stability, and semi-volatility, which favour long-range transport. Worldwide PCB production in 2000 was estimated at 1.3 million tonnes (Breivik et al. 2002).

20.2.2 DDTs

DDT and its metabolites (1,1-dichloro-2,2-bis ethane (DDD) and 1,1-dichloro-2,2-bis ethylene (DDE)) have been found in biota since the 1940s. After the war, DDT was used extensively as a pesticide on a variety of agricultural crops (i.e. cotton and peanuts) and to prevent the spread of diseases to humans by insects (e.g., malaria and typhus). The total global usage was estimated at 2.6 million tonnes up until 1992 (Voldner and Li 1995). DDT and related compounds are very persistent in the environment and their half-lives in soil range from 10 to 15 years. DDT has been declining in the temperate regions of the northern and hemisphere since the 1960s and especially since the mid-1970s when production and use were banned by many western nations (Mellanby 1992).

20.2.3 PBBs and PBDEs

PBBs, PBDEs and tetrabromobisphenol are the three main classes of brominated compounds used as flame retardant additives (Renner 2000). They are used at high volume in electric equipment such as computers, television sets, textiles (clothing), cars, airplanes and other applications (de Boer et al. 1998). Humans may absorb PBBs and PBDEs when they are emitted from electronic circuit boards and plastic computers and cabinets (Zelinski et al. 1993). PBBs and PBDEs show high lipophilicity, high resistance to degradation, and are expected to bioaccumulate effectively in the aquatic and terrestrial food chains. The annual usage of the flame retardants have drastically increased during the last few years. The annual world production in 1998 was estimated at 150 000 tonnes (Sellstrøm and Jansson 1995).

20.2.4 Perfluorinated alkyl substances (PFAS) and PFOS

In recent years environmental concern has arisen as PFAS have been reported in seabirds and marine mammals in the Arctic (Giesy and Kannan 2001; Smithwick et al. 2005; Verreault et al., submitted). The most pervasive PFAS reported in arctic biota has been PFOS (Kannan et al. 2001). PFOS is used as a refrigerant, surfactant, and as a component in pharmaceuticals, flame retardants, lubricant, paints, adhesives, cosmetics, paper coating, and insecticides. PFOS has been manufactured for over 50 years and has steadily increased its use. The annual US production was 3000 tonnes in 2000 (AMAP 2004).

20.2.5 PCN

PCNs exist as 75 congeners that are structurally similar to the PCBs and have many similar applications. These include electrical equipment, lubricants, solvents, dyes and sealants. They are also present as impurities in technical PCB and are formed during anthropogenic combustion processes. In the 1920s the worldwide annual production was approximately 9000 tonnes. One of the largest PCN producers voluntarily ceased production in the late 1970s. At the present, information about the world production is limited. Although the use of PCNs has declined over the past few decades, they are not prohibited in most countries and are still being used in many PCB-like applications (AMAP 2004).

20.3 Levels of POPs in marine food chains

The arctic environment is among the least polluted ecosystems in the world. The POP levels are low in most lower trophic level marine species from the Barents Sea area compared to the Baltic and the North Sea. The POP levels at lower trophic levels in the Barents Sea, are very similar to what is found in Alaska and Canada. High POP levels and possible biological effects are mainly associated with species at the top of the food webs (AMAP 1998, 2004). However, few studies to date have examined the potential effects of various POPs on the lower trophic level of the polar ecosystems.

There are large differences in POPs levels in Arctic animals. These differences may be attributed to several factors such as exposure, ability to metabolize contaminants, ability to excrete compounds, seasonal variation in body mass, age and sex (Bignert et al. 1993; Henriksen et al. 1996). For example, the sex difference in POP levels often seen in birds partly reflects the fact that female birds are able to deposit lipophilic compounds into the egg. In polar bears and other marine mammals, considerable amounts of POPs are also transferred to the foetus and via milk from the mother to the cub. This is one of the main reasons for lower levels of some POPs in sexually mature female polar bears than in males (Bernhoft et al. 1997).

Several detailed studies of the Arctic marine food webs and POP accumulation were carried out in the 1990s in different Arctic regions (e.g. Borgå et al. 2001; Fisk et al. 2001; Muir and Strachan 2003; Hoekstra et al. 2003). These are summarized along with factors of importance for POP flux and food webs in Borgå et al. (2004).

20.4 Marine invertebrates

Marine invertebrates provide the trophic link from phytoplankton to fish, seabirds and marine mammals in the arctic marine food web. These species do not only carry nutrients and energy, but also POPs. In general, marine invertebrates have low POP levels. In most marine species investigated at lower trophic levels in the Barents Sea area (Borgå et al. 2001; 2002 a-c), the POP levels are low compared to species at lower trophic levels from the sub-arctic and temperate areas (0.01-0.2 μg/g, lipid weight concentrations) (Borgå et al. 2001). In a comparison of the marine food web of the Barents Sea and the Canadian Arctic (Northwater Polynya), POP levels were relatively similar in copepods (*Calanus* spp.), euphausiids (*Thysanoessa* spp.), amphipods (*Themisto libellula*) and polar cod (*Boreogadus saida)* (Borgå et al. 2005a). This in contrast to what is found in higher trophic levels, where these POP levels are higher in the Barents Sea than in the Canadian and Alaskan marine food webs (Norstrom et al. 1998; Muir et al. 2000a; Borgå et al. 2005a). In the same species only hexachloro-cyclohexane (HCH) were higher in the Canadian Arctic marine food web than in the Barents Sea food web due to the proximity to point sources in eastern Asia. In all fish species from the Barents Sea the POP levels are lower than the levels found in the North- and Baltic Seas

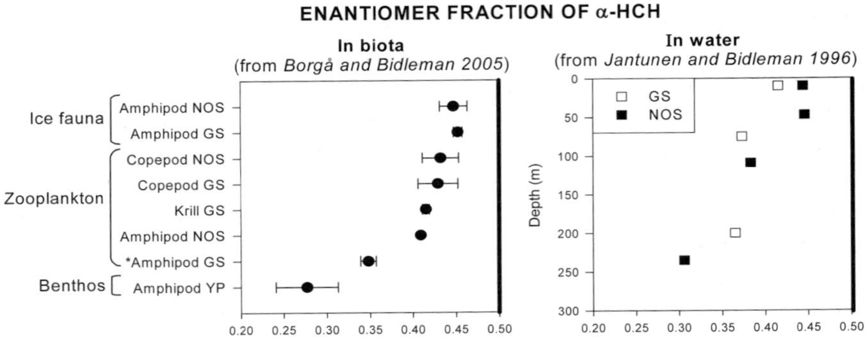

Fig. 20.1. Enantiomer fractions of α-HCH (mean ± 95 % confidence interval) in invertebrates and water from the Greenland Sea (GS), north of Svalbard (NOS and the Yermak Plateu (YP). * refers to predominantely pelagic amphipods that were collected with a bottom trawl. For details see Jantunen and Bidleman 1996 for water data, and Borgå and Bidleman 2005 for invertebrate data.

(AMAP 2004).

Despite low levels in invertebrates, they differ in POP levels according to diet, size and habitat (Borgå et al. 2002 a-c; Fisk et al. 2003). This has recently also been seen for chiral chlorinated pesticides in ice fauna, zooplankton and benthos (Borgå and Bidleman 2005) (Fig. 20.1), as well as in invertebrates from the Great Lakes (Wong et al. 2004). In addition, ice fauna differ in levels for some POPs reflecting the sea ice drift route across the Arctic Ocean (Borgå et al. 2002b; Borgå and Bidleman 2005).

Analysis of brominated flame retardant in polar cod has shown low levels compared to fish from other areas. Liver concentrations of PBDEs in herring (*Clupa harengus*) and cod (*Gadus morua*) were 15-60 times higher, respectively than the levels found in polar cod from Svalbard (Wolkers et al. 2004).

20.5 Seabirds

Seabird species living in the Arctic are contaminated by the same POPs as seabirds living in the south although their levels and proportions may occasionally differ to a certain extent (AMAP 1998; 2004). The POP levels in seabirds are mainly determined by their feeding habits and ability to biotransform contaminants (Borgå et al. 2005a). Common eider (*Somateria mollissima*), which feed on benthic organisms, and little auk (*Alle alle*), which mainly feed on copepods, have low POP levels (0.5-1.0 μg/g) (Savinova et al. 1995). However, in fish-eating species such as black-legged kittiwakes (*Rissa tridactyla*), common and Brünnich's guillemots (*Uria aalge* and *Uria lomvia*) and puffins (*Fratercula arctica*), the POP levels are somewhat higher (1.0-5.0 µg/g). The highest POP levels (1.0 to 40 µg/g) are found in herring gull (*Larus argentatus*), glaucous gull, great black-backed gull (*Larus marinus*) and the great skua (*Stercorarius skua*) (Savinova, Skaare and Gabrielsen, unpublished). The POP levels in gull species are 5-10 times higher than the other seabird species in the same area. In seabirds, PCBs comprise the majority of POP in all species, followed by DDT, chlordanes, HCB and HCHs (Savinova et al. 1995; Borgå et al. 2001). The high POP levels in herring gull, glaucous gull, great black-backed gull reflect their position on the top of the arctic food web. The levels of POP in different seabird species may also be explained by their migration pattern during the winter. Guillemots from low arctic colonies in Canada have shown higher PCB and HCB levels than guillemots from high arctic colonies (Braune and Donaldson 2000). Most seabird species from Svalbard, which migrate south to the North- or Norwegian Seas during the winter, have higher POP levels than seabirds that live in the Arctic throughout the winter (Borgå et al. 2005a; Savinova, Skaare and Gabrielsen, unpublished). A comparison of POP levels in 5 species of seabirds (black-legged kittiwakes, little auk, black guillemot (*Cepphus grylle*), common guillemot and glaucous gull), between the Barents Sea and the Canadian Arctic, show higher contaminant levels in species from the European Arctic (Barents Sea) (Borgå et al. 2005a).

Analysis of brominated flame retardants (e.g. PBDEs, hexabromocyclododecane (HBCD), PBBs), PFOS and PCN in seabirds have shown concentration in, e.g. glaucous gulls from Bjørnøya that were among the highest reported in any arctic seabirds, although generally lower, occasionally by many fold, compared to species from lower latitudes (Verreault et al. 2005a, 2005b). PFOS levels in glaucous gull plasma were comparable to levels of legacy POPs such as DDTs (Verreault et al., submitted).

20.5.1 Glaucous gulls

The glaucous gull has a circumpolar distribution in the Arctic. It is one of the largest gulls breeding in the Arctic and the only numerous avian predator in Svalbard, Frans Josef Land and Novaya Zemlya (Bakken and Tertitski 2000). The glaucous gulls breeding in the Barents Sea area do not migrate long distances. They winter mainly in the northern part of the Atlantic Ocean and remain there from November to March (Yudin and Firsova 1988). In Svalbard, the glaucous gull is both a predator and a scavenger (Løvenskiold 1964). On Bjørnøya, its diet consists mainly of eggs, chicks, adult birds, and crabs *(Hyas araneus)*, as well as offal from fishing boats (Bakken and Tertitski 2000).

PCBs and DDT are the most common POP measured in Glaucous gulls. PCBs and DDT constitutes on average 72 and 22 %, respectively, of the total pollution load measured in plasma (Verreault et al. 2005a). This is mainly due to the efficient bioaccumulation potential of PCBs and DDT and the ability of glaucous gulls to metabolize other POPs, although lower levels relative to marine mammals such as polar bear.

On Bjørnøya the sum PCB concentration in brain tissue of glaucous gulls found dead vary from 1 to 30 µg/g (wet weight (ww)), while healthy individuals vary from 0.5 to 10 µg/g (ww). In other tissue (muscle, liver and fat) the sum PCB varies from 0.5 to 23 µg/g (ww) (Gabrielsen et al. 1995; Sagerup et al. 2000). In blood from healthy glaucous gulls of normal body condition from Bjørnøya, the sum PCB were 0.6 µg/g (ww) in male gulls and 0.3 µg/g (ww) in female gulls (Bustnes et al. 2001; 2003; Verreault et al. 2004). At Bjørnøya, the POP levels vary within the population. Bustnes et al. (2001) studied two colonies, a few kilometers from each other, and found a large variation in POP concentration. Birds, which nested on the cliffs, had higher POP concentration than birds that nested at sea-level. While the birds on the cliffs were eating seabird eggs and chicks, the birds at sea-level fed more on fish, and thus were feeding at a lower trophic level. The concentration of sum PCB and sum DDT in glaucous gull were higher in the Barents Sea and Jan Mayen area than in Baffin Bay. The highest concentration was found around Frans Josef Land (Savinova, Skaare and Gabrielsen, unpublished data).

In glaucous gulls from Bjørnøya the levels of PCN ranged from 1.3 to 126 ng/g (lipid weight (lw)) in plasma and 1.8 to 162 ng/g (lw) in eggs (Verreault et al. 2005a). Of the PCB metabolites the hexa-chlorinated methylsulfonyl ($MeSO_2$)-PCB congener (range 13.5-551 ng/g (lw) was found in the plasma whereas the penta-chlorinated congeners 3'- and 4'-$MeSO_2$-CB101 (range 4.5-38.1 ng/g (lw)) dominated in the eggs from glaucous gulls (Verreault et al. 2005a). Of the PBDEs the plasma levels in glaucous gulls varied between 8.2 and 67.5 ng/g (ww) (Verreault et al. 2005b). The levels of PBDEs were 10 times lower than the concentration of sum PCB (sum 47 congeners). Of the brominated compounds that have been reported as naturally-occurring in the marine environment and/or metabolites of PBDEs, the methoxylated (MeO) and hydroxylated (OH)- PBDEs were found in plasma of glaucous gulls (Verreault et al. 2005b). The level of perfluorooctane sulfonate (PFOS) in plasma (48.1-349 ng/g (ww)) was the most pervasive perfuorinated alkyl substance (PFAS) found in glaucous gulls. The levels of PFOS in glaucous gulls are also the highest reported in any arctic seabird (Verreault et al., submitted).

20.6 Arctic fox

The arctic fox is an opportunistic feeder, which eats cached food, scavenged carcasses of seabirds, terrestrial birds, seals, Svalbard reindeer and Svalbard ptarmigans (Fuglei 2000). Some arctic foxes follow polar bears on the sea ice, feeding on remnants of seals killed by polar bears (Hiruki and Stirling 1989). The levels of PCBs in arctic fox show a great variation (1.0-45 μg/g), which probably reflects what they feed on (Wang-Andersen et al. 1993; Severinsen and Skaare 1997). Foxes sampled on the coast, which are feeding on marine species, have higher POP levels than foxes living inland, which feed on reindeer carcasses and terrestrial birds. Foxes in Canada, inland Iceland, and Alaska have lower POP levels than foxes along the coast of Iceland and on Svalbard (AMAP 2004). The reason for this may be that foxes from the coast of Iceland and Svalbard eat more marine species than foxes in Europe and in Canada.

20.7 Seals

The POP levels in different seal species from the Barents Sea area are low (10-50 times lower) when compared to seal species from the North Sea and the Baltic Sea. The PCB levels in ringed seal (*Phoca hispida*), harp seal (*Phoca groenlandica*), bearded seal (*Erignathus barbatus*) and Atlantic walrus (*Odobenus rosmarus*) are found at concentrations averaging 1.0-5.0 μg/g (ww) (AMAP 1998; 2004). The PCB levels in blubber of harp seals from the East Ice (Russian area) were three times higher (3.0 μg/g) (ww) than the levels in West ice (Greenland area) (Espeland et al. 1997; Kleivane et al. 1997). However, this comparison is confounded by a significant difference in blubber thickness between the two areas (less blubber in the East Ice seals), which could have increased the pollution levels in seals from the east. A geographical study of PCBs and DDTs in ringed seal blubber has shown higher levels - in samples from the Yenisey Gulf in the Russian Arctic, Svalbard and eastern Greenland compared to western Greenland and the Canadian Arctic (Muir et al. 2000). The ringed seal from Svalbard has PCB levels four times higher than seals from the western Canadian Arctic and Alaska (Wolkers et al. 1998; Muir et al. 2000). The highest levels of toxaphene, a chlorinated pesticide, have been found in harp seal collected east of Svalbard. The levels of toxaphene (Tox 26 and Tox 50) were 20 times higher than in ringed seal samples west of Svalbard (Wolkers et al. 1998; Wolkers et al. 2000).

In ringed seal, the levels of PBDEs were 10-20 times higher than Canadian seals. However, the levels were substantially lower than in seals from more southern latitudes (Wolkers et al. 2004).

20.8 Whales

The levels of POPs in different whale species also reflect their feeding habits. Beluga (*Delphinapterus leucas*), narwhal (*Mondon monoceros*) and harbour porpoise (*Phocoena phocoena*), which feed mainly on fish, have higher POP levels (5-6 μg/g) (ww) in their blubber than minke whales (*Balaenoptera acutorostrata*) (2-4 μg/g) (ww), which feed on krill and amphipods. Levels of POPs increase with age and there are large differences between males (highest level) and females. There are also geographical differences in the POP levels in whales. PCBs and DDTs increase from the west to the east in minke whales (Hobbs et al. 2002). In minke whales, the levels of POPs are also higher on the coast of Lofoten Island in northern Norway compared to the Svalbard and the Kola areas (Kleivane and Skaare 1998). In beluga whales most POPs are lower in southern Alaska than in eastern Canada and Svalbard (Wolkers 2002). The levels in recent measurements from the northeast Atlantic are 2-3 times lower than those made in the 1990s. This may be explained in part by changes in feeding habits to almost exclusively krill after the collapse of the capelin (*Mallotus villosus*) stocks in 1986 (AMAP 2004). Geographical differences in POP levels can also be explained by differences in migration pattern.

In killer whales (*Orcinus orca*) from Lofoten, which feed on herring, the highest PCB and DEE levels have been found of all whale species studied so far (Wolkers, pers. com.). The PCB levels were comparable to the levels found in polar bears from Svalbard.

Levels of PBDEs in beluga whales from Svalbard were 10-20 times lower than beluga whales from Canada. In addition, the levels in beluga whales were lower than in whales from more southern latitudes (Wolkers et al. 2004). The reason for higher levels in seals and whales in the European Arctic compared to the Canadian Arctic is probably due to a more effective PBDE transport from lower latitudes to the European Arctic.

Very high PCB levels (20-30 μg/g) (ww) have been found in harbour porpoise blubber samples from northern Norway. The levels of PCBs and DDTs are comparable to the Baltic Sea and North Sea and are the highest measured in any whale species from the Arctic (Berggren et al. 1999). The levels in harbour porpoises from Greenland are much lower. The reason for the high PCB levels in harbour porpoise from northern Norway is not fully understood. The levels of PCBs in pilot whales (*Globicephala melas*) from Faeroe Islands are higher than most other whales. Levels of other POPs are also comparably high (Dam and Bloch 2000). For example, the levels of brominated compounds in pilot whales are an order of magnitude higher than in other arctic marine mammals examined to date (van Bavel et al. 2001). Narwhal from west Greenland and the Canadian Arctic have similar POP levels. The levels in narwhal from Svalbard were considerably higher (Wolkers 2002).

20.9 Polar bears

Polar bears are widely distributed throughout the Arctic. They move south with the ice in the autumn and winter and then north as the pack ice melts in the spring and summer. Their primary prey are ringed seal and bearded seal (Stirling et al. 1982). Kleivane et al. (2000) also showed that polar bears from eastern Svalbard feed upon harp seals. Polar bears eat almost exclusively the blubber of seals. Polar bears have a very good capacity for oxidative biotransformation (e.g. via cytochrome P450 (CYP) 1A enzymes) of most POPs, as evidenced by high levels of PCB metabolites (i.e. OH-PCBs and $MeSO_2$–PCBs) found in plasma and tissues (Letcher et al. 1996; Sandau et al. 2000; Sandala et al. 2004; Verreault et al. 2005c).

Two studies (Andersen et al. 2001; Lie et al. 2003) investigating PCB and pesticide concentrations in polar bears from Svalbard, Frans Josef Land, Kara Sea, East Siberian Sea and Chukchi Sea (samples collected between 1987 and 1995) combined with earlier findings (Bernhoft et al. 1997; Norstrom et al. 1998) indicated that polar bears from Frans Josef Land and the Kara Sea have the highest sum PCB, sum chlordane and sum DDT levels among arctic animals. A decreasing trend was seen both east and west of these regions. At Svalbard, 35 mother/cub pairs were sampled between 1995 and 1998 (Lie et al. 2000) and the sum PCB geometric mean concentrations in plasma were 12 300 ng/g (lw) in cubs, 5820 ng/g (1w) in females with cubs, 6820 ng/g (1w) in yearlings, and 2945 ng/g (1w) in females with yearlings. In cubs, the sum PCB concentrations were significantly higher than in the other three groups. The reason for a higher sum PCB concentration in cubs was due to lactational transfer from their mothers.

In a recent circumpolar study (1996-2002) of polar bears from Alaska, Canada, Greenland and Svalbard the PCB and DDT levels were highest in the East Greenland and the Svalbard population (Fig. 20.2). In this study the levels of PCB and DDE levels were 6 and 3 times higher, respectively, in polar bears from east Greenland and Svalbard when compared to polar bears from Alaska and Canada (Verreault et al. 2005c). In polar bears from Svalbard the levels of PCNs ranged from 1.2 to 52 ng/g (lw) in fat samples. Compared to other marine species (e.g. the beluga whale, 0.04-0.4 ng/g (lw)) the PCN concentration is much higher (Gabrielsen et al. 2004). Of the PCB metabolites, a sum $MeSO_2$-PCB (ranging between 162 to 279 ng/g (lw)) was found in the fat samples of polar bears (Verreault et al. 2005c). The concentration of sum OH-PCB in Svalbard polar bear plasma from females ranged between 4.15 and 394 ng/g (ww) (Verreault et al. 2005b)(data for East Greenland polar bears see Sandala et al. 2004). PCB-metabolites are transferred from the polar bear females to the cubs via milk, resulting in concentrations that are approximately three times higher in the cubs than in the mother. While polar bears in the European Arctic usually have the highest levels of contaminants, the levels of PBDEs are low compared to Alaska, Canada and Greenland (Muir et al., in prep.).

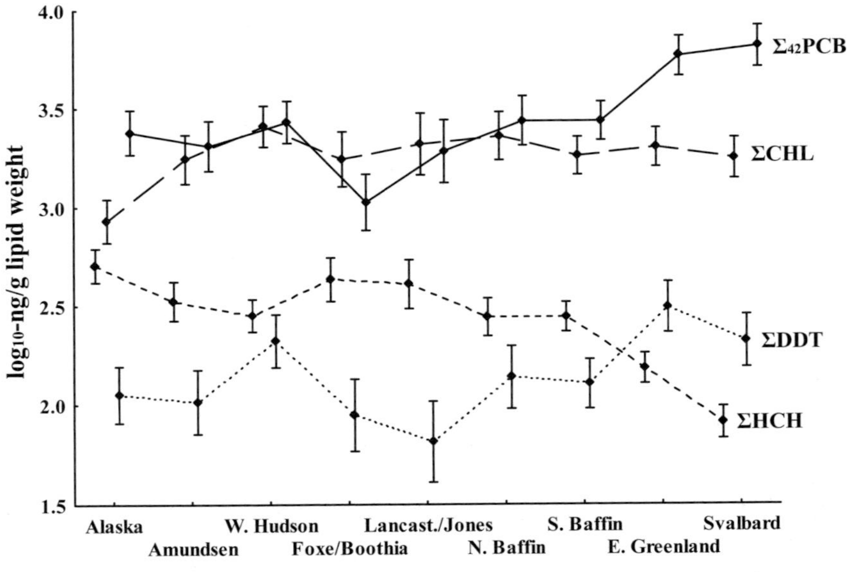

Fig. 20.2. Age-adjusted mean Σ (chlordanes) CHL, ΣHCH, ΣDDT, and Σ_{42}PCB concentrations ($\log_{10}$- transformed ng g^{-1} lipid weight) plotted with 95% confidence intervals (vertical bars) in adipose tissue of female polar bears from nine Arctic and Subarctic populations listed in order of longitude (Verreault et al. 2005c).

PBDEs in polar bear from Svalbard varied between 2.7 and 9.7 ng/g (ww). Of the brominated metabolites both the MeO and OH PBDEs (partly natural components in the marine environment, e.g. found in algae and sponges) were found in plasma of polar bears (Verreault et al. 2005b). Because the PBDE congener profile in the polar bear is comprised almost uniquely of PBDE 47, this may indicate that the polar bears have an effective metabolism of most PBDEs. The levels of PFOS (range 56.7 to 150 ng/g (ww)) were the most pervasive PFASs found in plasma from polar bears (Gabrielsen et al. 2004). In a geographical trend study, the PFOS concentration in liver samples of polar bears from south Hudson Bay and east Greenland were significant higher than in polar bears from Svalbard, high Arctic and the western Northwest Territories (Canada). The high concentration in polar bears from south Hudson Bay and east Greenland was suggested to be due to the proximity to sources in Europe and eastern North America (Smithwich et al. 2005).

20.10 Temporal trends in POPs in arctic seabirds and marine mammals

In general, the levels of most POPs- (i.e. PCB, DDT and HCB), which have been taken out of production and use in the 1970s and 1980s, show a decline. However, the decline has occurred with varying rates in different regions of the Arctic (AMAP 1998; 2004). This is also an explanation why animals differ in regional trends. Trend studies of DDT and PCB levels in seabird eggs and marine mammals (polar bears, seals and whales) from the Arctic and sub-Arctic show a clear reduction during the last 15-20 years (Braune and Donaldson 2000; Barrett et al. 1996; Henriksen et al. 2001; Verreault et al. 2005c; Muir et al. 2001; Stern and Addison 1999). In seabird eggs from northern Norway, there is a reduction in DDT and PCBs of 80-90 % between 1973 and 1993 (Barrett et al. 1996). In the Canadian Arctic, a similar reduction has been shown in eggs from guillemots, northern fulmars and black-legged kittiwakes (Braune and Donaldson 2000). In marine mammals (seals and whales), the reduction in DDT and PCBs are not great as those observed in seabirds (AMAP 2004). In polar bears from Svalbard, both PCB and HCB decreased significantly during the 1990s (Henriksen et al. 2001). The shape of the decline for polar bear suggests a levelling off in the latter part of the sampling period. Data from beluga also indicate that decline will be slow during the coming decade.

While PCBs and pesticides seem to be declining in the environment, the PBDE concentrations in arctic wildlife are increasing. Although the levels in the Arctic are still relatively low, a study in Canada showed a dramatic increase over a period of 15 years. In samples from blubber of ring seal and beluga, the levels increased by 3-9 times since the beginning of 1980s (Ikonomou et al. 2002). Generally, a doubling every five years seems to be the rule. Herring gull eggs from the Great Lakes in North-America show a similar increase as seen in ringed seal (Norstrom et al. 2002). The sources of environmental PBDE contamination include leakage from consumer products and industrial facilities that manufacture PBDEs as well as from disposal sites of products containing PBDEs. There are no temporal trend data from the European Arctic. If nothing is done to the "PCB of the 21st century" (i.e. proper restriction) the PBDEs may reach the same levels as PCBs in a few decades.

20.11 Effect studies

Ecological risk assessment of contaminants in arctic animals comprises assessments of exposures and effects, and risk characterization. In order to identify possible effects of contaminant exposure in free ranging animals from remote areas, two approaches are generally used. The first involves extrapolation and comparison. Possible effects are determined by comparing levels of contaminants in the species of interest to levels known to cause toxic effects in laboratory species (e.g., rats) or from observations on affected animals in the wild (AMAP 2004). The second approach investigates biological and potential toxic effects by studying biomarkers (indicators of biological responses) of contaminant exposure (AMAP 2004). Such studies may reveal subtle biological changes/disturbances associated with low and sub-lethal concentrations of contaminants. However, the significance of such effects for the health of the individual is often not obvious, and validity of extrapolating between biomarker responses measured in individuals and population level effects is not easily established. Thus, ecological risk assessment of contaminant exposure in free-ranging species will always be rather difficult without captive studies to demonstrate direct cause and effects (AMAP 2004).

Ecological risk assessment of contaminants in animals in the Arctic comprises assessments of exposures and effects, and risk characterization.

20.12 Effects of POPs in arctic animals

In field studies of Arctic free-living wildlife, it is difficult to prove a causal relationship between a suspected effect and specific contaminants. Thus, assessments of possible effects are based on associations between biological parameters and tissue contaminant concentrations. Some biological responses (or *biomarkers of exposure*) are quite specifically related to contaminant exposure (*e.g.* cytochrome P450 enzyme activities, accumulation of highly carboxylated porphyrins and thyroid hormone disruption), but are difficult to interpret in terms of consequences for the health of the individual, and populations (Peakall 1992). Other responses may be more directly coupled to the survival or reproduction of the individual, but influenced by a large variety of confounding factors.

Sometimes it is possible to compare the levels of POPs found in wildlife with effect thresholds derived from laboratory studies or environmental levels that are believed to have been implicated in observed effects (e.g., declines in bird populations). Such a process is not simple. For many arctic species, laboratory studies do not exist and comparisons should be made with caution since there are problems with extrapolating data across species. Also, unlike most laboratory studies, animals in the wild are exposed to a suite of different contaminants. They are also usually exposed to lower concentrations than laboratory animals. Furthermore, wild species are exposed to weathered mixtures of contaminants due to changes in composition of many POPs caused by abiotic degradation, metabolism, and filtering through the food web. Differences in species sensitivities to the effect of POPs also make it difficult to know which laboratory species best represent those in the Arctic. Arctic species differ from laboratory animals due to their fat dynamics, differences in life styles and life strategies, and differences in toxicokinetics (AMAP 2004).

In studies, which investigate biological and potential toxic effects by studying biomarkers of contaminant exposure, there are also limitations. It is not possible to determine causality, only that a statistical association has been found between a biomarker and the contaminant in question. Most POPs co-vary, and thus, it is not possible to state equivocally that the biomarker response has been caused by a particular contaminant. There may be other contaminants not analyzed, or biological parameters, that are just as important, or the response may be a result of synergistic, additive or antagonistic effects of contaminant mixtures. Biological variables such as age, sex, body condition, and presence of disease, or other stresses may also act as confounders, as they can cause similar biological effects as those seen from POPs. Therefore, for most reported biological effects in wildlife, the evidence for a causal link with a specific chemical contaminant is weak or non-existent. This is mainly due to the complexity of contaminant mixtures, the lack of data on chemical exposure, species sensitivity and mechanisms of action. Understanding the linkage between contaminants and health effects (e.g., reduced fitness or immunosuppression) is most likely to come from studies in laboratory animals. Crucial in establishing causal evidence for chemical-induced wildlife effects are semi-field or laboratory studies using the wildlife species of concern. Semi-field studies represent a useful approach to bridge the gap between the controlled conditions of laboratory experiments and environmental-exposure conditions in the field (AMAP 2004).

In the following assessment, results are presented in which associations have been reported between biological parameters and contaminant levels in glaucous gulls and polar bears from the Svalbard area. Levels of specific POPs are compared to no-observed adverse effect levels (NOAEL) or no-observed effect levels (NOEL), and the lowest observed adverse effect level (LOAEL) or lowest observed effect levels (LOEL) known to cause subtle effect in sensitive species. The purpose of these comparisons are to assess the likelihood that glaucous gull and polar bears may be at risk for effects of some POPs and to identify the contaminants associated with these effects.

20.12.1 Glaucous gulls

The NOEL range for sum PCB was 1.3 to 11.0 µg/g (ww) for effects related to reproductive success in seabirds. The LOEL range for various endpoints of reproductive success (hatching success, egg mortality, deformities, and parental attentiveness) ranged from 3.5 to 22 µg/g sum PCBs/g (ww) in eggs. For adults, sum PCB concentrations in brain tissue higher than 300 µg/g (ww) were associated with mortality. For dioxin-like compounds, the NOAEL range for reproductive effects was 1.5 to 200 pg toxic equivalency quotient (TEQ)/g (ww) in eggs, and the LOAEL range for various reproductive endpoints (deformities, hatching success, and mortality) ranged from 10 to 2200 pg TEQ/g (ww) in eggs (Barron et al. 1995; Giesy et al. 1994; Bosveld and van den Berg 1994). In eggs from peregrine falcons, DDE residues of 15 to 20 µg/g (ww) would result in eggshell thinning (Peakall et al. 1990).

In glaucous gull eggs from Svalbard/Bjørnøya, the sum PCB is close to or above NOEL and LOEL levels found for reproduction effects. Glaucous gull eggs from Bjørnøya are close to or above LOEL levels which have effects on hatching success in chickens and NOEL levels for hatching success in Foster terns (*Sterna forsteri*). The sum PCB in eggs from Bjørnøya is not above the levels that are related to increased egg mortality. The sum PCB found in liver from glaucous gulls from Bjørnøya is above levels reported to cause toxic effects in birds (AMAP 2004). The TEQ concentration was calculated to 2500 µg/g in glaucous gulls (Daelemans et al. 1992). These TEQ levels exceed all NOAELS and LOAELS for reproductive effects and LD_{50} in the range of other bird species (AMAP 2004). Sum DDT levels in eggs and liver exceed Canadian and U.S. Environmental Protection Agency (EPA) guideline levels for protecting wildlife. Thus glaucous gulls that prey on seabird eggs, chicks and adults have intake of sum PCB, sum DDT and TEQs high enough to cause effects.

20.12.1.1 Reproductive effects

Glaucous gulls from Bjørnøya with high levels of sum PCB were more often away from the nest and spent more time away than birds with low levels. Increased time away from the nest in glaucous gulls with high PCB levels may indicate that they needed more time to search for food than individuals with low PCB levels (Bustnes et al. 2001). Thus, chick survival may decrease for parents with elevated PCB exposure due to the fact that their parents spend less time protecting the chicks.

In glaucous gulls from Bjørnøya, the females with high POP levels (HCB, oxychlordane, DDE and PCBs) had a greater possibility for embryo mortality or unfertilized eggs than females with low POP levels. Also, the body condition at hatching was poorer in the first chick in a clutch (glaucous gulls normally lay 2-3 eggs) compared to the other chicks in females with high POP levels, suggesting that the female shunts a larger percentage of POPs to the first chick in the clutch. A negative relationship was also found between chick body condition (e.g., body mass) and the parents' blood concentration of HCB, β-HCH and PCB 28 for the second chick in a clutch. Except for the negative relationship found between the concentration of some POPs and date of egg laying no relationship was found toward other reproductive parameters (e.g.; clutch size, egg size, number of days incubating, predation of eggs and early chick survival) and the level of POPs (Bustnes et al. 2003).

Many POPs can disrupt endocrine function (Colborn et al. 1993). Female glaucous gulls from Bjørnøya, with high POP levels, produced more male chicks compared to females with low POP levels (Erikstad et al. 2005), strongly indicating that POPs are acting as endocrine disruptors in these birds.

In glaucous gulls from Bjørnøya, a significant relationship was found between levels of POPs and asymmetric wingfeathers. The effects were stronger for HCB than PCB and DDE (Bustnes et al. 2002). Asymmetric wingfeathers are an indication of stress and the observation of asymmetry may indicate that gulls with elevated POP concentrations are under increased stress during moulting.

20.12.1.2 Cytochrome P450 activity

In glaucous gull liver samples (Henriksen et al. 2000), a weak positive association was found between hepatic 7-ethoxyresorufin-O-dethylase (EROD) activity and PCB-levels. This may indicate enzyme induction by PCBs, but the EROD activities were low compared to other studies on fish-eating birds. Microsomal testosterone hydroxylase activity was only observed at the 6β-position and could not be related to levels of POPs. The low P450 associated enzyme activities in glaucous gulls suggests that they have a low capacity for metabolizing POPs, which may contribute to the high accumulation of POPs in this species.

In a laboratory study of glaucous gull chicks on Svalbard, a significantly higher level of CYP 1A enzymes was found in the liver in males from a group that were feed a diet of POPs (a natural diet they receive in the wild) compared to a group of males which were feed clean food (hen egg with little or no POPs) (Østby et al. 2005). Evaluation of blood POP levels demonstrated that the chicks fed a natural diet had higher POPs than the clean diet and there was also a positive correlation between blood POP levels and CYP 1A enzyme induction.

When making comparison of the TEQ value in glaucous gulls and several arctic seabird species (little auk, Brünnichs guillemot, kittiwake and black guillemot) the EROD levels were higher in little auk and lower in Brünnichs guillemot, kittiwake and black guillemot compared to glaucous gulls (Borgå et al. 2005b). Since the TEQ values in arctic seabirds were 100-200 times lower than the LOEL for CYP induction in common tern, it is assumed that most birds are below threshold levels for biological effects (Borgå et al. 2005b).

20.12.1.3 Effects on the hormone system

In male glaucous gulls from Bjørnøya a significant negative correlation was found between the PCB and HCB levels, and thyroxin (T4) levels and T4- to T3-ratios in the blood (Fig. 20.3). HCB, oxychlordane and PCB 118, 114 and 105 were the contaminants that had the strongest correlation to the decrease of the T4/T3-ratios in these glaucous gulls. The levels of T4 and T4/T3-ratio were lower in glaucous gulls from the colony that had highest levels of POPs when compared to the colony with low POP levels

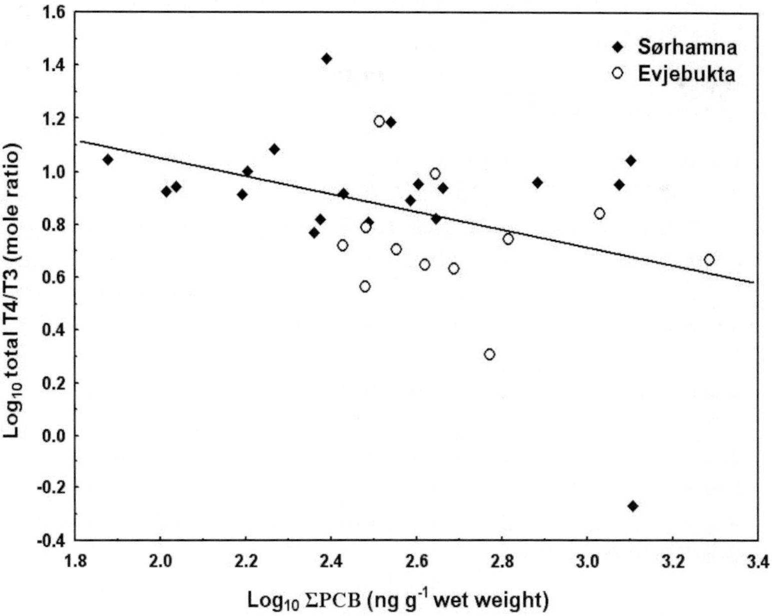

Fig. 20.3. Relationship between the total T4:T3 (thyroxin:triiodothyronine) ratios ($\log_{10}$ mole ratio) and blood levels of $\sum$ PCB ($\log_{10}$ ng/g wet wt), corrected for extractable plasma fat (%) and day of capture, for male glaucous gulls (n = 32) breeding in two colonies (Evjebukta and Sørhamna) at Bjørnøya (Norwegian Arctic) ($r = -0.40$, $p = 0.031$) (ng g^{-1} w.w.: nanograms per gram wet weight) (Verreault et al. 2004).

(Verreault et al. 2004).

20.12.1.4 Effects on the vitamin system

Many POPs, in particular 2,3,7,8-tetrachlorodibenzo-p-dioxin (TCDD) and dioxin-like PCBs, can interfere with vitamin A homeostasis in experimental animals (Zile 1992). Vitamin A is mainly stored as retinyl palmitate in lipid droplets of liver stellate cells (Blomhoff 1994). In herring gulls from colonies in eastern Canada, liver retinoid concentrations were inversely related to TCDD-concentrations in eggs from the same colonies (Spear et al. 1986; Anonymous 1991). In the sample of forty glaucous gulls from Bjørnøya, no significant relationships were found between liver retinoid concentrations and PCB levels (Henriksen et al. 2000). The hepatic vitamin A stores in glaucous gulls from Bjørnøya were larger than in herring gulls from contaminated locations in North America (Anonymous 1991) suggesting that the present contaminant levels do not influence vitamin A in gulls from Svalbard.

20.12.1.5 Genotoxic effects

In a laboratory study of glaucous gull chicks from Svalbard, a significant increase in levels of DNA-adducts (a measure of genetic mutations) was found in the group receiving a diet of POPs compared to a clean group. However, no clear relationship was found between levels of DNA adducts and POPs between the exposed and clean group (Østbye et al. 2005). The small sample size of birds included in this study may be one reason for the lack of statistical significance.

The frequency of chromosome abbreviation in lymphocytes was also higher in the POP-exposed group, when compared with the clean group. Although, no correlation was found for frequency of chromosome abbreviation and concentration of some of the POPs analyzed (Krøkje et al. 2005), suggesting that the present contaminant levels have genotoxic effects in gulls from Svalbard.

20.12.1.6 Effects on the immune system

Suppressed immune function has been associated with exposure to POPs in herring gulls from the contaminated Great Lakes area (Grasman et al. 1996). If establishment and/or survival of intestinal macroparasites are limited by host immune function, we would expect increased parasite intensities in animals with high organochlorine burdens, such as the glaucous gull. In a sample of 40 glaucous gulls from Bjørnøya in the western Barents Sea, numbers of intestinal macroparasites were compared with hepatic levels of selected POPs (Fig. 20.4) (Sagerup et al. 2000). After controlling for nutritional condition, no single parasite species was significantly associated with concentrations of PCBs or chlorinated pesticides. However, the intensity of all nematodes grouped together was positively correlated with all the 14 POPs measured, and significantly with 10 of them. The strongest correlations were with p,p'-DDT, Mirex, sum PCB, and PCB congeners 28, 118, 153, 138, 170, and 180. Although correlative, no immunological data were collected. However, these results suggest that POPs might affect immune function in the glaucous gull.

In glaucous gull from Bjørnøya a correlation was found between blood levels of POPs and increased levels of white blood cells. Females with high levels of HCB and oxychlordane have less possibility to initiate immune responses than glaucous gulls with low levels of these contaminants (Bustnes et al. 2004).

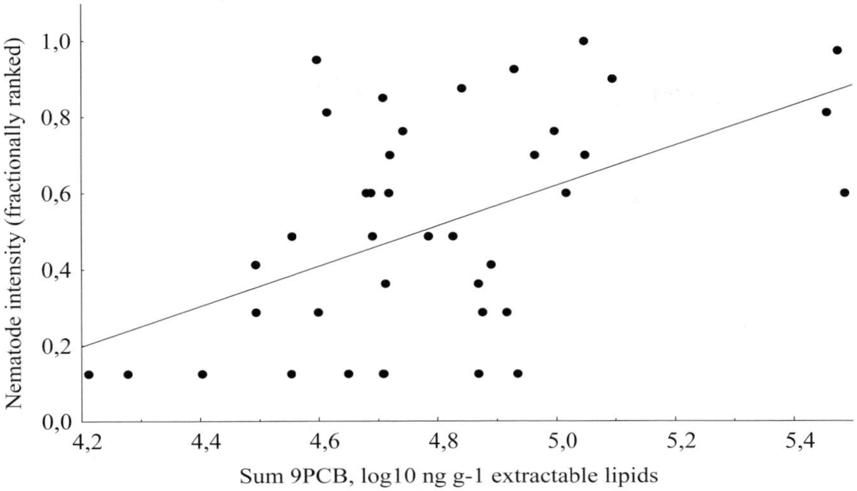

Fig. 20.4. Correlation between intestinal nematode infection intensity (fractionally ranked) and ∑ 9 PCB concentrations ($\log_{10}$ ng/g lipid weight, liver) in glaucous gulls from Bear Island (n = 40, $R^2 = 0.26$, P = 0.001) (Sagerup et al. 2000).

In a Svalbard laboratory study of glaucous gull chicks (56 days old chicks) fed a diet of POPs, a negative effect was found between POP levels in the diet and immunoglobulin (both IgG and IgM) concentrations when compared to a control group fed a clean diet (Larsen et al. 2004). The exposures of POPs through their diet appeared to cause a decrease in the immuno-competence and resistance of these chicks, which could impact their resistance to disease and infection.

20.12.2 Polar bears

In mink (*Mustela vison*), levels of 1230 ng/g (ww) of Aroclor 1254 in liver, was associated with impaired reproductive success (Platonow and Karstad 1973). Reduced growth and survival of mink kits were observed in female mink with 2000 ng/g (ww) Aroclor 1254 in liver tissue (Wren et al. 1987). An EC_{50} (effective concentration causing a response in 50 % of the treated organisms) in adult females for litter size was calculated to be 40 000 to 60 000 ng total PCB/g (lw) (approximately 1200 ng total PCB/g (ww) in muscle) and 2400 ng/g in muscle for kit survival (Leonards et al. 1995). Captive harbour seals (*Phoca vitulina*), exposed to PCBs via different fish diets, had reduced reproductive success at sum PCB levels of 25 000 ng/g (lw) in blood (Reijnders 1986). The EC_{50} for dioxin-like compounds was calculated to be 160 pg TEQ/g (ww) (5300-8000 pg/g (lw)) in mink muscle for litter size and 200 pg TEQ/g (ww) (6600-10000 pg/g (lw)) for kit survival (Leonards et al. 1995). Assessment based on subtle neurobehavioural effects in offspring of rhesus monkeys treated with PCBs and human mothers eating PCB contaminated fish, have resulted in an estimated LOAEL for effects on short term memory of 500-1000 ng/g (lw), and a NOAEL for effects on visual memory of 1000 ng/g (lw) in offspring or cord blood serum (Ahlborg et al. 1992). The LOAEL for immunosuppression is 21 000 ng sum PCB/g (lw) in rhesus monkeys (Tryphonas 1994).

In some polar bears from Svalbard, sum PCB concentrations are above the NOAEL and LOAEL levels for neurological-behavioural effects on their offspring, based on information studies of monkeys and humans and NOEL levels for kit survival in minks. The sum PCB in polar bears from Svalbard is also above NOEL and LOEL for reduced vitamin A in otters. Some individuals of polar bears above 3 years of age are above reported LOAEL levels for reduced immune function (based on information from monkeys). Some polar bears also have elevated levels of PCBs, which is correlated to low reproduction in seals and EC_{50} for reduced kit survival in mink (de Wit el at. 2004).

20.12.2.1 Reproductive effects

A negative effect on the reproduction and survival rate of young seals as result of PCB contamination has been shown in the Baltic (Olsson et al. 1992). Epi-zoological studies of polar bears indicate a reduced reproduction and survival of cubs at Svalbard compared to other arctic areas (Wiig 1995; Wiig et al. 1998), although the relationship between cub survival and POP levels has not been established. From laboratory studies on mammals, it is shown that high intake of PCB in a critical period may have an influence on the fetus and decrease in the survival of the offspring.

In Canada it was found that polar bear mothers with high levels of POPs (in the milk) had a greater chance for loosing the cubs than mothers with low POP levels. Mothers, which lost their cubs, had levels of PCBs, which were 3 times higher than females that did not lose their cubs (Nordstrom 1999).

In polar bears from Svalbard, the age composition in the population was different from the Canadian population. Female polar bears, which were more than 16 years old, constituted 12.7 % at Svalbard compared to 40.3 % in Canada (Wiig et al. 1998; Derocher et al. 2003), signifying that POPs may have an influence on the polar bear population at Svalbard.

It has also been hypothesized that PCBs and other POPs may be involved in the relatively high incidence of female pseudohermaphroditism in polar bears from Svalbard (Wiig et al. 1998). In polar bears from Svalbard, pseudohermaphroditism was found in 3 % of the total amount of female polar bears caught (Wiig et al. 1998). In female polar bears from Canada and Alaska, no such malfunction has been found (Andrew Derocher, pers. com.). Because POPs cause endocrine disruption, high POP levels in polar bears at Svalbard may be one reason for the observed pseudohermaphroditism.

20.12.2.2 Cytochrome P450 activity

Many planar halogenated hydrocarbons (e.g., dioxin-like compounds, which include some of the PCBs) have a common pattern of toxic effects that is associated with affinity to the aryl hydrocarbon (Ah) receptor and induction of isoforms from the CYP1A subfamily (Poland and Knutson 1982). Preliminary results from 13 polar bears from Svalbard show a positive correlation ($P = 0.026$) between CYP1A and total PCB concentration in white blood cells (Skaare et al. 2000).

In male polar bears from Canada, CYP 1A1 levels were correlated to mono-*ortho* and non-*ortho* substituted PCBs. CYP 2B was correlated to the concentration of chlordanes (mainly oxychlordane and nonachlor) and total mono- and non-*ortho*-substituted PCBs (de Wit et al. 2004).

20.12.2.3 Effects on the hormone system

The thyroid hormones are important for regulating metabolism and heat production. For arctic species thyroid hormones are very important for their adaptations to a cold climate. In young animals, the thyroid hormones are also important for the growth and development of the central neural system. Hormone data from polar bears at Svalbard (high POP levels) and Canada (low POP levels) show that the levels of thyroid hormones (e.g., T4) are negatively correlated to PCBs (Skaare et al. 2001). In polar bears from Svalbard, high levels of POPs, were negatively correlated to thyroid hormone levels (total T4/free T4) with POPs explaining 30 % of the variation in these ratios (Skaare et al. 2001). The correlation was strongest for sum PCBs and HCB. In another study of polar bears from Svalbard, a negative correlation was found between sum PCBs and total (t) T4, free (f) T4, fT3, tT4/tT3, tT3/fT3 and fT4/fT3. More thyroid hormone variables from female polar bears were affected when compared to males (Braathen et al. 2004). This may indicate that female polar bears are more vulnerable to effects of PCB on thyroid hormones than males.

Testosterone is an important male hormone, which is involved in sexual development. In males of polar bears from Svalbard, a negative relationship was found between the levels of sum PCB/sum pesticides and testosterone levels. The sum of PCB and sum of pesticides explained 57 % of the variation in testosterone levels (Oskam et al. 2003).

In female polar bears with cubs from Svalbard, a positive relationship was found between sum PCB and the level of the sex hormone progesterone (Haave et al. 2003). This may indicate that the exposure of PCB might disturb the timing of ovulation (release of eggs) in female polar bears and thereby impacting the possibility for fertilization. No relationship was found between sum PCB and estrogen in female polar bears from Svalbard (Haave et al. 2003).

Cortisol is involved in the body metabolism and is released during stress in animals. In polar bears from Svalbard a negative relationship was found between POP levels and the level of cortisol in the blood. The sum PCB and sum pesticides explained 27 % of the variation in cortisol levels (Oskam et al. 2004). This may indicate that the exposure of PCB can disturb metabolism and the release of stress hormones.

Female polar bears from Svalbard and the northern part of the Barents Sea, with and without cubs, had significantly higher progesterone levels than polar bears from other areas. However, no relationship was found between PCBs and progesterone levels within the Svalbard and Barents Sea bears (Haave et al. 2003).

Taken together, these findings strongly indicate that the present level of POPs may have an influence on the endocrine system on polar bears from Svalbard.

20.12.2.4 Effects on the vitamin system

Vitamin A homeostasis can be severely altered by exposure to POPs, and several vitamin A deficiency-like symptoms are associated with intoxication by poly-halogenated aromatic hydrocarbons (Zile 1992). Some PCB metabolites can disturb the formation of the protein complex responsible for retinol (vitamin A) and thyroxin transport (Brouwer et al. 1986). In harbour seals, which were fed fish from polluted waters, retinol and thyroid hormones in blood plasma were depressed, compared to seals fed fish from less polluted waters (Brouwer et al. 1989). In polar bears from Svalbard, the level of PCB, HCB and HCH was negatively correlated to the level of plasma retinol. POPs explained 12 % of the variation in retinol levels (Skaare et al. 2001). In another study from Svalbard, no relationship was found between the levels of PCB and retinol levels (Braathen et al. 2004).

When data from Svalbard (high POP levels) and Canada (low POP levels) were compiled and compared, a negative relationship was found between PCBs and retinol. A positive relationship was found between OH-PCBs and retinol levels (de Wit et al. 2004).

20.12.2.5 Effects on the immune system

Many studies have demonstrated adverse effects of PCBs and dioxin-like compound on the immune system of experimental animals (see *e.g.* review by Tryphonas 1994). In captive harbor seals fed fish highly contaminated with POPs, several measures of cell-mediated and humoral immune function were depressed compared to seals fed less contaminated fish (de Swart et al. 1996). Immunoglobulin G (IgG) is the most abundant class of anti-bodies in mammals, and consequently an essential part of the humoral immune system. In a sample of 52 polar bears from Svalbard, age- and sex-corrected levels of IgG were negatively correlated with sum PCB and HCB (Bernhoft et al. 2000). The PCB levels in bears were within in the range found to give immunotoxic effects in experimental animals (Bernhoft et al. 2000). In a subsequent study from Svalbard, which included more animals, the same negative relationship was found between IgG and PCBs. The sum PCB and sum pesticides explained 53 % of the variation in IgG levels (Lie et al. 2004). In polar bears from Svalbard, the ability to produce anti-bodies against flu- and rheo-virus and tetanus toxid after vaccination is 40-60 % explained of the sum PCB and sum of pesticides (Lie et al. 2004). The ability to produce anti bodies against *Mannheimia haemolytica* (earlier *Pasteurella* sp.) is 59 % explained by sum PCB and sum of pesticides (Lie et al. 2004). The lymphocyte function after *in vitro* stimulation by mitogens (PHA, Con A, PWM, LPS and PPD) and antigens (tetanus toxsoid and KLH) is 45-72 % explained by sum PCB and sum of pesticides (Lie et al. 2005). The association between POPs and IgG could indicate a contaminant-induced immunosuppression in Svalbard polar bears, with possible consequences for susceptibility to infectious diseases. This may indicate that the present contaminant levels may affect the health and the status for the polar bear at Svalbard.

20.13 Conclusion

The levels of POPs in the arctic environment are generally lower than found in more temperate regions. The present article shows that, while the levels of some POPs are decreasing (e.g., PCBs and DDTs), the levels of other POPs (brominated flame retardants and fluorinated) are increasing in the arctic animals. The main reason for this increase is the increase in the global production and use of these contaminants. The POP levels found in polar bears and glaucous gulls from these areas exceed the effect thresholds as demonstrated by laboratory and field studies, which indicate present POP level influence behavioural-, biochemical-, physiological- and immunological parameters. Today there is enough evidence to suggest that the contaminant levels affect the health of polar bears and glaucous gulls from Svalbard and Bjørnøya. Further monitoring and research is needed to get a better understanding of the relationship between POP exposure and biological effects in arctic animals.

Acknowledgements

I want to thank Birgit Braune, Canadian Wildlife Service, Ottawa, Myra Finkelstein, University of California, Santa Cruz, Katrine Borgå, Lisa Bjørnsdatter Knudsen, Jonathan Verreault and Kjetil Sagerup, The Norwegian Polar Institute, for their constructive comments and help on the manuscript. I also want to thank Prof. Dan Costa, University of California, Santa Cruz, for his hospitality while beeing on sabbatical at UCSC.

References

Ahlborg, UG, Hanberg, A, Kenne, K (1992) Risk assessment of polychlorinated biphenyls (PCBs). NORD 1992:26. Nordic Council of Ministers, Copenhagen, Denmark.

AMAP (1998) AMAP Assessment Report: Arctic Pollution Issues. Arctic Monitoring and Assessment Programme (AMAP) Oslo, Norway. xii+859 pp.

AMAP (2003) AMAP Assessment 2002: The influence of Global Change on Arctic Contaminant Pathways to within, and from the Arctic. Arctic Monitoring and Assessment Programme (AMAP) Oslo, Norway. xi+65 pp.

AMAP (2004) AMAP Assessment 2002: Persistent Organic Pollutants in the Arctic. Arctic Monitoring and Assessment Programme (AMAP) Oslo, Norway. xvi+310 pp.

Andersen, M, Lie, E, Derocher, AE, Belikov, SE, Bernhoft, A, Boltunov, AN,

Garner, GW, Skaare JU, Wiig Ø (2001) Geographic variation of PCB congeners in polar bears (*Ursus maritimus*) from Svalbard east to the Chukchi Sea. Polar Biol, 24:231-238.

Anonymous (1991) Toxic chemicals in the Great Lakes and associated effects. Minister of Supply and Services Canada.

Bakken, V, Tertitski, GM (2000) Glaucous gull (*Larus hyperboreus*). In; The status of marine birds breeding in the Barents Sea region (eds. Anker-

Nilssen, T, Bakken, V, Bianki, V, Golovkein, A, Strøm, H, Tatarinkova, L). Norsk Polarinstitutt Rapportserie, 113:94-96.

Barrett, RT, Skaare, JU, Gabrielsen, GW (1996) Recent changes in levels of pesistent organochlorines and mercury in eggs of seabirds from the Barents Sea. Environ Poll, 92:13 -18.

Barron, MG, Galbraith, H, Beltman, D (1995) Comparative reproductive and developmental toxicology of PCBs in birds. Comp Bioch Physiol, 112C:1-14.

Berggren, P, Ishaq, R, Zebühr, Y, Näf, C, Bandh, C, Broman, D (1999) Pattern and levels of organochlorines (DDTs, PCBs, non-ortho PCBs and PCDD/Fs) in male harbour porpoises (*Phocoena phocoena*) from the Baltic Sea, the Kattegat-Skagerrak Seas and the west coast of Norway. Mar Poll Bull, 38:1070-1084.

Bernhoft, A, Wiig, Ø, Skaare, JU (1997) Organochlorines in polar bears (*Ursus maritimus*) at Svalbard. Environ Poll, 96:1-16.

Bernhoft A, Skaare JU, Wiig Ø, Derocher AE, Larsen HJS (2000). Possible immunotoxic effects of organochlorines in polar bears (*Ursus maritimus*) at Svalbard. J Toxicol Environ Health A, 59:561-574.

Bignert, A, Göthberg, A, Jensen, S, Litzen, K, Odsjö, T, Olsson, M, Reutergårdh, L (1993) The need for adequate biological sampling in ecotoxicological investigations - a retrospective study of 20 years pollution monitoring. Sci Tot Environ, 128:121-139.

Blomhoff, R (1994) Transport and metabolism of vitamin A. Nut Rev, 52:S13-S23.

Blus, LJ (1995) Organochlorine pesticides. In: Handbook of ecotoxicology (eds. Hoffman, DJ, Rattner, BA, Nurton, GA, Cairns, J) Lewis Publishers, Boca Raton, USA, pp 275-300.

Bogan, JA, Bourne, WRP (1972) Organochlorine levels in Atlantic seabirds. Nature, 240:358.

Borgå, K, Bidleman, TF (2005) Enantiomer fractions of organic chlorinated pesticides in arctic marine ice fauna, zooplankton and benthos. Environ Sci Technol, 39:3464-3473.

Borgå, K, Gabrielsen, GW, Skaare, JU (2001) Biomagnification of organochlorines along a Barents Sea food chain. Environ Poll, 113:187-198.

Borgå, K, Gabrielsen, GW, Skaare, JU (2002a) Differences in contamination load between pelagic and sympagic invertebrates in the Arctic marginal ice zone: influence of habitat, diet and geography. Mar Ecol Prog Series, 325:157-169.

Borgå, K, Poltermann, M, Polder, A, Pavlova, O, Gulliksen, B, Gabrielsen, GW, Skaare JU (2002b) Influence of diet and sea ice drift on organochlorine bioaccumulation in arctic ice-associated amphipods. Environ Poll, 117:47-60.

Borgå, K, Gulliksen, B, Gabrielsen, GW, Skaare, JU (2002c) Size-related bioaccumulation and between-year variation of organochlorines in ice-associated amphipods from the Arctic Ocean. Chemosph, 46:1383-1392.

Borgå, K, Fisk AT, Hoekstra PF, Muir DCG (2004) Biological and chemical factors of importance in the bioaccumulation and trophic transfer of persistent organochlorine contaminants in Arctic marine food webs. Environ Toxicol Chem, 23:2367-2385.

Borgå, K, Gabrielsen, GW, Kleivane, L, Skaare, JU, Norstrom, R, Fisk, AT (2005a) Why do semi-circumpolar trends of OCs differ among Arctic trophic levels? Environ Sci Technol (accepted).

Borgå, K, Wolkers, H, Skaare, JU, Hop, H, Muir, DCG, Gabrielsen, GW (2005b) Bioaccumulation of PCBs in Arctic seabirds: influence of dietary exposure and congener biotransformation. Environ Poll, 134:397-409.

Bosveld, ATC, van den Berg, M (1994) Effects of polychlorinated biphenyls, dibenzo-p-dioxins and dibenzofurans on fish-eating birds. Environ Rev, 2:147-166.

Bourne, WRP, Bogan, JA (1972) Polychlorinated biphenyls in North Atlantic seabirds. Mar Poll Bull, 3:171-175.

Bowes, GW, Jonkel, CJ (1975) Presence and distribution of polychlorinated biphenyls (PCB) in arctic and subarctic marine food chains. J Fisheries Res Board Can, 32:2111-2113.

Braathen M, Derocher AE, Wiig Ø, Sørmo EG, Lie E, Skaare JU, Jenssen BM (2004) Relationships between PCBs and Thyroid Hormones and Retinol in Female and Male Polar Bears. Environ Health Perspect, 112:826-833.

Braune, BM, Donaldson, GM (2000) Trend in organochlorine concentrations in seabird eggs from the Canadian Arctic: 1975-1998. Workshop on persistent organic pollutants (POPs) in the Arctic: Human health and environmental Concerns. Rovaniemi, Finland, Jan. 2000, AMAP report 2000:1, abstract 27.

Breivik, K, Sweetman, A, Pacyna, JM, Jones, KC (2002) Towards a global historical emission inventory for selected PCB congeners- a mass balance approach. 1. Global production and consumption. Sci Tot Environ, 290:181-198.

Brouwer, A, Van den Berg, KJ, Blaner, WS, Goodman, DS (1986) Transthyretin (prealbumin) binding of PCBs, a model for the mechanism of interference with vitamin A and thyroid hormone metabolism. Chemosph, 15:1699-1706.

Brouwer, A, Reijnders, PJH, Koeman, JH (1989) Polychlorinated biphenyl (PCB)-contaminated fish induces vitamin A and thyroid hormone deficiency in the common seal (*Phoca vitulina*). Aq Toxicol, 15:99-106.

Bustnes, JO, Bakken, V, Erikstad, KE, Mehlum, F, Skaare, JU (2001) Patterns of incubation and nest site attentiveness in relation to organochlorine (PCB) contamination in Glaucous Gulls. J App Ecol, 38:791-801.

Bustnes, JO, Folstad, I, Erikstad, KE, Fjeld, M, Miland, ØO, Skaare, JU (2002) Blood concentration of organochlorines and wing feather asymmetry in glaucous gull. Funct Ecol, 16:617-622.

Bustnes, JO, Erikstad, KE, Skaare, JU, Bakken, V, Mehlum, F (2003) Ecological effects of organochlorine pollutants in the Arctic: a study of the glaucous gull. Ecol App, 13:504-515.

Bustnes, JO, Hanssen, SA, Folstad, I, Erikstad, KE, Hasselquist, D, Skaare, JU (2004) Immune function and organochlorine pollutants in Arctic breeding Glaucous Gull. Arch Environ Cont Toxicol (in press).

Colborn, T, vom Saal, FS, Soto, AM (1993) Developmental effects of endocrine-disrupting chemicals in wildlife and humans. Environ Health Perspect, 101:378-384.

Daelemans, FF, Mehlum, F, Schepens, PJC (1992) Polychlorinated biphenyls in two species of Arctic seabirds from the Svalbard area. Bull Environ Cont Toxicol, 48:828-834.

Dam, M, Bloch, D (2000) Screening of mercury and persistent organochlorine pollutants in lon-finned pilot whale (*Globicephala melas*) in the Faroe Islands. Mar Poll Bull, 40:1090-1099.

de Boer, J, Wester, PG, Klamer, HJC, Lewis, WE, Boon, JP (1998) Do flame retardants threaten ocean life? Nature, 394:28-29.

Derocher, AE, Wolkers, H, Colborn, T, Schlabach, M, Larsen, TS, Wiig, Ø (2003) Contaminants in Svalbard polar bear samples archived since 1967 and possible population effects. Sci Tot Environ, 301:163-174.

de Swart, RL, Ross, PS, Vos, JG, Osterhaus, ADME (1996) Impaired immunity in harbour seals (*Phoca vitulina*) exposed to bioaccumulated environmental contaminants: Review of a long-term feeding study. Environ Health Perspect Suppl, 104:823-828.

de Wit, C, Fisk, A, Hobbs, K, Muir, D, Gabrielsen, G, Kallenborn, R, Krahn, M, Norstrom, R, Skaare, J (2004) AMAP Assessment 2002. Persistent Organic Pollutants in the Arctic. Arctic Monitoring and Assessment Program report. Oslo, Norway. 310 pp.

EEA (2004) Arctic Environment: European perspectives. Why should Europe care? Environmental issue report No. 38. 58 pp.

Erikstad, KE, Moum, T, Bustnes, JO, Reiertsen, T (2005) Sex allocation effects from exposure to organic contaminant exposure in glaucous gull in the high Arctic (manuscript).

Espeland, O, Kleivane, L, Haugen, S, Skaare, JU (1997) Organochlorines in mother and pup pairs in two arctic seal species; Harp seal (*Phoca groenlandica*) and hooded seal (*Cystophora cristata*). Mar Environ Res, 44:315-330.

Fisher, BE (1999) Most unwanted persistent organic pollutants. Environ Health Perspect, 107:A18-A23.

Fisk, AT, Hobson, KA, Norstrom, RJ (2001) Influence of chemical and biological factors on trophic transfer of persistent organic pollutants in the Northwater Polynya marine food web. Environ Sci Technol, 35:732-738.

Fisk, AT, Muir, DCG, Hobbs, K, Borg, H, Braune, B, Burgess, N, Culhane, M, de Wit, C, Evans, M, Hickie, B, Hoekstra, P, Kuzyk, Z, Kwan, M, Lockhart, L, Macdonald, C, Norstrom, R, Outridge, P, Roach, P, Stern, G, Wayland, M, Stokker, Y (2003) Part I. In: Canadian Arctic Contaminants Assessment Report II: Contaminant Levels, Trends and Effects in the Biological Environment. (eds. Fisk, AT, Muir, DCG, Hobbs, K) Indian and Northern Affairs Canada, Ottawa, Canada.

Fuglei, E (2000) Physiological adaptations of the arctic fox to high Arctic conditions. Dissertation for the degree of Doctor Scientiarum. University of Oslo, Norway.

Gabrielsen, GW, Henriksen, EO (2001) Persistent organic pollutants in Arctic animals in the Barents Sea area and Svalbard: Levels and Effects. Memoirs of National Institute of Pol Res, Special Issue, 54:349-364.

Gabrielsen, GW, Skaare, JU, Polder, A, Bakken, V (1995) Chlorinated hydrocarbon in glaucous gulls (*Larus hyperboreus*) in the southern part of Svalbard. Sci Tot Environ, 160/161:337-346.

Gabrielsen, GW, Knudsen, LB, Verreault, J, Push, K, Letcher, RJ, Muir, DCG (2004) Halogenated organic contaminants and metabolites in blood and adipose tissue of polar bears (*Ursus maritimus*) from Svalbard. Report to the Norwegian Pollution Control Authority (SFT), Oslo, Norway. Report number 915/2004. 31 pp.

Giesy, JP, Kannan, K (2001) Global distribution of perfluorooctane sulfonate in wildlife. Environ Sci Technol, 35:1339–1342.

Giesy, JP, Ludwig, JP, Tillitt, DE (1994) Deformities in birds of the Great Lakes region:assigning causality. Environ Sci Technol, 28:128A-135A.

Giesy, JP, Feyk, LA, Jones, PD, Kannan, K, Sanderson, T (2003). Review of the effects of endocrine-disrupting chemicals on birds. Pure App Chem, 75:2287-2303.

Grasman, KA, Fox, GA, Scanlon, PF, Ludwig, JP (1996) Organochlorine-associated immunosuppression in prefledgling Caspian terns and herring gulls from the Great lakes: An Ecoepidemiological study. Environ Health Perspect Suppl, 104:829-842.

Haave, M, Ropstad, E, Derocher, AE, Lie, E, Dahl, E, Wiig, Ø, Skaare, JU, Jenssen, BM (2003) Polychlorinated Biphenyls and Reproductive Hormones in Female Polar Bears at Svalbard. Environ Health Perspect, 111:431-436.

Henriksen, EO, Gabrielsen, GW, Skaare, JU (1996) Levels and congener pattern of polychlorinated biphenyls in kittiwakes (*Rissa tridactyla*), in relation to mobilization of body-lipids associated with reproduction. Environ Poll, 92:27-37.

Henriksen, EO, Gabrielsen, GW, Trudeau, S, Wolkers, J, Sagerup, K, Skaare, JU (2000) Organochlorines and possible biochemical effects in glaucous gulls (*Larus hyperboreus)* from Bjørnøya, the Barents Sea. Arch Environ Cont Toxicol, 38:234-243.

Henriksen, EO, Wiig, Ø, Skaare, JU, Gabrielsen, GW, Derocher, AE (2001) Monitoring PCBs in polar bears: lessons learned from Svalbard. J Environ Monit, 3:493-498.

Hiruki, LM, Stirling, I (1989) Population-dynamics of the arctic fox, *Alopex lagopus*, on Banks Island, Northwest Territories. Can Field Nat, 103:380-387.

Hobbs, KE, Muir, DCG, Born, EW, Dietz, R, Haug, T, Metcalfe, T, Metcalfe, C, Øien, N (2002) Regional variation in levels and pattern of persistent organochlorines in North Atlantic minke whales. Abstract for the AMAP conference and workshop: Impact of POPs and Mercury on Arctic Environments and Humans, 20-24 January, 2002. Tromsø, Norway. 68 pp.

Hoekstra, PF, O'Hara, T, Fisk, AT, Borgå, K, Solomon, KP, Muir, DCG (2003) Trophic transfer of persistent organochlorines contaminants (OCs) within an Arctic marine food web from the southern Beaufort-Chukchi Seas. Environ Poll, 124:509-522.

Holden, AV (1972) Monitoring organochlorine contaminants in marine environment by the analysis of residues in seals. Marine pollution and sealife (ed. Ruivo, M) FAO, Fishing News Book, Surrey, pp 266-272.

Ikonomou, MG, Rayne, S, Addison, RF (2002) Exonential increase of the brominated flame retardants, polybrominated diphenyl ethers, in the Canadian Arctic from 1981-2000. Environ Sci Technol, 36:1886-1892.

Jantunen, LM, Bidleman, T (1996) Air-water gas exchange of hexachlorocyclohexanes (HCHs) and the enantiomers of alpha-HCH in arctic regions. J. geophys Res, 101, 28837-28846 (corrections: J geophys Res 1997, 102 :19279-19282).

Kannan, K, Franson, JC, Bowerman, WW, Hansen, KJ, Jones, PD, Giesy, JP (2001) Perfluorooctane sulfonate in fish-eating water birds including bald eagles and albatrosses. Environ Sci Technol, 35:3065–3070.

Kleivane, L, Skaare, JU (1998) Organochlorine contaminants in northeast Atlantic minke whales (*Balaenoptera acutorostrata*) Environ Poll, 101:231-239.

Kleivane, L, Espeland, O, Fagerheim, KA, Hylland, K, Polder, A, Skaare, JU (1997) Organochlorine pesticides and PCBs in the East ice harp seal (*Phoca groenlandica*) population. Mar Environ Res, 43:117-130.

Kleivane, L, Severinsen, T, Skaare, JU (2000) Biological transport and mammal to mammal transfer of organochlorines in arctic fauna. Mar Environ Res, 49:343-257.

Krøkje, Å, Bingham, C, Gabrielsen, GW, Husmo, R, Østby, L (2005) Genotoxic effects in Glaucous gull *(Larus hyperboreus)*. J Toxicol Environ Health (accepted).

Larsen, HJ, Gabrielsen, GW, Sagerup, K (2004) Effekter av persistente organiske miljøgifter på immunrespons hos polarmåke. Report to the Norwegian Pollution Control Authority (SFT), Oslo, Norway. Report number 918/2004. 18 pp.

Leonards, PEG, de Vries, TH, Minnaard, W, Stuijfzand, S, de Voogt, P, Cofino, WP, van Straalen, NM, van Hattum, B (1995) Assessment of experimental data on PCB-induced reproduction inhibition in mink, based on an isomer- and congener-specific approach using 2,3,7,8-tetrachlorodibenzo-p-dioxin toxic equivalency. Environ Toxicol Chem, 14:639-652.

Letcher, RJ, Norstrom, RJ, Lin, S, Ramsay, MA, Bandiera, SM (1996) Immunoquantitiation and microsomal mono-oxygenase activities of hepatic cytochrome P4501A and P4502B and chlorinated hydrocarbon contaminant levels in polar bears (Ursus maritimus). Toxicol App Pharm, 137:127-140.

Lie, E, Derocher, AE, Wiig, Ø, Skaare, JU (2000) Polychlorinated biphenyls in mother/offspring pairs of polar bears (*Ursus maritimus*) at Svalbard. Organoh Comp, 49:457-460.

Lie, E, Bernhoft, A, Riget, F, Belikov, SE, Boltunov, AN, Derocher, AE, Garner, GW, Wiig, Ø, Skaare, JU (2003) Geographical distribution of organochlorine pesticides (OCPs) in polar bears (*Ursus maritimus*) in the Norwegian and Russian Arctic. Sci Tot Environ, 306:159-170.

Lie, E, Larsen, HJS, Larsen, S, Johansen, GM, Derocher, AE, Lunn, NJ, Norstrom, RJ, Wiig, Ø, Skaare, JU (2004) Does high organochlorine (OC) exposure impair the resistance to infection in polar bears (*Ursus maritimus*)? Part I: Effect of OCs on the humoral immunity. J Toxicol Environ Health A, 67:555-582.

Lie, E, Larsen, HJS, Larsen, S, Johansen, GM, Derocher, AE, Lunn, NJ, Norstrom, RJ, Wiig, Ø, Skaare, JU (2005) Does high organochlorine (OC) exposure impair the resistance to infection in polar bears (*Ursus maritimus*)? Part II: Effect of OCs on mitogen- and antigen-induced lymphocyte proliferation. J Toxicol Environ Health, Part A, 68:457-484.

Løvenskiold, HL (1964) Avifauna Svalbardensis. Norsk Polarinstitutt Skrifter, 129:1-460.

Macdonald, RW, Barrie, LA, Bidleman, TF, Diamond, ML, Gregor, DJ, Semkin,
RG, Strachan, WMJ, Li, YF, Wania, F, Alaee, M, Alexeeva, LB, Backus, SM, Bailey, R, Bewers, JM, Gobeil, C, Halsall, CJ, Harner, T, Hoff, JT, Jantunen, LMM, Lockhart, WL, Mackay, D, Muir, DCG, Pudykiewicz, J, Reimer, KJ, Smith, JN, Stern, GA, Schroeder, WH, Wagemann, R, Yunker, MB (2000) Contaminants in the Canadian Arctic:5 years of progress in understanding sources occurrence and pathways. Sci Tot Environ, 254:93-234.

McFarland, VA, Clarke, JU (1989) Environmental occurrence, abundance, and potential toxicity of polychlorinated biphenyl congeners -considerations for a congener-specific analysis. Environ Health Perspect, 81:225-239.

Mellanby, K (1992) The DDT story. British Crop Protection Council, Farnham. 113 pp.

Muir, DCG, Wagemann, R, Hargrave, BT, Thomas, DJ, Peakall, DB, Norstrom, RJ (1992) Arctic marine ecosystem contamination. Sci Tot Environ, 122:75-134.

Muir, D, Braune, B, De March, B, Norstrom, R, Wagemann, R, Lockhart, L, Hargrave, B, Bright, D, Addison, R, Payne, J, Reimer K (1999) Spatial and temporal trends and effects of contaminants in the Canadian Arctic marine ecosystem: a review. Sci Tot Environ, 230:83-144.

Muir, D, Innes, S, Cleement, M, Riget, F, Dietz, R, Kleivane, L, Skaare, J, Severinsen, T , Nakata, H, Tanabe, S (2000) Circumpolar trends of PCBs and organochlorine pesticides in the Arctic marine environment inferred from levels in ringed seals. Workshop on persistent organic pollutants (POPs) in the Arctic: Human health and environmental Concerns. Rovaniemi, Finland, jan. 2000, AMAP Report 2000:1, abstract 49.

Muir, D, Fisk, A, Kwan, M (2001) Temporal trends of persistent organic pollutants and metals in ringed seals from the Canadian Arctic. In: Synopsis of research conducted under the 2000-2001 Northern Contaminant Program (ed. Kalhok, S.) Ministry of Indian Affairs and Northern Development, Ottawa, Canada, pp 208-214.

Muir, D, Strachan, W (2003) Persistent organic pollutants in seawater. In; Sources, occurence, trends and pathways in the physical environment. The Canadian Arctic Contaminants Assessment Report II (eds. Bidleman, T, Macdonald, R, Stow, J) Indian and Northern Affairs Canada, Ottawa, Canada, pp 92-99.

Muir, DCG, Backus, S, Derocher, AE, Dietz, R, Evans, T, Gabrielsen, GW, Nagy J, Norstrom, RJ, Sonne, C, Stirling, I, Taylor, MK, Letcher, RJ Brominated flame retardants in Polar Bears (Ursus maritimus) from Svalbard, Greenland, Alaska and the Canadian Arctic (in prep).

Norstrom, RJ (1999) Trends and effects of contaminants in polar bears. In: Synopsis of Research Conducted under the 1999-2000 Northern Contaminants Program (ed. Kalhok, S) Ministry of Indian and Northern Affairs, Ottawa, Canada, pp 173-181.

Norstrom, RJ, Belikov, SE, Born, EW, Garner, GW, Malone, B, Olpinski, S, Ramsay, MA, Schliebe, S, Stirling, I, Stishov, MS, Taylor, MK, Wiig, Ø (1998) Chlorinated hydrocarbon contaminants in polar bears from eastern Russia, North America, Greenland, and Svalbard: Biomonitoring of Arctic pollution. Arch Environ Cont Toxicol, 35: 354-367.

Norstrom, RJ, Simon, M, Moisey, J, Wakeford, B, Weseloh, DVC (2002) Geographical distribution (2000) and temporal trends (1981-2000) of brominated diphenyl ethers in Great Lake herring gull eggs.
Environ Sci Technol, 36:4783-4789.

Olsson, M, Karlsson, B, Ahnland, E (1992) Seals and seal protection - summary and comments. Ambio 21: 606-606.

Oskam IC, Ropstad E, Lie E, Derocher AE, Wiig Ø, Dahl E, Larsen S, Wiger R, Skaare JU (2003) Organochlorines affect the major androgenic hormone, testosterone, in male polar bears (*Ursus maritimus)* at Svalbard. J Toxicol Environ Health A, 66:2119-2139.

Oskam IC, Ropstad E, Lie E, Derocher AE, Wiig Ø, Dahl E, Larsen S, Wiger R, Skaare JU (2004) Organochlorines affect the steroid hormone cortisol in freerenging polar bears *(Ursus maritimus)* at Svalbard, Norway. J Toxicol Environ Health A, 67:957-977.

Peakall, DB (1992) Animal biomarkers as pollution indicators. Chapman & Hall Ecotoxicology Series 1. Chapman & Hall, London.

Peakall, DB, Noble, DG, Elliott, JE, Somers JD, Erickson, G (1990) Environmental contaminants in Canadian peregrine falcons, *Falco peregrinus*: A toxicological assessment. Can Field Nat, 104:204-254.

Platonow, NS, Karstad, LS (1973) Dietary effects of polychlorinated biphenyls in mink. Can J Comp Med, 37:391-400.

Poland, A, Knutson, JC (1982) 2,3,7,8-tetrachlorodibenzo-p-dioxin and related halogenated aromatic hydrocarbons: Examination of the mechanism of toxicity. Ann Rev Pharm Toxicol, 22:517-554.

Reijnders, PJH (1986) Reproductive failure in common seals feeding on fish from polluted coastal waters. Nature, 324:456-457.

Renner, R (2000) What fate for brominated fire retardants? Environ Sci Techn, 34:222A-226A.

Sagerup, K, Henriksen, EO, Skorping, A, Skaare, JU, Gabrielsen, GW (2000) Intensity of parasitic nematodes increases with organochlorine levels in the glaucous gull. J App Ecol, 37:1-9.

Sandala, GM, Sonne-Hansen, C, Dietz, R, Muir, DCG, Valters, K, Bennett, ER, Born, EW, Letcher, RJ (2004) Hydroxylated and methyl sulfone PCB metabolites in adipose and whole blood of polar bears (*Ursus maritimus*) from East Greenland. Sci Total Environ, 331:125-141.

Sandau, CD, Meerts, IATM, Letcher, RJ, Mcalees, AJ, Chittim, B, Brouwer, A, Norstrom, RJ (2000) Identification of 4-hydroxyheptachlorostyrene in polar bear plasma and its binding affinity to transthyretin: a metabolite of octachlorostyrene? Environ Sci Technol, 34:3871-3877.

Savinova, TN, Polder, A, Gabrielsen, GW, Skaare, JU (1995) Chlorinated hydrocarbons in seabirds from the Barents Sea area. Science Tot Environ, 160/161:497-504.

Sellstrøm, U, Jansson, B (1995) Analysis of tetrabromobisphenol A in a product and environmental-samples. Chemosp, 31: 3085-3092.

Severinsen, T, Skåre, JU (1997) Levels of heavy metals and persistent organic components in some terrestrial animals from Svalbard. The AMAP International Symposium on Environmental Pollution of the Arctic, Tromsø, Norway, June 1-5, 1997. Extended abstracts, Volume 1:407-409.

Skaare, JU, Bernhoft, A, Derocher, AE, Gabrielsen, GW, Goksøyr, A, Henriksen, E, Larsen, HJS, Lie, E, Wiig, Ø (2000) Organochlorines in top predators at Svalbard - occurrence, levels and effects. Toxicol Lett 112:103-109.

Skaare, JU, Bernhoft, A, Wiig, Ø, Norum, KR, Haug, E, Eide, DM, Derocher, AE (2001) Relationships between plasma levels of organochlorines, retinol and thyroid hormones from polar bears (*Ursus maritimus*) at Svalbard. J Toxicol Environ Health A, 62:227-241.

Smithwick, MM, Muir, DCG, Mabury, SA, Solomon, KR, Sonne, C, Born, EW, Dietz, R, Derocher, AE, Evans, T, Gabrielsen, GW, Letcher, RJ, Martin, JW, Nagy, J, Stirling, I, Taylor, MK (2005) A circumpolar study of perfluorinated acids in polar bears (*Ursus maritimus*). Environ Sci Technol (in press).

Spear, PA, Moon, TW, Peakall, DB (1986) Liver retinoid concentrations in natural populations of herring gulls (*Larus argentatus*) contaminated by 2,3,7,8-tetrachlorodibenzo-p-dioxin and in ring doves (*Streptopelia risoria*) injected with a dioxin analogue. Can J Zool, 64:204-208.

Stern, GA, Addison, RF (1999) Temporal trends of organochlorines contaminants in southeastern Baffin beluga. In: Synopsis of research conducted under the 1998/99 Northern Contaminant Program (ed. Kalhok, S.) Ministry of Indian Affairs and Northern Development, Ottawa, Canada, pp 203-212.

Stirling, I, Kingsley, MCS, Calvert, W (1982) The distribution and abundance of seals in the eastern Beaufort Sea 1974-1979. Occasional Paper No. 7. Canadian Wildlife Service, Ottaewa, ON, Canada. 25 pp.

Tryphonas, H (1994) Immunotoxicity of polychlorinated biphenyls: Present status and future considerations. Experim Clin Immunogenetics, 11:149-162.

van Bavel, B, Dam, M, Tysklind, M, Lindstrom, G (2001) Levels of polybrominated diphenyl ethers in marine mammals. Organoh Comp, 52:99-103.

Verreault, J, Skaare, JU, Jenssen, BM, Gabrielsen, GW (2004) Effects of organochlorine contaminants on thyroid hormone levels in Arctic breeding glaucous gulls, *Larus hyperboreus*. Environ Health Perspect 112:532–537.

Verreault, J, Letcher, RJ, Muir, DCG, Chu, S, Gebbink, WA, Gabrielsen, GW (2005a) New organochlorine contaminants and metabolites in plasma and eggs of glaucous gulls (*Larus hyperboreus*) from the Norwegian Arctic. Environ Toxicol Chem (in press).

Verreault, J, Gabrielsen, GW, Chu, S, Muir, DCG, Andersen, M, Hamaed, A, Letcher, RJ (2005b) Flame retardants and methoxylated and hydroxylated PBDEs in two Norwegian Arctic top–predators: glaucous gulls and polar bears. Environ Sci Technol (in press).

Verrault, J, Muir, DCG, Norstrom, RJ, Fisk, AT, Lunn, NF, Stirling, I, Gabrielsen, GW, Derocher, AE, Sandala, GM, Gebbink, W, Braune, B, Wakeford, B, Taylor, M, Nagy, J, Branigan, M, Obbard, M, Letcher RJ (2005c) Geographical and temporal trends of chlorinated hydrocarbon contaminants and metabolites in polar bears (*Ursus maritimus*) from Svalbard, Greenland, Alaska and Canadian Arctic populations. Sci Tot Environ (in press).

Verreault, J, Houde, M, Gabrielsen, GW, Berger, U, Haukaas, M, Letcher, R, Muir, DCG. Perfuorinated alkyl substances in plasma, tissues and eggs of glaucous gulls (*Larus hyperboreus*) from Svalbard, Arctic Norway. Environ Sci Technol (submitted).

Voldner, EC, Li, YF (1995) Global usage of toxaphene. Chemosph, 27:2073-2078.

Wang-Andersen, G, Skaare, JU, Prestrud, P, Steinnes, E (1993) Levels and congener pattern of PCBs in Arctic fox, *Alopex lagopus*, in Svalbard. Environ Poll, 82:269-275.

Wiig, OE (1995) Distribution of polar bears (*Ursus maritimus*) in the Svalbard area. J Zool, London, 237:515-529.

Wiig, OE, Derocher, AE, Cronin, MM, Skaare, JU (1998) Female pseudohermaphrodite polar bears at Svalbard. J Wildl Dis, 34:792-796.

Wolkers, J (2002) POPs in beluga whales and narwhal. Unpublished data. Norwegian Polar Institute, Polar Environmental Centre, Tromsø, Norway.

Wolkers, J, Burkow, IC, Lydersen, C, Dahle, S, Monshouwer, M, Witkamp RF (1998) Congener specific PCB and polychlorinated camphene (toxaphene) levels in Svalbard ringed seals (*Phoca hispida*) in relation to sex, age, condition and cytochrome P450 enzyme activity. Sci Tot Environ, 216:1-11.

Wolkers, H, Burkow, IC, Lydersen, C, Witkamp, RF (2000) Chlorinated pesticide concentrations, with an emphasis on the polychlorinated camphenes (toxaphenes), in relation to cytochrome P450 enzyme activities in harp seals (*Phoca groenlandica*) from the Barents Sea. Environ Toxicol Chem, 19:1632-1637.

Wolkers, H, van Bavel, B, Derocher, AE, Wiig, Ø, Kovacs, KM, Lydersen, C, Lindsrom, G (2004) Congener-Specific accumulation and food chain transfer of polybrominated diphenyl ethers in two Arctic food chains. Environ Sci Technol, 38:1667-1674.

Wong, CS, Mabury, SA, Whittle, DM, Backus, SM, Teixeira, C, Devault, DS, Bronte, CR, Muir, DCG (2004) Organochlorine compounds in Lake Superior: Chiral polychlorinated niphenyls and biotransformation in aquatic food web. Environ Sci Technol, 38:84-92.

Wren, CD, Hunter, DB, Leatherland, JF, Stokes, PM (1987) The effects of polychlorinated biphenyls and methylmercury, singly and in combination, on mink. II: Reproduction and kit development. Environ Cont Toxicol, 16:449-454.

Yudin, KA, Firsova, LV (1988) The glaucous gull. Birds of the USSR. Moscow: Nauka, pp. 165-174 (in Russian).

Zelinski, V, Lorenz, W, Bahadir, M (1993) Brominated flame retardants and resulting PBDD/F in accidental fire residues from private residences. Chemosph, 27:1519-1528.

Zile MH (1992) Vitamin A homeostasis endangered by environmental pollutants. Proc Soc Exper Biol Med, 201:141-153.

Østby, L, Gabrielsen, GW, Krøkje, Å (2005) CYP1A induction and DNA adduct formation in Glaucous gulls (*Larus hyperboreus*), fed with environmentally contaminated gull eggs. Ecotoxicol Environ Safety, 62:000-000.

21 Arctic health problems and environmental challenges in Greenland.

Gert Mulvad[1,2], Henning Sloth Petersen[1,2] and Jørn Olsen[1,3]

[1]Commission for Scientific Research in Greenland. [2]Center of Primary Health Care, Nuuk, Greenland. [3]The Danish Epidemiology Science Centre, Department of Epidemiology and Social Medicine, University of Aarhus, Aarhus C, Denmark, e-mail: GM@gh.gl.

21.1 Introduction

The traditional diet in Greenland is to a large extend based upon marine animals and fish (Deutch 2004). Today the Greenlandic diet is a mixture of traditional food and imported food, this is the way it has been for some generations. Due to weather conditions most of fresh food come from wild animals or fish. Greenland has a production of lamb and a limited supply of vegetables but most produced foods are imported from outside. A large part of the diet still stem from seafood, fish or sea animals, but imported fabricated foods are expected to continue to take over an increasing part of their energy consumption (Deutch 2004, Mulvad 1996).

In any community it is important how food is produced, how it is prepared and how it is consumed. These things are important to the individual and for the way people come together. That is why food is more than just getting the necessary nutrition, it is also essential for social life and the ways families function. Since food is increasingly imported and come in full or almost full fabricated forms the quality of nutrition changes. Along with this public health may be affected as well as the social aspect of eating and preparing the meals. Still the traditional diet is very important to the population culturally and financially. It is also of importance in order to get sufficient nutrients, because in many places imported food is available mostly in poor quality (Deutch 2004, Pars 2000).

High levels of long range transported contaminants to the Arctic have been documented by (AMAP 2003). In Greenland high contents of organic contaminants are found in people (Hansen et al 2002, Deutch 1998, Deutch et al. 2000) and the pollution has reached a level of concern. However, the replacement of traditional food by substandard, imported food is of significant concern.

Pollutants that accumulate in this part of the food chain can best be studied in the Inuit population. Other dietary studies focusing upon a diet rich in fat, like n-3 fatty acids, have been subject to study (Dyerberg 1975). However, much more could be done. The partly isolated Inuit population with its ethnic background provides new opportunities for genetic studies, as well as studies on the health impact of unique social circumstances, light and extreme cold weather.

Greenland is now in a transition between a disease pattern characterized by acute diseases, mainly infections, and chronic diseases, diabetes and cardiovascular diseases. This transition took place in Europe many decades ago but now we have a chance to study the process with modern technology. Organization and logistic of the health care system in sparsely populated Arctic regions is of great interest. Further more Greenland has public health problems which more than anything require a long term strategy for prevention.

21.2 Contaminants, diet and health effect in the Arctic

21.2.1 Organic environmental contaminants in the Arctic

A common problem for some environmental contaminants is that they have natural half lives of many years. They are also long range transported and can cause problems for generations. Contaminants examined by the Arctic Monitoring and Assessment Program, (AMAP), are PCBs, pesticides such as DDT, HCB, HCH, chlordane, dieldrin and toxaphene, and heavy metals. The persistent organic pollutants (POPs), are accumulated in the food chain in fatty tissue, creating the highest levels in orsoq/blubber from seals and tooth whales. Barleene whales, however, have a lower place in the food chain and are therefore less contaminated. The highest levels are also found in the oldest individuals. Humans and polar bears are at the top of the food chain, and are thus highly exposed to these chemicals (NERI 2004). Generally the levels are low in fish from Greenland and it is not a problem to eat Greenlandic fish. In the Baltic there has been found levels in seals up to 100 times higher than along the coasts of Greenland, but in the countries around the Baltic, seals are not eaten.

Table 21.1. Chlorinated organic contaminants in adipose tissue. (Mulvad, Pedersen et al.)

	Greenland	Canada	USA	Finland
	n = 42			
Fatty acids	1993	1983	1983-84	1983
PCB(mg/kg)	15,8	2,1	1,2	0,3
DDT(ug/kg)	4450	2630	1670	330
HCB(ug/kg)	752	10	31	20

Currents in the sea and in the air spread the contaminants unevenly so that seals and people along the East coast of Greenland are more exposed to most POPs (Deutch 1998, Deutch et al. 2000).

21.2.2 Heavy metal in the Arctic

In Greenland, research on mercury is extensive. Mercury is a heavy metal that has always been in the environment, it also constitutes a pollution without borders. Mercury levels in blood from Greenlandic people are some of the highest measured in human samples (Hansen et al. 1990).

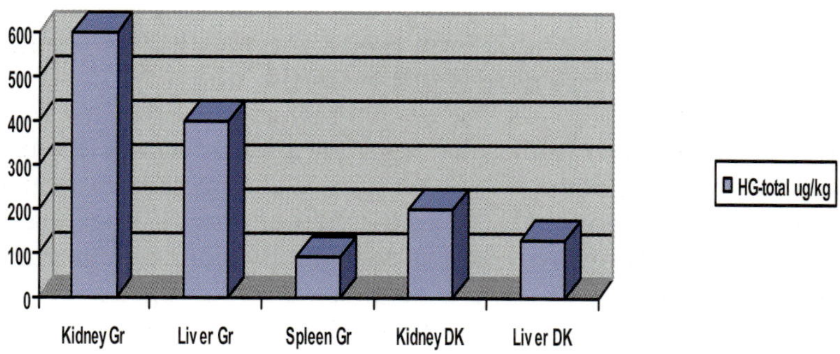

Fig. 21.1. Mercury in organs from Greenland(Gr) and Denmark(DK). Mercury level are some of the highest measured in human samples. (Mulvad, Pedersen, Hansen et al. 1996)

Lead is a well known contaminant that may stem from hunting bullets (NERI 2004) or from petrol and gasoline. Cadmium comes mostly from smoking cigarettes. The high smoking frequency in Greenland creates rather high blood levels in Greenlandic people in general. Heavy metals are accumulated on their way through the food chain and are also accumulated in human tissues.

21.2.3 The possible effect of the contaminants?

It is not dangerous for adults to eat the traditional food and the change from traditional diets to westernized MacDonalization may be a poor option concerning health. It is in relation to pregnant women and especially unborn babies that contaminants cause concern.

Clinical studies have shown an effect on the ability to learn and on memory, as a function of exposure to e.g. mercury early in life (Weihe et al. 2004). Some surveys indicate that the hormone balance and the immune system may be affected, i.e. the ability to fight infections (AMAP 2003). Some of these effects may perhaps be counteracted by the fatty acids in sea mammals.

As the woman accumulates contaminants throughout her life, contaminants will be present in the body even if the diet is changed during pregnancy (Deutch 1998). Some surveys also suspect POPs for affecting the quality of sperm (Abell et al. 2000).

21.2.4 Omega-3 fatty acid a gift from the sea

The traditional Greenlandic diet, where sea mammals are dominant, provides heat, energy and a healthy diet, at least for physical active people (Pars 2000, Dyerberg et al. 1975); the well known fat from sea mammals protects against coronary heart diseases and possibly also immune and inflammatory diseases.

A great amount of evidence from epidemiological studies and clinical trials support a theory of protective effect against coronary heart disease for fish consumption and intake of marine omega-3 fatty acids (Jul et al. 1994, Pedersen 2000, Pedersen et al 2003). Biological pathways for this risk reduction include membrane stabilization in the cardiac myocite, inhibition of platelet aggregation, favorable modifications of the lipid profile, decrease in blood pressure and reduction of the inflammatory response of the endothelium. Results from epidemiological studies suggest a threshold effect for the consumption of fish and omega-3 fatty acids.

Table 21.2. Relative contrations of fatty acids in plasma phospholipids among Inuit Women from Greenland, aged between 49 and 65 years, and women from Quebec, Canada. (Cote, Mulvad, Pedersen et al, 2004)

Fatty acids	Greenland Women (N=153) Mean	Quebec Women (N=93) Mean
Eicosapentaenoic: EPA (C20:5 n-3)	4,7	0,6
Docosahexaenoic: DHA (C22:6 n-3)	7,2	1,4
EPA+DHA	12,0	2,0
PUFA, n-3 series	13,8	2,8
PUFA, n-6 series	22,8	28,3
EPA/AA ratio	0,90	0,09
n-3/n-6 ratio	0,67	0,10
MUFA, n-9 series	18,0	15,7
Saturated	45,0	53,0

Relative concentrations are expressed as the percentage of total acids in plasma phospholipids.
This number represents all women who were within the same group of age

Risk reduction is especially important for cardiac sudden death. Nevertheless, protection against non-fatal coronary heart disease has also been observed (Pedersen et al 2003). Menstrual discomfort in Danish women reduced by dietary supplements of seal oil capsules. Shown in a trial done in Aarhus University (Deutch et al. 2000). Omega 3 fatty acid may also help prevent pre-term births (Olsen et al. 2002). Selenium is an important vitamin which limits the effect of mercury. A number of other vitamins and trace elements are well represented in the traditional diet and at the same time it is a diet which is high in protein and low in sugar. It is a diet well suited for physical active hunters living in the cold Arctic environment.

21.2.5 Health impact of light and extreme cold weather provides opportunities in Greenland of special studies.

Vitamin D is obtained from dietary sources and from endogenous synthesis in the skin. Dietary sources rich in vitamin D are fat fish and sea mammals. Ordinary meat, milk and eggs contain less vitamin D and vegetables are void of vitamin D. The endogenous synthesis occurs in the skin under ultraviolet light B (UVB) stimulation. The endogenous production depends on sun exposure, age, clothing, skin pigmentation, and use of sun protection. The exposure to UVB depends again on latitude, solar height, absorption in ozone layer and atmosphere, and reflection from clouds.

In Greenland, protective clothing is customary, and summer is short with a low solar zenith altitude Thus, yearly exposure to UVB-sunlight is limited. However, the traditional Inuit diet is rich in sea mammals that contains large quantities of vitamin D. During the last decades, significant cultural changes have occurred in Greenland. Today, many Greenlanders are living on a westernized Danish diet that is low in natural dietary sources containing vitamin D Furthermore, Danish food is not fortified with vitamin D. Changes from a traditional- to a westernized-fare are associated with a reduced vitamin D status in Greenlanders, especially in winter time. (Rejnmark et al. 2004). The influence of age, gender, latitude, season, diet and ethnicity on plasma 25-hydroxy-vitamin D 25 OHD was studied showing that in addition to ethnicity (Danes versus Greenlanders), 25 OHD levels were influenced by age, season (summer > winter), and diet (a traditional Inuit diet > westernised diet).

A longitudinal study on osteoporosis was conducted in Nuuk, in September 2002. The objective was to evaluate risk factors of osteoporosis and changes in parameters of ultrasound densitometry, after two years, among perimenopausal Inuit women who previously participated in a cross section study. After two years, the prevalence of osteoporosis has doubled. Moreover, the study suggests an association between smoking and change in stiffness, and mono-ortho PCB congeners concentration was strongly and negatively associated with bone stiffness (Cote et al. 2003).

21.2.6 Ethnic background and genetic influence

The first known immigration was by Eskimos who came from the west more than 6.000 years ago, and since then there has been 4 immigrations by Eskimos and their descendants - Inuit - the latest only about 150 years ago. From the east came the Vikings 1.000 years ago, but they disappeared after a few hundred years, perhaps without leaving any genetic trace.

Until the 1960s, only 2-3% of the Greenlandic population were non-Greenlanders, so a major genetic influence on the Inuit tribe has only taken place during the last decades, where especially in the 1970s and 1980s a stream of laborers, employers and academics invaded the country and with an overweight of younger virile men that made significant marks not only on the color of the kids today, but also resulted in relative depletion of Greenlandic females. In spite of common beliefs these bureaucrats and

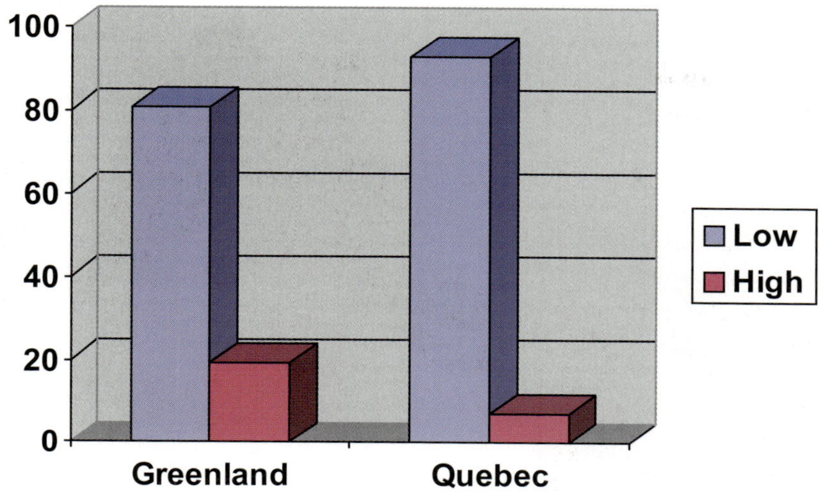

Fig. 21.2. Prevalence of risk of osteoporotic fractures among Inuit women from Nuuk, Greenland (N=153) and women from Quebec City Canada (N=2972). (Cote, Mulvad, Pedersen et al. 2004)

craftsmen were more successful in spreading their genes than the Vikings. More than 20% of the population were foreigners at that time. Since the late 1980's, the ratio has slowly decreased. Looking at the diversity in samples of Greenlandic Inuit shows 58% of the y chromosomes have been assigned to European origin in contrasts with a complete absence of European mitochondrial DNA, it indicates a male-biased European admixture. (Bosch et al. 2003).

21.2.7 Other diseases

Greenland is now in a transition between a disease pattern characterized by acute diseases, mainly infections and chronic diseases, diabetes and cardiovascular diseases. After World War 2 measles epidemic, tuberculosis, sexual transmitted diseases and hepatitis has been described. The lifestyle of the Inuit is in the same period undergoing a rapid transformation, like in most other countries that are not isolated from the rest of the world. These transitions also influence health parameters in Greenland. Infections are still frequent, especially acute respiratory infection, HIV, chlamydia,

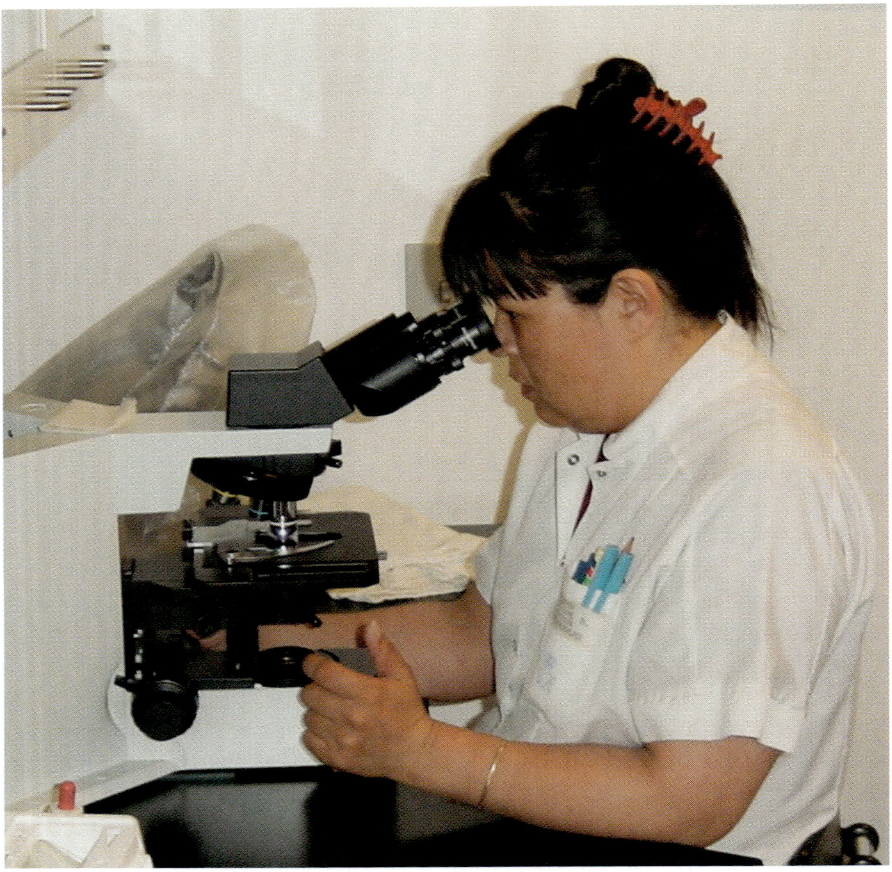

Fig. 21.3. Infection are still frequent in Greenland

tuberculosis and hepatitis (Koch 1999, Homoe 2001). We also se an increase in the food born infections due to the increase of imported food.

The effect of the western lifestyle, obesity and central fat pattern are associated with several cardiovascular risk factors including diabetes, hypertension and dyslipidemia. Central fat patterns and obesity are more prevalent among the Inuit compared to the Danish population (Bjerregaard et al 2002, Jorgensen et al. 2002).

21.3 Organization and logistics of the health care system

For obvious reasons, experience in how to establish an efficient and competent public health system in sparsely populated Arctic regions is of great interest in the northern part of North America as well as in northern Europe and Russia. It is also a well known fact that technology from one area cannot necessary be transferred from one area to the next. Since most of health technology is developed by means of research in technically advanced settings we face a shortage a knowledge for how best to modify and transfer this technology to quite different settings.

Greenland is the largest island in the world covering 2.2 million sq. kilometers of which 90% is covered by ice, and 10% is a narrow strip of land along the coastline. The climate is Arctic. About 56.000 inhabitants are settled in 18 towns and 60 settlements along a coastline of 4.000 km. The towns have 600-6.000 inhabitants with the exemption of the capital, Nuuk, that has 14.000 inhabitants. About 20% of the total population live in settlements with 50 to 500 inhabitants.

21.3.1 Health Care Centers

Health care takes place in 17 district medical centers, one for each town including its settlements. Nuuk has a central hospital for specialized treatment, and one of the University Hospitals in Copenhagen (Rigshospitalet) is used for more intensive care or specialized treatment. The district medical centers are basically autonomous units. Depending on the population, there are 1-5 physicians, nurses, midwifes, health care assistants, lab-technicians, translators, secretaries and others. Typically, the doctors and nurses are Danish and the rest Greenlanders. In the smallest district, a doctor and a nurse are the only medically trained personnel. A district has from 1 to 10 surrounding settlements to look after, and the distance can be 300 kilometers. Boats and dog sledges can be used for transport, but in many emergency cases only helicopter is an option, and a helicopter is often placed far away. The cost for an emergency transportation can be very high.

The advantage of the health care system is the very close contact between the health care staff and the patients. The staff members are themselves part of the local communities and usually know their patients well. This contact is key to optimize immunization programs pr treatment modalities for venereal diseases, diabetes and mental diseases. Also the close relation to the social care system, police and institutions has advantages. The staff should be in a good position to implement prevention and more comprehensive public health programmers when such programmers are formulated by the health care authorities.

21.3.2 Telemedicine

To strengthen the connections between remote units to the main hospital and to specialists outside Greenland, the implementation of telemedicine is now of growing importance. Telemedicine is the use of information technology for electronic transmission of information, pictures, sound and other health-related data that is needed to make appropriate diagnosis and treatment plans. Based on the experience already obtained with telemedicine in Greenland, , there are good reasons to implement telemedicine on a

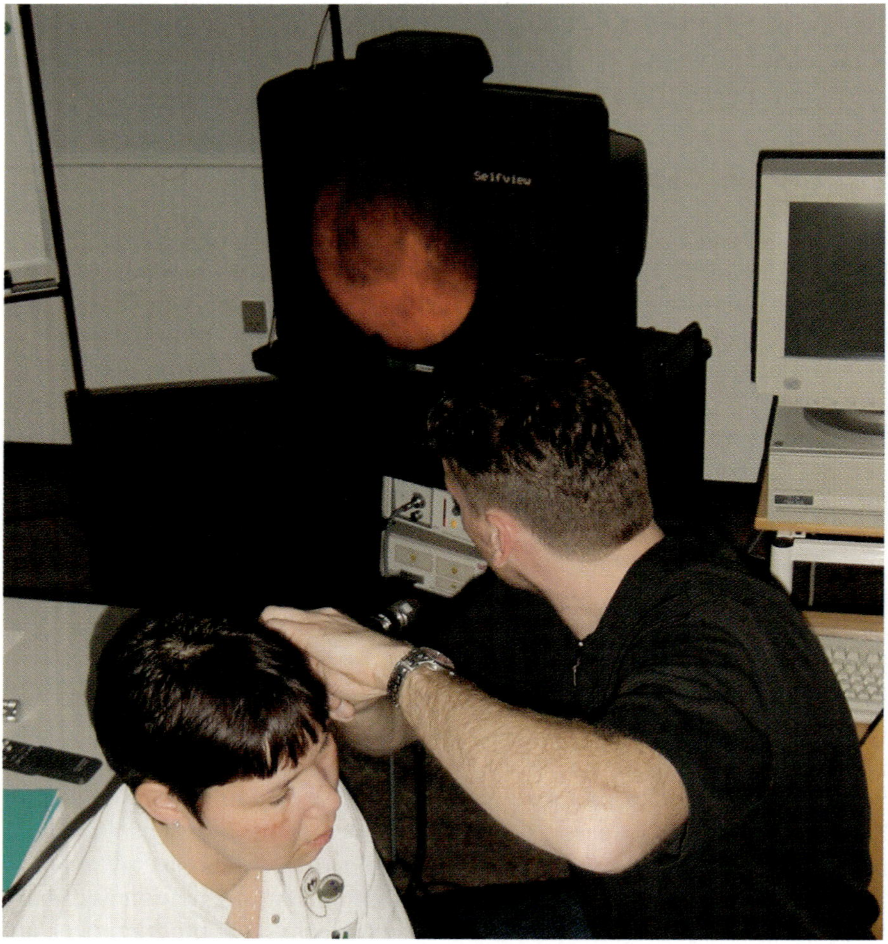

Fig. 21.4. Telemedicine is important between remote units.

larger scale (Stensgaard 2000).

It is important that the results are registered and analyzed currently throughout the period in order to ensure that the experience gained will support the continued use of telemedicine in Greenland and also be published to the benefit of others. Video-consultations can be established and X-ray photos and electro- cardiograms can also be transmitted. sonography of heart, dermatology and psychiatry are other areas. A common record system will also help provide updated information about each patient.

21.3.3 Strategy for improved health

The major characteristic of recent health and social problems in Greenland are (compared to Denmark): low mean lifetime expectancy, high infant mortality, increasing rates of diabetes, cardiovascular diseases and cancer, high rate of infectious diseases (tuberculosis, HIV, hepatitis B virus infection, Helicobacter pylori infection, meningitis), high rate of suicide, high rate of lethal accidents, high rate of legal abortions, domestic violence, tobacco, alcohol and drug abuse, mental health are stressed by unstable family relationships, contamination of the traditional diet (Chief Medical Officer 2000). The health of the population does not match the amount of money spend on health care. Health expenditure/GDP (per cent) Greenland: 9,2, Denmark: 8,2 Norway: 7,6. (Health Statistic in the Nordic Countries 2001).

Social studies and public health science have to take into consideration the conditions under which people live, e.g. health standards and housing, schools and education as well as social relations should be studied further. We need research in Greenland for the people not only on the people, and with the people.

Lifestyle diseases, the incidence of dental diseases, mental health, various forms of cancer, as well as infectious diseases are important in relation to the well-being and health of the population (Lynge 1997, Bjerregaard et al. 2002).

Social studies and studies on public health are to a great extent designed to accommodate the specific need for research in Greenland (Bjerregaard et al. 2003, Commission for Scientific Research in Greenland 2003), covered by this strategy within the framework of the effort.

To promote Health, well-being and economy of the Greenlandic population, it is suggest that, within this broad framework, social studies and public health science together encourage research which includes health hazards as an integral part of cultural aspects as well as research on social behavior in general. Since many of the problems related to problems that

Fig. 21.5. Children in Greenland - Greenland for children.

accumulate during a life time, a life course aspect should be implemented. Many results of studies indicate that prevention should start early, at the time when susceptibility is shaped or programmed and when habits are still modifiable. The effort should, in a broad sense, concentrate on the conditions under which children grow up under the theme: Children in Greenland- Greenland for children.

21.4 Conclusions

Global pollution has reached a level where the environment is affected and there is concern for people and animals that seat on the top of the food chain. Replacement of traditional food by substandard, imported food is also of significant concern. Any food recommendation must be structured to the community and take into consideration the available sources of food. For this reason it is important to keep monitoring the pollution that accumulates in the food chains and to take this information into consideration together with other available information concerning nutrition. These recommendations have to be well balanced and to incorporate all health aspects of diet and the social and economic consequences of our diet and how we get the food we need to eat. The pattern of disease will be increasingly influenced by chronic diseases being developed over long periods of time. The causes for this may be found early in life; during pregnancy, when all organs are formed, and during early childhood, where habits and

Table 21.3. Estimated daily intake of traditional and imported foods in East Greenland in 2001 by semi-quantitative food frequency questionnaire (n=180). (Deutch, Pedersen 2004)

Food	Daily intake
Seal meat and blubber	41g/d
Polar bear	23g/d
Walrus	14g/d
Whale meat and blubber	37g/d
Fish	78g/d
Reindeer, hare, muskox	50g/d
Imported food	638g/d

circumstance may have lifelong influence. In Greenland too, it is necessary to understand the social, cultural and environmental causes better in order to ensure a better preventive effort.

The conditions for children form the basis for the future of Greenland both economically, socially and for health. The position of the family and the definition of its responsibilities are central matters in this process. Time has come to concentrate the effort where it will have the greatest and longest lasting effect. A research program like this one should focus on families who are just starting to have children. Research should include genetic, social and cultural aspects on one hand and disease, behavior and cognitive development on the other. Research should have a longitudinal angle and should provide possibilities of trying out theory through interventional projects. Time has come for the policy makers to think ahead and to make decisions that span more than time periods of election.

Acknowledgements

The present papers has been supported by The Commission for Scientific Research in Greenland, Sygekassens Helsefond, Denmark, Greenland Home Rule.

References

Abell A, Ernst E and Bonde JP (2000) Semen quality and sexual hormones in greenhouse workers. Scand J Work Environ Health. 26(6):492-500

AMAP (2003) AMAP Assessment 2002: Human Health in the Arctic. AMAP secretariat, Oslo, Norway

Annual Report from the Chief Medical Officer in Greenland (2000)
Bjerregaard P and Curtis T (2002) Cultural change and mental health in Greenland: the association of childhood conditions, language and urbanization with mental health and suicidal thoughts among the Inuit of Greenland. Social science and Med 54. 33-48
Bjerregaard P, Jorgensen ME, Andersen S and Mulvad G et al. (2002) Decreasing overweight and central fat patterning with Westernization among the Inuit in Greenland and Inuit migrants. Int. J. of Obesity 26, 1503-1510.
Bjerregaard P, Mulvad G and Olsen J (2003) Studying Health in Greenland. Int. J. of Circumpolar Health 62:1. 5-16
Bosch E, Rosser ZH, Norby S, Lynnerup N and Jobling MA (2003) High level of male-biased Scandinavian admixture in Greenlandic Inuit shown by Y-chromosomal analysis. Hum Genet 112:253-363
Commission for Scientific Research in Greenland (2003): Strategi for dansk-grønlandsk polarforskning 2003-2007 (Danish).
Côtè S, Dodin S, Blanchet C, Mulvad G, Pedersen HS and Dewailly E (2003). A longitudinal study of bone mass among perimenopausal inuit women of Greenland; ICCH12 (2003) Nuuk (abstract).
Deutch B (1998) Biostatistik og epidemiologisk vurdering af levevis og miljømedicinske data fra Grønland. MPH thesis. University of Aarhus (Danish).
Deutch B and Hansen JC (2000) High Human plasma levels of organochlorine compounds in Greenland. Regional differences and lifestyle effects. Dan Med Bull 47(2):132- 7
Deutch B, Jorgensen EB and Hansen JC (2000) Menstrual Discomfort in Danish Women Reduced by Dietary Supplements of Omega-3 PUFA and B12 (Fish Oil and Seal Oil Capsules). Nutrition Research vol 20 NO. 5, 621-631
Deutch B, Pedersen HS and Hansen JC (2004) Dietary composition in Greenland 2000, plasma fatty acids and persistent organic pollutants. Sci Total Environ. 20;331(1-3):177-88
Dyerberg J, Bang HO and Hjorne N (1975) Fatty acids composition of the plasma lipids in Greenland Eskimos. Am J Clin Nutr. 28:958-66
Hansen JC, Reiersen LO and Wilson S (2002) Arctic Monitoring and Assessment Programme (AMAP), strategy and results with focus on the Human health assessment under the second phase of AMAP, 1998-2003. Int J Circumpolar Health 61(4):300-18
Hansen JC, Tarp U and Bohm J (1990) Prenatal exposure to methyl mercury among Greenlandic polar Inuits. Arch Environ Health 45(6):355-8
Health Statistics in the Nordic Countries (2001) NOMESCO Nordic Medico Statical Committee.
Homøe P (2001) Otitis Media in Greenland. Int J of Circ Health vol. 60 suppl. 2, 1-54. Disputats. Universitet of Copenhagen.
Jorgensen M, Bjerregaard P and Borch-Johnsen K (2002) Diabetes and impaired glucose tolerance among the inuit population og Greenland. Diabetes Care 25:1766-1771

Jul E, Mulvad G and Pedersen HS et al. (1994) The Relationship between a low Rate of Ischaemic Heart Disease and the traditional Greenlandic Diet with High Amounts of Monounsaturated and N-3 Polyunsaturated Fatty Acids.. Arctic Med. Research vol 53; Suppl, 2 pp 282-284

Koch A (1999) A longitudinal community based study of respiratory tract infections in Greenlandic children: disease burden and risk factors. Ph. D. thesis, University of Copenhagen

Lynge I (1997) Mental disorders in Greenland. Past and present. Meddr Grønland, Man and Society 21:1-73

Mulvad G and Pedersen HS et al. (1996) The Inuit diet. Fatty acids and antioxidants, their role in ischemic heart disease, and exposure to organochlorines and heavy metals. An international study. Arct Med Res 55 Suppl. 1:20-24

National Environmental Research Institute (NERI) (2004) Contaminants in the traditional Greenland diet. Ministry of the Environment, Denmark, NERI Tehnical Report, No. 492

Olsen SF and Secher NJ (2002) Low consumption of seafood in early pregnancy as a risk factor for preterm delivery: prospective cohort study. BMJ 23;324(7335):447

Pars T (2000) Forbruget af traditionelle grønlandske fødevarer I Vestgrønland.. Ph.D. thesis, University of Copenhagen. (Danish)

Pedersen HS (2000) Det grønlandske obduktionsprojekt. Atherosclerose og intracerebral hæmorrhagi i relation til fedtsyrasammensætningen i fedtvæv og validering af dødsårsagsmønsteret. Ph.D. thesis, University of Aarhus. (Danish)

Pedersen HS, Mulvad G, Newmann WP and Boudreau DA (2003) Atherosclerosis in coronary arteries and aorta among Greenlanders: An autopsy study. Atherosclerosis Vol 170 Issue 1 Pages 93-103

Rejnmark L, Joergensen ME and Pedersen MB et al. (2004) Vitamin D insufficiency in Greenlanders on a Westernized fare: ethnic differences in calcitropic hormones between Greenlanders and Danes. Calcified Tissue International 74:255-263

Stensgaard T (2000) Udbygning og styrkelse af det telemedicinske samarbejde i de nordiske og tilgrænsende områder. DIVS 2000:820, Nordisk Ministerråd, København (Danish)

Weihe P, Hansen JC and Murata K et al. (2002) Neurobehavioral performance of Inuit children with increased prenatal exposure to methylmercury. Int J Circumpolar Health. 61(1):41-9

Index

Printing: Krips bv, Meppel
Binding: Stürtz, Würzburg

Islamic Finance, Governance and Regulation

Islamic financial institutions (IFIs) are often criticised due to insufficient regulatory provisions, thus lowering clients' confidence in them. Additionally, few countries have well-established legislation relating to the Islamic finance industry, while others are only now developing preliminary regulations to incorporate Islamic finance within their own financial systems. This book offers a comprehensive understanding of Islamic financial regulation, outlining the regulatory landscape in various jurisdictions, examining how it supports the development and foundation of IFIs and addressing the regulation and supervision of IFIs from a global perspective. The book begins by introducing Islamic finance and its current position in the international banking system, then goes on to review the historical developments of the field. It discusses the key financial regulators, their functions and the regulations that have been implemented to foster the growth of Islamic finance. It includes cases from both Eastern and Western countries. It thoroughly examines how Shariah principles are integrated into the regulatory framework of Islamic finance. Specifically, the book examines the laws and arrangements established in different countries regarding the governance of the Islamic finance industry as well as the legal criteria for setting up and governing IFIs. The focus is on exploring diverging laws and their impact on developing Islamic Finance. This will serve as an academic and practitioner guide to understanding the preliminary regulations required for the establishment, governance and management of IFIs. Also, it offers practical insights and guidance for professionals working in the Islamic finance industry, including bankers, lawyers, regulators and policymakers.

Hakimah Yaacob is an Associate Professor at the Faculty of Islamic Economics and Finance, University Islam Sultan Sharif Ali (UNISSA), Brunei Darussalam.

Razali Mat Zin is a Professor at the Faculty of Islamic Economics and Finance, University Islam Sultan Sharif Ali (UNISSA), Brunei Darussalam.

Qaisar Ali is a lecturer at the School of Business and Management, Bandung Institute of Technology, Indonesia.

Islamic Business and Finance Series

Series Editor: Ishaq Bhatti

There is an increasing need for western politicians, financiers, bankers, and indeed the western business community in general to have access to high quality and authoritative texts on Islamic financial and business practices. Drawing on expertise from across the Islamic world, this new series will provide carefully chosen and focused monographs and collections, each authored/edited by an expert in their respective field all over the world.

The series will be pitched at a level to appeal to middle and senior management in both the western and the Islamic business communities. For the manager with a western background the series will provide detailed and up-to-date briefings on important topics; for the academics, postgraduates, business communities, manager with western and an Islamic background the series will provide a guide to best practice in business in Islamic communities around the world, including Muslim minorities in the west and majorities in the rest of the world.

Islamic Finance and Sustainable Development
A Global Framework for Achieving Sustainable Impact Finance
Edited by Mohd Ma'Sum Billah, Rusni Hassan, Razali Haron and Nor Razinah Mohd. Zain

Islamic Finance in the Modern Era
Digitalization, FinTech and Social Finance
Edited by Hussain Mohi-ud-Din Qadri and M. Ishaq Bhatti

The Islamic Economic System
Cultural Context in a Global Economy
Muhammad Awais, Ali Osman Ozturk, Omar Khalid Bhatti and Nazima Ellahi

Islamic Finance, Governance and Regulation
Global Perspectives
Hakimah Yaacob, Razali Mat Zin and Qaisar Ali

For more information about this series, please visit: www.routledge.com/Islamic-Business-and-Finance-Series/book-series/ISLAMICFINANCE

Islamic Finance, Governance and Regulation

Global Perspectives

Hakimah Yaacob, Razali Mat Zin and Qaisar Ali

Routledge
Taylor & Francis Group
LONDON AND NEW YORK

First published 2025
by Routledge
4 Park Square, Milton Park, Abingdon, Oxon OX14 4RN

and by Routledge
605 Third Avenue, New York, NY 10158

Routledge is an imprint of the Taylor & Francis Group, an informa business

British Library Cataloguing-in-Publication Data
A catalogue record for this book is available from the British Library

ISBN: 978-1-032-74810-8 (hbk)
ISBN: 978-1-032-74813-9 (pbk)
ISBN: 978-1-003-47103-5 (ebk)

DOI: 10.4324/9781003471035

Typeset in Times New Roman
by KnowledgeWorks Global Ltd.

Contents

Figures and tables

Figures

Tables

Acknowledgement

All praise to Allah (s.w.t), the most Gracious and most Merciful, by whose grace and blessing this work has been completed. We express our sincere indebtedness and gratitude to our beloved families, friends and colleagues for their support, patience, perseverance and encouragement.

Preface

Islamic Finance, Governance and Regulation: Global Perspectives covers regulatory provisions for Islamic finance in 47 jurisdictions. The book explains the regulatory framework and supervision of Islamic Financial Institutions (IFIs) from global perspectives. We anticipate that the book will stimulate forward thinking in shaping the regulation and supervision of Islamic finance. It is fervently established that the rationality and objectivity of Islamic Finance are losing features and characters under the spheres of the conventional financial environment.

This book outlines the regulatory landscape in various jurisdictions and examines how it supports the development and foundation of IFIs. To explore the regulatory environment of the selected countries, the authors have discussed the main financial regulators, their functions and the promulgated regulations by these regulators to foster the growth of Islamic finance. This book covers the jurisdictions from both Eastern and Western countries offering an insight into whether the financial regulatory infrastructure of non-Islamic countries may support establishing IFIs as well as critically analyses the feasibility of Islamic finance to embed in these countries.

This book is expected to serve as an academic and practitioner guide to understand the preliminary regulations required for the establishment, governance and management of IFIs. IFIs are often criticised due to insufficient regulatory information lowering clients' confidence in these institutions. This book fulfils the lack of information on regulatory arrangements and discusses several regulatory regimes in countries with well-established Islamic finance industries. It also outlines the regulatory system supportive of further developing and establishing an Islamic finance industry.

Abbreviations

AALCO	Asian-African Legal Consultative Organisation
AAOIFI	Accounting & Auditing Islamic Financial Institutions
ABS	Australian Bureau of Statistics
ADR	Alternative Dispute Resolution
AfDB	African Development Bank
AMAF	Africa Microfinance Action Forum
AML-CFT	Anti-money Laundering-Combating the Financing of Terrorism
AMU	Arab Maghreb Union
ANIE	Asian Newly Industrialized Economies
APRA	Australian Prudential Regulation Authority
ARCIFI	Arbitration and Reconciliation Centre for Islamic Financial
ASEAN	Association of South East Asian Nations
ASFIM	Association des Sociétés de Gestion et Fonds d'Investissement
BAMA	Bank Al Muamelat Assahiha
BAMUI	Indonesia Mu'amalat Arbitration Body
BBA	Bay' Bithaman Ajil
BCBS	Basel Committee for Banking Supervision
BCBS	Basel Committee on Banking Supervision
BCEAO	Central Bank of West African States
BCP	Basel Core Principles
BESA	Bond Exchange of South Africa
BICRA	Banking Industry Country Risk Assessment
BNM	Bank Negara Malaysia
BOC	Bank of China
BOJ	Bank of Japan
BOK	Bank of Korea
BOT	Bank of Thailand; Bank of Taiwan
BPM	Banque Populaire de Mauritanie
BRIC	Brazil, Russia, India and China
BRVM	Bourse Regionale des Valeurs Mobilieres
BSI	Banking System Indicator
CAD	Canadian dollar

CBC	Central Bank of the Republic of China (Taiwan)
CBL	Central Bank of Libya
CBN	Central Bank of Nigeria
CBOS	Central Bank of Sudan
CBSS	Central Bank of South Sudan
CCHS	Check Clearing House System CHF Swiss Franc
CDB	China Development Bank
CDC	British Colonial Development Corporation
CDRN	Commercial Dispute Resolution News
CEMAC	Central African Economic and Monetary Union
CFC	Casablanca Finance City
CGAP	Consultative Group to Assist the Poor
CIFS	CBC Interbank Funds-Transfer System
COMESA	Common Market for Eastern and Southern Africa
COSUMAF	Commission for Market Surveillance of the Central African
CPI	Consumer price index
DBSA	Development Bank of Southern Africa
DBUs	Domestic banking units
DGBAS	Directorate-General of Budget, Accounting and Statistics of the Executive
DMO	Debt Management Office
DRAM	Dynamic random access memory ECB European Central Bank
D-SIBs	Domestic Systemically Important Banks
DSX	Douala Stock Exchange
DVB	Delhi Vidyut Board
EAC	East African Community
ECOWAS	Economic Community of West African States
EFSA	Egyptian Financial Supervision Authority
EMPEA	Emerging Markets Private Equity Association
ESM	European Stability Mechanism
ETF	Exchange-traded Fund
ETFs	Exchange-traded funds
EU	European Union
EUR	Euro
FAO	Food and Agriculture Organization
FDI	Foreign Direct Investment
FDIC	Federal Deposit Insurance Corporation
FED	Federal Reserve System
FEWACCI	Federation of West African Chambers of Commerce and
FIBE	Faisal Islamic Bank of Egypt
FOMC	Federal Open Market Committee FSA Financial Services Agency, Japan
FSAP	Financial Sector Assessment Program
FSB	Financial Stability Board
FSC	Financial Services Commission

FSC	Financial Supervisory Commission
FSF	Financial Stability Forum
FSIs	Financial soundness indicators F
FSSA	Financial System Stability Assessment
FX	Foreign exchange
G20	Group of Twenty
GBP	British Pound GDP Gross domestic product
GCC	Gulf Cooperation Council
GDP	Gross domestic products
GDP	Gross Domestic Product
GEF	Global Environmental Fund
GIIPS	Greece, Ireland, Italy, Portugal, and Spain
GNP	Gross national product GPD Generalized Pareto distribution
GSE	Ghana Stock Exchange
G-SIBs	Global Systemically Important Banks
GTSM	Gretai securities market
HKMA	Hong Kong Monetary Authority
HPA	Hire Purchase Act
HSBC	Hong Kong and Shanghai Banking Corporation
IADI	India Africa Diamond Institute
IAIEPA	India-Africa Institute of Education, Planning and Administration
IAIFT	India Africa Institute of Foreign Trade
IAIIT	India-Africa Institute of Information Technology
IAIS	International Association of Insurance Supervisors
IAS	International Accounting Standards
ICBC	Industrial and Commercial Bank of China
ICD	Islamic Corporation for the Development of the Private Sector
ICP	Insurance Core Principles
IFAAS	Islamic Finance Advisory & Assurance Services
IFAD	International Fund for Agricultural Development
IFC	International Finance Corporation
IFRS	International Financial Reporting Standards
IFRSs	International Financial Reporting Standards
IFSB	Islamic Financial Services Board
IIFM	International Islamic Financial Market
IILM	International Islamic Liquidity Management Corporation
IIRA	International Islamic Rating Agency
IMF	International Monetary Fund
COSUMAF	Commission for Market Surveillance of the Central African
IOSCO	International Organization of Securities Commissions
IRS	Interbank Remittance System JCIC Joint Credit Information Center
IRTI	Islamic Research & Training Institute
IsDB	Islamic Development Bank

ISFD	Islamic Solidarity Fund for Development
JGBs	Japanese government bonds JPY Japanese Yen
J-REITs	Japan Real Estate Investment Trust
KLRCA	Kuala Lumpur Regional Centre for Arbitration
KRW	Korean Won
KSE	Khartoum Stock Exchange
LCR	Liquidity coverage ratio
LDCs	Less developed countries
LOHAS	Lifestyles of Health and Sustainability
LTRO	Longer-term Refinancing Operation
LTV	Loan to value
MAS	Monetary Authority of Singapore
MDG	Millennium Development Goals
MOF	Ministry of Finance
MOF	Ministry of Finance
MoU	Memorandum of Understanding
MUI	Majelis Ulama Indonesia
NADRAS	National Alternative Dispute Resolution Advisory Council
NLSA	National Legal Service Authority
ODR	Online Dispute Resolution
OFC	Offshore Financial Centers
ROSCs	Reports on Observance of Standards and Codes
SAC	Shari'ah Advisory Council
SJT	Summary Jury Trial
SROs	Self-Regulatory Organizations
SS	Financial Supervisory Service, South Korea
UNICITRAL	UNCITRAL United Nations Commission on International Trade Law

1 Global Finance and Islamic Financial System

1.1 Introduction

This chapter lays the foundation of the book and introduces Islamic finance (IF) and its current position in the world banking system. The historical developments of the Islamic banking system are reviewed and related to the current Islamic financial system in different jurisdictions. This chapter is divided into four subsections. The evolution and significance of worldwide central banks are discussed in section 1.2 followed by a brief explanation of fractional reserve system and the discussion on the need to standardise the financial system in sections 1.3 and 1.4. The authors critically review the significance and developments in global regulatory frameworks and their relation to the Islamic financial industry.

1.2 Background of Global Islamic Financial Regulations

This section analyses the background of the establishment of central banks in each jurisdiction. Therefore, elaborative provisions are necessary for clarity to some extent. This section sheds light on the historical development of banks in each country until their current establishment. All the relevant acts are clearly mentioned particularly in relation to the regulator's mandate, scope of reference and objectives of its establishment. Having almost the same objectives in most jurisdictions, central banks or the monetary authority of the country primarily focus on financial stability (Buiter, 2014). Initially, the establishment of central banks was to provide financial assistance to the government. Later on, the functions of central banks extended to control the conduct of financial institutions to ensure safe and sound practices in the industry. These include fair, responsible and professional business conduct of Islamic financial institutions. With the passage of time, the regulatory objectives of each central bank were not limited to only financial stability but also to society, customers and investors. In order to protect the interests of their stakeholders, the regulations became stringent. Central banks shall exercise their powers and perform the functions in a way that they consider most appropriate for the purpose of meeting the regulatory objectives.

This chapter also attempts to highlight the significance of central banks in their role to ensure compliance to the international regulatory framework as well as adherence to the local laws and standards. Particularly, the authors cover the banking

DOI: 10.4324/9781003471035-1

standards adoption up until dispute resolution mechanism. Based on the availability of reports, some cases are discussed. However, most cases involving arbitrations are unavailable and limited due to the inclusion of Shariah governance framework in jurisdictions like Malaysia and Brunei based on the available laws.

In some jurisdictions, contracts associated with Riba, gharar and maisir are unenforceable in their court of law. The parties need to resort to other dispute resolution mechanisms to enforce the contract. Unlike other jurisdictions, Malaysia imposes a strict observance and legal obligation in Shariah compliance. Malaysia is the first country that penalised for a breach of Shariah in the civil courts and, consequently, recognised a breach of Shariah as equivalent to a breach of law. The introduction of the Islamic Financial Services Act 2013 (IFSA 2013) has emerged as a new interpretation of Shariah muamalat law, which is now being heard and argued in front of civil court judges (Engku Ali and Oseni, 2017).

Regulatory Framework is defined as "the existence of the necessary infrastructure which supports the control, direction or implementation of a proposed or adopted course of action, rule, principle or law." Regulatory frameworks come in two forms: the centralised framework and localised framework. In order to understand the regulatory frameworks, the authors refer to the historical root, which is the Bretton Woods Agreement and it is an important part of world financial history (Bordo, 1993a). Also, the creation of the International Monetary Fund (IMF) and the valuation of gold and foreign exchange rates remain important to this day. The agreement also made currencies convertible for trade and other current account transactions (Bordo, 1993b). Since the collapse of the Bretton Woods System, IMF members are free to choose any form of necessary exchange arrangement (except pegging their currency to gold) and allow the currency to float freely, pegging it to another currency or a basket of currencies, adopting the currency of another country, participating in a currency bloc, or forming part of a monetary union.

The Bretton Woods Agreement resulted in the establishment of the IMF and the World Bank in July 1944. The goal of the agreement was to establish a framework for economic cooperation and development that would lead to a more stable and prosperous global economy. While this goal remains central to both institutions, their work is constantly evolving in response to new economic developments and challenges (Phillips, 1983). Unlike in the gold standard or Bretton Woods, "modern" money is not backed by or redeemable for gold and hence the term fiat money. This brings the issue of seigniorage, which is the benefit one gets from the first use of fiat money, i.e., the free purchasing power that new money, not backed by gold or anything with intrinsic value, carries with it. It is submitted that the authors argue that the usage of fiat money is not a problem in the financial system (Tymoigne, 2014). Comparing fiat money with gold dinar is unjustifiable since fiat money is regulated by the authority of each jurisdiction. People used gold and silver as the main currency of exchange previously due to its precious values and unregulated. However, this statement needs to be stated in subtle. The usage of fiat money, however, needs to be free from any speculation and backed by financial stability. Issues and challenges of regulated and regimented interest are discussed in detail in the below sections.

1.3 Evidence of the Fractional Reserve Banking System

According to Rothbard (1995), the sequential introduction of the incorrect fractional reserve and financial intermediation theories of banking have led us further away from the truth. Such research should focus on the role of interested parties, especially that of internationally active banks, central banks and privately funded think tanks, in influencing academic discourse. It is worrying, for instance, that the topic of bank credit creation has been a virtual taboo for the thousands of researchers at the world's central banks over the past half-century (Turner, 2017).

1.4 Evolution of Standardisation in the Banking System

There are several factors that led the world banking system to adopt a standardisation within its routine operation (Kerwer, 2005). This book submits that the standardisation in banking began in the 1930s after the Bank of International Settlement was established followed by the Bretton Wood Agreement in 1944.

The collapse of Bank Heuss Herstatt in 1974 was a good bargaining factor to standardise the banking practices in the world (Yaacob & Rani, 2018). The Committee of Banking Regulations and Supervisory Practices (Basel) was initially established by the Central Bank Governors of the Group of Ten (G10) countries at the end of 1974 in the aftermath of serious disturbances in international currency and banking markets (notably the failure of Bankhaus Herstatt in West Germany) (Goodhart, 2011). Since its inception, the Basel Committee has expanded its membership from the G10 to 45 institutions from 28 jurisdictions. Starting with the Basel Concordat, first issued in 1975 and revised several times, the Committee has established a series of international standards for bank regulation. Among them, the most notable is the landmark publications of the accords on capital adequacy, commonly known as Basel I, Basel II and, most recently, Basel III. The Core Principles for Effective Banking Supervision (Core Principles) are the *de facto* minimum standard for sound prudential regulation and supervision of banks and banking systems.

The minimum standards originally issued by the Basel Committee on Banking Supervision (BCBS) in 1997 are used by countries as a benchmark for assessing the quality of supervisory systems and for identifying future work to achieve a baseline level of sound supervisory practices (King and Tarbert, 2011). The Core Principles are also used by the IMF and the World Bank, in the context of the Financial Sector Assessment Programme (FSAP), to assess the effectiveness of countries' banking supervisory systems and practices (Ozdemir, 2018). The revised Core Principles define 29 principles that are needed for a supervisory system to be effective. Those principles are broadly categorised into two groups: the first group (Principles 1–13) focuses on the powers, responsibilities and functions of supervisors, while the second group (Principles 14–29) focuses on prudential regulations and requirements for banks. The original Principle 1 has been divided into three separate principles, while new principles related to corporate governance, and disclosure and transparency, have been added. The Core increased from 25 to 29 Principles (Bank of International Settlements, 2020).

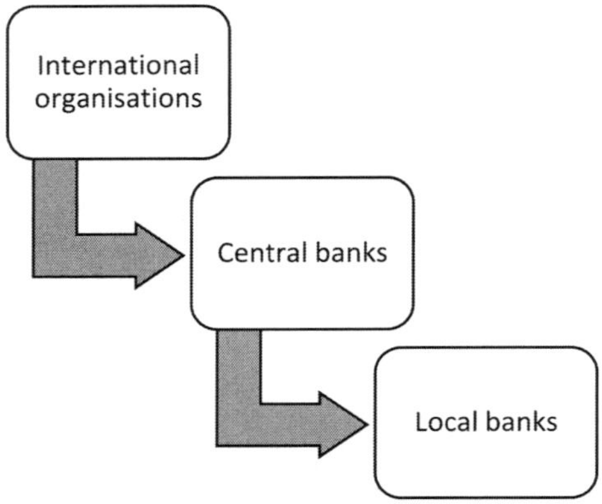

Figure 1.1 The three layers relationship.

Figure 1.1 shows the implementation of the fractional reserve system, which is required under the BCBS and adopted by several jurisdictions discussed in this book. The Central Banks Acts used in different jurisdictions are also analysed in this book. It is notable from Figure 1.1 that the statutory reserve is required by the Central Banks Acts of these jurisdictions. The statutory reserve refers to the fractional reserve requirement to be applied in the banking system. Eventually, the "money multiplier" demonstrates that a sum of money introduced into the monetary system will increase the money supply by more than its notional value.

According to the World Bank Working Paper, in dual financial systems with fairly developed conventional money markets, Islamic banks evolve in an interest rate-dominant environment (Khatat, 2016). Due to arbitrage between conventional and Islamic financial systems, there tend to be spillovers from conventional interest rates to Islamic banks' funding costs, to returns of Profit-Sharing Investment Accounts (PSIAs) as well as to costs of Islamic credit.

In most jurisdictions that apply dual banking systems, there is often no IF equivalent to the money market or government securities yield curves that can serve as references to price Islamic banks' credit. As a result, some Islamic banks tend to rely on conventional interest rates to price their *Murabaha* and Ijarah contracts (Khatat, 2016).

The regulatory framework of IF falls under the same structures of conventional banking (Figure 1.2). The institutions at Layer 1 stand as a silent regulator as their characteristics influence softly. The regulators in Layer 1 consist of the World Bank, the IMF, the International Organization of Securities Commissions (IOSCO), the Bank of International Settlement (BIS), the Financial Services Board (FSB), the Financial Action Task Force (FATF), etc. The bank provides a regulatory and supervisory framework for prudential supervision and a fair and transparent financial market. The regulatory framework also promotes financial integrity and aims to prevent money laundering and terrorist financing. According to Yaacob and Rani (2018), the relationship between how the banking system works is best demonstrated through three layers as shown in Figure 1.1. The incorporation of standards issued by international organisations will be distributed to central banks. Central

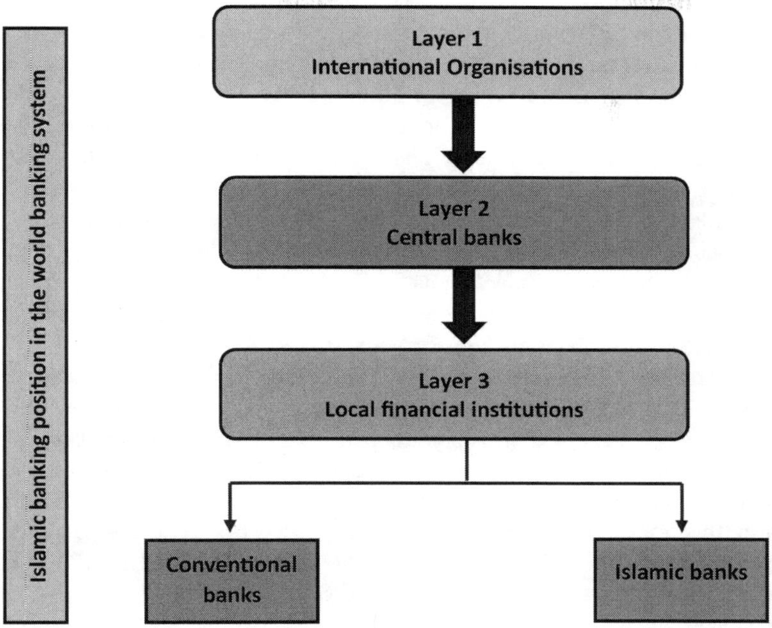

Figure 1.2 The connection between Layer 1, Layer 2 and Layer 3 in the context of financial institutions.

banks will act as an agent to disseminate information, standards, guidelines and any written directives from international organisations. Standardisation is meant to create harmonised practices in the financial system. Undeniably standardisation is needed toward globalisation.

The three tiers relationship appears to have coherent effects of standardisation and practices. Eventually, all the local institutions embed the same standards. By incorporating the standards into the local jurisdictions or local institutions, the monitoring and reviewing processes require compliance. These compliance requirements demand large compliance toward the standards.

Figure 1.2 also indicates the connection of Layers 1, 2 and 3. The second Layer lies down to the central banks around the world. The central banks are the institutions that execute the policies and mandates from Layer 1. From Layer 2, the line goes down to Layer 3. Layer 3 consists of local banks in each jurisdiction.

The implementation of regulations and standards from Layer 1 goes down to Layer 2 at the central banks level and passed down to Layer 3 at the financial institutions level consisting of conventional and Islamic banking as shown in Figure 1.2.

Based on Figure 1.2, Islamic banking is positioned as equal to the conventional banking systems. The only difference is they are safeguarded by strict supervision by the Central Bank or the monetary authority to ensure Shariah products, and any Shariah-related matters are fully complied with Islamic Shariah law. However, Islamic banking compliance requirements are higher than those of conventional banking. For example, the establishment of conventional banking and Islamic banking must comply with the provisions of the Companies Act 1965[1] in addition to the licensing requirement of the central banks under the licensing requirement under the Malaysian Central Bank of Malaysia Act 2009. It is also worth mentioning that the board of directors' duties and obligations are statutorily spelled out in the Act. In addition, these obligations are read together with the Corporate Governance Standards issued on the 3 August 2016.[2] This Standard[3] applies to licensed banks, licensed investment banks, licensed Islamic banks, licensed insurers, licensed takaful operators and financial holding companies. This policy document sets out the central bank's expectations for financial institutions to have in place effective corporate governance arrangements consistent with the long-term viability of a financial institution (Bank Negara Malaysia, 2024).

The board responsibilities outlined in this policy document should be read together with section 56 of the Financial Services Act 2013 (FSA) and section 65 of the Islamic Financial Services Act 2013 (IFSA). The board must have a board charter that sets out the mandate, responsibilities and procedures of the board and the board committees, including the matters reserved for the board's decision. In addition, this policy document is issued pursuant to– (a) sections 47(1) and 58(4) of the Financial Services Act 2013 (FSA); and (b) sections 29(2), 57(1) and 67(4) of the Islamic Financial Services Act 2013 (IFSA).

However, the liabilities of the Board of Islamic financial institutions (IFIs) further added up in the Shariah Governance Framework (SGF) for their ultimate accountability and responsibility for the overall Shariah governance, namely to perform diligent oversight over the effective functioning of the Financial Institution's SGF, to ensure that these guidelines are effective and are commensurate with the size, complexity and nature of its business.

Apart from the SGF, the Board is responsible for ensuring compliance with every guideline issued. For example, the Late Payment Guidelines require the attestation by the Board of Directors that all requirements of the guidelines have been met including governance requirements in paragraph 7 of the Guidelines. Paragraph 7 states that the Board of Directors (Board) must ensure that the IFI has satisfied all the requirements of the guidelines, including putting in place appropriate controls and safeguards to monitor and determine the *ta'widh*, *gharamah* and combined rates, and ensuring capability of the information system and the IFI's compliance with Shariah rulings. On top of these guidelines, the Board is statutorily responsible for ensuring compliance with other guidelines such as the Introduction of New Product Guidelines and Istisna's Guideline. For example, the Istisna Guideline issued by Bank Negara Malaysia states that the roles and responsibilities of the Board with respect to istisna' transaction must include the following:

1 Set the IFI's business strategy and risk appetite with regard to the application of the istisna' contract;
2 Approve the IFI's internal policies and oversee the internal procedures for effective risk management to ensure compliance with the relevant laws and regulations;

3 Establish appropriate systems to implement the application of istisna' contract consistent with Shariah requirements;
4 Ensure that the IFI has adequate and qualified personnel with sufficient knowledge and competency on the concept, application and risks associated with the istisna' contract; and
5 Ensure independent reviews are conducted regularly to assess compliance with the policy documents issued by the bank and relevant internal policies established by the IFI.

It is inferred that the above statutory duties and responsibilities as double compliance for the IFIs. On top of compliance with legal requirements, the IFIs require another compliance in Shariah. The existence of layers in the world banking system and the connection between those layers have created different layers of compliance (Figure 1.3).

Based on Figure 1.3, the Standards Basel Core Principles (SBCB) states the Central Bank has to impose penalties for non-compliance of the Basel requirement. The assessment methodology can be used in multiple contexts: (1) self-assessment performed by banking supervisors themselves; (2) IMF and World Bank assessments of the quality of supervisory systems, for example, in the context of FSAP; (3) reviews conducted by private third parties such as consulting firms; or (4) peer review conducted, for instance, within regional groupings of banking supervisors.[4]

In order to achieve full objectivity, compliance with the Core Principles is best assessed by suitably qualified external parties consisting of two individuals with strong supervisory backgrounds who bring varied perspectives to provide checks

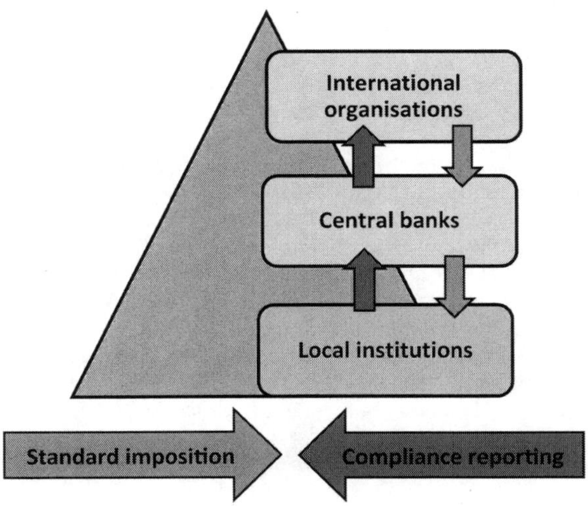

Figure 1.3 The connection between compliance and reporting between Layers 1, Layer 2 and Layer 3.

and balances; however, experience has shown that a recent self-assessment is a highly useful input to an outside party assessment.

Based on Figure 1.3, the Accounting and Auditing Organization for Islamic Financial Institutions (AAIOFI) and IFSB are not within the framework of the banking systems. Imposing Islamic banks to comply with the standards of these two organisations may lead to double compliance. For example, in Malaysia and Brunei, their Shariah standards are issued strictly by the Shariah Advisory Council (SAC) of Malaysia and the Shariah Financial Supervisory Board (SFSB) Brunei. Non-compliance with their SAC and SFSB may result in legal penalties.

Notes

1 Within Malaysian context.
2 Within Malaysia Context.
3 Or interchangeably used with the Policy Documents.
4 So far, the assessment involving 150 countries. The Basel Committee on Banking Supervision has issued guidelines for performing self-assessments: Conducting a supervisory self-assessment– practical application, Basel, April 2001. The regular reports by the IMF and the World Bank on the lessons drawn from assessment experiences as part of FSAP exercises constitute a useful source of information which has been used as an input to improve the principles.

References

BIS, Bank of International Settlements. (2020). The Basel Core Principles - Executive Summary. https://www.bis.org/fsi/fsisummaries/bcps.pdf [Accessed 13 March 2023].

BNM, Bank Negara Malaysia. (2024). Corporate Governance. https://www.bnm.gov.my/documents/20124/963937/PD-Corporate_Governance.pdf [Accessed 10 April 2024].

Bordo, M. D. (1993a). The Bretton Woods international monetary system: A historical overview. *In A retrospective on the Bretton Woods system: Lessons for international monetary reform* (pp. 3–108). University of Chicago Press.

Bordo, M. D. (1993b). *The gold standard, Bretton Woods and other monetary regimes: An historical appraisal. National Bureau of Economic Research, Working Paper* No: 4310. https://www.nber.org/system/files/working_papers/w4310/w4310.pdf [accessed 27 March 2024].

Buiter, W. H. (2014). *The role of central banks in financial stability: How has it changed?* Pp. 11–56. World Scientific. https://doi.org/10.1142/9789814449922_0002.

Engku Ali, E. R. A., & Oseni, U. A. (2017). Towards an effective legal and regulatory framework for Islamic financial transactions: Major initiatives of the Central Bank of Malaysia. *International Journal of Law and Management*, *59*(5), 652–672.

Goodhart, C. (2011). *The Basel Committee on banking supervision: A history of the early years* 1974–1997. Cambridge University Press.

Kerwer, D. (2005). Rules that many use: Standards and global regulation. *Governance*, *18*(4), 611–632. https://doi.org/10.1111/j.1468-0491.2005.00294.x.

Khatat, M. E. H. (2016). *Monetary policy in the presence of Islamic banking*. International Monetary Fund.

King, P., & Tarbert, H. (2011). Basel III: An overview. *Banking & Financial Services Policy Report*, *30*(5), 1–18.

Ozdemir, B. (2018). Evolution of risk management from risk compliance to strategic risk management: From Basel I to Basel II, III and IFRS 9. *Journal of Risk Management in Financial Institutions*, *11*(1), 76–85.

Phillips, R. (1983). The role of the International Monetary Fund in the post-Bretton Woods era. *Review of Radical Political Economics*, *15*(2), 59–81. https://doi.org/10.1177/048661348301500203

Rothbard, M. (1995). Fractional reserve banking. *The Freeman*, *45*, 624–27.

Turner, A. (2017). *Between debt and the devil: Money, credit, and fixing global finance*. Princeton University Press.

Tymoigne, E. (2014). Modern money theory, and interrelations between the treasury and central bank: The case of the United States. *Journal of Economic Issues*, *48*(3), 641–662.

Yaacob, H., & Rani, A. N. B. H. A. (2018). *Regulated and regimented interest in the financial and economy system: Issues and challenges. In Proceedings of the 1st International Conference on Islamic Economics, Business, and Philanthropy (ICIEBP 2017) - Transforming Islamic Economy and Societies,* (pp. 10–16). Science and Technology Publication. https://www.scitepress.org/Papers/2017/70759/70759.pdf [accessed 02 January 2024].

2 Islamic Financial Regulations in ASEAN

2.1 Introduction

This chapter outlines financial regulations in popular Southeast Asian economies namely Malaysia, Indonesia, Brunei, Singapore, Philippines and Thailand. The authors explain currently established regulations for the growth of Islamic financial institutions (IFIs) in countries having well-established Islamic finance industries as well as review how different financial regulations support the establishment of Islamic financial systems in countries such as Singapore, the Philippines and Thailand. The chapter presents a detailed overview of different regulators and their role in promoting and strengthening Islamic finance industries in Malaysia, Indonesia and Brunei.

2.2 Malaysian Islamic Finance Industry and Regulatory Provisions

Malaysia applies a dual banking system. The recognition of the dual banking system is written clearly in the Central Bank of Malaysian Act (CBMA) 2009. Islamic banking in Malaysia started in 1963 with the establishment of Lembaga Tabung Haji and Bank Islam Malaysia Berhad (BIMB) in 1983. In the same year, the law to regulate Islamic banking was issued known as Islamic Banking Act 1983. BIMB started to fully operate as an IFI on 1 July 1983 and was listed on the Kuala Lumpur Stock Exchange in 1991. The Interest-Free Banking Scheme was introduced in 1993 and later replaced with the Islamic Banking Scheme. SAC was established in 1997 (Ariff, 1989). BIMB was licensed under the Islamic Banking Act (IBA) 1983 with 10 years exclusivity. The Government Investment Act (GIA) was enacted in 1983 to facilitate the liquidity management of the BIMB. Shariah Advisory Body (SAB) must be set up before a license can be granted to Islamic banks as a requirement for licensing (Section 3 of IBA 1983).

In 1984, Takaful was introduced as an alternative to conventional insurance and to facilitate Islamic banking activities. In 1993, the interest-free Banking Scheme known as *Skim Perbankan Tanpa Faedah* (SPTF) was introduced by the Central Bank of Malaysia (CBM) (Alam et al., 2023). This was initiated by the three pioneering banks (UMBC, Maybank, and Bank Bumiputra Malaysia Berhad). SPTF was applied through Islamic windows by the participating conventional banks. In 1996, Section 124 was added to the Banking and Financial Institution Act (BAFIA)

DOI: 10.4324/9781003471035-2

1989 to accommodate the establishment of Islamic windows within conventional banks. In 1998, the Islamic Banking Scheme (Skim Perbankan Islam [SPI]) was introduced replacing SPTF. In 1999, Bank Muamalat Berhad was set up as the second bank licensed under the IBA 1983. In 2003, Section 13B was inserted into the IBA 1983. Simultaneously, Section 16B was inserted into the CBMA 1958. The effects of this new insertion were as follows;

1 Reference to SAC was made as advisory;
2 Section 13B of IBA 1983 allows the Islamic banks to seek advice from the SAC in shariah matters;
3 Section 16B of CBMA 1958 makes it optional and non-compulsory for the court or the arbitrator to refer to any directives or SAC on shariah matters;
4 Should a reference be made by the court, the SAC's ruling will bind the arbitrator and remain persuasive in court.

The International Islamic Banking Business (IIBB) began in 2005 and granted approval to a foreign Islamic bank Kuwait Finance House (KFH) to establish in Malaysia. IIBB was introduced in parallel to the trade liberalisation policy. Conventional banks participating in the Islamic banking business were encouraged to apply Islamic subsidiary concepts to avoid legal impediments in their activities. In the same year, the Islamic subsidiary concept was first applied by RHB Islamic. In 2009, the CBMA was enacted replacing the previous Central Bank Act 1958 (Kunhibava, 2012). Reference to SAC and any rulings by SAC are made compulsory upon courts and arbitrators. After decades of being regulated under the post-judgment interest through the High Court Rules, the Court Rules Act 2012 recognised the Inclusion of *Ta'widh* and *Gharamah* provision replacing the imposition of post-judgment interest on the Islamic financial contracts. The Minister has appointed 1 March 2013 as the date of starting operation of the Subordinate Courts (Amendment) Act 2010 [A1382]. The lower court's monetary jurisdiction is increased up to one RM1 million for sessions court. Retail cases involving Islamic Financial contracts are to be heard by the lower courts.

2.2.1 Regulators

There are three regulators governing financial activities in Malaysia. First is the CBM (Bank Negara Malaysia) to govern the Takaful and Islamic banking. Second, the Securities Commission governs the Islamic Capital Market and other securities issuances. The third regulator is the Labuan Offshore aimed at governing all the financial activities in Labuan (Ariff & Rosly, 2011). The Securities Commission is regulated under the Securities Commission Malaysia Act 2015. Islamic Capital Market is regulated under the Capital Market Services Act 2007 amendment 2012. Labuan Offshore is regulated under the Labuan Financial Services Authority Act 1996 (Act 545)]. Labuan Islamic Financial and Securities Services Act (LFSA) 2010 was introduced to regulate the market conduct. The role and function of these regulators are briefly discussed below.

2.2.1.1 Bank Negara Malaysia

Bank Negara Malaysia (the CBM) *aka* BNM is a statutory body that started operating on 26 January 1959 (Sheng, 1989). BNM is governed by the CBMA 2009. The role of BNM is to promote monetary and financial stability which aims to provide a conducive environment for the sustainable growth of the Malaysian economy. BNM's monetary policy stance is to maintain price stability while remaining supportive of growth. It is also responsible for the stability of the financial system. This is achieved by developing a sound, resilient, progressive and diversified financial sector which serves to support the sectors of the real economy. It also plays an important function in implementing initiatives to deepen and strengthen the financial markets, including the foreign exchange market.

BNM has played a significant developmental role in developing the financial system infrastructure and advancing the financial inclusion agenda (Bank Negara Malaysia, 2022). This is to ensure all economic sectors and segments of society have access to financial services. In addition, it also oversees the nation's payment systems infrastructure which emphasises the efficiency and security of the financial systems. As the banker and adviser to the Government, BNM provides advice on macroeconomic policies and the management of public debt. It is also the sole authority in issuing the national currency and in managing the country's international reserves.

2.2.1.2 Securities Commission

The Securities Commission Malaysia (SC) is a statutory body entrusted with the responsibility of regulating and systematically developing Malaysia's capital markets. It has direct responsibility for supervising and monitoring the activities of market institutions and regulating all persons licensed under the Capital Markets and Services Act 2007 (Mohamed Sadique et al., 2010). Established on 1 March 1993 under the Securities Commission Act 1993, the SC is a self-funding statutory body with investigative and enforcement powers. Prior to the establishment of the SC, the capital market was regulated by six government bodies, namely:

- Capital Issues Committee (CIC), Ministry of Finance;
- Panel on Takeovers and Mergers, Prime Minister's Department;
- Foreign Investment Committee, Prime Minister's Department;
- Companies Commission of Malaysia;
- Ministry of International Trade and Industry (MITI); and
- Bank Negara Malaysia (BNM/Central Bank)

It reports to the Minister of Finance and its accounts are tabled in Parliament annually. The SC's many regulatory functions include:

1 Supervising exchanges, clearing houses and central depositories;
2 Registering authority for prospectuses of corporations other than unlisted recreational clubs;
3 Approving authority for corporate bond issues;

4 Regulating all matters relating to securities and derivatives contracts;
5 Regulating the take-overs and mergers of listed corporations, public companies, and entities as specified by the SC;
6 Regulating all matters relating to unit trust schemes;
7 Licensing and supervising all licensed persons;
8 Encouraging self-regulation; and
9 Ensuring proper conduct of market institutions and licensed persons.

2.2.2 Islamic Banking Laws in Malaysia

On 30 June 2013, the Financial Services Act 2013 (FSA) and the Islamic Financial Services Act 2013 (IFSA) came into effect by substituting and repealing the Banking and Financial Institutions Act 1989, the Insurance Act 1996, the Payment Systems Act 2003, the Exchange Control Act 1953, the Islamic Banking Act 1983 and the Takaful Act 1984 (Ariff, 1989). The objectives of these Acts were to empower BNM to counter future risks related to financial stability in the financial sector, increase consumer protection, promote competition in the broader financial services sector and step forward towards global trends in financial regulations.

2.2.2.1 Islamic Financial Services Act 2013

In order to regulate the Islamic banking industry, Malaysia introduced the Islamic Financial Services Act 2013. This Act replaced the Islamic Banking Act 1983 and the Takaful Act 1984. This is the first Act introduced in the Malaysian Parliament under the Federal Law applying shariah as the governing authority in the banking system. The Act also introduced several Arabic words pertaining to Islamic banking products such as mudharabah, musharakah, ijarah, musharakah mutanaqisah, wakalah, etc. IFSA 2013 brought a new dimension to defining shariah and recognising the growth of Islamic banking in Malaysia.

2.2.2.2 Financial Services Act 2013

FSA 2013 was introduced on the same date as IFSA 2013. The key difference between the FSA and the IFSA lies in the introduction of a new Part IV of the IFSA to strengthen shariah governance whereby IFIs shall ensure end-to-end shariah compliance with regard to their policies, procedures and operations (Kamaruddin et al., 2020). The wording carries verbatim in nature, however, with the additional shariah in IFSA for all acts and operations. The Act allows the operation of Islamic windows. Once they have operated as a window, the unit will automatically be regulated under the IFSA 2013.

2.2.2.3 Central Bank Malaysia Act 2009

The CBM derives its power from the CBMA 2009. Their power of supervising and monitoring any financial activities including Islamic banking and conventional is embedded in the Act (Hasan, 2010). The amendment was made to the CBM.

Recently, Malaysia emerged as one of the first countries allowing digital currency by virtue of Malaysia among the first countries that regulate digital currencies. The Anti-Money Laundering and Counter Financing of Terrorism (AML/CFT) – Digital Currencies (Sector 6) that was issued by the Financial Intelligence and Enforcement Department which refers to the digital representation of value that (1) functions as a medium of exchange and (2) is interchangeable with any money (including through the crediting or debiting of an account) but excluding electronic money, as defined under the FSA 2013 [Act 758] and the IFSA 2013 [Act 759], issued by an approved issuer of electronic money under those Acts. The term uses the word "digital currencies" and not virtual currencies. By excluding e-money, the Malaysian definition falls under type 3 as proposed by the ECB. Malaysia adopts a model of the Financial Action Task Force (FATF). The virtual currency is divided into two basic types: convertible and non-convertible virtual currency. Although the paper uses "non-convertible" and "closed," and "convertible" and "open" as synonyms, it should be emphasised that the notion of "convertible currency" does not in any way imply an ex officio convertibility (e.g. in the case of the gold standard), but rather a de facto convertibility (e.g. because a market exists). Thus, a virtual currency is "convertible" only as long as some private participants make offers and others accept them since the "convertibility" is not guaranteed at all by law.

2.3 Negara Brunei Darussalam

2.3.1 Islamic Banking System in Brunei Darussalam

Brunei Darussalam's banking system is characterised by a dual banking system consisting of Islamic banks and conventional banks (Ali, 2018). The banking system consists of eight banks comprising one Islamic Bank and seven conventional banks, as well as an Islamic Trust Fund which is set up under its statute. Out of the eight banks, three are international, three are regional and two are domestic. In addition, there are three licensed finance companies, two are conventional and one is Islamic. The three finance companies are wholly owned subsidiaries of three licensed banks in Brunei Darussalam. There are four offshore banks operating within the framework of the International Banking Order of 2000. There is only one licensed pawnbroker in Brunei Darussalam.

2.3.2 The Regulators

His Majesty the Sultan of Brunei Darussalam has consented to the Autoriti Monetari Brunei Darussalam Order, 2010 which commenced on 1 January 2011. The Autoriti Monetari Brunei Darussalam Order, 2010, provides for the establishment of Autoriti Monetari Brunei Darussalam (AMBD), its Board and matters connected to the objects, operation, administration, functions, powers and duties of AMBD including relationships between AMBD and the government, AMBD and banks and financial institutions, as well as consequential and related amendments to other written laws that govern the activities supervised by AMBD.

His Royal Highness Prince Haji Al-Muhtadee Billah, the Crown Prince and Senior Minister at the Prime Minister's Office in his capacity as the Chairman of AMBD consented to officiate the Establishment of AMBD and chaired the First Meeting of the AMBD Board of Directors, on 10 January 2011 at AMBD Head Office located in Bandar Seri Begawan.

The growth of Islamic finance in Brunei can be categorised into three types (Shahid Ebrahim & Kai Joo, 2001). The capital market, Takaful and banking. Applying the dual banking system equipped Brunei with strategic equilibrium. Bruneian banks have a reputation for high capital adequacy ratios (CAR). In 2015, the banking sector's average CAR was 21.5%, which is well above the minimum regulatory requirement of 10%. Current minimum (quantum of) capital requirements for locally incorporated banks and overseas incorporated banks are BND100 million and BND1 billion, respectively. These serve to ensure that banks operating in Brunei Darussalam are financially robust. The quality of Brunei Darussalam's banks is reflected by their unanimous investment-grade credit ratings (Brunei Darussalam Central Bank, 2016). 2016 Statistics show that the financial sector contributes to 3% of total GDP (Zulhazmi, 2015). Recent statistics by the Department of Economic Planning and Development show that the financial sector's share of GDP is around 5.1%. Building upon the platform developed and leveraging the myriad of opportunities in Asia – the world's fastest-growing region – *Wawasan 2035* envisages that, by 2035, the financial sector's contribution to GDP will have expanded to 8% of GDP (Brunei Darussalam Central Bank, 2016). To maintain their existing credit ratings, Bruneian-owned and incorporated banks need to be demonstrably well-capitalised and maintain CAR at levels above the minimum international threshold (8% of risk-adjusted exposures) (Brunei Darussalam Central Bank, 2016). The growth of the regulatory framework of Islamic finance or finance generally in Brunei can be divided into three phases as follows:

1 The first phase from 1935 to 1980 (45 years)
2 The second phase from 1981 to 2011(30 years)
3 The third phase from 2011 to preset

According to a Spanish traveller, Antonio Pigafetta, the greatness and glory of the Brunei Empire in the 15th century was due to the expansion of the Brunei Empire and was in line with the development of the Brunei foreign trade mainly to meet the demand of the Chinese and European traders to Borneo Forest products and so on (Wright, 1977). In general, these developments can be divided into four phases. The development of Muamalat in Brunei Darussalam dates back to the codification of Brunei Canon Law under the sovereignty of Sultan Hassan (1582–1598) which contains 46 Sections with six parts (Venardos, 2010). The Canon Law of Brunei marks a new milestone in financial transactions in Brunei. Founded as the apex of the law, the code delineated permission to conduct trade but forbade usury. The prohibition of usury is the ruling adopted from Islamic teaching. The code also explains the sales of land, mortgages and bankruptcy.

Islamic banking was first introduced in Brunei Darussalam in the early 1990s and has since seen tremendous growth. In 1993 (Abdul' Ali et al., 2014), the Islamic Bank of Brunei (IBB) Berhad was established. Baiduri Bank was established later in 1994. In 2015, it held a significant role in Brunei Darussalam's banking industry with total assets of BND8.94 billion and deposits totaling BND7.34 billion which accounted for 52.4% and 51.5% of the total market share. The operations of the licensed banks are governed by the provisions of the Banking Order 2006 and the Islamic Banking Order 2008. All fully licensed banks and finance companies operating in Brunei Darussalam are mandatory members of the Brunei Darussalam Deposit Protection Corporation which was established in January 2011 (Hassan et al., 2022). In 2005, the largest Islamic bank in Brunei was established known as Bank Islam Brunei Darussalam (BIBD). In 2006, Brunei issued a shariah Financial Supervisory Board (SFSB) Order (Yaacob et al., 2021). As per Section 3 of the SFSB Act 2006, the establishment of the Board known as the Syariah Financial Supervisory Board shall be the authority for the ascertainment of the Laws of Islam for the purposes of Islamic banking business, takaful business, Islamic financial business, Islamic development financial business and any other business which is based on Syariah principles and which is supervised and regulated by the authority. The SFSB is regulated by the Ministry of Finance (Yaacob et al., 2021). The functions of the Board outlined in Figure 2.1 are as follows:

1 The Board shall have such functions as may be determined by the authority.
2 Every financial institution may consult the SFSB on shariah matters relating to the Islamic banking business, takaful business, Islamic financial business, Islamic development financial business and any other business that is based on shariah principles.
3 The relationship between the SFSB and the SAB is best described in Figure 2.4.

The governance of IFIs is supervised by the Ministry of Finance which may issue any written directives in relation to Islamic banking business, takaful business,

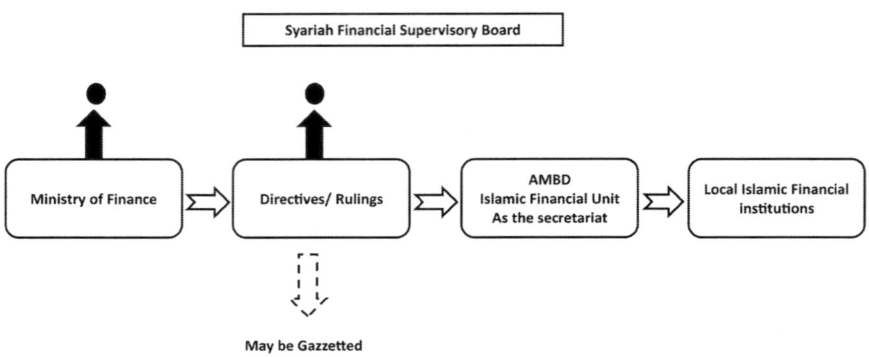

Figure 2.1 Hierarchy of Shariah rulings in Islamic finance.

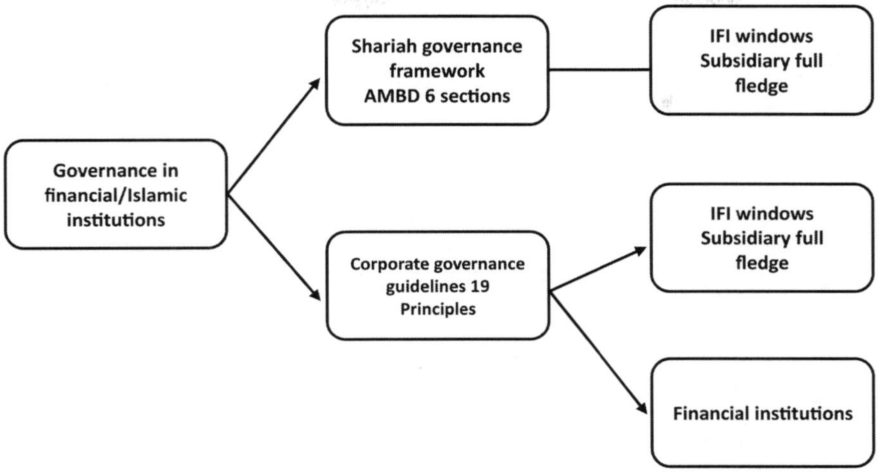

Figure 2.2 Governance systems in Islamic Banks Brunei.

Islamic financial business, Islamic development financial business and any other business with the advice of the Board (Haqqi, 2017). The governance mechanism used by the Ministry of Finance Brunei Darussalam is shown in Figure 2.2. Any financial institution which fails to comply with or contravenes any written directives issued under subsection (1) is guilty of an offence and liable on conviction to a fine not exceeding \$500,000 and to a further fine not exceeding \$5,000 for every day during which the contravention continues (Hassan et al., 2022).

Islamic banks in Brunei adopt dual governance systems, i.e., compliance with the corporate governance Guideline and the shariah governance framework. The tiers of shariah governance are represented in Figure 2.3.

Any proceedings in the court or arbitration regarding Islamic banking business, takaful operation, Islamic banking development or any business conducted following shariah principles should be referred to the respective Authority. The arbitrator and the court may consider the Board's rulings while seeking guidance related to the shariah merits of the case. This means there are two ways of reference whenever involves court or arbitration proceedings as envisaged by the Order are as follows:

- Reference to any written Directives of the Ministry of Finance
- References to the Ministry of Finance for a ruling from the Board.

The Syariah Financial Supervisory Board may, with the approval of the Ministry of Finance, cause any decision of the Board to be published by notification in the Gazette The systematic decision-making process about Islamic finance cases used by the shariah courts is sketched in Figure 2.4.

The establishment of BIBD was made possible by virtue of the Brunei Darussalam Companies Act, Chapter 39 ("the Act"). The operation of BIBD as an Islamic

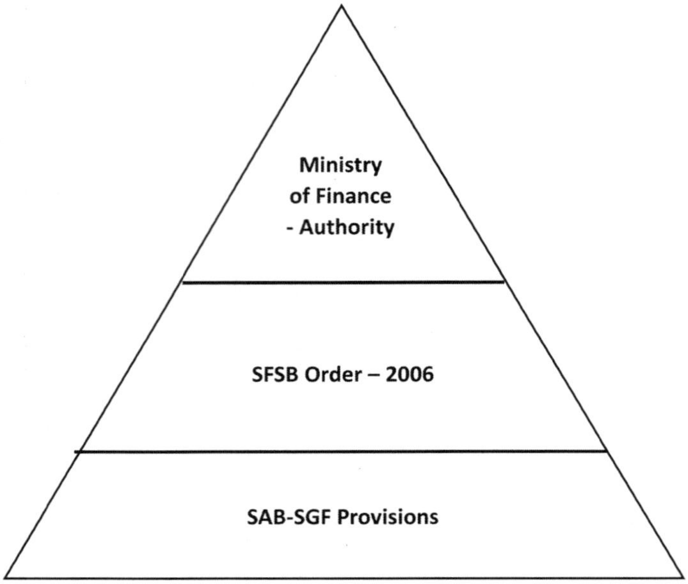

Figure 2.3 The dual governance system of IFIs in Brunei Darussalam.

bank is made in line with the Islamic Banking Order 2008. The Islamic Banking Order 2008 does not impose stringent requirements in monetary form as a penalty for shariah non-compliance. The interpretation of Hukum Shara' is tightened with strict interpretation in Section 2 of the Islamic Banking Order 2008. This interpretation is pivotal to avoid any plebeian to shariah.

The bank is principally engaged in the provision of Islamic banking business as allowed under the Islamic Banking Order, 2008 and shariah principles as advised by the SAB. The subsidiaries are principally engaged in the provision of investment banking, Islamic hire-purchase, stockbroking, asset and fund management, leasing and management services. In the year 2014, continuous successful implementation of several key-consumer products occurred which were based on the *Tawarruq* concept in delivering shariah-based financial solutions to customers.

Figure 2.4 The procedures and observance on Qaul Mu'tamad.

At the same time, BIBD has enhanced the subsidiary of BIBD At-Tamwil's consumer products, from the *AlIjarah Thumma Al-Bai'* (AITAB) shariah concept to *Al-Ijarah Al-Muntahiyah Bitamleek* (IMBT). The growth of Islamic banking in Brunei in phases three and four is summarised in Table 2.1.

Table 2.1 Phases of Islamic banking development in Brunei.

29 September 1991	Tabung Amanah Islam Brunei (TAIB)
1992	The International Bank of Brunei Berhad (IBB)
13 Jan 1993	Islamic Bank of Brunei Berhad
Undated	The government abolished the practice of paying interest in the government loan Scheme for the construction of houses.
1994	Baiduri Bank Berhad
1995	Money-Changing and Remittance Businesses Act, Chapter 174
March 1995	Development Bank of Brunei Berhad (DBB) was initially established as a conventional bank
4 April 2000	The Sultan decreed the DBB to operate on Islamic banking principles.
1 July 2000	Successful conversion of DBB to the Islamic Development Bank of Brunei (IDBB) as the second Islamic bank in Brunei
2000	International Banking Order
2000	The International Business Companies Order, 2000 (S 56/00)
2000	The International Limited Partnerships Order, 2000 (S 45/00)
2000	The International Trusts Order, 2000 (S 55/00)
2001	The Mutual Funds Order, 2001.
2001	The Securities Order, 2001
February 2001	IDBB opened its first subsidiary company, Takaful IDBB Sdn Bhd
2002	the International Insurance and Takaful Order, 2002 (S 43/02)
2002	IFSB established
2002	The Pawn Broking Oder
1 July 2003	Revised Finance Companies Act, Cap 89, 1972
2003	International Business Companies Order 2003
2004	Currency and Monetary Order-repealed the Currency Act 1967
1 February 2004	Brunei Currency Board was renamed as Brunei Currency and Monetary Board. Two new polymer notes for circulation consisting of B$50 and B$100 were introduced. The reverse design featured the nature theme as in the first polymer series of B$1, B$5 and B$10.
2005	BIBD, Brunei's largest bank and flagship Islamic financial institution, was formed in 2005 through the merger of IBB and IDBB
2006	Syariah Financial Supervisory Board Order was enacted
2006	*Banking Act 1957 Repealed by Section 131 of the Banking Order, 2006 [S 45/2006] w.e.f. 04-03-2006*
2006	Banking Order
2006	Hire Purchase Order
2006	Insurance Order
25 April 2006	In exercise of the power conferred by subsection (1) of Section 11 the Minister of Finance, with the approval of His Majesty the Sultan and Yang Di-Pertuan, hereby appoints the 4 March 2006 as the date on which the Banking Order, 2006 shall commence.

(*Continued*)

Table 2.1 (Continued)

2008	Islamic Banking Order
2010	Authority Monetari Brunei Darussalam (AMBD) Order 2010 and effective on 1 January 2011.
2010	Deposit Protection Order, 2010 (DPO, 2010)
2011	Brunei Darussalam Deposit Protection Corporation (BDPC) to administer the deposit protection scheme aimed at protecting depositors against the loss of their deposit in the unlikely event the member institution (banks/finance companies) is unable to meet its obligations to depositors.
2011	Autoriti Monetari Brunei Darussalam (AMBD) was established, adopting the AMBD Order 2010
2011	Anti-Terrorism Order, 2011 came into force
2012	The establishment of the Credit Bureau (CB) Unit and the launching of the Credit Report Service
2012	Criminal Asset Recovery Order, 2012, came into force
2013	Anti-Terrorism (Terrorist Financing) Regulations, 2013 came into force
2013	Securities Markets Order, 2013, came into force which repealed the Securities Order, 2001 and the Mutual Funds Order, 2001.
2013	Brunei Darussalam Islamic Investment Summit
2013	The Chairmanship of the ASEAN Central Bank Governors' Meeting
2013	Centre for Islamic Banking, Finance and Management (CIBFM) came under the purview of AMBD
2014	The National Real-Time Gross Settlement System was launched
2014	The Chairmanship of the IFSB Council
2014	Brunei Darussalam became a member of the Egmont Group of Financial Intelligence Units
2014	The Chairmanship of the ASEAN Insurance Regulator's Meeting (AIRM)
2014	Self-Inquiry & Dispute Resolution (SIDR) introduced by AMBD
Guideline No: ITS/ G-1/2014	Guideline on Registration of General Insurance Agents
Guideline No: ITS/ G-3/2014	Guideline on Fit and Proper Criteria for Key Responsible Persons in Insurance and Takaful
2015	Securities Markets Regulations, 2015, came into force
2015	Payment and Settlement Systems (Oversight) Order, 2015; Payment and Settlement Systems (Finality and Netting) Order and Bills of Exchange Act (amendment) came into force
2015	Commenced National Risk Assessment (NRA) in Brunei Darussalam
2016	Automated Clearing House goes live
2016	Collateral Registry – Secured Transaction Order, 2016
2016	AMBD signed the International Organization of Securities Commissions (IOSCO) Multilateral Memorandum of Understanding (MoU) concerning Consultation and Cooperation and the Exchange of Information

Source: (Authors', 2024).

2.3.2.1 Historical Development of AMBD

The establishment of AMBD signifies yet another important milestone in the development of the financial sector in Brunei Darussalam. It also underscores the commitment of the Government of His Majesty the Sultan and Yang Di-Pertuan Negara Brunei Darussalam in achieving and maintaining a sound and dynamic financial system; by implementing appropriate measures and embarking on necessary reforms in the financial sector to support the objectives of national economic development.

AMBD is a statutory body, acting as the central bank of Brunei Darussalam which undertakes several core functions, chief of which is the formulation and implementation of monetary policies, the regulation and supervision of financial institutions as well as currency management.

Four divisions previously under the Ministry of Finance merged to form AMBD, namely:

1 The Financial Institutions Division or FID
2 The Brunei Currency and Monetary Board or BCMB
3 The Brunei International Financial Center or BIFC
4 Part of the Research and International Division or RID

The principal objectives of the AMBD are as follows:

1 To achieve and maintain domestic price stability;
2 To ensure the stability of the financial system, in particular by formulating financial regulation and prudential standards;
3 To assist in the establishment and functioning of efficient payment systems and to oversee them;
4 To foster and develop a sound and progressive financial services sector.

2.3.2.2 Monetary Policy Framework in Brunei

The Monetary Policy Framework in Brunei Darussalam is based on a currency board system with the Brunei dollar anchored to the Singapore dollar at par. Brunei Darussalam also has a Currency Interchangeability Agreement (CIA) with Singapore that was signed in 1967 and this agreement continues to remain in effect between AMBD and the Monetary Authority of Singapore (MAS) (Lee & Azali, 2010).

Banking and specialised markets are primarily responsible for licensing, supervision and regulation of banks and finance companies (both Islamic and conventional), offshore banks, money changers and money remittance companies and pawnbrokers. In fulfilling its mandate, the division administers and enforces the following legislation:

- Banking Order, 2006
- Islamic Banking Order, 2008
- International Banking Order, 2000
- Finance Companies Act, Chapter 89

- Hire Purchase Order, 2006
- Pawnbrokers Order, 2002
- Money-Changing and Remittance Businesses Act, Chapter 174

In discharging its responsibilities in the above regard, it strives to benchmark itself to the relevant international standards such as the BCP set by the BCBS and those of the IFSB.

2.3.2.3 Establishing an Islamic Bank in Brunei Darussalam

In considering whether to grant a licence, the Authority shall have regard to:

a The need to protect the public interest or the interests of depositors, customers, other creditors and potential depositors, customers and other creditors; and
b The need to protect the security, reputation and economic interests of Brunei Darussalam in relation to financial matters
  i The Authority shall not grant a licence unless it is satisfied that:
  ii The applicant enjoys a high reputation in the financial community or will be able to do so if a licence is granted;
  iii Every person who is to be a director, controller, or manager of the business is a fit and proper person to hold that position;
  iv The applicant will carry on banking business in Brunei Darussalam and, in carrying on the business, will at all times maintain adequate accounting and other records and adequate systems of control of its business and records.

The statistics of the minimum capital paid to establish an Islamic bank in Brunei Darussalam are presented in Table 2.2.

In the case of the minimum paid-up capital, the foreign bank branches in Brunei Darussalam (Both conventional and Islamic banks) Net Head Office Funds should be B$30 million.

Islamic Finance has experienced rapid growth over the past decade. It has emerged as an effective tool for financing development worldwide, including in non-Muslim countries (Saiti et al., 2022). Islamic finance has already been mainstreamed within the global financial system namely, Nigeria, South Africa and Kenya through the existence of Islamic banking and takaful firms in these

Table 2.2 Minimum capital requirements for establishing an Islamic bank in Brunei Darussalam.

Banks	*Minimum paid-up capital*
Head Office in Brunei Darussalam	B$100 million
Head Office outside Brunei Darussalam	B$1,000 million
Conventional banks	B$500 million
Islamic banks	

Source: (BDCB, 2024).

countries. Brunei Darussalam should also take advantage of this opportunity to diversify its economy.

The rapid development of the Islamic finance industry in Brunei Darussalam is very encouraging as being supported by His Majesty's aspiration to be the hub for Islamic finance (Damit, 2007). The first Islamic trust fund was the establishment of *Perbadanan Tabung Amanah Islam Brunei (TAIB)* in 1991. This was the starting point for Brunei's Islamic finance industry with a decree of His Majesty the Sultan of Brunei Darussalam on 25 September 1990 in a meeting of the Islamic Council.

> If we are not able to do something, it does not mean that we just sit still and do nothing. If we are unable to because we do not have skilled labor or the expertise, and yet we are in dire need to meet the demands, say for example, of Fardhu Kifayah, then what is wrong with following the international practice or traditions of nations by the mutual exchange of information or undergoing training or, if necessary, the hiring of experts on temporary employment with us to help us make our projects succeed the way we plan and desire? For example, [in] the matter concerning the Islamic financial system, the most popular nowadays is the Islamic bank. It is no longer a dream of imagination, but we are informed that there are already more than fifty Islamic commercial banks in Islamic countries. This is also one of the obligations of Fardhu Kifayah for each of the said Islamic countries, including our State of Brunei Darussalam…
>
> (Siddique, 1992)

In 2023, Brunei Darussalam was ranked 9th in the latest Global Islamic Economy Indicator (GIEI), the indicator evaluated the relative strengths of the ecosystem that a country has to support the development of the Islamic economy which also includes Islamic finance. Brunei Darussalam's Islamic financial ecosystem now comprises Islamic banks, takaful operators, Islamic investment dealers, Sukuk and other ancillary services (Islamic Financial Services Board, 2023).

There are efforts underway to deepen the Islamic financial market in Brunei Darussalam, which will be further explained in this chapter. The finance sector in Brunei Darussalam is one of the sources of economic production, which contributes to GDP growth.

2.3.2.4 Establishment of SFSB

To strengthen the Islamic finance industry, a proper legal and regulatory framework is needed. Therefore, the SFSB Order was introduced in 2006 to enhance shariah governance and supervision and accords the board the power to make shariah rulings on Islamic transactions. The first Islamic banking legislation to regulate the Islamic banking industry is the Islamic Banking Act of Chapter 168, an act to provide for the licensing and regulation of Islamic banking business. In 2008, the Ministry of Finance introduced the Banking Order 2008, which intended to expand the Islamic banking sector by strengthening the supervision and regulation of Islamic banking activities. This also aims to attract foreign participants to strengthen the

Table 2.3 List of the financial institutions supervised by the Central Bank of Brunei Darussalam.

Types of Institutions	*No. of institutions*	*Name of institutions*
Islamic Bank	1	BIBD
Islamic Trust Fund	1	Perbadanan TAIB
Islamic Finance Companies	1	BIBD At Tamwil Berhad
General Takaful	2	Insurans Islam TAIB General Takaful Sdn Bhd Takaful Brunei Am Sdn Bhd
Family Takaful	2	Insurans Islam TAIB Family Takaful Sdn Bhd Takaful Brunei Keluarga Sdn Bhd
SME Bank	1	Bank Usahawan
Islamic Window	1	Standard Chartered

Source: (Authors, 2024).

industry as a whole and to nurture local knowledge and expertise. In the same year, Takaful Order 2008 was introduced. The order is a new comprehensive legislation to govern the activities of the domestic takaful industry and provide the legal platform for a level playing field between conventional insurers and takaful operators to carry out Family Takaful and General Takaful.

2.3.2.5 Central Bank

In 2010, the foundation for the establishment of a strong regulator for the Bruneian Islamic finance industry was created. The Autoriti Monetari Brunei Darussalam Order 2010 was passed to regulate the financial institutions (Table 2.3) in Brunei. Since the establishment of the AMBD, matters concerning Islamic finance have been looked and regulated by them. AMBD is the central bank in Brunei Darussalam and is responsible to oversea the operations of banking including Islamic and conventional segments and insurance industries. It is also responsible to reduce redundancy and bureaucracy.In the same year, the Centre for Islamic Banking, Finance and Management was introduced under the purview of AMBD as it plays an important role in developing the human capacity in the areas of Islamic banking, finance and management. Furthermore, the Memorandum of Understanding was signed between Perbadanan TAIB and ICD to establish TAIB Bank in 2016. Hence currently, the Ministry of Finance Brunei supervises the following Islamic financial institutions.

2.3.2.6 Pawnbrokers

Currently, pawnbroking in Brunei is regulated under the Pawn Brokers Order Ch63. The basic criteria to apply licensing for pawnbrokers in Brunei Darussalam is as follows:

- To be conducted in accordance with Hukum Syara;
- Shall apply to all persons whether or not professing the Islamic religion;

- Any person may apply to the Licensing Officer for a license, on payment of an application fee of B$100;
- Licenses shall be granted annually and a fee of B$5,000 shall be charged for every license;
- To provide security to the satisfaction of and in a manner determined by the Authority amounting to B$100,000;
- No person can carry on the business of a pawnbroker without a license.

2.3.2.7 Money Changers and Remittance Businesses

The money changers and remittances are currently governed under the Money Changers and Remittance Businesses 1995, Ch174. In considering any application for a licence the Authority may require to be satisfied as to:

- The good character of the applicant or, if the applicant is a company, the general character of the management of the company;
- The financial condition of the applicant; and
- Whether the public interest will be served by the granting of the licence.

2.3.2.8 Capital Market and Services

The Capital Market Unit administers the Securities Markets Order, 2013 (SMO) and its supplementary legislation the Securities Markets Regulations, 2015 (SMR) (Hamdan & Hoon, 2019). The SMO and SMR provide a legal and regulatory framework for the licensing and supervision of market operators such as securities exchanges, clearing houses, trading facilities and credit rating agencies; capital market intermediaries such as dealers, fund managers and investment advisers; and securities products and the collective investment schemes.

The types of licences available for the capital market intermediaries are Capital Market Services Licence and Representative's Licence which allow licensees to conduct the following regulated activities:

- Dealing and arranging deals in investments as principal or agent;
- Managing investments including collective investment schemes and establishing, operating, or winding up a collective investment scheme;
- Giving or offering investment advice in his capacity as an investment adviser or a financial planner;
- Using computer-based systems for giving investment instructions; and
- Safekeeping and administration of assets including custodial services.

In order to facilitate the comprehensive determination of a borrower's creditworthiness, AMBD has, pursuant to Section 42A of the AMBD Order, 2010, established the Credit Bureau. The Credit Bureau is a unit under the Regulatory Department of AMBD and provides the service of collecting, collating and consolidating credit information in the form of a credit report.

Table 2.4 List of financial institutions and their parent companies.

Baiduri Bank Berhad	Malayan Banking Berhad (Maybank)
Baiduri Finance Berhad	United Overseas Bank Limited (UOB)
Bank Islam Brunei Darussalam (BIBD)	RHB Bank Berhad
BIBD At-Tamwil Berhad	Standard Chartered Bank
Hongkong and Shanghai Banking Corporation Limited (HSBC)	Perbadanan Tabung Amanah Islam Brunei (TAIB)
HSBC Finance (B) Berhad	

Source: (BDCB, 2024).

Since September 2012, this service has been available to all licensed banks (Table 2.4), finance companies and the Islamic Trust Fund in this country, who are Members of the Credit Bureau:

2.3.2.9 Dispute Resolution

Before 29 September 1959, the duties of Legal Adviser to the Brunei Government were performed by the Attorney General of Sarawak, who was assisted by a Deputy Legal Adviser stationed in Brunei (Skully, 1984). When the Constitution was proclaimed on 29 September 1959, the appointment of Attorney General of Brunei came into being and from that date, the Attorney General of Sarawak ceased to be the Legal Adviser to the Brunei Government. The Commercial court became fully operational in February 2016. The first case was registered on 2 February 2016 and as of 23 November 2016, the court has a record of 155 cases filed. Approximately 70% of the cases filed are suits arising out of breaches under banking and financial services. The remaining 30% deals with disputes under the sale and delivery of goods, breaches in development projects, breaches in tenancy agreements and sale and purchase of commodities. Out of the 155 cases filed, no appearance and/or defence has been filed in 77 cases resulting in the entry of Judgment in Default of Appearance and/or Judgment in Default of Defence.

The functioning of the Credit Bureau is presented in Figure 2.5. AMBD provides a comprehensive dispute resolution mechanism. Any information that the customer believes, satisfies any of the following elements in the Disputable Information of the Self-Inquiry Report:

1 The credit facility, guarantee, security or cheque dishonour either does not belong to the Subject or has been reported more than once or has erroneous information;
2 The closed credit facility as on the Report Date is still reported as active; or
3 Demographic (including individual or company profile), employment and relationship information either does not belong to the Subject or is erroneously reported. For the avoidance of doubt, error shall not include any contention raised by the Subject involving issues relating to the application of banking internal processes and methodologies such as interest calculations or levy of charges.

Based on Figure 2.5, the boundaries of regulatory growth in Islamic banking and finance in Brunei Darussalam are considerably fast. The first phase of growth

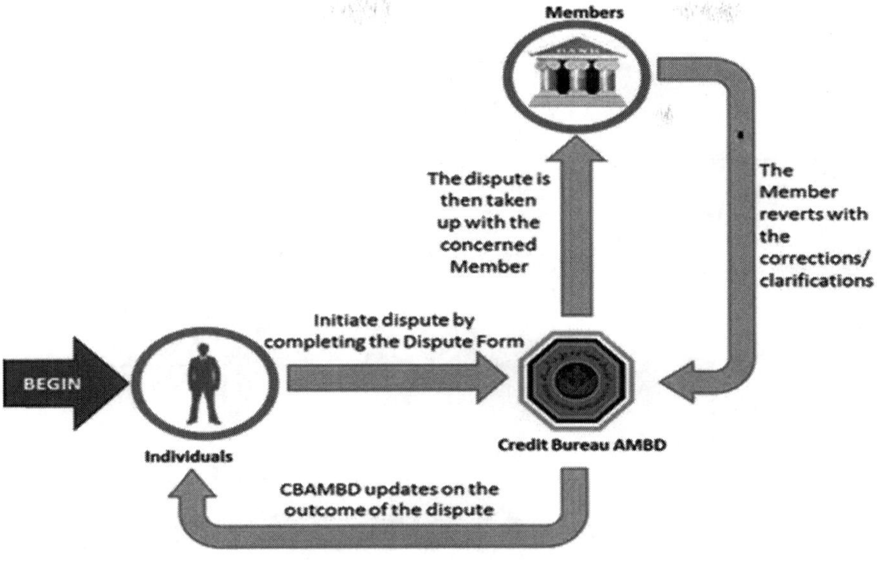

Figure 2.5 The functioning of Credit Bureau of Brunei Darussalam.

ranges from strengthening the position of currency and monetary identity. By the separation of 1967, Brunei started to hail as the main player in the banking industry resulting from the rapid and robust growth in the oil and gas industry. However, there are still improvements to be made as follows:

- The need to establish Muamalat Court for dispute settlement. Currently, all disputes go to the civil courts. shariah muamalat issues are unique and require interpretations and recognition in principles. Malaysian established the Muamalat Court in 2014 but still faces difficulties among judges who are incompetent in understanding shariah.
- In order to boost rapid growth in Islamic finance, AMBD should issue shariah standards and Guidelines directing industries on the Modus operandi of each contract. Compliance with shariah is taken seriously in shariah, and therefore need guidelines in order to speed up growth.
- Products should be diversified and modified vehemently to suit the culture and customs of banking in faith in Brunei.
- The muqtadha al-aqd is an important element to ensure the products are in line with shariah. Therefore, it is pivotal to understand that muqtadha al-aqd is followed and recognised in legal documents.

2.3.2.10 Shariah Standards Issuance

As of to date, Brunei has issued several Guidelines and shariah Standards as follows:

Guidelines No.IFAU/G/1/2018: Guidelines on the Islamic Product Approval Process Under Section 14 of The Syariah Financial Supervisory Board Order 2006.

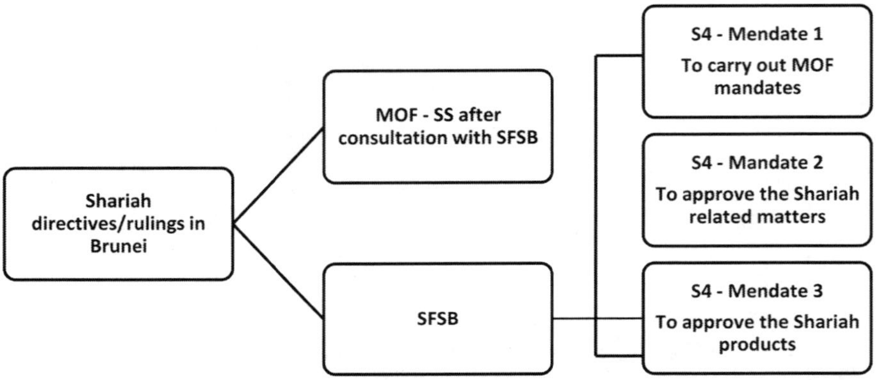

Figure 2.6 Issuance of Shariah ruling.

Notice No.IFAU/N/1/2018: Notice to Financial Institution On Syariah Governance Framework

Guidelines No.IFAU/G/2/2018: Guidelines on Syariah Governance Framework for Financial Institution in Brunei Darussalam.

Figure 2.6 outlines the steps and legal guidelines to issue the shariah guidelines in Brunei Darussalam.

2.3.2.11 Shariah Audit

Based on Figure 2.6, shariah directives are issued through SFSB and the Ministry of Finance after consultation with SFSB.

2.4 Indonesia

The Republic of Indonesia is the world's largest archipelagic nation, consisting of over 17,508 islands stretching 5,120 km from east to west and 1,760 km from north to south. Indonesia's islands cover a total of about 2 million square kilometres, with territorial waters extending to more than double the land area (Bidani & Ravallion, 1993).

2.4.1 The Regulators

As the central bank of Indonesia, Bank Indonesia is vested with a mandate for the regulation and development of Islamic banking in the country stipulated in Act No. 10 of 1998 concerning Banking and Act No. 23 of 1999 amended by Act No. 3 of 2004 concerning Bank Indonesia (Sari et al., 2016). As part of this role, Bank Indonesia is internationally active as a member of international organisations concerned with Islamic banking and financial services (Tanuwidjaja & Choy, 2006). Regulation and development of Islamic banking are vital to the creation of a sound Islamic banking system that will contribute to Indonesia's economic development. This is especially important in view of the gap between potential demand

and supply-side availability of Islamic banking products and services in Indonesia. This gap represents an opportunity for local and international investors to play a role in the Islamic banking industry (Pratiwi, 2016).

Bank Indonesia as an independent central bank was initiated when a new Central Bank Act, the UU No. 23/1999 on Bank Indonesia, was enacted on 17 May 1999 and was amended with UU No.3/2004 on 15 January 2004 (Sari et al., 2016). The Act confers its status and position as an independent state institution and free from interference by the Government or any other external parties. As an independent state institution, Bank Indonesia is fully autonomous in formulating and implementing each of its tasks and authority as stipulated in the Act.

- External parties are strictly prohibited from interfering with Bank Indonesia's implementation of its tasks, and
- Bank Indonesia has the duty to refuse or disregard any attempt of interference in any form by any party.

In its capacity as the central bank, Bank Indonesia has one single objective of achieving and maintaining stability of the Rupiah value. The stability of the value of the Rupiah comprises two aspects: one is the stability of the Rupiah value against goods and services and the other is the stability of the exchange rate of the Rupiah against other currencies. The first aspect is reflected by the rate of inflation, and the second aspect is reflected by the development of the Rupiah exchange rate against other currencies.

The prescribed single objective is intended to establish a target that is clear and to be achieved by Bank Indonesia within its limits and responsibilities. This way, the achievement of the objective of Bank Indonesia will be easily measurable.

The legal basis for the operation of the Indonesian banking system is set out in Act No. 7 of 1992 concerning Banking (amended by Act No. 10 of 1998). Within the institutional framework, Indonesian banks are divided into commercial banks and rural banks and are also separated as conventional and Islamic banks based on the nature of their business.

2.4.2 *Islamic Banking*

In view of providing a wider banking services alternative to the Indonesian economy, the development of Islamic banking in Indonesia is implemented under a dual banking system in compliance with the Indonesian Banking Architecture (API) (Rizvi et al., 2020). Islamic banking and conventional banking systems jointly and synergically support a wider public fund mobilisation in the framework of fostering the financing capability of national economic sectors.

The Blueprint of Islamic Banking Development in Indonesia was prepared by Bank Indonesia in 2002 to provide guidance to stakeholders of Islamic banking and to set the position and vision of Bank Indonesia in developing Islamic banking in the country (Abduh & Azmi Omar, 2012). In the process of preparing this Blueprint, various aspects have been taken into consideration such as the actual condition of the national Islamic banking industry including related tools, development trends of the Islamic banking industry within an international scale as well and system development of national Islamic finance that has started to be materialised and inseparable from wider architectural landscape such as API and Indonesian Financial System Architecture (ASKI) including international best practices formulated by international Islamic financial institutions such as IFSB (Islamic Financial Services Board), AAOIFI and IIFM.

The development of Islamic banking was directed to provide the highest benefits to the public and to give optimal contribution to the national economy (Nastiti & Kasri, 2019). Consequently, its development path is always referred to other strategic plans, such as API, ASKI as well as Medium Term National Development Plan (RPJMN) and Long-Term National Development Plan (RPJPN). Therefore, the policy in developing Islamic banking is a part and an activity supporting the achievement of large-scale strategic planning in development at the national level.

The Blueprint of Islamic Banking in Indonesia defines the vision, mission and target of Islamic banking development as well as strategic initiatives with clear priorities in response to the main challenge and in reaching the target for the next ten years by achieving significant market share of Islamic banking through the advancement of the role of Islamic banking in the national, regional and international financial activities with other Islamic financial sectors.

In the condition of its integration, in the short term, the phase of Islamic banking development is more directed towards servicing the huge potential of the domestic market. In other words, Indonesian Islamic banking must become a domestic player with international quality in services and performance.

Subsequently, the Islamic banking system envisioned by Bank Indonesia is modern Islamic banking with open and universal characteristics and inclusive to all Indonesian people without any exception (Marlina et al., 2021). It refers to the banking system presenting applicative forms of Islamic economic concepts that are wisely formulated in the current context of problems faced by Indonesia within the consciousness of the historical socio-cultural condition of the country. Only by doing that, the aspiration development of Islamic banking will always be recognised and accepted by Indonesian people as a part of the solutions to the various problems encountered by this country.

In July 2007, Indonesia issued the Investment Guide to Islamic Banking in Indonesia. The Investment Guide to Islamic Banking is envisaged as a useful information tool on Indonesia's potential, investment mechanisms and requirements for potential investors (Millar & Anwar, 2009). Bank Indonesia remains ready to provide facilitation for parties interested in seizing the opportunity to enter Indonesia's rapidly expanding Islamic banking industry.

2.4.3 Islamic Banking Laws

The legal framework for the Islamic banking system in Indonesia is provided in Act No. 7 of 1992 (amended by Act No. 10 of 1998) which accommodates the existence of a dual banking system in Indonesia. In addition, Act No. 23. of 1999 (amended by Act No. 3 of 2004) concerning Bank Indonesia provides the legal basis for Bank Indonesia to perform the regulation and oversight of the banking system, including Islamic banks, and to issue central bank instruments complying with sharia principles.

Bank Indonesia has taken a strategic step for the long-term development of Islamic banking by preparing the Islamic Banking Blueprint (until 2015), which sets out short-, medium- and long-term initiatives aligned with the API for 2004–2015.

The development of the Islamic banking system in Indonesia must uphold prudent banking and compliance with Sharia principles (Sari et al., 2016). The infrastructure for compliance with sharia principles is now in place. The Indonesian Council of Ulamas (MUI) has established the National Sharia Board (DSN) as the sole institution for issuing fatwas on Islamic financial instruments in Indonesia and the appointment of Sharia Supervisory Boards (DPS) at banks. These boards are responsible for ensuring that banks conducting business based on Sharia principles comply with these principles in their operations, products and services. In addition to its active role at home in the development and provision of Islamic banking infrastructure, Bank Indonesia is also active in international Islamic banking organisations such as the Islamic Financial Services Board (IFSB) and Accounting, and Auditing Organization for Islamic Financial Institutions (AAOIFI). Bank Indonesia's involvement includes active participation in working groups for deliberation of the Guidance for Islamic Financial Institutions, some aspects of which have also been applied in the regulations issued by Bank Indonesia.

2.4.3.1 Otoritas Jasa Keuangan (OJK)

On 31 December 2013, Bank Indonesia, also known as the Central Bank, officially transferred its banking microprudential supervisory powers to the Financial Services Authority ("OJK"). The OJK now has the authority to monitor individual banks, including the supervision of licensing matters and the day-to-day operations of banks (Syah, 2020). Bank Indonesia will retain macroprudential supervisory powers and will monitor the banking system as a whole. After this latest transfer of power, the OJK has taken over full supervision of capital markets and financial institutions in Indonesia, including banks and non-banking financial institutions such as insurance companies, financing companies and retirement funds. Supervisory powers over capital markets and non-banking financial institutions were transferred from the Capital Market and Financial Institutions Supervisory Agency ("Bapepam-LK") to the OJK on 31 December 2012, exactly one year prior to the transfer of banking microprudential supervision from Bank Indonesia (Wardani & Basri, 2020). The transfer of power from Bapepam-LK and Bank Indonesia to OJK was mandated in Article 55 of Law No. 21 of 2012 regarding the OJK.

Bank Indonesia will continue to carry out macroprudential functions and deal with systemic threats to the industry. The central bank will still supervise monetary policy, payment systems and other macroprudential issues. Below are several legislations empowered and regulated under OJK:

OJK Regulation Regarding Assessment on Level of Bank Soundness for Sharia Commercial Banks and Sharia Business Unit (Peraturan OJK Tentang Penilaian Tingkat Kesehatan bank Umum Syariah dan Unit Usaha Syariah) OJK Regulation Number 8/POJK.03/2014 regarding assessment on the level of bank soundness for Sharia commercial banks and Sharia business units is issued in order to improve the effectiveness of evaluation on the level of soundness in banks in order to deal with changes in the complexity of business and risks profile that may come from banks and their subsidiaries, and also with changes in the approach of evaluation on level of soundness that is implemented internationally which can influence current approach of such evaluation.

Therefore, banks are obligated to maintain and improve their level of bank soundness by implementing prudential principles, Sharia principles and risk management in carrying out their business activities. Additionally, in taking the responsibility to sustain bank business, directors and board of commissioners are responsible for upholding and monitoring the level of bank soundness as well as taking necessary measures to sustain and improve the level of bank soundness.

Sharia commercial banks are obligated to conduct assessments on the level of bank soundness, both individually and by consolidation. As for Sharia business units, they are obligated to conduct assessments on the level of bank soundness individually. In order to provide implementation guidelines for POJK, OJK issued one OJK Circular Letter for Sharia Banks, namely SEOJK No. 10/SEOJK.03/2014 concerning Bank Rating of Sharia Commercial Banks and Sharia Business Units. The circular letter contains more technical details concerning soundness assessments of sharia banks and sharia business units along with the obligation for Sharia commercial banks and sharia business units to report their risk profile. A few popular regulations of Bank Indonesia are tabulated in Table 2.5.

Supervision of the sharia capital market is based on Regulation No. II.K.1 concerning the Criteria and Issuance of a Sharia-Based Securities List where the Issuer is required to submit a Sharia-Based Securities List on 31 May and 21 November. OJK initiated surveillance of one Investment Management Company, specifically the mandatory report submitted based on the report of the Issuer of the Sharia-Based Securities List (DES) on 31 May and 21 November 2014. Based on the outcome of the surveillance, the report was in accordance with prevailing regulations.

2.4.4 Shariah Non-Bank Financial Industry

During 2014, OJK approved 59 insurance products, consisting of 10 sharia general insurance products, 39 sharia life insurance products and 10 sharia bancassurance products (Kharisma et al., 2022). In addition, OJK also granted new business licenses to one sharia conventional finance company and one sharia guarantee company. Furthermore, OJK also undertook fit and proper tests for 39 directors and commissioners in the sharia insurance subsector and 13 directors and commissioners in the special sharia financial services and sharia financing subsector in order to ensure the competence and credibility of directors and commissioners at sharia nonbank financial institutions.

Table 2.5 Bank Indonesia's major regulations.

Title	*Description*
Bank Indonesia Regulation Number 13/23/PBI/2011	Bank Indonesia Regulation Number 13/23/PBI/2011 Concerning The Implementation of Risk Management in Islamic Commercial Banks and Islamic Business Units
Bank Indonesia Regulation Number 13/13/PBI/2011	Bank Indonesia Regulation Number 13/13/PBI/2011 Concerning Asset Quality Rating For Islamic Commercial Banks And Islamic Business Units
Bank Indonesia Regulation Number 13/9/PBI/2011	Bank Indonesia Regulation Number 13/9/PBI/2011 Concerning Amendment to Bank Indonesia Regulation Number 10/18/PBI/2008 Concerning Restructuring of Financing for Islamic Banks and Islamic Business Units
Bank Indonesia Regulation Number 13/5/PBI/2011	Bank Indonesia Regulation Number 13/5/PBI/2011 Concerning Legal Lending Limit for Islamic Rural Banks
Bank Indonesia Regulation Number 11/33/PBI/2009	Bank Indonesia Regulation Number 11/33/PBI/2009 Concerning The Implementation of Good Corporate Governance by Islamic Commercial Banks and Islamic Business Units
Bank Indonesia Regulation Number 11/23/PBI/2009	Bank Indonesia Regulation Number 11/23/PBI/2009 Concerning Islamic Rural Banks
Bank Indonesia Regulation Number 11/15/PBI/2009	Bank Indonesia Regulation Number 11/15/PBI/2009 Concerning Conversion of Business Activities from Commercial Bank to Sharia Bank
Bank Indonesia Regulation Number 11/10/PBI/2009	Bank Indonesia Regulation Number 11/10/PBI/2009 Concerning Sharia Business Units
Bank Indonesia Regulation Number 11/3/PBI/2009	Bank Indonesia Regulation Number 11/3/PBI/2009 Concerning Sharia Commercial Bank
Bank Indonesia Regulation Number 10/32/PBI/2008	Bank Indonesia Regulation Number 10/32/PBI/2008 Concerning Sharia Banking Committee

Source: http://www.ojk.go.id/en/kanal/perbankan/regulasi/regulasi-perbankansyariah/Default.aspx

In order to expand Non-Bank Financial Services to society, OJK granted a sharia business unit license to one sharia insurance company and two sharia finance companies and gave permission for the establishment of a new branch office to seven companies consisting of five sharia insurance companies and two sharia finance companies in order to expand public access to nonbank financial services. Moreover, OJK also registered eight marketing offices.

2.4.4.1 Mediation in Disputes

As part of the implementation of the Dispute Resolution Road Map in the Financial Services Sector, OJK actively and continuously supports and assist the industry. Those commitments focused on increasing the capacity of industry, not only in developing internal dispute resolution in internal Financial Services Providers but also the establishment of Alternative Dispute Resolution Institutions as an external dispute resolution in the financial services sector (Hadad et al., 2011). In 2014, OJK organised four mediator certification programmes, three adjudication and arbitrage workshops, and one settlement of disputes in the financial services sector based on sharia workshop. The series of capacity-building programmes followed by 172 industries will be prepared for operationalisation of Alternative Dispute Resolution Institution.

To prepare for an Alternative Dispute Resolution Institution (LAPS), OJK issued an OJK Circular Letter containing assessment guidelines for alternative dispute resolution institutions, and intensively communicated, assisted and coordinated with associations in banking, financing, guarantee, venture capital and pawn brokers sector that currently do not have LAPS. OJK assessed three LAPS in 2014 to ensure they met the requirements set. The three LAPS are the Capital Market Arbitration Board (BAPMI), the Indonesian Institute of Insurance Mediation and Arbitration (BMAI), and the Pension Funds Mediation Board (BMDP). In October 2014, the venture capital sector established the Indonesian Institute of Ventura Arbitration (BAVI), while other sectors such as banking, guarantee, financing and pawn brokers are still in the process of preparing the deed of establishment of LAPS.

2.5 Philippines

In 1939, as required by the Tydings-McDuffie Act, the Philippine legislature passed a law establishing a central bank. As it was a monetary law, it required the approval of the United States president (Takagi, 2016). However, President Franklin D. Roosevelt disapproved of it due to strong opposition from vested interests (Hutchcroft, 1998). A second law was passed in 1944 during the Japanese occupation, but the arrival of the American liberalisation forces aborted its implementation.

Shortly after President Manuel Roxas assumed office in 1946, he instructed then Finance Secretary, Miguel Cuaderno, Sr. to draw up a charter for a central bank (Nagano, 1999). The establishment of a monetary authority became imperative a year later as a result of the findings of the Joint Philippine-American Finance Commission chaired by Mr. Cuaderno. The Commission, which studied Philippine financial, monetary and fiscal problems in 1947, recommended a shift from the dollar exchange standard to a managed currency system. A central bank was necessary to implement the proposed shift to the new system.

Immediately, the Central Bank Council, which was created by President Manuel Roxas to prepare the charter of a proposed monetary authority, produced a draft. It was submitted to Congress in February 1948. By June of the same year, the newly proclaimed President Elpidio Quirino, who succeeded President Roxas, affixed his signature on Republic Act No. 265, the Central Bank Act of 1948. The establishment of the Central Bank of the Philippines was a definite step towards national sovereignty. Over the years, changes were introduced to make the charter more responsive to the needs of the economy. On 29 November 1972, Presidential Decree No. 72 adopted the recommendations of the Joint IMF-CB Banking Survey

Commission which made a study of the Philippine banking system. The Commission proposed a programme designed to ensure the system's soundness and healthy growth. Its most important recommendations were related to the objectives of the Central Bank, its policy-making structures, the scope of its authority and procedures for dealing with problem financial institutions.

2.5.1 Financial Regulators

The Regulator is the Bangko Sentral NG Pilipinas. The power is mandated in the New Central Bank Act (RA 7653). Art 3 states that the responsibility and Primary Objectives of the Bank are as follows;

- The Bangko Sentral shall provide policy directions in the areas of money, banking and credit. It shall have supervision over the operations of banks and exercise such regulatory powers as provided in this Act and other pertinent laws over the operations of finance companies and non-bank financial institutions performing quasi-banking functions, hereafter referred to as quasi-banks and institutions performing similar functions.
- The primary objective of the Bangko Sentral is to maintain price stability conducive to a balanced and sustainable growth of the economy. It shall also promote and maintain monetary stability and the convertibility of the peso.

On 15 January 2013, the Bangko Sentral ng Pilipinas (BSP) released Circular No. 781 which provides the implementing guidelines on the revised risk-based capital adequacy framework, particularly on the minimum capital and disclosure requirements (Abola et al., 2000). The new framework which took effect in 2014 applies to all universal and commercial banks (U/KBs) and their subsidiary banks and quasi-banks. This framework allowed a reasonable transition period for banks to comply with the new minimum requirements. While the international rules allow staggered implementation, the BSP adopted the capital reforms in full, in recognition of the strong capital position of the banking industry.

The BSP implements new minimum capital ratios of 6.0% Common Equity Tier 1 (CET1) ratio, 7.5% Tier 1 ratio and 10.0% Total Capital Adequacy Ratio (CAR). A capital conservation buffer (CCB) of 2.5%, comprised CET1 capital was also prescribed.

In recognition of the distinct structure of foreign bank branches (FBBs) which operate under the U/KB license, a calibrated Basel III framework was issued under Circular No. 822 dated 13 December 2014. The reform highlighted the role of Permanently Assigned Capital which is the CET1 equivalent for FBBs. Shortly after, R.A. No. 10641 amending R.A. No. 7721 was approved to further liberalise the entry of foreign banks. To implement this law, the BSP issued Circular No. 858 dated 21 November 2014, providing the new capital computation for FBBs.

In October 2014, Circular No. 856 Basel III Framework for Dealing with Domestic Systemically Important Banks (DSIBs) was issued to address systemic risk and interconnectedness by identifying banks that are deemed systemically important within the domestic banking industry. Minimum buffers composed of CET1 capital shall be required of systemically important banks starting January 2017.

2.5.2 Islamic Banking in the Philippines

At present, there is only one Islamic bank in the Philippines, the Al-Amanah Islamic Investment Bank of the Philippines. Al-Amanah Bank was established in 1973 (Hadji Latif, 2021). While it has been over four decades since the creation of the first Islamic Bank, Islamic banking itself has not grown in large part because of legal constraints. The Al-Amanah Charter created the Bank but not a framework for Islamic banking per se. No such enabling law has so far been passed. In fact, the General Banking Law of 2000 defines Islamic banks as specifically pertaining to Al-Amanah Bank only. The GBL does not provide for the creation of other Islamic banks.

Based on the World Bank Report, The Philippines Development Plan 2011–2016 identified the development of an Islamic finance framework as one of the priorities (Tagoranao et al., 2022). Interest in developing Islamic finance in the Philippines demonstrated in recent years by the Government, legislators and the private sector is driven by four main factors. The first is the desire to address the needs of the Philippines Muslim population. The second is ASEAN integration. As a member of ASEAN and a neighbour to Indonesia and Malaysia, the Philippines looks to expand economic links with neighbouring countries and sees regional and global Halal markets and linked financing as an important component of its ASEAN integration strategy (Estrada et al., 2010). The third is the opportunity to tap international financial markets for both resource mobilisation and investment diversification through Sukūk and equity markets. And the fourth is the need to find a solution for Al Amanah Bank – an existing Islamic bank with a record of poor performance. Islamic finance, or to be precise Islamic banking, in the Philippines is presently defined by the law adopted in 1990 establishing Al-Amanah Islamic Investment Bank of the Philippines. Al-Amanah Bank is a sole Islamic bank, state owned, and in its present form not financially viable. The existing framework for Islamic banking in the Philippines must be understood in the context of the history of conflict in Mindanao. The ongoing peace process, ASEAN integration and globalisation of Islamic financial markets offer an opportunity to change course, move forward and develop effective Islamic finance in the Philippines.

The primary law is the Republic Act No. 6848, otherwise known as the "Charter of Al Amanah Islamic Investment Bank of the Philippines," and the Manual of Regulations for Banks. The Manual contains regulations on Islamic banking approved by the Monetary Board through Resolutions 161 and 244 in 1996. The regulations cover several issues, including the purpose of the bank, functions of the *Shari'ah* advisory council, powers of the bank, return on investment funds, authorised operations, investment risk funds, as well as accounting and training of banking personnel. The Islamic banking system is regulated by Bangko Sentral in

Pilipinas, the first and only Islamic financial institution is Amanah Islamic Investment Bank, which was established in 1990.

2.5.3 Islamic Banking Laws in the Philippines

Legal and regulatory: the charter of AAIIBP is the only piece of legislation guiding Islamic banking in the Philippines. The Charter does not address key issues such as tax neutrality for Islamic and conventional products, resulting in higher costs for Islamic products. In addition, the unavailability of deposit guarantee/insurance by the Philippine Deposit Insurance Corporation has raised apprehensions over Islamic deposit accounts offered by the bank (Komijani & Taghizadeh-Hesary, 2018).

A draft bill amending RA 6848, known as the Charter of Al-Amanah Islamic Investment Bank of the Philippines and providing for the regulation and organisation of an expanded Islamic banking system in the Philippines with a short title of "The Philippines Islamic Finance Act" was submitted to Congress in August 2015 (HB5989) and Senate in January 2016 (SB3150). The Bill seeks to strengthen Al-Amanah Bank to service a broader market while enabling both the establishment of other Islamic banks and engagement in Islamic banking arrangements by conventional banks under the supervision and regulation of BSP (Venardos, 2010). The proposed bill also mandates the government provide programmes for increased consumer awareness and capacity building.

The bill includes an important provision on tax neutrality requiring the Government to ensure neutral tax treatment between Islamic banking transactions and equivalent conventional banking transactions. The bill empowers BIR to issue policies and guidelines to ensure tax neutrality. The bill provides a basis for deposit insurance for Islamic banks through the establishment of an insurance mechanism on takaful principles. The drafting of the bill included active consultations among BSP, NCMF, Amanah Bank, DBP, PDIC, BIR and other key agencies. The drafting was informed by international practice, including from the World Bank. There is generally broad-based support for the bill among the regulators and business community and it is expected to be passed into law.

The effective implementation of the law will depend on the issuance of appropriate implementing regulations and guidance by BSP, BIR and other concerned agencies as well as the establishment of the necessary institutional support through bank supervision and bank resolution processes, Shari'ah Councils and accounting and auditing. BSP is working on identifying specific steps to enable effective implementation of the provisions of the proposed bill if passed into law. BSP has already initiated work on drafting supervisory guidance within the existing legal framework for minimum prudential standards and expectations relative to Shari'ah-compliant debt-type Sukūk investments of universal and commercial banks. The authorities intend to build on the international experience available to inform the development of regulatory framework and supervisory standards for Islamic finance in the Philippines. The BSP also plans to strengthen the capacity of the supervisory staff in terms of their understanding of Islamic finance concepts and supervisory practices to design and implement appropriate supervisory processes for Islamic banking in the Philippines. Such capacity-building efforts would potentially involve engaging qualified international experts to work with BSP supervisory staff to develop implementing regulations, principles and supervision manual, attending training courses on Islamic finance supervision, and working with qualified experts to fine-tune supervisory practices once Islamic banks or windows are authorised to operate in the Philippines.

2.6 Singapore

2.6.1 Financial Regulators

2.6.1.1 Monetary Authority of Singapore (MAS)

The MAS is governed by the MAS Act, which confers MAS powers to issue legal instruments for the regulation and supervision of financial institutions. In addition, MAS also has frameworks and guidelines in place on topics that cut across various classes of financial institutions (Parrado, 2004).

MAS applies the same set of admission criteria when considering an application by a conventional bank and an Islamic bank to operate in Singapore. MAS' primary concern, when admitting new players, is the safety and soundness of the new institution. While the unique features of Islamic banking may alter the source and extent of risks, MAS does not expect the risk profile of an Islamic bank to be fundamentally different from its conventional banking counterparts (Wilson, 2015). Fundamentally, MAS expects all banks, Islamic or conventional, to remain focused on their core banking business.

2.6.1.2 Unified Regulatory Framework

A conventional bank with existing operations in Singapore intending to conduct Islamic banking business should notify MAS if any such plans are being made and before commencing Islamic banking activities (Montes & Giap, 1999). Every bank should also ensure that they are well-managed and possess the necessary risk management capabilities to offer Islamic banking services.

As part of the single regulatory framework, a bank carrying out Islamic banking activities will be required to comply with the same set of rules and regulations as any other bank in Singapore, namely the Banking Act, Banking Regulations, Notices and Directives (Chow & Pei, 2018). These rules will include the need to maintain eligible assets (MAS Notice 640), maintain sufficient liquidity buffers (MAS Notice 613), keep ample provisions (MAS Notice 612), observe large exposure limits (Banking Act, Section 29), limit property-related exposures (Banking Act Section 33), put in place strict anti-money laundering controls (MAS Notice 626) and comply with minimum regulatory capital requirements. (MAS Notice 637). Every bank should also observe any general principles laid out in guidelines closely (e.g. Guidelines on Risk Management Practices).

2.6.2 Establishing Islamic Bank in Singapore

The nature of the risk faced by an Islamic bank in Singapore is unique and relatedly to shariah compliance risk. In addition to managing the risks faced by conventional banks, such as credit, market and operational risks, an Islamic bank also has to ensure that it is in compliance with shariah rulings as this carries a significant reputational risk to the bank (Delle Foglie et al., 2023). As a prudential regulator, MAS does not prescribe what constitutes shariah compliance nor endorse specific shariah rulings. Nevertheless, MAS expects Islamic banks to take into account shariah compliance matters and to manage this compliance risk as part of their overall risk management process. Nothing in these guidelines should be construed as expressing an opinion on shariah acceptability.

2.6.2.1 Single Capital Framework

MAS Notice 637 sets out the minimum CAR for all banks incorporated in Singapore, and the rules they must follow in calculating these ratios (Pradhan et al., 2023). The capital requirements in MAS Notice 637 will apply to Islamic financial products on the same basis as they apply to other banking products. The examples used in these guidelines are for illustrative purposes only and are not intended to be definitive or exhaustive. The actual application of the requirements in MAS Notice 637 will vary depending on the circumstances of each case. A bank incorporated in Singapore is responsible for making its assessment of the risks associated with each product and the capital rules that are applicable.

Islamic banks commonly fund their activities through the acceptance of non-interest-bearing deposits, such as wadiah and qard hassan deposits. Banks in Singapore are allowed to offer such non-interest-bearing deposits as these deposits satisfy the legal definition of a "deposit" set out in the Banking Act.

2.7 Thailand

2.7.1 The Regulator

2.7.1.1 Bank of Thailand (BOT)

According to BOT Act B.E.2485 as amended by B.E.2551, the functions and roles of BOT are as follows:

1 Print and issue banknotes and other security documents.

   The BOT prints and issues banknotes and other security documents under the enforcement of the Currency Act and has sole rights to print and issue banknotes in the Kingdom (Lerskullawat, 2023).

2 Promote monetary stability and formulate mmonetary policies.

   The BOT implements monetary policy as specified by the Monetary Policy Committee as follows: mobilising the deposits, determining the interest rate for loans to financial institutions, trading foreign exchange and exchanging for future cash flow, borrowing foreign exchange in order to maintain the monetary stability, borrowing money in order to implement the monetary policy, trading securities as necessary and exchanging for the future cash flow in order to control the money supply in the country's financial system and borrowing or lending the securities with or without returns.

3 Manage the BOT's assets

The BOT manages its assets (excluding the assets within the currency reserve according to the Currency Act) and invests such assets for returns by realising the security, liquidity, return on asset and management risks.

4 Provide banking facilities to the government and act as the registrar for the government bonds

The BOT provides banking facilities to the government in terms of depository and lending facilities for the Ministry of Finance, acts as the custodian for the government, acts as the representative of the government for investment in assets and FX, trades and transfers the bill of exchange, securities, and share certificate and controls and oversees FX. In addition, the BOT may provide banking facilities to the state enterprise or other government agencies. Moreover, the BOT acts as the registrar for government bonds by acting as the government representative in purchasing and selling government bonds, paying principal and interest or acting as the registrar of state enterprises, specialised financial institutions or other government agencies.

5 Provide banking facilities for the financial institutions.

The BOT provides banking facilities for financial institutions by acting as lender of the last resort for the financial institutions, acting as the custodian for the financial institutions and ordering the financial institutions to report or explain the assets, liabilities or contingent liabilities.

6 Establish or Support the establishment of the payment system

The BOT establishes or supports the establishment of a payment system, and electronics clearing system, and administers such systems for safety and efficiency.

7 Supervise and examine the financial institutions

The BOT supervises, examines and analyses the financial status and performance, and risk management system of the financial institutions in order to promote financial institutions stability.

8 Manage the country's foreign exchange rate under the foreign exchange system and manage assets in the currency reserve according to the Currency Act.

9 Control the foreign exchange according to the exchange control act.

2.7.2 Islamic Banking in Thailand

The history of the Islamic financial system in Thailand started with the establishment of a cooperative society, Pattani Islamic Saving Cooperative, that operated based on shariah in 1987 (Thippayana et al., 2023). By the end of 2001, four other Islamic saving cooperatives were established in Southern Thailand, i.e. Ibnu Affan Saving Cooperative (Pattani), As-Siddiq Saving Cooperative (Songkla), Saqaffah

Islam Saving Cooperative (Krabi) and AlIslamiah Saving Cooperative (Phuket). Islamic banking products and services were first introduced to Muslims in Thailand with the implementation of "Islamic Window" by GSB in 1998. A similar concept was also introduced by BAAC in 1999. The Krung Thai Bank was the first bank to set up an "Islamic Branch" in 2001. This branch offers a full range of basic banking products and services based on Islamic principles. Another key milestone for Islamic banking in Thailand was the enactment of the Islamic Bank of Thailand Act 2002. This law paved the way for the establishment of the first full-pledge Islamic bank, the Islamic Bank of Thailand in 2003. This bank, which is located in Bangkok, plans to open branches in Yala, Pattani, Narathiwat and Songkla.

Currently, two bodies regulate the financial institutions in Thailand, The Ministry of Finance and The Bank of Thailand (the central bank). With regard to the existing institutions that offer Islamic banking products and services, they are regulated by two different authorities (Bank of Thailand, 2024). For example, both GSB and BAAC are under the supervision of the Ministry of Finance through the Government Savings Bank Act, B.E. 2489 (1946) and The Bank for Agriculture and Agricultural Cooperatives Act, B.E.2509 (1966). However, Krung Thai Bank, which was established under the Commercial Banking Act B.E. 2505 (1962), is within the ambit of the Bank of Thailand.

2.7.3 Islamic Banking Laws

In Thailand, Islamic banks are governed under the Islamic Bank of Thailand Act 2002, which comes under the jurisdiction of the Ministry of Finance. Since BOT is responsible for monetary policy and regulating commercial banks in Thailand, BOT is in a better position to supervise the operations and activities of Islamic banks. Hence, it is recommended that the falls within the ambit of this central bank.

The government's intention was to set up an Islamic Bank under the government's jurisdiction which has its own sets of law (Harahap et al., 2023). The Ministry of Finance was then assigned to draft "The Islamic Bank of Thailand Act" which was later approved by the parliament in October 2002.

Islamic Bank of Thailand is set up under the Islamic Bank of Thailand Act 2002 and regulated by the Ministry of Finance. The Bank commenced operations in June 2003 with a paid-up capital of 1 billion baht and had its headquarters also the first branch in Klongton. The Bank continued to open more branches, especially in Bangkok and Southern provinces and by the end of 2005 the Bank had a total of nine branches. The expansion of the Bank's business continued through the acquisition of shariah Banking Services of Krung Thai Bank PCL in November 2005. It was conducted under the government's policy and that generated the number of branches increased from 18 to 27. To facilitate the growing business and increase competitive effectiveness the Bank relocated the headquarters from Klongton to Asoke, a prime business area, in August 2005.

During the initial stage of incorporation, IBank focused solely on Muslim market segmentation. However, IBank believes the shariah banking system is able to serve anyone and thus creates and develops more innovative and competitive products, embarks on rebranding exercise and provides the best services to customers. The Bank promotes ethical culture, social responsibility, Islamic principles and good governance through the operational framework resulting in IBank as be differentiated and new alternative for everyone regardless of religion. Currently, IBank is a state-owned enterprise and has 82 branches throughout the country.

References

Abduh, M., & Azmi Omar, M. (2012). Islamic banking and economic growth: The Indonesian experience. *International Journal of Islamic and Middle Eastern Finance and Management*, *5*(1), 35–47.

Abdul 'Ali, M.J.A., Tan, T.S., & Lee, Y.C. (2014). Brunei: Islamic Banking and Finance. *In The Islamic Finance Handbook (eds. Thiagaraja, S., Morgan, A. Tebbutt, A., & G. Chan, G*. John Wiley & Sons Singapore Pte. Ltd. https://doi.org/10.1002/9781118936863.ch04.

Abola, V. A., Neri, E. Jr, & Supangan, R. (2000). The Banking system and the savings rate in the Philippines. In Masuyama, S., Vandenbrink, D., & Yue, C. H. *Restoring East Asia's dynamism*. Institute of Southeast Asian Studies: Singapore.

Alam, A., Fianto, B. A., Ratnasari, R. T., Ahmi, A., & Handayani, F. P. (2023). History and development of takaful research: A bibliometric review. *SAGE Open*, *13*(3).

Ali, Q. (2018). Service quality from customer perception: Comparative analysis between Islamic and conventional bank. *Journal of Marketing and Consumer Research*, *43*, 70–82.

Ariff, M. (1989). Islamic banking in Malaysia: Framework, performance, and lessons. International Journal of Economics, *Management and Accounting*, *2*(2).

Ariff, M., & Rosly, S. A. (2011). Islamic banking in Malaysia: Unchartered waters. *Asian Economic Policy Review*, *6*(2), 301–319. https://doi.org/10.1111/j.1748-3131.2011.01208.x

Bank of Thailand. (2024). Financial Sector Assessment Program. https://www.bot.or.th/content/dam/bot/supervision/fsap/FSAPeng.pdf [Accessed 17 March 2024].

BDCB, Brunei Darussalam Central Bank. (2016). Brunei Darussalam Financial Sector Blueprint 2016 – 2025. https://www.bdcb.gov.bn/financial-sector-development/financial-sector-blueprint [Accessed 25 August 2023].

Bidani, B., & Ravallion, M. (1993). A regional poverty profile for Indonesia. *Bulletin of Indonesian Economic Studies*, *29*(3), 37–68.

BNM, Bank Negara Malaysia. (2022). Financial Sector Blue Print 2022–2026. https://www.bnm.gov.my/documents/20124/5915429/fsb3_en_book.pdf [Accessed 13 May 2023].

Chow, H. K., & Pei, S. F. (2018). Financial sector in Singapore. In Volz, V., Morgan, P. J. Yoshino, N. *Handbook of Banking and Finance in Asia*, pp. 165–178. Routledge: London.

Damit, M. Y. H. (2007). Brunei Darussalam: Towards a new era. *Southeast Asian Affairs*, *1*, 101–113.

Delle Foglie, A., Boukrami, E., Vento, G., & Panetta, I. C. (2023). The regulators' dilemma and the global banking regulation: The case of the dual financial systems. *Journal of Banking Regulation*, *24*(3), 249–263.

Estrada, G. B., Park, D., & Ramayandi, A. (2010). Financial development and economic growth in developing Asia. *Asian Development Bank Economics Working Paper*, Paper No. 233.

Hadad, M. D., Agusman, A., Monroe, G. S., Gasbarro, D., & Zumwalt, J. K. (2011). Market discipline, financial crisis and regulatory changes: Evidence from Indonesian banks. *Journal of Banking & Finance*, *35*(6), 1552–1562.

Hadji Latif, S. D. (2021). Knowledge and attitudes towards Islamic banking in the Philippines. *Journal of Islamic Accounting and Business Research*, *12*(2), 169–185.

Hamdan, M., & Hoon, C. Y. (2019). Brunei Darussalam. Southeast Asian Affairs, 85–102.

Haqqi, A. R. A. (2017). Regulatory regimes of Islamic banking in ASEAN Economic Community (AEC): A uniting force in the industry of the region. *Advances in Social Sciences Research Journal*, *4*(13). 24,39.

Harahap, B., Risfandy, T., & Futri, I. N. (2023). Islamic law, Islamic finance, and sustainable development goals: A systematic literature review. *Sustainability, 15*(8), 6626.
Hasan, Z. B. (2010). Sharia governance in Islamic financial institutions and the effect of the Central Bank of Malaysia Act 2009. *Journal of International Banking Law and Regulation*, 3. https://ssrn.com/abstract=2196827 [Accessed 22 September 2023].
Hassan, M. A. D., Besar, M. H. A. H., & Jaidin, J. H. (2022). An analysis of the legal and governance framework for protection and promotion of the rights of the Islamic bank, customers, and community: A case study of Brunei Darussalam. *Journal of Islamic Finance, 11*(2), 84–94.
Hutchcroft, P. D. (1998). *Booty capitalism: The politics of banking in the Philippines*. Cornell University Press.
IFSB, Islamic Financial Services Board. (2023). Islamic Financial Services Industry Stability Report -2-23 - Navigating a Challenging Global Financial Condition. https://www.ifsb.org/wp-content/uploads/2023/10/Islamic-Financial-Services-Industry-Stability-Report-2023_En.pdf [Accessed 20 August 2023].
Kamaruddin, M. I. H., Hanefah, M., Shafii, Z., & Salleh, S. (2020). Comparative analysis on Shariah governance in Malaysia: SGF 2010, IFSA 2013 and SGPD 2019. *Journal of Public Administration and Governance, 10*(1), 110–131.
Kharisma, B., Nurmandi, A., Muallidin, I., Kurniawan, D., & Loilatu, M. J. (2022). The role of the financial services authority (OJK) in preventing illegal fintech landing in the COVID-19 pandemic in Indonesia. In *International conference on human-computer interaction (*pp. 568–577). Cham: Springer International Publishing.
Komijani, A., & Taghizadeh-Hesary, F. (2018). *An overview of Islamic banking and finance in Asia*. Routledge Handbook of Banking and Finance in Asia, 505–518.
Kunhibava, S. (2012). Islamic banking in Malaysia. *International Journal of Legal Information, 40*(1–2), 191–201.
Lee, C., & Azali, M. (2010). Currency linkages among ASEAN. *The Singapore Economic Review, 55*(03), 459–470.
Lerskullawat, A. (2023). Financial development and asset price reaction to monetary policy: Case study of Thailand. *International Journal of Economic Policy in Emerging Economies, 18*(3-4), 269–282.
Marlina, L., Rusydiana, A. S., Hidayat, P., & Firdaus, N. (2021). Twenty years of Islamic banking in Indonesia: A biblioshiny application. *Library Philosophy and Practice, 4999.*
Millar, R., & Anwar, H. (Eds.). (2009). *Islamic finance: A guide for international business and investment.* London: GMB Publishing Ltd.
Mohamed Sadique, R. B., Roudaki, J., Clark, M. B., & Alias, N. (2010). Corporate fraud: an analysis of Malaysian securities commission enforcement releases. World Academy of Science, Engineering and Technology, pp. 1213–1222. https://hdl.handle.net/10182/4211 [accessed 05 February 2024].
Montes, M., & Giap, T. K. (1999). Developing the financial services industry in Singapore. In Masuyama, S., Vandenbrink, D., Chia Siow Yue, C. H. *East Asia's financial systems: evolution and crisis* (pp. 231–59). Institute of Southeast Asian Studies, Singapore.
Nagano, Y. (1999). Politics and Philippine Banking during the American period. *Philippine Political Science Journal, 20*(43), 61–82.
Nastiti, N. D., & Kasri, R. A. (2019). The role of banking regulation in the development of Islamic banking financing in Indonesia. *International Journal of Islamic and Middle Eastern Finance and Management, 12*(5), 643–662.
Parrado, E. (2004). *Singapore's unique monetary policy*: *How does it work?. IMF Working Paper No. 2004/010.*
Pradhan, R. P., Nair, M. S., Arvin, M. B., & Hall, J. H. (2023). Institutional quality, financial development and sustainable economic growth among lower income countries. In *Natural resources forum* (Vol. 47, no. 3, pp. 435–483). Oxford, UK: Blackwell Publishing Ltd.
Pratiwi, A. (2016). Islamic banking contribution in sustainable socioeconomic development in Indonesia: An epistemological approach. *Humanomics, 32*(2), 98–120.

Rizvi, S. A. R., Narayan, P. K., Sakti, A., & Syarifuddin, F. (2020). Role of Islamic banks in Indonesian banking industry: An empirical exploration. *Pacific-Basin Finance Journal, 62*, 101117.

Saiti, B., Ardo, A. A., & Yumusak, I. G. (2022). Why non-Muslims subscribe to Islamic banking? *Qualitative Research in Financial Markets, 14*(2), 247–269.

Sari, M. D., Bahari, Z., & Hamat, Z. (2016). History of Islamic bank in Indonesia: Issues behind its establishment. *International Journal of Finance and Banking Research, 2*(5), 178–184.

Shahid Ebrahim, M., & Kai Joo, T. (2001). Islamic Banking in Brunei Darussalam. *International Journal of Social Economics, 28*(4), 314–337.

Sheng, A. (1989). Bank restructuring in Malaysia. Financial Regulation (World Bank, 1992), 195–238.

Siddique, S. (1992). *Brunei Darussalam 1991: The non-secular nation. Southeast Asian Affairs, 91-100*. ISEAS - Yusof Ishak Institute: Singapore.

Skully, M. T. (1984). *Financial Institutions and Markets in Southeast Asia: A study of Brunei*. Indonesia, Malaysia, Philippines, Singapore and Thailand: Springer.

Syah, R. (2020). Dynamic Control Financial Supervision (OJK) for Growth Customer Behavior using KYC System.

Tagoranao, D. M. S., Gamon, D. A. D., & Tagoranao, J. S. (2022). The contributions of Islamic social financial institutions in realizing sustainable economic recovery for the Muslim communities in the Philippines. *Al-Risalah Journal: Academic Biannual Refereed Journal, 6*(1), 1–21.

Takagi, Y. (2016). Central banking as state building: Policymakers and their nationalism in the Philippines, 1933–1964 (Vol. 17). Nus Press.

Tanuwidjaja, E., & Choy, K. M. (2006). Central Bank credibility and monetary policy in Indonesia. *Journal of Policy Modeling, 28*(9), 1011–1022.

Thippayana, P., & Thinnam, N. (2023). Credit-using behavior of Muslims in non-Muslim country: A study of Nakhon Si Thammarat, Thailand. *Cogent Business & Management, 10*(1), 2179457.

Venardos, A. M. (2010). *Current issues in Islamic banking and finance: Resilience and stability in the present system*. World Scientific.

Wardani, V. P., & Basri, M. M. (2020). Literature review on the role of the financial services authority (OJK) in oversighting Sharia investment. *Journal of Islamic Economic Laws, 3*(2), 159–170.

Wilson, P. (2015). Monetary policy and financial sector development. *The Singapore Economic Review, 60*(03), 1550031.

Wright, L. (1977). Brunei: An historical Relic. *Journal of the Hong Kong Branch of the Royal Asiatic Society, 17*, 12–29.

Yaacob, H., Yaacob, A., Basir, K. H., & Ali, Q. (2021). Enabling legal environment of Islamic finance and social finance in Brunei Darussalam: A demand for improvement. In *Impact of Zakat on sustainable economic development* (pp. 186–195). IGI Global.

Zulhazmi, A. (2015). Islamic finance in Brunei well on track for 2016 and beyond. https://www.islamicfinancenews.com/islamic-finance-in-brunei-well-on-track-for-2016-and-beyond.html [Accessed 14 July 2023].

3 Islamic Financial Jurisdictions in South Asia

3.1 Introduction

This chapter expounds on the development of Islamic finance (IF) in South Asian countries (Bangladesh, India, Pakistan and Sri Lanka). The chapter strives to sketch an accurate picture of the chronological development of IF and different regulations established for the governance of Islamic financial institutions in these developing economies. Some of these jurisdictions namely Pakistan and Bangladesh are among the pioneers of IF since these countries continue patronaging and developing their IF industries. It is also observed that IF has a huge potential in both Muslim and non-Muslim-dominant countries as there is a clear demand for Islamic financial products and claims that existing financial regulations are supportive of establishing and developing the IF industry in the country.

3.2 Bangladesh

The regulator for the banking sector is the Bangladesh Bank, which is the central bank of Bangladesh. The establishment was based on the Presidential Decree on the Bangladesh Bank Order, 1972 (President's Order No. 127 of 1972). It was necessary to establish a central bank in Bangladesh to manage the monetary and credit system of Bangladesh with a view to stabilising domestic monetary value and maintaining a competitive external par value of the Bangladeshi Taka towards fostering growth and development of the country's productive resources in the best national interest (Hossain, 1988).

3.2.1 Banking Laws

The Banking Companies Act 1991 contains provisions for Islamic banking activities, following an amendment in 1993 (Sarker, 2000). Notwithstanding that the law specifically recognises Islamic finance contracts such as mudharabah and musharakah, generally the same banking laws as applied to conventional banks are imposed on Islamic financial institutions. The first Islamic financial institution in Bangladesh was the Islamic Bank Bangladesh Ltd. which was established in 1983 (Iqbal et al., 2006). To date, there are four other full-fledged Islamic financial

DOI: 10.4324/9781003471035-3

institutions, with some banks operating as Islamic windows. The law imposes a strict adherence to Shariah compliance and falls under the responsibility of the board of directors of the respective banks to ensure that the activities of the banks and their products are Shariah compliant. The Board of the Islamic banks/Subsidiary company/Conventional commercial banks having Islamic branches, therefore, be constituted with directors having requisite knowledge and expertise in Islamic Jurisprudence. The Board may form an independent Shariah Supervisory Committee (SSC) with experienced and knowledgeable persons in Islamic Jurisprudence (Ullah, 2014). However, the Board shall be responsible for any lapses/irregularities on the part of the SSC.

In November 2009, the Central Bank issued Guidelines for Conducting Islamic Banking in Bangladesh including the criteria for setting up full-fledged Islamic Bank (Alam et al., 2019). The guideline has been prepared based on the Banking Companies Act 1991, Companies Act 1994 and Prudential Regulations of Bangladesh Bank. However, this guideline should be treated as supplementary, not a substitute, to the existing banking laws, rules and regulations (Alam et al., 2020). In case of any point not covered under this guideline as also in case of any contradiction, the instructions issued under the Banking Companies Act and Companies Act will prevail.

The guideline defines "Shariah" as rules and regulations that have their origin in the holy Qur'an and Sunnah to govern all aspects of human life. The definition of Islamic bank is interesting as "Islamic bank" means such a banking company or an Islamic banking branch(es) of a banking company licensed by Bangladesh Bank, which follows the Islamic Shariah in all its principles and modes of operations and avoids receiving and paying of interest at all levels. The definition is elaborative as compared to the Islamic Financial Services Act (2013) in Malaysia as well as the Brunei Islamic Banking Order (2008). "Islamic Banking Business" refers to the banking business whose goals, objectives and activities comply the principles of Islamic Shariah and no part of the business either in form and substance contain any elements not approved by Islamic Shariah.

The following will be the broad criteria for consideration of setting up of the scheduled Islamic Commercial Bank:

1 The proposed bank company will be a public limited company and a minimum of 50% share shall be offered to the public.
2 All the financial transactions of the banking company shall be conducted based on the principles of Islamic Shariah.
3 The ways and means of resource mobilisation, expansion and nature of business transactions have to be stated in the application form.
4 The applicant(s) shall indicate expertise and other facilities available to them for ensuring the operation of their Islamic Banking business as per Islamic Shariah.
5 The banking company, to commence business, shall raise a minimum paid-up capital of Tk.2.00 billion and shall maintain the required capital adequacy ratio at all times, as prescribed by the Bangladesh Bank.

6 The minimum shareholding stake of each sponsor shall be Tk. 2.5 million, and the maximum shall be 10% of the proposed bank's total share capital. This ceiling of 10% applies to any individual, company or family member, either severally and jointly or both. The term "family" includes the spouse, father, mother, son, daughter, and brother of the individual or anyone dependent on that individual.
7 The sponsors' share shall not be transferred before the expiry of three years from the commencement of the business without permission from Bangladesh Bank.
8 The ceiling of 10% may be relaxed in case of a bank set up as a joint venture with a foreign financial institution or banking company.
9 Application shall stand disqualified if any of the sponsors and/or directors, their spouses or firms:

- has been convicted by a court of Law for the commission of any criminal offence inside or outside Bangladesh;
- has been involved in any illegal activity, particularly illegal banking business, receipt of deposits, financial dealings and other business;
- has failed to settle liability/meet obligations to banks and other financial institutions.

10 They shall furnish names of the banks/financial institutions along with the names of the branches with which they have had dealings. Bank reports are also required to be submitted;

- has failed to pay taxes/duties. They shall indicate their Tax Identification Numbers (TIN);
- is in the opinion of the sanctioning authority enjoys adverse reputation regarding integrity and performance.

3.3 India

3.3.1 Financial Regulators

The Reserve Bank of India was established on 1 April 1935, in accordance with the provisions of the Reserve Bank of India Act, of 1934 (Mohan, 2006). The Central Office of the Reserve Bank was initially established in Calcutta but was permanently moved to Mumbai in 1937. The Central Office is where the Governor sits and where policies are formulated (Reddy, 2006).

Though originally privately owned, since nationalisation in 1949, the Reserve Bank has been fully owned by the Government of India. The Preamble of the Reserve Bank of India describes the basic functions of the Rseserve Bank as:

> to regulate the issue of Bank Notes and keeping of reserves with a view to securing monetary stability in India and generally to operate the currency and credit system of the country to its advantage.

3.3.1.1 Main Functions

Monetary Authority: Formulates, implements and monitors the monetary policy (Mohan, 2008).

Objective: To maintaining price stability and ensuress adequate flow of credit to productive sectors.

Regulator and supervisor of the financial system: Prescribes broad parameters of banking operations within which the country's banking and financial system functions.

Objective: To maintain public confidence in the system, protect depositors' interest and provide cost-effective banking services to the public.

Manager of Foreign Exchange: Manages the Foreign Exchange Management Act, 1999.

Objective: To facilitate external trade and payment and promote orderly development and maintenance of the foreign exchange market in India.

Issuer of currency: Issues and exchanges or destroys currency and coins not fit for circulation.

Objective: To give the public an adequate quantity of supplies of currency notes and coins and in good quality.

Developmental role: Performs a wide range of promotional functions to support national objectives.

Related Functions: Banker to the Government: Performs merchant banking functions for the central and the state governments; also acts as their banker.

Banker to banks: Maintains banking accounts of all scheduled banks.

3.3.2 Banking Laws

The Reserve Bank of India has established a few important regulations to supervise the financial activities in the country. Some of these Acts are as follows

- Reserve Bank of India Act, 1934
- Public Debt Act, 1944/Government Securities Act, 2006
- Government Securities Regulations, 2007
- Banking Regulation Act, 1949
- Foreign Exchange Management Act, 1999
- Securitisation and Reconstruction of Financial Assets and Enforcement of Security Interest Act, 2002 (Chapter II)
- Credit Information Companies(Regulation) Act, 2005
- Payment and Settlement Systems Act, 2007
- Payment and Settlement Systems Regulations, 2008 and Amended up to 2011 and BPSS Regulations, 2008
- The Payment and Settlement Systems (Amendment) Act, 2015 - No. 18 of 2015
- Factoring Regulation Act, 2011
- Other relevant Acts
- Negotiable Instruments Act, 1881

- Bankers' Books Evidence Act, 1891
- State Bank of India Act, 1955
- Companies Act, 1956/Companies Act, 2013
- Securities Contract (Regulation) Act, 1956
- State Bank of India Subsidiary Banks) Act, 1959
- Deposit Insurance and Credit Guarantee Corporation Act, 1961
- Banking Companies (Acquisition and Transfer of Undertakings) Act, 1970
- Regional Rural Banks Act, 1976
- Banking Companies (Acquisition and Transfer of Undertakings) Act, 1980
- National Bank for Agriculture and Rural Development Act, 1981
- National Housing Bank Act, 1987
- Recovery of Debts Due to Banks and Financial Institutions Act, 1993
- Competition Act, 2002
- Indian Coinage Act, 2011: Governs currency and coins
- Banking Secrecy Act
- The Industrial Development Bank (Transfer of Undertaking and Repeal) Act, 2003
- The Industrial Finance Corporation (Transfer of Undertaking and Repeal) Act, 1993

3.3.3 Islamic Banking in India

In relation to the establishment of an Islamic bank in Kerala under the Kerala State Industrial Development Corporation (KSIDC), which is the designated agency for the formation of the bank, would hold an 11% stake in the proposed bank. According to government officials, it would be registered as a non-banking finance company, before being transformed into a full-fledged Shariah-compliant bank (Fasih, 2012). The project proposed to raise an initial capital of Rupees 500 crore (crore is 10 million in the Indian numbering system) from leading non-resident Indians (NRIs) and Indian business houses. Purely based on Shari'ah principles, the bank would avoid interest-based business activities. The proposed Kerala-based bank would invest funds in infrastructure projects and two instruments, bay' al-salam (deferred delivery) and istisna, have been identified for such investments. The bank would invest all its funds in wealth-generating investments and distribute profits to its shareholders. It would also set apart a social fund and provide interest-free loans to Gulf returnees to set up businesses or small-scale ventures.

In 2010, however, a rather more immediate problem reared its head, when Dr Subramaniam Swamy, president of India's Janata party and a former government minister, succeeded in putting the project on hold, issuing a writ in the High Court arguing that the involvement of government agencies in setting up an Islamic bank runs contrary to the secular principles enshrined in the Indian constitution (Beg, 2016). In February 2011, the Kerala High Court dismissed the writ, observing that they had no objection to KSIDC carrying on a business that was in accordance with Shari'ah law in addition to complying with the laws of the country

(D:\2024B\yaccob\author reviewed\21Shahid & Raj, 2019). They also stated that, although the institution was based on religious principles, its motive was not to propagate religion and the state's participation in it was based on purely commercial reasoning. High Court argues that the involvement of government agencies in setting up an Islamic bank runs contrary to the secular principles enshrined in the Indian constitution. The Reserve Bank of India (RBI) has proposed the opening of an "Islamic window" in conventional banks for the "gradual" introduction of Sharia-compliant or interest-free banking in the country.

Both the Centre and RBI have been exploring the possibility of introducing Islamic banking for a long to ensure financial inclusion of those sections of society that remain excluded due to religious reasons.

3.4 Pakistan

Section 3 of the State Bank of Pakistan Act states that "As soon as may be after the commencement of this Act, steps shall be taken to establish, in accordance with the provisions of this Act, a bank to be called the State Bank of Pakistan or Bank Daulat-e-Pakistan, for the purposes of taking over, as from the first day of July 1948, the management of the currency from the Reserve Bank of India, and carrying on the business of Central Banking." Section 3(2) further states that "the Bank shall be a body corporate by the name of State Bank of Pakistan or Bank Daulat-e-Pakistan, having perpetual succession and a common seal, and shall by the said name sue and be sued" (Arby, 2004).

The Islamic banking system is regulated by the State Bank of Pakistan (SBP), and within the purview of the State Bank of Pakistan Act 1956, the Policies for Promotion of Islamic Banking, detailed criteria for Setting Up Scheduled Islamic Commercial Bank based on Principles of *Shari'ah* in the Private Sector, detailed Criteria for Setting up of scheduled Islamic banking subsidiaries by existing commercial banks, guidelines for opening of standalone branches for Islamic banking by existing banks, as well as the fit and proper criteria for appointment of *shari'ah* advisors (Ayub, 2021). In Pakistan, there is a long history of attempts to Islamise the banking sector, notably since the President of Pakistan's announcement in February 1979 that interest was to be removed from the economy within three years (Rammal & Parker, 2013). A special Ordinance was passed in 1980 authorising *mudharabah* companies. A landmark decision in 1991 by the Federal *Shari'ah* Court called for the present financial system to be radically changed to bring it into conformity with the *Shari'ah*. As a result, the SBP is increasing its efforts to comply with that decision.

Accordingly, interest was eliminated from the operations of specialised financial institutions, including the House Building Finance Corporation, the National Investment Trust and the Mutual Funds of Investment Corporation of Pakistan, in 1979, and for the commercial banks from 1981 until 1985 (Zafar & Sulaiman, 2020). However, it is only in recent years that Islamic financial institutions have taken off. The first full-fledged Islamic financial institution was Al-Meezan Bank, which was established in 2002, while the first full-fledged foreign Islamic financial institution was Emirates Global Islamic Bank, which was established in 2007. Notwithstanding the above, conventional banks may operate Islamic windows via separate subsidiaries or at separate dedicated branches only.

Another market would be the insurance and capital market, which is regulated by the Securities and Exchange Commission of Pakistan (SECP) (Majeed & Zainab, 2017). The relevant law is the Securities and Exchange Ordinance 1969, and also the Takaful Rules 2005, which provide the framework for takaful operations in Pakistan. They allow SECP to set up a central *Shari'ah* board while requiring each takaful operator to have its own *Shari'ah* board and ensure that *Shari'ah* audits are carried out from time to time. The first takaful operator, Pak-Kuwait Takaful, was licensed in 2006, while the second, Takaful Pakistan Ltd., was licensed in 2007.

3.4.1 Financial Regulators

The State Bank of Pakistan (SBP) is incorporated under the State Bank of Pakistan Act, of 1956, which gives the Bank the authority to function as the central bank of the country. The SBP Act mandates the Bank to regulate the monetary and credit system of Pakistan and to foster its growth in the best national interest to secure monetary stability and fuller utilisation of the country's productive resources.

SBP's fsunctions are mainly governed by:

- SBP Act, 1956 (as amended up to 19 November 2015)
- Banking Companies Ordinance, 1962
- Foreign Exchange Regulations Act, 1947
- Payment System and Electronic Fund Transfer Act, 2007

Under the State Bank of Pakistan Order 1948, the Bank was charged with the duty to "regulate the issue of Bank notes and keeping of reserves to secure monetary stability in Pakistan and generally to operate the currency and credit system of the country to its advantage." The scope of the Bank's operations was considerably widened in the State Bank of Pakistan Act 1956, which required the Bank to "regulate the monetary and credit system of Pakistan and to foster its growth in the best national interest to secure monetary stability and fuller utilisation of the country's productive resources". Under financial sector reforms, the State Bank of Pakistan was granted autonomy in February 1994. On 21 January 1997, this autonomy was further strengthened by issuing three Amendment Ordinances (which were approved by the Parliament in May 1997) namely, the State Bank of Pakistan Act, 1956, the Banking Companies Ordinance, 1962 and the Banks Nationalisation Act, 1974. The changes in the State Bank Act gave full and exclusive authority to the State Bank to regulate the banking sector, to conduct an independent monetary policy and to set limits on government borrowings from the State Bank of Pakistan. The amendments in Banks Nationalisation Act abolished the Pakistan Banking Council (an institution established to look after the affairs of NCBs) and institutionalised the process of appointment of the Chief Executives and Boards of the nationalised commercial banks (NCBs) and development finance institutions (DFIs), with the State Bank having a role in their appointment and removal (Ahmed et al., 2013). The amendments also increased the autonomy and accountability of the Chief Executives and the Boards of Directors of banks and DFIs.

3.4.1.1 Functions

The traditional functions, which are generally performed by central banks almost all over the world, may be classified into two groups: (1) the primary functions including the issue of notes, regulation and supervision of the financial system, bankers' bank, lender of the last resort, banker to Government, and conduct of monetary policy; and (2) the secondary functions including the agency functions like management of public debt, management of foreign exchange, etc., and other functions like advising the government on policy matters and maintaining close relationships with international financial institutions.

The non-traditional or promotional functions, performed by the State Bank, include the development of a financial framework, institutionalisation of savings and investment, provision of training facilities to bankers and provision of credit to priority sectors. The State Bank also has been playing an active part in the process of Islamisation of the banking system. The main functions and responsibilities of the State Bank can be broadly categorised as follows:

3.4.2 Islamic Banking Laws

Below are several laws regulating the banking industry in Pakistan:

3.4.2.1 SBP Act

An Ordinance to provide for the establishment of SBP Banking Services Corporation.

- The Financial Institutions (Recovery of Finances) Ordinance, 2001
- Banking Companies Ordinance 1962 (as modified up to 02 April 2011)
- Banks Nationalization Act 1974 (as modified up to 30 June 2007)

3.4.2.2 F.E. Manual

- National Accountability Bureau (NAB) Ordinance 1999
- Microfinance Institutions Ordinance, 2001
- Payment Systems and Electronic Fund Transfer Act, 2007

3.4.2.3 Prudential Regulations

- Anti-Money Laundering and Combating the Financing of Terrorism (AML/CFT) Regulations & Guidelines

3.4.2.4 Pakistan Coinage Act

- Pakistan Coinage Act Amendment
- The Negotiable Instruments Act, 1881
- Electronic Transactions Ordinance, 2002
- Credit Bureaus Licensing Criteria
- Credit Bureaus Regulations
- Credit Bureau Act 2015
- Credit Bureaus Amendment Act 2016
- Exchange Companies Manual 2016
- Foreign Exchange Regulation Act, 1947
- Financial Institutions (Secured Transactions) Act, 2016

3.4.3 Dispute Resolution

Section 82A of the Banking Companies Ordinance 1962 states that the Appointment of Mohtasib shall be appointed by the President in consultation with the Governor of the State Bank of Pakistan (Khan et al., 2022). The Banking Mohtasib shall be a person of high integrity and unimpeachable banking or legal credentials who is not a shareholder of a banking company or financial institution and is not, and has not been, a bank defaulter. The jurisdiction of the Banking Mohtasib in relation to banking transactions shall be to:

1. enquire into complaints of banking malpractices;
2. perverse, arbitrary or discriminatory actions;
3. violations of banking laws, rules, regulations or guidelines;
4. inordinate delays or inefficiency;
5. corruption, nepotism or other forms of maladministration.

3.4.4 Banking Muhtasib

The Banking Mohtasib shall have the power and responsibility:

1. to entertain complaints from customers, borrowers, banks or from any concerned body or organisation;
2. to facilitate the amicable resolution of complaints after giving hearings to the complainant and the concerned bank;
3. to receive evidence on affidavit;
4. to issue a commission for the examination of witnesses.

In the event that complaints cannot be resolved by consent, to give a finding which shall be acted upon in the manner set out in the Act, a complaint shall be made on solemn affirmation or oath in writing addressed to the Banking Mohtasib. The complaint shall set out the full particulars of the transaction complained of and the name and address of the complainant. All reports submitted by the Banking Mohtasib shall be published and released to the public unless he directs otherwise for reasons to be recorded.

3.5 Sri Lanka

Before the establishment of the Central Bank, functions relating to central banking were conducted by the Currency Board System that was set up under the Paper Currency Ordinance No.32 of 1884 (Wijewardena, 2007). After gaining political independence, the Currency Board System was considered inadequate and unsuitable for meeting the needs of a developing country and an independent nation. Therefore, in July 1948, the Government of Ceylon requested the United States Government for technical expertise to set up a central bank, which resulted in Mr. John Exter, an economist from the Federal Reserve Board of the United States being appointed to carry out this task.

The Exter Report on the rationale and the legal framework for the Central Bank was presented to the Government of Ceylon in November 1949 and led to its formation (Weerasinghe, 2018). The Central Bank of Ceylon was established by the Monetary Law Act (MLA) No.58 of 1949 and commenced operations on 28 August 1950. It was renamed the Central Bank of Sri Lanka in 1985.

The Central Bank was given wide powers to administer and regulate the entire money, banking and credit system of the country (Gunawardena, 2005). The Central Bank was also given the sole right and authority to issue currency and it also became the custodian of the international reserves of the country. The objectives of the Central Bank as specified in the MLA in 1949 were as follows:

1 The stabilisation of domestic monetary values (maintenance of price stability).
2 The preservation of the par value or the stability of the exchange rate of the Sri Lankan Rupee (maintenance of exchange rate stability).
3 The promotion and maintenance of a high level of production, employment and real income in Sri Lanka.
4 The encouragement and promotion of the full development of the productive resources of Sri Lanka

3.5.1 Regulator

The Regulator for banking activities including Islamic banking is the Central Bank of Sri Lanka. The Central Bank's focus and functions have evolved since its formation, in response to the changing economic environment. In keeping with trends in central banking, the objectives of the Central Bank were streamlined by amending the Monetary Law Act (MLA) in 2002, to enable it to pursue its core objectives and to free it of the multiple objectives that were originally assigned to it. The Central Bank has two core objectives:

1 Maintaining economic and price stability
2 Maintaining financial system stability

3.5.2 Banking Laws

The financial system of the country is regulated by several legislative enactments (Ranawaka et al., 2023). The key Acts relevant to the regulatory role of the Central Bank are the Monetary Law Act, the Local Treasury Bills Ordinance, the Registered Stocks and Securities Ordinance, the Exchange Control Act, the Banking Act, the Finance Business Act, the Finance Leasing Act, Payment and Settlement Systems Act and Financial Transactions Reporting Act. The revised Banking Act No 30 of 1988, as amended in 2005, allows both commercial banks and specialised banks to operate on a Shariah-compliant basis, including "the acceptance of a sum of money in any manner or form from any person for a fixed period of time for investment in a business venture of the bank on the basis that profits or losses of the venture will be shared with the person from whom such money is accepted in a manner determined at the time the money is accepted." This landmark legislation came in after years of intensive discussions and lobbying by Amana Investments, the pioneer Islamic services provider in the country, with the Central Bank of Sri Lanka (Mihular, 2014).

There are 13 licensed insurance companies in Sri Lanka including one Takaful operator (Amãna Takaful). The total premium collected in 2004 amounted to US$239.11 million, while the total assets of the insurance companies amounted to US$705.57 million. The insurance penetration level measured in terms of total gross premium as a percentage of GDP is around 1.5% and the insurance density or per capita premium amounts to US$14.80. This offers a promising growth opportunity for strong insurance companies and banks that can offer bancassurance products. The Takaful concept in insurance is seeing increasing market acceptance, and two of the largest insurance operators have announced their intention to offer Takaful products to the market in 2006.

3.5.2.1 Monetary Law Act (MLA) No. 58 of 1949

An Act to establish the monetary system of Sri Lanka and the Central Bank to administer and regulate the system and to confer and impose upon the Monetary Board of the Central Bank powers, functions and responsibilities necessary for the purpose of such administration and regulation, and to provide for connected matters (Perera & Jayawickrema, 2014).

- Local Treasury Bills Ordinance No. 8 of 1923
- An ordinance to provide for the borrowing of money by the issue of Treasury Bills in Sri Lanka
- Registered Stock and Securities Ordinance No. 7 of 1937

An ordinance to make provision for the creation and issue of registered stock, government promissory notes and bearer bonds for the purpose of raising loans in Sri Lanka.

Exchange Control Act No. 24 of 1953: An Act to make provisions conferring powers, and imposing duties and restrictions, in relation to gold, currency, payments, securities, debts, and the import, export, transfer and settlement of property, to authorise the Central Bank to administer the provisions aforesaid, and to provide for matters connected therewith.

Banking Act No. 30 of 1988: An Act to provide for the introduction and operation of a procedure for the licensing of persons carrying on banking business; for the regulation and control of matters relating to the business of banking; and to provide for matters connected therewith or incidental thereto.

Finance Business Act No. 42 of 2011: An Act to provide for control and supervision of finance companies and to provide for matters connected therewith or incidental thereto.

Finance Leasing Act No.56 of 2000: An Act to provide for the regulation and monitoring of finance leasing businesses; to specify the rights and duties of lessors and lessees and suppliers of equipment; and for matters connected therewith or incidental thereto.

Payment and Settlement Systems Act No 28 of 2005: An Act to provide for the regulation of payment, clearing and settlement systems; for the disposition of securities on the books of the Central Bank; for the regulation of providers of money services; for the electronic presentment of cheques; and to provide for matters connected therewith or incidental thereto.

Financial Transactions Reporting Act No. 6 of 2006: An Act to provide for the collection of data relating to suspicious financial transactions to facilitate the prevention, detection, investigation and prosecution of the offences of money + of terrorism respectively; to require certain institutions to undertake due diligence measures to combat money laundering and the financing of terrorism; to identify the authority which will be responsible for monitoring the activities of all institutions to whom this Act applies; and to provide for matters connected therewith or incidental thereto.

Microfinance Act No 6 of 2016: An act to provide for the licensing, regulation and supervision of companies carrying on microfinance business; the registration of non-governmental organisations accepting limited savings deposits as microfinance non-governmental organisations; for the setting up of standards for the regulation and supervision of microfinance non-governmental organisations and micro-credit non-governmental organisations and to provide for matters connected therewith or incidental thereto.

References

Ahmed, R., Siddiqui, K., & Mufti, I. (2013). Islamic banking in Pakistan–Problems and prospects. *Asian Journal of Research in Banking and Finance*, *3*(7), 42–72.

Alam, M. K., Ab Rahman, S., Mustafa, H., Shah, S. M., & Hossain, M. S. (2019). Shariah governance framework of Islamic banks in Bangladesh: Practices, problems and recommendations. *Asian Economic and Financial Review*, *9*(1), 118.

Alam, M. K., Miah, M. S., Siddiquii, M. N., & Hossain, M. I. (2020). The influences of board of directors and management in Shariah governance guidelines of the Islamic banks in Bangladesh. *Journal of Islamic Accounting and Business Research*, *11*(9), 1633–1647.

Arby, M. F. (2004). *State Bank of Pakistan: Evolution, functions and organization. Munich Personal RePEc Archive*. https://mpra.ub.uni-muenchen.de/13614/ [accessed 11 January 2024].

Ayub, M. (2021). The corporate governance for Islamic financial Institutions in Pakistan: Need for dynamic regulatory policies. *In Monetary policy, Islamic finance, and Islamic corporate governance: An international overview* (pp. 37–60). Emerald Publishing Limited.

Beg, S. A. N. A. (2016). Prospects, problems and potential of Islamic banking in India. *International Journal of Accounting and Financial Management Research*, *6(*3), 9–20.

Fasih, F. (2012). Inclusive growth in India through Islamic banking. *Procedia-Social and Behavioral Sciences*, *37*, 97–110.

Gunawardena, C. A. (2005). *Encyclopedia of Sri Lanka*. Sterling Publishers Pvt. Ltd.

Hossain, M. (1988). *Credit for alleviation of rural poverty: The Grameen Bank in Bangladesh* (Vol. 65). Research Report of the International Food Policy Research Institute.

Iqbal, M., Molyneux, P., & Conermann, S. (2006). Thirty years of Islamic banking. History, performance and prospects. *Bankhistorisches Archiv*, *32*(2), 155–158.

Islamic Banking Order. (2008). *Negara Brunei Darussalam. Brunei Darussalam Central Bank, Bandar Seri Begawan.* https://www.bdcb.gov.bn/docs/regulatory/banking/islamic-banking-order-2008.pdf [accessed 02 March 2024].

Islamic Financial Services Act. (2013). *Laws of Malaysia. Bank Negara Malaysia, Kuala Lumpur.* https://www.bnm.gov.my/documents/20124/8102422b-e6dd-d149-8db0-e3637e89ed5c [accessed 13 February 2024].

Khan, H. A., Khan, F., & Barua, S. J. (2022). Alternative dispute resolution and sustainability of economic development for the Consumer's benefit: An Islamic moral economy perspective. *Asian Journal of Law and Policy*, *2*(2), 99–111.

Majeed, M. T., & Zainab, A. (2017). How Islamic is Islamic banking in Pakistan? *International Journal of Islamic and Middle Eastern Finance and Management*, *10*(4), 470–483.

Mihular, R. (2014). *Sri Lanka: Putting the pieces together. In Thiagaraja, S., Morgan, A., Tebbutt, A., & Chan, G. The Islamic finance handbook* (pp. 461–476). John Wiley & Sons: Singapore. https://doi.org/10.1002/9781118936863.ch27.

Mohan, R. (2006). Evolution of central banking in India. Reserve Bank of India. https://www.rbi.org.in/scripts/BS_SpeechesView.aspx?Id=285 [accessed 12 January 2024].

Mohan, R. (2008). The role of fiscal and monetary policies in sustaining growth with stability in India. *Asian Economic Policy Review*, *3(*2), 209–236.

Perera, R., & Jayawickrema, V. (2014). Monetary policy rules in practice: Evidence for Sri Lanka. *Munich Personal RePEc Archive, No. 95584*. https://mpra.ub.uni-muenchen.de/95584/ [accessed 20 January 2024].

Rammal, H. G., & Parker, L. D. (2013). Islamic banking In Pakistan: A history of emergent accountability and regulation. *Accounting History*, *18*(1), 5–29.

Ranawaka, P., Khatibi, A., Azam, S. F., & Tham, J. (2023). An overview of corporate governance laws and regulations applicable to the Banking industry in Sri Lanka. *International Journal of Research Publications*, *130*(1), 268–292.

Reddy, Y. V. (2006). Economic reforms and changing role of the Reserve Bank of India. India in a Globalising World: Some Aspects of Macroeconomy, Agriculture, and Poverty: *Essays in Honour of CH Hanumantha Rao*, 81.

Sarker, A. A. (2000). Regulation of Islamic banking in Bangladesh: Role of Bangladesh bank. *International Journal of Islamic Financial Services*, *2*(1), 45–56.

Shahid, M., & Raj, J. (2019). Islamic banking in India: An overview. *ZENITH International Journal of Multidisciplinary Research*, *9*(6), 246–252.

Ullah, H. (2014). Shari'ah compliance in Islamic banking: An empirical study on selected Islamic banks in Bangladesh. *International Journal of Islamic and Middle Eastern Finance and Management*, *7*(2), 182–199.

Weerasinghe, P. N. (2018). Evolution of monetary and exchange rate policy in Sri Lanka and the way forward. *Staff Studies*, *48*(1).

Wijewardena, W. A. (2007). *Central Banking Nearly six decades after John Exter (No. 62)*. Central Bank of Sri Lanka.

Zafar, M. B., & Sulaiman, A. A. (2020). Islamic banking in Pakistan: Emergence, growth, and prospects. *In Growth and emerging prospects of international Islamic banking* (pp. 61–77). IGI Global.

4 Islamic Finance in Central Asia

4.1 Introduction

This chapter covers regulatory arrangements for Islamic financial institutions in Central Asian countries such as Afghanistan, Azerbaijan, Kyrgyzstan, Russia and Uzbekistan. The chapter focuses on these central Asian economies as these are major economies in the region containing a large Muslim population with a clear demand for Islamic financial services. Hence, the regulators of these economies have established certain regulations to develop Islamic finance (IF) in their respective countries. The authors present an overview of different regulators and supportive laws for the development of IF in these jurisdictions.

4.2 Afghanistan

The Islamic Republic of Afghanistan is a country located in South and Central Asia where Muslims account for 99% of its total population of 31.6 million (Bhatnagar & Shahab Ahmed, 2021). The decades of war have destroyed the infrastructure and nearly every industry in Afghanistan. However, after the end of the Taliban regime in 2001, the transitional democratic government had scarce financial resources to rebuild and reform a stable infrastructure in the country (Rubin, 2000). The international community generously contributed to restoring the framework, infrastructure and standard of living in Afghanistan. During the restructuring phase, different profit and non-profit organisations donated money in different currencies which allowed reviving of the eroded banking sector. Table 4.1 shows the key developments during the evolution of conventional and Islamic banking in Afghanistan.

During the restoration of the banking sector, Islamic banking paved its way in Afghanistan between 2008 and 2009. The Da Afghanistan Bank (DAB; Central Bank of Afghanistan) took an initiative with the coordination of technical support of the Shariah Council comprised of well-known local religious scholars to draft the first Islamic banking law (Schoon, 2017). The initial draft of Islamic banking law to introduce Islamic banking was sent to parliament for approval. During the approval phase, approximately US$33 billion was circulating in the Afghanistan economy and most of this circulating cash was deposited in the Kabul Bank along

DOI: 10.4324/9781003471035-4

Table 4.1 Events of financial developments in Afghanistan.

Date	*Event*
2003	Emergence of Banking Law
2004	Emergence of DAB law
2008–2009	The first draft of Islamic banking was sent for government approval
February 2010	Kabul Bank started to offer Islamic banking services
2015	A regulatory framework was issued for Islamic banks
2018	Islamic Bank of Afghanistan was issued license

Source: (Authors, 2024).

with other 16 banks. In February 2010, Kabul Bank started an Islamic banking section and started to offer Islamic banking services in its all branches through Islamic banking windows.

Currently, seven Afghani banks (New Kabul Bank, Afghan Bank Mille, Afghan United Bank, Maiwand Bank, Ghazanfar Bank, Afghanistan International Bank and Azizi Bank) offer limited Islamic banking services through Islamic banking windows (Disli et al., 2023). These Islamic banking windows offer limited Islamic banking products such as musharakah, mudarabah and ijarah financing. These services are deemed insufficient based on the fact that there is a high demand for Islamic banking products and the majority of Afghan Muslims (57%) are unbanked and do not engage with conventional banks due to interest factors (Hamdard, 2011).

Ulamas and religious scholars actively promote Islamic banking in Afghanistan which signifies the need to establish a full-fledged Islamic bank to cater to the demand of Islamic banking services. DAB has shown a positive interest in developing a progressive environment by supporting a regulatory framework for the growth of Islamic banks.

DAB issued a regulatory framework for Islamic banks in 2015 to address the rules based on standards issued by the Bahrain-based Accounting and Auditing Organisation for Islamic Financial Institutions. Under these guidelines, DAB officially issued the first Islamic banking license to Bakhtar Bank and renamed it as Islamic Bank of Afghanistan (IBA) in 2018. IBA offers mudaraba savings account and Ijara financing products under Article 2.2 of the Afghanistan Bank Law Moreover, IBA also aimed to launch wealth management and digital banking services in the near future.

DAB issued detailed procedures and guidelines for Islamic banking regulation, products outline and Shariah parameters. Despite a supporting legal framework for Islamic banking, there are practical barriers to the actual growth of Islamic banks.

The lack of human capacity is a major challenge for Islamic banks as most of the Islamic bankers lack the required skill and knowledge to effectively deliver the actual Islamic banking services to customers (Rostan et al., 2021). Even though Kabul Bank has established a training centre to contribute to the capacity building, it will take time for employees to develop the required skills as the Islamic banking industry is still in its nascent stage in this country.

Additionally, Islamic banks need to extend their product range such as takaful and sukuk issuance to attract foreign investment and enhance public awareness. The active government support, demand for Islamic banking and gradually developing infrastructure are optimistic gestures for the growth of IF in Afghanistan.

4.2.1 Regulators

The banking laws and regulations in Afghanistan emerged during peace restoration in the country, particularly the lending activities were introduced in the banking system (Pavlović & Charap, 2009). The most important development in the banking regulation was the emergence of the DAB Law in 2004. Initially, the international community and donors generously contributed to assist in strengthening DAB to perform regulation, supervision and monitory operations. The banking law evolved in September 2003 which laid the foundation of banking supervision in Afghanistan. This law granted a mandate to the Central Bank to act as the main regulating and monitoring authority of commercial banks.

The financial institutions in Afghanistan can be divided into licensed Banks and non-bank institutions. These financial institutions are licensed under Article 6(1) of the Law of Banking in Afghanistan. As of March 2019, there are 14 banks and 3 non-bank institutions in Afghanistan. The banks are further divided into:

- State-owned (Bank Millie Afghan, Pashtany Bank, New Kabul Bank),
- Eight private (Azizi Bank, Afghanistan International Bank, Islamic Bank of Afghanistan, Maiwand Bank, Afghan United Bank, Ghazanfar Bank, Arian Bank)
- Three branches of foreign banks (Habib Bank Limited, National Bank of Pakistan, Bank Alfalah Ltd.).

The non-banks institutions include Electronic Money Institutions, FXDS and PSPS.

These financial institutions are regulated under the banking laws of Afghanistan which are as follows:

4.2.1.1 Financial Leasing Law (Statute Law)

This Law is disseminated on 17 March 2014 under Articles 10 and 14 of the Constitution of Afghanistan for the regulation of financial lease affairs. The main objective of this law is to promote small and medium-level investments through financial leases and provide job opportunities through this law under Article 2 (Disli & Jalaly, 2024).

4.2.1.2 Da Afghanistan Bank Law

Article 1 (1.1) describes the establishment of the Central Bank DAB in 2003 under this law. Article 2 describes that the main function of DAB is to maintain domestic price stability. DAB enjoys the power of being a financial regulator for all the licensed institutions (Article 3). The highest policy and decision-making body (Article 6) in DAB is known as the Supreme Council and it will be charged by a Governor who shall be responsible for internal control and audit of the DAB's operation and other staff-related issues.

4.2.1.3 Afghanistan Banking Law

This law was formulated under Article 79 of the Constitution after the approval of the Cabinet of the Islamic Republic of Afghanistan under resolution number 17. Article 2 of this law describes the main purpose of this law which gives mandate to government to regulate the banking affairs under articles 10 and 11 of the Constitution of Afghanistan.

4.2.1.4 Counter Financing of Terrorism

This law was enacted under Article 7 of the constitution of the Islamic Republic of Afghanistan to implement the International Convention for the Supersession of Financing of Terrorism. Article 2 of this law prohibits the provision of funds or property for terrorist acts, terrorist organisations or terrorists.

4.2.1.5 Anti-money Laundering and Proceeds of Crime

This law was passed under Article 7 of the Constitution to protect and promote the financial integrity of Afghanistan. The key role of this law under Article 5 is to fight against the use of financial institutions and designated non-financial businesses and professions (DNFBPS).

4.2.1.6 Negotiable Instrument Law

This law was enacted in February 2007 for the facilitation of business affairs to create easy access to credit for businesses that are looking to expand commerce and investment under Article 2.

4.2.1.7 Secured Transaction of Movable Property Law

This law was enacted on 31 May 2009 to pave the way for the recovery of debt using Movable property through mutual trust between parties involved in a banking transaction under Article 2.

4.2.1.8 Law of Mortgage on Immovable Property

This law was enacted in June 2009 to regulate the affairs concerning the mortgage of immovable property in banking transactions and determine the rights and obligations of involved parties.

4.2.2 Financial Regulatory Authorities

There are two key regulatory authorities to regulate the financial institutions in Afghanistan:

4.2.2.1 Da Afghanistan Bank

The Royal Decree of King Zahir in 1940 established DAB. Initially, its operations were limited to printing money and covering budget deficits. Article 2 of the DAB law describes that the bank's primary objective is to maintain domestic price stability. DAB is a key contributor in the formulation of internal economic policies to promote sustainable economic growth with prejudice to its primary objectives. Additionally, it is responsible to:

Structure the Monetary policy of Afghanistan under the Article 62 of DAB Bank Law, 2003
Formulate currency policy and Afghanistan currency arrangements
Effectively manage official foreign exchange reserves of Afghanistan
Print, mint and issue coins and banknotes
Act as a fiscal policy advisor to the state
License, regulate and supervise banks, foreign exchange dealers, money services providers, payment system operators, securities services providers and securities transfer system operators.

4.2.2.2 Financial Supervision Department

The financial supervision department (FSD) was established in late 2003 after the enactment of DAB and Banking Law to discharge DAB's regulatory responsibilities. This department is responsible for on-site and off-site financial reviews of the banks on a quarterly basis. Currently, FSD has 165 professional staff members who supervise the financial system of Afghanistan. The key objective of FSD is to ensure the financial stability of all licensed banks that provide both Islamic and conventional banking services.

4.3 Azerbaijan

4.3.1 Regulators

After gaining its independence in 1991, Azerbaijan transformed its economy from a central planning economy to a free market economy. The existing communist system in those days was not suitable for the creation of banking protocols; therefore, Azerbaijan adopted new procedures to create two different types of banking systems known as central and commercial banking (Mikayılov & Muxtarov, 2013).

In the initial phase of independence, political instability and the Armenian hostile invasion deeply affected the country's economy. After regaining its political stability in 1994, Azerbaijan started to implement relevant policies to recover from economic crises. These policies included a legal framework for states and banks which were one of the cornerstones of its economy. During this era, the banking sector started to develop and provide significant support in the development of the real sector (Mikayılov & Muxtarov, 2013).

In the beginning of the free market economy, the banking sector was not fully regulated and it was not internally controlled and audited, which resulted in the emergence of several local and foreign banks. However, soon new legislation was created and implemented on banks, especially on founding capital amount requirement. The government of Azerbaijan continues to create new legislation to ensure the stability of the banking sector and provide protection to the public. The central bank terminated the licenses of 11 banks in 2016 and another 2 in the subsequent year. Currently, there are 2 state-owned and 28 private licensed banks in Azerbaijan.

4.3.2 Financial Regulations

Some of the laws implemented in the banking sector are as follows:

4.3.2.1 Constitution of the Republic of Azerbaijan

This law established the Central Bank of Azerbaijan and regulates relevant statutory acts along with international treaties to which the Republic of Azerbaijan is a signatory.

4.3.2.2 Civil Code of Azerbaijan

It was enacted in 2000, and all credit unions, Central banks, National banks, non-bank credit and insurance institutions in Azerbaijan are regulated under the Civil Code of Azerbaijan. Additionally, all microfinance non-bank credit institutions (NBCIs) are registered as limited liability companies (LLCs) and are regulated under Articles 87–96 of the Civil Code of Azerbaijan (Isazade, 2010).

4.3.2.3 Humanitarian Organizations Guidelines of 2009

Non-bank credit organisations established with humanitarian status are regulated by the guidelines issued in 2009. These guidelines provide a platform for local and international non-bank credit organisations to transform into either non-commercial or commercial entities.

These banking legislations are implemented by the regulatory authorities of Azerbaijan charted in Figure 4.1.

4.3.2.4 Central Bank of the Republic of Azerbaijan

Article 14 of the Constitutional Law of the Republic of Azerbaijan provides the legal basis for an independent banking system and the circulation of money (Mammadov & Ahmadov, 2021). A presidential decree in 1992 established a national bank of the Republic of Azerbaijan known as the Central Bank. The Civil Code of

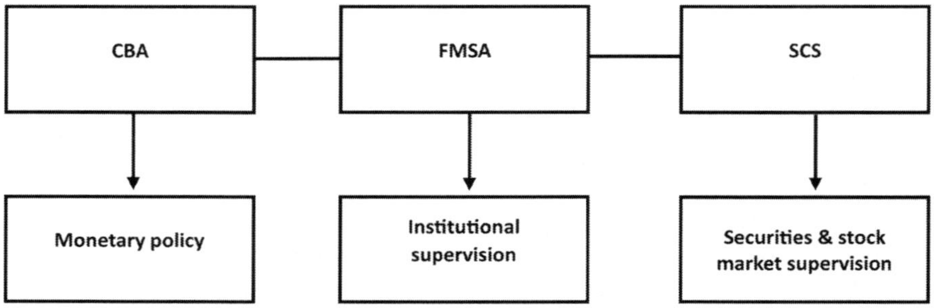

Figure 4.1 Key regulators of the Azerbaijan banking industry.

Azerbaijan (Article 2) provides it the mandate of being an independent authority. Item II of Article 19 of the Constitution of the Republic of Azerbaijan declares it the sole property of the state. The main objectives of the Central Bank are to:

- Set and implement the monetary and exchange rate policy (Article 5)
- Organise money circulation (Item II of Article 19 of Constitutional Law)
- Regularly set and announce an official exchange rate of Manat
- Regulate and develop centralised interbank and other licensed payment systems
- Maintain and manage gold and foreign exchange reserves

4.3.2.5 Financial Markets Supervision Authority (FMSA)

The financial markets are regulated and maintained through an executive authority known as FMSA. Article 6 of the Civil Code of Azerbaijan grants mandate to FMSA reserve an exclusive mandate to issue and revoke banking licenses, and permits to open branches, departments and representative offices.

4.3.2.6 State Committee for Securities

The security and stock markets are regulated through the State Committee for Securities (SCS) under the Auspices of the President of Azerbaijan. The Committee was established in 2008; however, there is no indication of its activities affecting the banking sector.

4.3.3 Islamic Banking

The Republic of Azerbaijan is a country in the South Caucasus region, located at the crossroads of Southwest Asia and Southeastern Europe. Muslims account for 99% of its total population of 9.9 million which indicates the need to establish IF in the country. Despite having a large Muslim population, IF industry in Azerbaijan is still at an embryonic stage. However, the government has started to make significant efforts to promote and establish IF in the country (Gresh, 2007).

On the commercial scale, Kovsarbank attempted to launch comprehensive Islamic banking services in 2008. However, the Central Bank of Azerbaijan revoked its license within a few months of operation and ultimately bank had to terminate its operation (Mamedov & Gasimov, 2021). After the initial ignition of IF and the financial crises of 2008, the government started to realise the resilience and the need for Islamic banks. This time, the country's largest state-owned bank, the International Bank of Azerbaijan (IBA) started to offer Islamic banking services through Islamic banking windows. The Islamic banking assets of IBA rose to USD$520 million at the end of year 2014 as compared to USD$160 million in 2013. Nonetheless, the bank had to terminate the Islamic banking department due to structural changes and get rid of distressed assets and poor management.

Even though there are significant growth indicators such as a high human development rate on par with Europe, a low unemployment rate and a growing economy yet, there is no Islamic bank in Azerbaijan to provide comprehensive Islamic financial services. Altogether, six banks (IBA, Kauthar Bank, the Islamic Development Bank Group, Tauran Bank, Azeri Bank and Nikoil Bank) provide limited Islamic banking services in the country.

Recently, global Islamic financial services providers and local authorities have made significant efforts to develop Islamic banking services in Azerbaijan. The private sector arm of the Islamic Development Bank (IDB) Group, the Islamic Corporation for the Development (ICD), aims to launch Islamic banking services in Azerbaijan and acts as a stimulator to initiate changes in the country's banking legislation to make room for Islamic banking (AzerNews, 2017).

ICD and the Association for Azerbaijan-Arab Countries Cooperation (ACCA) signed a memorandum of understanding (MOU) to cooperate and facilitate information and expertise exchange for the effective development of Islamic banking. Both institutions agreed to organise joint investment forums, professional training programmes, market research, workshops, publications, study tours, qualifications and many more. Similarly, two other Islamic financial institution, AlHuda Centre of Islamic Banking and Economics (CIBE) and Studying of Economic Resources (SER) signed an agreement to promote and develop Islamic banking and finance in Azerbaijan and the rest of the Caucasus countries. The agreement aims to create an exclusive strategy for research, product development, Shariah advisory and capacity building initiatives for the sustainable development of IF in Azerbaijan and neighbouring regions (CIBE, 2017).

Despite conserving serious efforts for the expansion of IF in the country, without support from local regulatory authorities, it is vague to expect Islamic banking to fully develop in Azerbaijan. The absence of a supporting legal environment to safeguard and recognise Islamic banking, unfriendly conditions of the Central bank, absence of an interest-free financial marketplace, lack of linkage institutions to provide information and support to new entrants, talent shortage and low awareness and understanding are alarming factors to halt Islamic banking development in the country.

The plethora of opportunities created by flourishing IF in the country may positively affect the government of Azerbaijan to create and enable a supporting playing field for Islamic banks. IF can be labelled as an asset-based or participatory finance to create a wider acceptance. IF can help in the infrastructure development of Azerbaijan by acting as an alternative financing mechanism such as Sukuks and by attracting foreign investors from GCC and Asia.

4.4 Kyrgyzstan

In the Kyrgyz Republic, a bank is a financial institution established to attract deposits from individuals and legal entities and other funds, and allocation on its behalf, based on terms of collectability, repayment, maturity and implementation of settlements in accordance with the client's instructions Kloc & Woodward, 1995). The Bank is a legal entity, established as a joint stock company and implementing its activity in accordance with the legislation of the Kyrgyz Republic on the basis of license, issued by the National (Central) Bank of the Kyrgyz Republic (hereinafter referred to as the Bank of Kyrgyzstan). The term "Bank" or a word combination with this term may be used in a firm name and for advertising purposes only by institutions with the right to carry out banking operations in compliance with this Law and the Law of the Kyrgyz Republic "On the National Bank of the Kyrgyz Republic." Requirements to a bank, prescribed by this Law, can be made to other financial institutions licensed by the Bank of Kyrgyzstan in the manner prescribed by the Kyrgyz Republic legislation and regulations of the Bank of Kyrgyzstan.

4.4.1 Regulators

The National Bank of the Kyrgyz Republic performs the following basic functions (Yadav, 2016):

1 Develops, determines and implements the monetary policy in the Kyrgyz Republic;
2 Exercises regulation and supervision over activities of banks and finance and credit institutions licensed by the National Bank of the Kyrgyz Republic;
3 Develops and conducts the uniform currency policy;
4 Has the sole responsibility to issue banknotes;
5 Promotes effective functioning of the payment system;
6 Establishes rules for conducting banking operations, accounting and reporting for the banking system.

4.4.2 Islamic Banking Laws

In Kyrgyz, the Law on Banking Transactions and Operations based on Islamic Financing Principles and the Regulation of Implementation of Islamic Financing Principles (Draft 2007) provides for, among others, the licensing of Islamic financial institutions. It imposes the requirement for an Islamic financial institution to have its own *Shari'ah* board and allows Islamic financial institutions to follow the AAOIFI *Shari'ah* Standards (Nagimova, 2022). It also specifies a range of *Shari'ah*-based products to be offered by Islamic financial institutions. The whole Islamic banking system is regulated by the National Bank of the Kyrgyz Republic.

"Islamic bank" operates in accordance with the Islamic principles of banking and financing at the availability of the required license of the Bank of Kyrgyzstan. The bank, having an "Islamic window" represents a bank, rendering the services in accordance with the Islamic principles of banking and financing within the frames of the corresponding division ("Islamic window") if the required license issued by the Bank of Kyrgyzstan is available. Islamic principles of banking and financing are the regulations of bank operations implementation in accordance with the standards of Shariah in the order, established by regulatory acts of the Bank of Kyrgyzstan. "Islamic window" is a corresponding division of the bank, implementing operations in accordance with the Islamic principles of banking and financing. Standards of Shariah are the whole set of regulations for the performance of economic or any other activity, developed and approved by international organisations, establishing the standards of Islamic banking management (Accounting and audit organisation for Islamic financial institutions and the Council of Islamic Financial Services). Shariah Council is a body of the bank, performing operations in accordance with the Islamic principles of banking and financing. This Council is responsible for compliance of the Bank transactions (contracts) in accordance with the standards of Shariah.

4.4.3 Islamic Banks in Kyrgyzstan

Article 13 of the Licensing of Banking Transactions states as follows;

1 A bank shall receive the status of a legal entity from the moment of the state registration in compliance with the legislation of the Republic.
2 To start its activity, the bank shall receive the license of the Bank of Kyrgyzstan for the implementation of banking activity for the fee amount that is established in compliance with the legislation of the Kyrgyz Republic.
3 The Bank of Kyrgyzstan is the only institution on the territory of the Kyrgyz Republic, authorised to issue licenses for performing banking operations, including operations conforming with the Islamic principles of banking and financing. The license for banking operations is given in the order, established by the Bank of Kyrgyzstan. The information about banks, having the license of the Bank of Kyrgyzstan, is provided in the mass media once a year.
4 A bank, its affiliates and representative missions shall not be subject to registration in the local government agencies.
5 Upon consideration of the documents provided by Article 14 of the present Law, the Bank of Kyrgyzstan may give preliminary consent to issue the license to a bank.

That means that the license will be given to the applicant only after it determines the staff structure, pays the necessary minimum capital, presents the state registration certificate and performs other requirements provided by the legislation of the Kyrgyz Republic and the Bank of Kyrgyzstan.

The license shall be issued to the applicant after they receive preliminary permission and the Bank of Kyrgyzstan makes sure the following are implemented:

1 A state registration certificate is submitted;
2 The required minimum authorised capital is fully paid;
3 Bank premises and equipment meet requirements prescribed by the Bank of Kyrgyzstan to ensure the security of deposits;
4 The founders selected a bank's independent external auditor for the forthcoming financial year;
5 The bank selected its personnel in compliance with the requirements of the Bank of Kyrgyzstan and tested information systems for carrying out banking activity;
6 The bank met other requirements prescribed by the current legislation.

- Licenses issued by the Bank of Kyrgyzstan shall have unlimited terms of effect and shall be effective on all the territory of the Kyrgyz Republic.
- Licenses issued by the Bank of Kyrgyzstan shall be nominal (shall not be alienated) and shall not be subject to transfer to third parties.

7 All banking transactions may be carried out only if such transactions are expressly provided in the license.
8 B transactions performed without the license of the Bank of Kyrgyzstan shall be considered void. Unlicensed banking activity shall entail the liability in accordance with the legislation of the Kyrgyz Republic.

4.5 Russia

Article 75 of the Constitution of the Russian Federation establishes a special legal status for the Bank of Russia, gives it the exclusive right to issue currency (Part 1) and protect the ruble and ensures its stability, which is the main function of the Bank of Russia (Part 2). The status, purposes, functions and powers of the Bank of Russia are also spelled out in Federal Law No. 86-FZ, dated 10 July 2002 (Perotti, 2002). "On the Central Bank of the Russian Federation (Bank of Russia)," and other federal laws.

4.5.1 Regulators

According to Article 3 of the Federal Law "On the Central Bank of the Russian Federation (Bank of Russia)," the goals of the Bank of Russia are to protect the ruble and ensure its stability, promote the development and strengthen the Russian banking system, ensure the stability and development of the national payment system, develop the financial market of the Russian Federation and ensure its stability.

A key element of the legal status of the Bank of Russia is its independence, which implies, above all, that the Bank of Russia is a special public and legal institution that has the exclusive right to issue currency and organise its circulation (Subbotina, 2009). The Bank of Russia is not a body of state power, but its powers are, in effect, the functions of a body of state power, because their implementation implies the use of state compulsion. The Bank of Russia performs the functions and exercises the powers stipulated by the Constitution of the Russian Federation and the Federal Law "On the Central Bank of the Russian Federation (Bank of Russia)" independently from the federal bodies of state power, regional authorities and local governments. Its independent status is codified in Article 75 of the Constitution of the Russian Federation and Article 1 and Article 2 of the Federal Law 'On the Central Bank of the Russian Federation (Bank of Russia)'.

The legislative powers of the Bank of Russia imply its exclusive right to issue regulations binding on the federal bodies of state power, regional authorities, local governments and all legal entities and private individuals on matters placed within its competence by the Federal Law "On the Central Bank of the Russian Federation (Bank of Russia)" and other federal laws. According to Article 104 of the Constitution of the Russian Federation, the Bank of Russia does not have the power to initiate legislation, but it is involved in the law-making process not only by virtue of issuing its own regulations but also owing to the fact that the drafts of federal laws and statutory acts of federal bodies of executive power concerning the performance by the Bank of Russia of its functions must be submitted to the Bank of Russia for consideration and approval (Vysman, 1993).

The Bank of Russia is a legal entity. Its authorised capital and other property are federal property. Nevertheless, the Bank of Russia has both proprietary and financial independence. It exercises its powers to own, use and manage its property, including international reserves, in compliance with the purposes and according to the procedure established by the Federal Law "On the Central Bank of the Russian Federation (Bank of Russia)." Bank of Russia property may not be seized or encumbered without its consent unless the federal law stipulates otherwise. The financial independence of the Bank of Russia implies that it covers its expenses from its own income. The Bank of Russia may defend its interests in court, including international courts, the courts of foreign states and courts of arbitration.

The state is not liable for the Bank of Russia's obligations, just as the Bank of Russia is not liable for the state obligations, unless they have assumed such obligations or unless federal laws stipulate otherwise (Åslund, 2012). The Bank of Russia is not liable for the obligations of credit institutions and non-credit financial institutions, while the credit institutions and non-credit financial institutions are not liable for the obligations of the Bank of Russia, except the cases in which the Bank of Russia or credit institutions and non-credit financial institutions assume such obligations.

The Bank of Russia is accountable to the State Duma of the Federal Assembly of the Russian Federation (hereinafter referred to as the State Duma), which appoints and dismisses the Bank of Russia Governor (on the proposal of the President of the Russian Federation) and members of the Bank of Russia Board of Directors (on the proposal of the Bank of Russia Governor, agreed with the President of the Russian Federation), sends and recalls its representatives in the National Financial Board within its quota and considers the guidelines for the single state monetary policy and Bank of Russia annual report and takes decisions on them. On the proposal of the National Financial Board, the State Duma may take the decision to get the Audit Chamber of the Russian Federation to audit the financial and economic activities of the Bank of Russia and its divisions and establishments. In addition, the State Duma holds parliamentary hearings on the Bank of Russia's activities with the participation of its representatives and hears the reports by the Bank of Russia Governor about the Bank of Russia's activities in the course of presenting the annual report and guidelines for the single state monetary policy.

4.5.2 Banking Laws

The law to regulate banking activities is under the Federal Law on Banks and Banking Activities. The Central Bank of Russia performs its functions in compliance with the Constitution of the Russian Federation, Federal Law "On the Central Bank of the Russian Federation (Bank of Russia)" and other federal laws. According to Article 75 of the Constitution of the Russian Federation, the principal function of the Bank of Russia is to protect the ruble and ensure its stability (Vernikov, 2007). The Bank of Russia is the sole issuer of currency. Pursuant to Article 4 of the Federal Law "On the Central Bank of the Russian Federation (Bank of Russia)," the Bank of Russia performs the following functions:

- In collaboration with the federal government, it elaborates and implements a single state monetary policy;
- In collaboration with the federal government, it elaborates and implements the policy towards developing the financial market of the Russian Federation and ensuring its stability;
- It is the sole issuer of cash and organiser of cash circulation;
- It approves the graphic symbol of the ruble as a sign;
- It is the creditor of last resort for credit institutions and it organises the credit institution refinance system;
- It sets the settlement rules in the Russian Federation;
- It exercises supervision and oversight of the national payment system;
- It sets the rules for conducting banking operations;
- It services budget accounts of all levels of the Russian budget system, unless the federal laws stipulate otherwise, by effecting settlements at the instruction of the authorised bodies of executive power and government extra-budgetary funds, which are assigned the task of organising the execution of and executing the budgets;
- It efficiently manages the Bank of Russia's international reserves;
- It takes the decision on the state registration of credit institutions, issues banking licences to credit institutions and suspends and revokes them;
- It makes decisions on the state registration of non-governmental pension funds;
- It supervises the activities of credit institutions and banking groups;
- It exercises regulation, control and supervision over the activities of non-credit financial institutions in compliance with federal laws;
- It registers securities issues and prospectuses and registers reports on the results of securities issues;

- It exercises control and supervision over the compliance by issuers with the requirements of federal legislation on joint-stock companies and securities;
- It exercises regulation, control and supervision over corporate governance in joint-stock companies;
- It conducts independently or at the instruction of the Russian Government all types of banking operations and other transactions necessary for the performance of Bank of Russia functions;
- It organises and exercises foreign exchange regulation and foreign exchange control pursuant to federal legislation;
- It sets the procedure for effecting settlements with international organisations, foreign states and legal entities and private individuals;
- It approves industry accounting standards for credit institutions, the Bank of Russia, and non-credit financial institutions, the chart of accounts for credit institutions and the procedure for its application, the chart of accounts for the Bank of Russia and the procedure for its application;
- It sets and publishes official exchange rates of foreign currencies against the ruble;
- It takes part in the compiling of Russia's balance of payments forecast and organises the compiling of Russia's balance of payments;
- It takes part in the development of the methodology for compiling Russia's financial account within the national account system and organises the compiling of Russia's financial account;
- It keeps official statistical records of direct investments to and from Russia in compliance with federal legislation;
- It establishes independently the statistical methodology of direct investments to and from Russia, the list of respondents approves the procedure for their submitting of primary statistical data on direct investments, including the methods of federal statistical review;
- It analyses and makes forecasts for the situation in the Russian economy and publishes the corresponding materials and statistical data;
- It pays compensation for household deposits with bankrupt banks uncovered by the compulsory deposit insurance system in the cases and according to the procedure established by the federal law;
- It is the depository of the IMF funds in the Russian currency and it conducts operations and transactions provided by the Articles of Agreement of the International Monetary Fund and the agreements with the International Monetary Fund;
- It exercises control over compliance with the requirements of federal legislation on countering the illegal use of insider information and market manipulation;
- It protects the rights and legitimate interests of shareholders and investors in the financial markets, insurers, insured persons and beneficiaries recognised as such in accordance with the insurance legislation, and also insured persons in the system of compulsory pension insurance, depositors and participants of a non-governmental pension fund in the system of non-governmental pension insurance;
- It performs other functions in compliance with federal laws.

The legal regulation of banking activities shall be provided by the Constitution of the Russian Federation, this Federal Law, the Federal Law on the Central Bank of the Russian Federation (Bank of Russia), other federal laws and Bank of Russia regulations.

Art 75 of the Constitution of the Russian Federation states as follows:

- The monetary unit of the Russian Federation is the ruble. The Central Bank of the Russian Federation is the sole issuer of currency. The introduction and issuance of other currencies in the Russian Federation are prohibited.
- To protect the ruble and ensure its stability is the principal function of the Central Bank of the Russian Federation, which it performs irrespective of other public authorities.

The system of federal taxes and general principles of taxation and levies in the Russian Federation are stipulated by federal laws. Government loans are issued in the procedure specified by federal law and placed on a voluntary basis.

4.5.3 Islamic Banking in Russia

The first bank to implement the Islamic financial system in Russia was the Badr-Forte Bank in Moscow. It was licensed in 1991 and started using IF methods in 1997.

The Badr-Forte Bank, its Russian name being International Commercial Bank, is the only structure in the banking sphere guided by Islamic business rules. Licensed by the Russian Federation Central Bank in 1991, it has been using Islamic funding methods since 1997 (Hidayat et al., 2020). Significantly, it is not called Islamic, yet its charter says that the bank has the right "to act according to Russian and international laws by applying Islamic economic banking technologies which do not contradict the banking laws of Russia." The Badr-Forte Bank is a member of the General Council of Islamic Banks under the IDB: its operations have been recognised at the international level as corresponding to the Shari'a. On 14 August 1992, it was announced that a United Islamic Commercial Bank Inc. was established by KamAZ, ElAz, the West Siberian Metallurgical Combine, the Islamic Cultural Center in Moscow and the All-Russia Tatar Cultural and Educational Center. The initiative of the leaders of the Union of Industrialists and Entrepreneurs of Russia was supported by the administration of the Kemerovo Region. Administration Head Aman Tuleev was asked to be the bank's president.

This structure was set up for practical rather than religious reasons: It aimed to develop the scientific and technical potential of the CIS countries and encourage business activities of enterprises by attracting foreign capital primarily from the Muslim states. Business communities of Saudi Arabia, the UAE, Turkey and the Central Asian states showed a certain interest in the project, which was not realised – the bank never went into operation.

In December 2006, the Central Bank of Russia suddenly stripped the bank of its license on account of continuous non-compliance to standard banking regulations and the Russian federal law on countering the legalisation of illegal earnings (money laundering) and the financing of terrorism. Since then, many other Islamic banking organisations were set up, but most eventually closed down: Amal Financial House, a subsidiary of Bulgar Bank in Yaroslavl, lost its license in 2017, Ellips Bank in Nizhniy Novgorod and Express Bank in Dagestan, established in 2011, were also branded as inefficient and eventually got liquidated in 2013.

They were not banks, but Islamic "windows" inside conventional banks. Since Sharia laws prohibit acceptance of specific interest or fees, these "windows" helped bypass Russian law that states that banks must always charge fees.

The law states that every first transaction of each client requires a notification from a bank supervisor. Badr-Forte bank, according to the regulator, did not fully comply with this law and had suspicious transactions to the tune of 33.9 billion rubles (about $550 million).

In December 2016, TASS news agency reported that Russia and Iran are exploring the establishment of an Islamic bank as the two countries expand their economic cooperation. The entire banking system in Iran must abide by Islamic principles such as "a ban on interest and gambling, which foreign firms would need to follow if they are to grab business ranging from long-term financing to insurance." The creation of a Russian-Iranian Islamic bank was discussed. No timeframe or potential size was given for the proposed bank, which would be the first such institution operating in the two countries.

Later, in 2016, Tatagroprombank launched The Partnership Banking Center, which operated under Sharia (Islamic laws), and many analysts believed in its success. The centre also hoped that investments would start flowing to Kazan, the so-called Muslim capital of Russia, which is the capital of the internal republic of Tatarstan.

The Partnership Banking Center started operations in March 2016 as a subsidiary of LLC Tatagroprombank. The centre says it provides a wide range of banking services irrespective of the religious beliefs of clients. It works with both individuals and companies and has an agreement with the Islamic Development Bank in Saudi Arabia.

On March 9, the Russian State Duma (lower house of parliament) decided to reject the bill that would let Islamic banking fully and legally operate in Russia. Rais Suleimanov from the National Strategy Institute believes that one of the reasons for the rejection was the vague statement from Deputy Dmitriy Savelyev who brought in the bill, according to a report on EADaily.

On 24 September 2004, the State Duma passed amendments to the Land Code which transferred, free of charge, land under the buildings and other facilities used for religious and charity purposes to the religious organisations that owned them. Those religious organisations that do not own the buildings in which they function will receive the land under them gratis for the entire period they continue using the buildings. These amendments were intended to favour the Russian Orthodox Church; however, other confessions profited from them as well. When summing up the results, those who took part in the seminar "Waqf and Its Prospects in Russia Today" recommended that the Council of the Muftis of Russia set up, together with the IDB and the General Secretariat of the Waqfs of Kuwait, a work group to study whether the principles of waqf property can be adjusted to Russia's legislation, as well as render assistance, methodologically and organisationally, to create a single register of waqfs in Tatarstan.

4.6 Uzbekistan

4.6.1 Regulators

Article 1 of the Law of the Republic of Uzbekistan "About the Central Bank of the Republic of Uzbekistan" states that the Central Bank is a legal entity and is the exclusive property of the state (Akimov & Dollery, 2006). It is enacted by Article 124 of the Constitution that the banking system of the Republic of Uzbekistan is headed by the Central Bank of the Republic.

4.6.1.1 Central Bank of Uzbekistan

Central Bank is an economically independent institution exercising its costs from its own revenues.

The provisions reflect the norms of international practice, under which the Central Bank is established as a legal entity belonging to the state, but the financial activities are carried out by the Central Bank itself (Akimov & Dollery, 2009). The Central Bank is a public body, which in the framework of its competence develops guidelines for the single state monetary policy, has the exclusive right to issue banknotes in circulation in the form of notes and coins as legal tender, is the body of currency regulation and control, acts as a banker, advisor and fiscal agent of the government of the Republic of Uzbekistan, performs the functions of supervision and regulation of banks and other lending institutions. The properties of the Central Bank are its cash and other tangible assets, the cost of which is reflected in the balance sheet of the Central Bank.

The main purpose and objectives of the Central Bank of the Republic of Uzbekistan are (Younas & Turdialiev, 2021);

- Under the Law of the Republic of Uzbekistan "About the Central Bank of the Republic of Uzbekistan," the main purpose of the Central Bank is to ensure the stability of the national currency.
- The main objectives are the formation, adoption and implementation of monetary policy and policy on currency regulation;
- Organisation and provision of effective payment system in the Republic of Uzbekistan;
- The licensing and regulation of banking activities, the activities of credit unions, micro-credit organisations and pawn shops, the supervision of banks, credit unions, micro-credit organisations and pawnshops, licensing of production of forms of securities;

- Storage and management of official foreign exchange reserves of the Republic of Uzbekistan, including the provisions of the Government under the agreement;
- Organise jointly with the Ministry of Finance the cash execution of the state budget.

To ensure the stability of the national currency, the Central Bank of the Republic of Uzbekistan solves its basic tasks, the essence of which are as follows:

- Implementing concerted monetary and currency policy, the Central Bank exercises supervision over the total money supply in circulation, the stability of the exchange rate of the national currency against foreign currencies, and promotes maintaining public confidence in the banking system.
- Varying the amount of the money supply through its credit and exchange business, the Central bank affects the dynamics of prices of goods and services, the total volume of production, exports and imports, impacts on employment levels.

The Central Bank is the body of currency regulation and currency control and is responsible to:

- issue normative documents on foreign exchange regulation, mandatory for all persons on the territory of the Republic of Uzbekistan;
- issue and revoke licenses for foreign currency transactions to banks, other businesses and individuals, supervise and regulate their activities;
- set limits on open currency position and other economic measurement data for banks;
- establish a procedure for determining the rate of the national currency against foreign currencies;
- own and monitors the international reserves of the Republic of Uzbekistan.

The task of organising and providing a settlement system determines the main role of the Central Bank to ensure an efficient payment system in the country. Payments on the territory of the Republic of Uzbekistan are carried out in the form of cash and non-cash payments.

The Central Bank of the Republic of Uzbekistan enacts obligatory normative documents setting the rules of accounts. Uniform application of these rules provides the continuity and timeliness of accounts throughout the country.

To ensure an effective system of settlements the Central Bank constantly works to introduce new tools and calculations based on modern technology and develops the system of accounting and reporting in accordance with international standards for the provision of the financial and economic information of consumers – shareholders, investors, depositors, auditors, creditors and foreign companies, organisations and banks.

One of the main objectives of the Central Bank of the Republic of Uzbekistan is the banking regulation and supervision of banks and other lending institutions. This is necessary to maintain the stability of the banking system and protect the interests of shareholders, depositors and creditors.

Among the most important means to achieve this objective is the licensing of banks and other lending institutions. The purpose of licensing is to grant an access to the market for banking and credit services offered by financially sound organisations operated by persons of impeccable reputation. This ensures public confidence in the banking and financial system.

The Central Bank contributes to the maintenance of international reserves at a level sufficient for the implementation of the monetary and currency policy of the Republic of Uzbekistan, as well as settling with international transactions.

In 1998, the banks acquired enough power to be able to work in compliance with Basel recommendations (Akimov, 2015). Based on the Basel recommendations on banking supervision and international practice, they established the statutory acts regulating the activity of commercial banks within the framework of international standards of financial statements. There it was used the principles of the advanced revelation of the issues on commercial bank balance and taking measures to stave off running out of capital as well as decrease banking liquidity for the purpose of supporting financial steadiness of commercial banking organisations. In 1999–2000, further liberalisation of international monetary policy was adopted. Uzbekistan has introduced the mechanism of foreign exchange sales in the over-the-counter foreign exchange market. Such a mechanism proposes the purchase of exchange by authorised banks from their clients on a contractual basis at a free exchange rate on the basis of supply and demand in the foreign exchange market.

References

Akimov, A. (2015). The political economy of financial reforms in authoritarian transition economies: A comparative study of Kazakhstan and Uzbekistan. In *Development and financial reform in emerging economies* (pp. 125–147). Routledge.

Akimov, A., & Dollery, B. (2009). Financial development policies in Uzbekistan: An analysis of achievements and failures. *Economic Change and Restructuring*, *42*, 293–318.

Akimov, A. V., & Dollery, B. (2006). Uzbekistan's financial system. An evaluation of twelve years of transition. *Problems of Economic Transition*, *48*(12), 6–31.

Åslund, A. (2012). The enigmatic Russian banking system: An introduction. *Eurasian Geography and Economics*, *53*(2), 244–249.

AzerNews. (2017). *ICD inks MOU to introduce Islamic banking in Azerbaijan*, 18 May 2017. https://www.azernews.az/business/113318.html [accessed 12 March 2023].

Bhatnagar, S., & Shahab Ahmed, Z. (2021). Geopolitics of landlocked states in South Asia: A comparative analysis of Afghanistan and Nepal. *Australian Journal of International Affairs*, *75*(1), 60–79.

CIBE, Center of Islamic Banking and Economics. (2017). Press release, May 15, 2017, Lahore. https://www.findevgateway.org/sites/default/files/announcement/press_release_azerbaijan_5.2017.pdf [Accessed on March 11, 2019].

Disli, M., Hatam, A. K., & Jalaly, S. (2023). Faithful finance: Unlocking Banking Islamization in Afghanistan. *Economies*, *11*(7), 189.

Disli, M., & Jalaly, S. (2024). The performance of Islamic and conventional microfinance loans in Afghanistan: The Taliban and beyond. *Emerging Markets Review*, *59*, 101104.

Gresh, G. F. (2007). The rise of Islamic banking and finance in Central Asia. *Al Nakhlah*, *9*.

Hamdard, A. K. (2011). Practical aspects of Islamic banking in Afghanistan. *The daily Afghanistan*. http://outlookafghanistan.net/topics.php?post_id=326 [Accessed on March 15, 2019).

Hidayat, S. E., Rafiki, A., & Svyatoslav, S. (2020). Awareness of financial institutions' employees towards Islamic finance principles in Russia. *PSU Research Review*, *4*(1), 45–60.

Isazade, A. R. (2010). Azerbaijan: Legal status of business entities. *Journal of International Trade Law and Policy*, *9*(2), 193–212.

Kloc, K., & Woodward, R. (1995). The Banking system in Kyrgyzstan. *Russian & East European Finance and Trade*, *31*(6), 73–94.

Mamedov, Z. F., & Gasimov, A. (2021). Islamic financial system: key trends and prospects. *Economic and Social Development: Book of Proceedings*, 459–465.

Mammadov, I., & Ahmadov, F. (2021). Financial development and economic growth: Evidence from Azerbaijan. *WSEAS Transactions on Business and Economics*, *18*, 237–252.

Mikayılov, Q., & Muxtarov, Ş (2013). Azerbaycan bankacılık sisteminin finansal analizi: 2006-2011. *Journal of Qafqaz University*, *1*(1), 75–83.

Nagimova, A. Z. (2022). Islamic Finance in the post-Soviet Central Asia and Transcaucasia: The evidence from Kyrgyzstan. *Journal of Islamic Accounting and Business Research*, *13*(8), 1197–1215.

Pavlović, J., & Charap, J. (2009). Development of the Commercial Banking System in Afghanistan: Risks and Rewards. *IMF Working Paper*, WP/09/150.

Perotti, E. (2002). Lessons from the Russian meltdown: The economics of soft legal constraints. *International Finance*, *5*(3), 359–399.

Rostan, P., Alami, S., Mohammad, E., Rahmani, A., & Azizi, S. (2021). Challenges of islamic banking in least developed countries: The case of Afghanistan. JOEEP: *Journal of Emerging Economies and Policy*, *6*, 5–19.

Rubin, B. R. (2000). The political economy of war and peace in Afghanistan. *World Development*, *28*(10), 1789–1803.

Schoon, N. (2017). The implementation of Islamic Financial Services in Afghanistan: A case study. Critical issues and challenges in Islamic *Economics and Finance Development*, 97–114.

Subbotina, N. (2009). Challenges that Russian banks face implementing the AML regulations. *Journal of Money Laundering Control*, *12*(1), 19–32.

Vernikov, A. (2007). *Russia's banking sector transition: Where to*?. BOFIT Discussion Paper No. 5/2007. http://dx.doi.org/10.2139/ssrn.1001669 [accessed 21 January 2024].

Vysman, I. (1993). The new Banking legislation in Russia: Theoretical adequacy, practical difficulties, and potential solutions. *Fordham Law Review*, *62*, 265.

Yadav, R. (2016). Financial sector reforms and commercial banks performance in Kyrgyzstan: 1991–2015. *Management*, *5*(28), 1–13.

Younas, A., & Turdialiev, M. A. (2021). Special features of international financial centres and its establishment in Uzbekistan. *The American Journal of Political Science Law and Criminology*, *3*(08), 17–21.

5 Islamic Financial Jurisdictions in East Asia

5.1 Introduction

Although Islam is not a dominant religion in East Asian countries (Hong Kong, Japan and South Korea), Islamic banking services are rising in the region due to the increasing Muslim population and the popularity of Islamic financial services. To cater to this escalating demand, the regulators of these countries have developed certain regulations to accommodate the increasing demand for Islamic financial services. This chapter reviews different regulators and banking laws of these economies and analyses the relevance of these regulations in creating Islamic financial services.

5.2 Hong Kong

5.2.1 Regulators

Hong Kong has one of the highest concentrations of overseas banking institutions. There are 70 famous banking giants operating out of the world's 100 leading banks (Hong Kong Monetary Authority, 2024). The banking system of Hong Kong's deposit-taking institutions is maintained through a three-tier system. The authorised institutions are categorised into three institutions namely licensed banks, restricted banks and deposit-taking companies. The authorised financial institutions are:

- Licensed Banks

  These banks operate current and saving accounts, accept deposits of any size and maturity and pay and collect cheques drawn by or paid in by customers. As of February 2019, under this category, there are 152 licensed banks in Hong Kong.
- Restricted Licence Banks

  These banks primarily engage in merchant banking and capital market activities. These banks are allowed to take deposits of any maturity of HK$500,000 and above. Under this category, 18 locally and overseas incorporated banks are included.
- Deposit-taking Companies

  Most of these companies are owned or associated with banks. These companies perform specialised activities such as consumer financing and securities

DOI: 10.4324/9781003471035-5

business. They are allowed to take deposits of HK$100,000 or above with an original term of maturity of at least three months. As of February 2019, there are 16 locally incorporated deposit-taking companies operating in Hong Kong.

- Local Representative Offices

  Overseas representative banks are allowed to establish their offices in Hong Kong but are not allowed to engage in banking business. The main role of these offices is to liaise work between the bank and its customers in Hong Kong. There are 48 local representative offices of overseas banks in Hong Kong.

The legal framework for banking in Hong Kong is covered under the following legislation:

- Banking Ordinance (Cap 155, 155L, 155M, 155Q, 155R) (Hong Kong Monetary Authority, 2024)
- Supervisory Policy Manual
- Securities and Future Ordinance (Hong Kong e-Legislation, 2019)
- Companies Ordinance (Hong Kong e-Legislation, 2020)
- Bills of Exchange Ordinance
- Common Law (Ng, 2009)
- Money Lenders Ordinance

The regulatory framework of Hong Kong is a sectoral structure that operates through a tier system. Figure 5.1 outlines the regulatory structure of Hong Kong's financial industry.

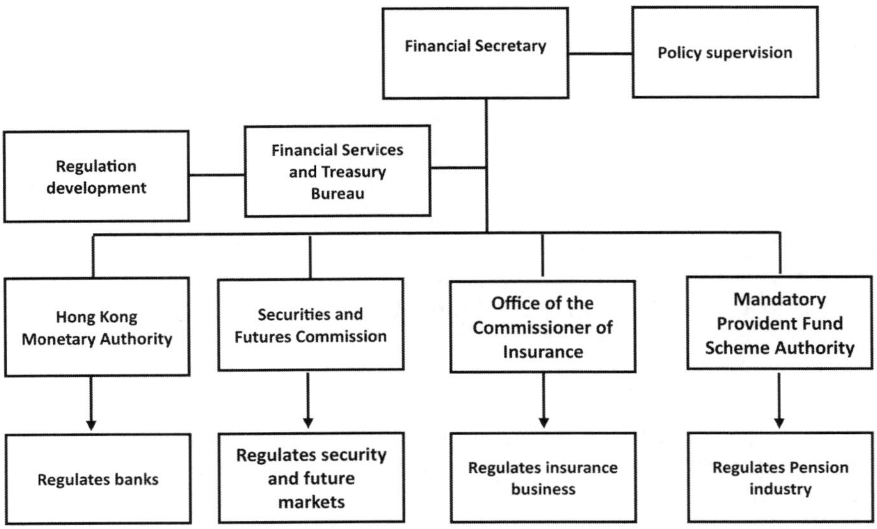

Figure 5.1 Three-tier regulatory system of Hong Kong.

Under the first tier, the Financial Secretory oversights the overall policy and the Financial Services and Treasury Bureau (FSTB) translates these policies into regulation. Under the second tier, specialised regulatory agencies such as the Hong Kong Monetary Authority (HKMA), Security and futures Commission (SFS), Office of the Commissioner of Insurance (OCI) and Mandatory Provident Fund Schemes Authority (MPFA) regulate banks, security and future markets, insurance business and pension industry, respectively. Under the third tier, a self-regulatory organisation oversights the members' activities (Arner et al., 2010).

Additionally, there are other bodies known as the Financial Reporting Council (FRC), Companies Registry, Hong Kong Deposit Protection Board (HKDPB), Independent Commission, Financial Intelligence Unit (FIU) and the Consumer Council which play a significant role in accounting and auditing standards for listed companies, Companies Ordinances, deposit insurance scheme, fight against corruption, promoting financial Intelligence and consumers protection (Chan et al., 2011).

5.2.1.1 Hong Kong Government

The government of Hong Kong focuses on monitoring a range of aspects of public finance, monetary and financial affairs under articles 106 and 113 of the Basic Law of Hong Kong instead of involving in the day-to-day regulation of the financial system. The Financial Secretary is responsible for the monetary system, exchange fund, public finance, financial system and the status of the international financial centre. FSTB implements the policies of public finance in the financial system set by the Financial Secretary and the International Financial Centre ensures that the country's regulatory regime is regularly updated to cater to investors' demands.

5.2.1.2 Hong Kong Monetary Authority

HKMA acts as a central bank and regulatory authority at the same time. It is responsible for administering the official monetary policy, ensuring banking stability, acting as an international financial centre and managing exchange funds (Kwan & Lui, 1999). HKMA is also responsible for regulating the activities of authorised institutions (licensed banks, restricted licensed banks and deposit-taking companies). Under the Banking Ordinance, it is responsible to monitor banking business, defined to include only deposit-taking and cheque-related services hence lending business is not covered by the Banking Ordinance. HKMA is responsible for endorsing the issued guidelines by HKAB for the Code of Banking Practice.

5.2.1.3 Insurance Authority (IA)

It was established in 2015 under the Amendment of Insurance Ordinance 2015. It is an independent regulator which does not depend on the Government. The IA replaced the regulatory functions and Office of the Commissioner of Insurance (previously the Government department). IA was empowered by the Insurance Ordinance on 26 June 2017 to protect the interests of policyholders and promote the general stability of the insurance industry. IA focuses to authorise and supervise insurers and is expected to take the mandate of existing insurance intermediaries from the three self-regulatory organisations and implement a new statutory regulatory and licensing regime within two years (Hong Kong e-Legislation, 2022).

5.2.1.4 Securities and Futures Commission (SFC)

This commission was established in 1989 after the shocks of the market crises of 1987 in Hong Kong (Godfrey, 2018). It is an independent legislative body responsible for the regulation of securities and future markets. SFC was established under the SFC Ordinance and subsidiary legislation whose key functions are to promote fairness, efficiency, competitiveness, transparency and orderliness of the securities and future industry and to offer protection and education to the public investing in financial products. Recently, its investor education mandate was transferred to the Investor Education Centre.

5.2.1.5 Mandatory Provident Fund Schemes Authority

This statutory body was established in 1998 under the Mandatory Provident Fund Schemes Ordinance to facilitate the workforce in Hong Kong for the accumulation of savings for their retirement. The key functions of the MPFA include oversight of the operation of occupational retirement (ORSO) schemes.

5.2.2 Islamic Banking in Hong Kong

Hong Kong is a Buddhist-dominant territory with a population of approximately 7.4 million and has a small Muslim community of 300,000 members. Despite its small Muslim population, the Hong Kong Monetary Authority (HKMA) has made no secret of its aspirations to establish an Islamic finance hub, especially in the capital markets segment. HKMA issued a policy statement in 2007 to launch Islamic banking through the issuance of sukuk which became the first initiative at the government level to introduce Islamic finance in Hong Kong.

The launching of sukuks was the result of amendments in the financial system which was dependent on the mandatory approval of the Inland Revenue Ordinance and the Stamp Duty Ordinance (Ibrahim, 2015). After a waiting period of six years, the government offered its inaugural sukuks in 2014 under the Government Bond Programme. In the year 2015, HKMA issued a second sukuks of US$1 billion for five years and currently several local firms such as Airport Authority, metro operator MTR Corp and Hong Kong Mortgage Corp (HKMC) are involved in sukuks issuance.

Islamic banking in Hong Kong is offered through full-fledged Islamic banks, Islamic banking windows and cooperative financial institutions. Malaysia's Hong Leong Bank became the first bank in 2008 to obtain a license to offer comprehensive Islamic banking services in Hong Kong. The bank focuses on tapping the markets of mainland China and West North Asia through this platform leveraging Hong Kong's status as an international financial centre (Reuters, 2008). Hong Kong continues to be an attractive base for funds and asset managers. The leading Islamic financial services providers such as Malaysia, Indonesia and Pakistan have established banks in Hong Kong to tap mainland China and Asia's financial sources. The below-listed banks provide Islamic financial services in Hong Kong:

- Hong Leong Bank
- International Islamic Bank for Investment and Development (IIBID)
- National Bank of Pakistan (NBP)
- Standard Chartered Islamic banking
- The Islamic Bank of Asia Limited
- DBS Bank
- CIMB Islamic Bank
- Bank Mandiri
- BNI group
- Bank Islam

Another method to offer Islamic banking services in Hong Kong is by establishing Islamic banking windows as it is an effective and cost-efficient method. Any institution looking to establish an Islamic banking window in Hong Kong is not required a separate authorisation under the Banking Ordinance. However, it is mandatory to propose a realistic and feasible business plan and must discuss it with HKMA. The discussion between HKMA and the institution includes the following topics:

- The risk management process of the institution
- The policies, procedures and controls to ensure the Shariah compliance of products and services
- The preparation of contractual documentation to minimise the legal disputes
- The proper disclosure of risk and return to depositors

The government of Hong Kong actively promotes the development of Islamic finance, especially sukuk market. In line with its agenda of promoting Islamic finance, the HKMA amended its laws in 2013 and 2014 to provide a tax structure for sukuk and so far; three sukuks have been issued under the Government Bond Programme. HKMA jointly organised an Islamic finance conference with Bank Negara Malaysia for talent development and the Treasury Markets Association has also organised various training sessions and seminars on Islamic finance.

Historically, Hong Kong has been a fundraising centre for Chinese enterprises, and now it needs to maintain its position through innovation and new product development. Islamic finance strong existence could be a great opportunity for Hong Kong to continue attracting these enterprises (Komijani & Taghizadeh-Hesary, 2018). The investors from Islamic world are keen on investment opportunities in Asia particularly in mainland China in order to diversify their portfolios. Offering Shariah-compliant investment opportunities to these investors will enable them to gain entry into the Chinese market and at the same time mainland Chinese issuers can also make use of this platform to target a wealthy investor base in the Islamic world.

Despite making significant strides to establish its platform for Islamic finance offering, there are several areas of concern that create doubts about the growth and future of Islamic finance in Hong Kong. Some of these concerns are as follows:

- Tax issues
- Lack of knowledge
- Risk management
- Adherence to Islamic laws
- Contractual documentation
- Disclosure of risk and return to depositors

5.3 Japan

5.3.1 Regulators (The Bank of Japan)

There are two principal regulators in Japan. The Financial Services Authority and the Bank of Japan (BoJ). The power to the Financial Services Authority is delegated by the Prime Minister (Hoshi & Ito, 2004). The on-site and off-site inspection of investment banking activities is performed by the Securities and Exchange Surveillance Commission. BoJ also has supervisory authority over banks, based primarily on its contractual agreements and transactions with them. The regulator's powers as prescribed in the Banking Act include receipt of various reports, the ability to carry out on-site inspections (where a bank must, in practice, disclose any and all information it holds to the regulator) and the power to make orders of business improvement and suspension (Vittas & Kawaura, 1995).

BoJ is the central bank of Japan. It is a juridical person established based on the Bank of Japan Act (Act No 89, 1997 hereinafter the Act) and is neither a government agency nor a private corporation. The Act sets the Bank's objectives "to issue banknotes and to carry out currency and monetary control" and "to ensure smooth settlement of funds among banks and other financial institutions, thereby contributing to the maintenance of stability of the financial system."

The Act also stipulates the Bank's principle of currency and monetary control as follows: "Currency and monetary control by the BoJ shall be aimed at achieving price stability, thereby contributing to the sound development of the national economy". It is important to note that the Japanese regulatory framework regulates commercial banking activities and investment banking activities separately.

5.3.1.1 Regulator's Objectives

The purpose of the BoJ or the Central Bank of Japan is:

- To issue banknotes and to carry out currency and monetary control.
- To ensure smooth settlement of funds among banks and other financial institutions, thereby contributing to the maintenance of stability of the financial system.

- Currency and monetary control by the BoJ shall be aimed at achieving price stability, thereby contributing to the sound development of the national economy.

The BoJ's autonomy regarding currency and monetary control shall be respected.

The BoJ shall endeavour to clarify to the citizen the content of its decisions, as well as its decision-making process, regarding currency and monetary control.

The BoJ shall, taking into account the fact that currency and monetary control are components of overall economic policy, always maintain close contact with the government and exchange views sufficiently, so that its currency and monetary control and the basic stance of the government's economic policy shall be mutually compatible.

Article 5 (1) explains the public nature of the BoJ's Business and its autonomy. In light of the public nature of its business and property, the BoJ shall endeavour to conduct its business in a proper and efficient manner. In implementing this Act, due consideration shall be given to the autonomy of the BoJ's business operations (Walter, 2006).

5.3.2 Islamic Banking and Takaful Laws

Under the Japan Financial Services Agency's (JFSA) 2007, a "Plan for Strengthening the Competitiveness of Japan's Financial and Capital Markets" was introduced. Islamic finance was given a highlight. In 2008, JFSA amended rules to allow subsidiaries of Japanese banks to conduct Islamic finance transactions, with foreign subsidiaries later allowed to take Islamic deposits, but the rules are seen as restrictive.

The regulator is considering allowing banks to provide Islamic products in the domestic market for the first time and will present the results of a consultation on rule changes later this month. Islamic products require multiple transfers of title of the underlying asset, and so can present regulatory challenges for new jurisdictions in areas such as tax.

The most significant regulatory reforms influencing Islamic finance in 2009 were amendments to the Ordinance for Enforcement of the Banking Act/Insurance Business Act, under which certain types of Islamic finance transactions were for the first time added to the scope of business that the subsidiaries of banks/insurance companies may conduct (Suzuki & Miah, 2018). Second, tax reforms have reduced Japanese taxation risks for foreign fund managers of certain investment funds on their income and capital gains from the disposition of funds. Financial Services Agency (J-FSA) made an attempt to improve the infrastructure for Islamic finance in Japan by submitting a request for tax reform for the fiscal year 2010 (Tax Reform Request) with a view to implementing necessary tax measures in favour of Islamic finance transactions. The Tax Reform Request followed the relaxation of financial regulations in 2008, which allowed subsidiaries of Japanese banks and insurance companies to provide certain Islamic finance services in such forms as Murabaha or Ijarah. These changes and developments demonstrate a high level of interest in Islamic finance among both public and private sectors and growing expectations for the development of Japanese legal infrastructure in the area of Islamic finance.

5.3.3 The Banking Laws in Japan

The principal legislation regulating banks in Japan is as follows:

- The Banking Act (Act No. 59 of 1981). The Banking Act, together with the orders and ordinances issued thereunder, primarily governs the following matters in respect of banks: licensing, organisation and governance, business scope, customer protection, prohibited acts, capital adequacy, subsidiaries and shareholders, accounting, disclosure, and mergers and acquisitions. By March 2008, the regulatory framework of Basel II was fully introduced into Japanese banking laws through amendments to the FSA administrative notice17 including, inter alia, the internal ratings-based approach and the advanced measurement approach. Following Basel II's introduction, Basel III is also being introduced into Japanese banking laws.
- Financial Instruments and Exchange Act. The key legislation regulating banks' provision of securities-related services is the Financial Instruments and Exchange Act (Act No. 25 of 1948, as amended; the FIEA), the orders and ordinances issued thereunder and the Guidelines. In September 2013, the FSA announced its new monitoring policy to enhance its supervisory oversight function by further integrating its on-site inspection and off-site supervision into a continuous and seamless monitoring process. This new monitoring policy may lead to significant changes in the relationship between banks and the FSA.

5.3.3.1 Requirements for Establishing a Bank in Japan

- The license must be obtained from the FSA to conduct a banking business.
- A foreign bank can establish either a local subsidiary or a local affiliate in the form of a joint-stock company (kabushiki kaisha), established as a foreign bank branch in Japan, and obtain a banking licence for such a bank branch. Banks in Japan are only permitted to engage in Banking Businesses, businesses incidental to Banking Businesses and such other businesses expressly permitted under the Banking Act and other legislation.

The FSA clarified its supervisory guidelines with respect to matters to be monitored, such as the circumstances of fund transfers within the foreign bank group (including transfers to and from its head offices and branch accounts), the foreign bank branch's assets in Japan, deposit types provided by the foreign bank branch and the foreign bank branch's manner of treating deposits in Japan. The FSA also clarified that these points of consideration are applicable not only to criteria for licensing but also to the daily monitoring of foreign bank branches:

- Effective from April 2014, foreign bank branches, in principle, are required to maintain assets (in the form designated by a Cabinet Order) equal to or more than ¥2 billion at all times in Japan, which is equal to the required minimum capital amount of the local entity banks;
- Foreign bank branches in Japan are required to maintain at all times assets equal to the minimum capital amount that local banks are required to maintain (i.e., JPY 2 billion);

- Foreign bank branches are now also required to explain certain matters to customers (such as the fact that deposits in foreign bank branches are not covered under Japan's deposit insurance system); and
- The penalty for breach of an order to maintain assets in Japan, which may be issued by the FSA to a bank (including a foreign bank branch) has been augmented.

5.3.3.2 Dispute Resolution

The Arbitration Law (Act No. 138 of 1 August 2003) came into force in March 2004. This Arbitration Law is applicable to arbitral proceedings whose place of arbitration is in Japan. The Arbitration Law generally adopts the UNCITRAL Model Law (prior to its 2006 amendment). An arbitration agreement is defined as an agreement by the parties to leave the resolution of all or certain civil disputes which have already arisen or which may arise in the future in respect of a defined legal relationship (irrespective of whether contractual or not) to one or more arbitrators, and to accept the award made therefore. The International Commercial Arbitration Committee, the former body of JCAA, was established in 1950 within the Japan Chamber of Commerce and Industry, with the support of six other business organisations including the Japan Federation of Economic Organizations, the Japan Foreign Trade Council and the Federation of Banking Associations of Japan, to serve as an organisation to settle commercial disputes and promote international trade, thereby contributing to the development of the Japanese economy. Although in practice most appointed arbitrators are legal professionals (such as professors, former judges and attorneys), no such qualification is required under Japanese law. Non-lawyers of other countries may act as arbitrators in arbitral proceedings in Japan (Ramseyer, 1988). The JCAA maintains a list of arbitrators that includes more than 100 lawyers, foreign lawyers and professors located in Japan and elsewhere. The list of arbitrators, however, merely exists as a consultative resource and is not binding on parties to an arbitration administered by the JCAA.

5.4 South Korea

There are several Korean companies that are seen as the keenest issuers of sukuk including oil refiners that import crude oil from the Middle East (Wan Ahmad et al., 2019). State affiliated energy groups with business in the region and airline operators. Companies such as GS Caltex, Korean Air Hyundai, Samsung and many others are reportedly exploring the possibility of raising funds from the sukuk market. The main reason for choosing Islamic finance is due to the strategic alternatives of funding for Korean issuers and enhancing ties with Islamic countries from "trade" towards "finance" Prior to the year 1997.

5.4.1 Financial Regulators in South Korea

There were four regulators in the Korean Financial System – the Office of Bank Supervision (OBS), the Securities Supervisory Board (SSB), the Insurance Supervisory Board (ISB) and the Non-bank Supervisory Authority (NSA). After the Asian Financial Crisis, the South Korean regulatory structure was reformed (Choi, 1993). The Financial Services Commission (FSC) was established in 1998 and has the authority to license and supervise all financial institutions in Korea including non-banking financial institutions. To facilitate the supervision process, all four regulators were consolidated to form the Financial Supervisory Service (FSS) in January 1999 following the FSA model in the United Kingdom. FSS functions as the executive arm of FSC and is responsible for the actual supervision and inspection of financial institutions. In addition, the SFC was placed under FSC to oversee the securities and futures market. With this consolidation, the Bank of Korea's (the central bank) primary role is to focus on the monetary policy of Korea. Looking into this regulatory reform is important as it highlights which organisation would have the power to make policy changes in Korea to facilitate the introduction and operation of Islamic Finance in Korea. As the FSC is the decision-making body, it would have the authority to recommend the policy changes needed.

The Bank of Korea is the 59th regulatory body to join the IFSB, bringing total membership to 184, joining the likes of the central banks of Luxembourg and Japan and the monetary authorities of Hong Kong and Singapore (Ko, 2014). The move could augur stronger links between South Korea and Islamic finance hubs in southeast Asia.

5.4.2 Banking Laws

The regulatory framework in Korea has been reformed. In early 1997, the Presidential Committee on Financial Reform was established. In December 1997, an Act on the establishment of a financial Supervisory Organisation was also passed (Shim & Park, 2016). Korea's interest in Islamic Finance is quite recent and this was fuelled mainly by the interest of Korean companies to tap into the Islamic financial market. Since funding in the traditional market (the United States and Europe) was drying up due to the financial crises, Korean companies are looking for alternative funding sources. Even prior to the financial crises.

In January 2013, CIMB became the first ASEAN financial institution to obtain an equities brokerage license in South Korea since 2002, enabling it to serve both domestic and international clients (Shaikh et al., 2020). CIMB Korea is now in the position to offer its global clients access to the South Korean equity market and to catalyse business and investment flows between South Korea and the Asia Pacific. CIMB Korea also provides corporate advisory services and research coverage to both domestic and international clients.

South Korea has had a long and frustrating journey trying to pass legislation favourable to Islamic financial transactions but to no avail. Prepared and submitted for the first time in September 2009, the Ministry of Strategy and Finance was unsuccessful in getting the amendment to the Special Tax Treatment Control Act, containing provisions that would neutralise the existing double tax treatment of Sukuk transactions, approved by the National Assembly's Strategy and Finance Committee.

Amendments were introduced again in February 2011 only to be rejected again due to strong opposition from Korean Christian groups, and there have been no signs from the government in attempting to re-introduce a bill that would create a level playing field for Shariah financial transactions.

Takaful is not yet available in the South Korean market nor provided by any local entities; however, the Korea Trade Insurance Corporation (K-sure) in November 2016 signed a reinsurance agreement with the Islamic Corporation for the Insurance of Investment and Export Credit ((ICIEC)) to build mutual business opportunities. Under the agreement, the (ICIEC) will reinsure commercial and political risks related to medium/long-term export credit transactions which involve Korean exporters and contractors selling capital goods and services to (ICIEC) member countries (Baber, 2020). The (ICIEC) will also extend reinsurance support to K-sure for Korean investors investing in (ICIEC) member countries.

References

Arner, D. W., Hsu, B. F., & Da Roza, A. M. (2010). Financial regulation in Hong Kong: Time for a change. *Asian Journal of Comparative Law*, *5*, 1–47.

Baber, H. (2020). Financial inclusion and FinTech: A comparative study of countries following Islamic finance and conventional finance. *Qualitative Research in Financial Markets*, *12*(1), 24–42.

Chan, R. S., Lau, C. K., & Ng, A. W. (2011). Compliance and value relevance of audit committees: Evidence from Hong Kong. *Journal of Financial Reporting and Accounting*, *9(*1), 74–97.

Choi, B. S. (1993). *Financial policy and big business in Korea: The perils of financial regulation. The politics of finance in developing countries*, 23–54.

Godfrey, A. M. (2018). Hong Kong Securities and Futures Commission (SFC) highlights operating, control and compliance deficiencies in asset management industry. *Journal of Investment Compliance*, *19*(1), 63–68.

Hong Kong Monetary Authority. (2024). The Three-Tier Banking System. https://www.hkma.gov.hk/eng/key-functions/banking-stability/banking-policy-and-supervision/three-tier-banking-system.shtml [Accessed 14 April 2024].

Hong Kong e-Legislation. (2019). Exchange Fund Ordinance – Cap 66. https://www.elegislation.gov.hk/hk/cap66!en.assist.pdf?FILENAME=Assisted%20Monolingual%20PDF%20(English).pdf&DOC_TYPE=K&PUBLISHED=true [Accessed 10 April 2024].

Hong Kong e-Legislation. (2022). Cap 41 – Insurance Ordinance. https://www.elegislation.gov.hk/hk/cap41 [Accessed 10 April 2024].

Hong Kong e-Legislation. (2020). Cap 19 – Bills of Exchange Ordinance. https://www.elegislation.gov.hk/hk/cap19 [Accessed 20 April 2024].

Hoshi, T., & Ito, T. (2004). Financial regulation in Japan: A sixth year review of the Financial Services Agency. *Journal of Financial Stability*, *1*(2), 229–243.

Ibrahim, M. H. (2015). Issues in Islamic banking and finance: Islamic banks, Shari'ah-compliant investment and sukuk. *Pacific-Basin Finance Journal*, *34*, 185–191.

Ko, D. W. (2014). Financial regulatory reforms in Korea responding to the 2008 global financial crisis: Are they following the global trends? *Asia Pacific Law Review*, *22*(2), 33–55.

Komijani, A., & Taghizadeh-Hesary, F. (2018). An overview of Islamic banking and finance in Asia. *In Routledge Handbook of Banking and Finance in Asia (eds. Volz, U., Morgan, P., & Yoshino, N.* Routledge: London.

Kwan, Y. K., & Lui, F. (1999). Hong Kong's currency board and changing monetary regimes. In *Changes in exchange rates in rapidly developing countries: Theory, practice, and policy issues* (pp. 403–436). University of Chicago Press.
Ng, K. H. (2009). *The common law in two voices: Language, law, and the postcolonial dilemma in Hong Kong*. Stanford University Press.
Ramseyer, J. M. (1988). Reluctant litigant revisited: Rationality and disputes in Japan. *Journal of Japanese Studies*, *14*(1), 111–123.
Reuters, T. (2008). Hong Leong Bank launches Islamic banking biz in HK. https://www.reuters.com/article/idUKHKG26641820080818/ [Accessed 17 September 2023].
Shaikh, I. M., Qureshi, M. A., Noordin, K., Shaikh, J. M., Khan, A., & Shahbaz, M. S. (2020). Acceptance of Islamic financial technology (FinTech) banking services by Malaysian users: An extension of technology acceptance model. *Foresight*, *22*(3), 367–383.
Shim, J., & Park, J. H. (2016). Public participation and trust in government: The case of the Korean financial regulatory agency. *Public Performance & Management Review*, *40*(1), 1–22.
Suzuki, Y., & Miah, M. D. (2018). *Dilemmas and challenges in Islamic finance. Looking at equity and microfinance*. Routledge.
Vittas, D., & Kawaura, A. (1995). Policy-based finance, financial regulation, and financial sector development in Japan (Vol. 1443). World Bank Publications.
Walter, A. (2006). From developmental to regulatory state? Japan's new financial regulatory system. *The Pacific Review*, *19*(4), 405–428.
Wan Ahmad, W. M., Hisham Hanifa, M., & Hyo, K. C. (2019). Are non-Muslims willing to patronize Islamic financial services? *Journal of Islamic Marketing*, *10*(3), 743–758.

6 Islamic Financial Regulations in the Middle East

6.1 Introduction

Islamic financial regulations in Middle Eastern countries are discussed in detail in this chapter. A detailed overview of the financial regulators, diverse regulations and their role in promoting the development and growth of Islamic finance (IF) in these countries is presented and reviewed. The chapter highlights the difference in regulatory and governance structure of countries with equal demand for Islamic financial services. This will allow comparing the regulatory difference between developing and underdeveloped Islamic financial jurisdictions and how these differences in implementing Islamic financial regulations may impact the performance and growth of Islamic financial industries. From a regional coverage perspective, this chapter is exclusive as it exhibits almost all developing and developed countries in the Middle East.

6.2 Bahrain

6.2.1 The Regulator

The Central Bank of Bahrain (CBB) is a public corporate entity established by the CBB and Financial Institutions Law 2006 (Ramady, 2015). It was created on 6 September 2006. The CBB is responsible for maintaining monetary and financial stability in the Kingdom of Bahrain. It succeeded the Bahrain Monetary Agency, which had previously carried out central banking and regulatory functions since its establishment in 1973 (shortly after Bahrain secured full independence from Great Britain).

The CBB inherits the Bahrain Monetary Authority's 33-year track record and wide range of responsibilities. It implements the Kingdom's monetary and foreign exchange rate policies, manages the government's reserves and debt issuance, issues the national currency and oversees the country's payments and settlement systems (Sarea, 2019). It is also the sole regulator of Bahrain's financial sector, covering the full range of banking, insurance, investment business and capital markets activities. The CBB's wide scope of responsibilities allows a consistent policy approach to be undertaken across the whole of the Kingdom's financial sector.

DOI: 10.4324/9781003471035-6

It also provides a straightforward and efficient regulatory framework for financial services firms operating in Bahrain.

6.2.2 Islamic Banking Law

Bahrain is one of the earliest and most active jurisdictions to introduce Islamic banking in its financial system. The regulator for its Islamic banking, insurance and capital markets is the CBB, and the primary legislation is the Prudential Information and Regulations for Islamic Banks (PIRI) 2000 which covers areas such as capital adequacy, asset quality, management of investment accounts, corporate governance and liquidity management (Ali et al., 2024). The non-profit organisation namely Accounting and Auditing Organisation for Islamic Financial Institutions (AAOIFI) was established in 1991 and since then it has issued standards on accounting, auditing, governance and *Shari'ah* practices, with which players in Bahrain are required to comply. The first Islamic financial institution was Bahrain Islamic Bank which was established in 1979. There are now more than 30 institutions licensed in Bahrain, including full commercial banks, investment banks and offshore banking units.

6.3 Egypt

The Central Bank of Egypt (CBE) was established by Presidential Decree No. 64 of the Year 2004 Promulgating the Statute of the CBE (Awad, 2008). The CBE has successfully completed the first wave of its banking reform program (started in 2004 and ended in December 2008). In January 2009, CBE initiated the second wave of its reform program scheduled to end by 2011 (extended up to 31 March 2012). This wave includes several pillars including applying Basel II in the Egyptian banking sector in cooperation with the European Union.

6.3.1 The Regulator

The main regulator of the Islamic financial system is CBE. The main piece of legislation is Law No. 88 of 2003 relating to the Central Bank Law and the Banking and Currency Sectors, and the Executive Regulations issued by Presidential Decree No. 101 of 2004 (Al-Mashat & Billmeier, 2008). In this regard, all banking and financial activities in Egypt, including those of the Islamic financial institutions, are regulated by the Central Bank. CBE is an autonomous regulatory body, assuming the authorities and powers vested therein by Law No. 88 for 2003, and the Presidential Decree No. 65 for 2004. This part of the website highlights the core purposes of the Bank and some of the work that is undertaken to achieve them. Other parts of the website provide more information about each of the Bank's activities. The main objectives and functions of the CBE are as follows:

- Realizing price stability and ensuring the soundness of the banking system.
- Formulating and implementing the monetary, credit & banking policies.

- Issuing banknotes and determining their denominations and specifications.
- Supervising the banking sector.
- Managing the foreign currency and international reserves of the country.
- Regulating the functioning of the foreign exchange market.
- Supervising the national payments system.
- Recording and following up on Egypt's external debt (public and private).

However, separate permits may be required from the Capital Market Authority to perform banking activities relating to capital markets, i.e. the creation of investor funds. The Financial Regulatory Authority is a public Authority, having a legal status, established under Law 10 of the year 2009. The Authority is responsible for supervising and regulating non-banking financial markets and instruments, including the Capital Market, the Exchange, and all activities related to Insurance Services, Mortgage Finance, Financial Leasing, Factoring and Securitization (Handy, 2001). Financial Regulatory Authority's role is to regulate the market and ensure its stability and competitiveness to attract more local and foreign investments "The mandate of the Authority also includes limiting inconsistency risks and addressing problems arising from applying different supervisory rules."

The Financial Regulatory Authority has replaced the Egyptian Insurance Supervisory Authority, the Capital Market Authority and the Mortgage Finance Authority in application of the provisions of the supervision and regulation of Insurance law no. 10 of 1981, the Capital Market law no. 95 of 1992, the Depository and Central Registry law no. 93 of 2000, the Mortgage Finance law no. 148 of 2001, as well as other related laws and decrees that are part of the mandates of the above authorities. The authority is also the admin authority for companies to establish Islamic banking law.

Egypt has three full-fledged Islamic domestic banks and an absence of any international bank offering Islamic products. The first experiment with IF was Mit Ghamr Savings Bank (1963–1967). A leading full-fledged Islamic financial institution is the Faisal Islamic Bank of Egypt, which was established in 1979. Faisal Islamic Bank of Egypt (FIBE) was founded in 1979 as the first bank in Egypt to operate entirely in accordance with Sharī'ah principles (Galal Abdullah Mouawad, 2009). Despite the bank's recent successes, the CBE's refusal to allow the opening of new branches has prevented it from expanding its network. This has reportedly led to the overcrowding of FIBE's existing branches. Al Bakara Bank Egypt is a subsidiary of the Bahrain-based Al Baraka Banking Group and was previously known as the Egyptian Saudi Finance Bank. The National Development Bank of Egypt was restructured into an Islamic bank in 2010 after the Abu Dhabi Islamic Bank, the second-largest Islamic bank in the Emirates, acquired a 51% stake for US$28 million under the Egyptian government's privatisation drive (Fayed, 2013).

6.3.2 Basel II Adoption

Basel II Program Memorandum of Understanding (MOU) was signed in 2008 between CBE and European Central Bank (ECB); in cooperation with seven European National Central Banks (NCBs) namely, Bulgarian National Bank,

Czech National Bank, Deutsche Bundesbank, Bank of Greece, Banque de France, Banca d' Italia and National Bank of Romania. The Program duration was three years starting in January 2009 (Ghanem, 2017). A dedicated task force has been set up inside the CBE to manage all aspects leading to the implementation of the Basel II framework. Also, an EU resident project coordinator from Banque de France joined the program. The program phases included the capacity building for the specialized task force and banking supervision executives on Basel II issues, intense coordination with the banking system based on discussion papers that were issued on the significant topics related to Basel II, quantitative impact studies (QIS) to assess the consequences of future Basel II regulations, set up of a new data-warehouse framework to enhance data collection and finalization of the fine-tuning of Basel II regulations.

The program of implementing Basel II in the Egyptian banking sector has ended successfully; it has been decided that banks will be required to apply Basel II pillar I regulations starting from December 2012 and June 2013 according to the end of the financial year of each bank.

According to the Bank of International Settlement Report 2015, CBE followed two core principles for the implementation of Basel II: simplicity and communication. Simplicity was required to remain consistent with the varied levels of sophistication in banks' information and control systems and to ensure a smooth transition from existing regulations; standardised approaches are the logical consequence (Abdel-Baki, 2012). Communication was a core factor of success for a new regulatory framework. CBE's implementation strategy focused on the Standardised Approach and its related issues for credit and market risks and the Basic Indicator for operational risk.

In addition, some internal treatments were adopted to suit the Egyptian banking environment while still adhering to the conservative principles of the Basel II framework. Pillar 1 – Credit risk: In 2010, the Draft Regulation Published (Discussion Paper) was introduced to the market including a brief overview of the Internal Ratings-Based approach (IRB) with a definition for both the Foundation and the Advanced IRB while the final rule in force included only the Standardised Approach to give banks room to fully digest this approach before moving to the more advanced approaches (Fouad et al., 2022). Pillar 1 – Operational risk: In 2010, the Draft Regulation Published (Discussion Paper) was introduced to the market including a detailed overview of the Standardised/Alternative Standardised Approach as well as the Advanced Measurement Approaches while the final rule in force included only the Basic Indicator Approach to give banks room to fully digest this approach before moving to the more advanced approaches. Pillar 2 – As part of Pillar 2, the Supervisory Review and Evaluation Process (SREP) is already enforced at CBE within the different departments of the Supervision and Control sector.

Through its supervisory tools, the CBE's different supervisory departments, including the offsite and onsite supervision, take the necessary actions for an adequate and thorough review and evaluation of the Egyptian banking sector including foreign branches operating in Egypt (El-Faham, 2020). However, the Internal Capital Adequacy Assessment Process (ICAAP) as part of the SREP was introduced to the market in 2013. As for the risks not captured in Pillar 1, namely, interest rate risk in the banking book, concentration and liquidity risks, Draft Regulations (Discussion Papers) capturing these risks were published in 2011. The Final Rule in that regard is postponed for the time being until the Egyptian banking sector digests the new Pillar 1 framework. Pillar 3 – In December 2008, CBE introduced the new International Financial Reporting Standards (IFRS), which were applied to all banks, in addition to the CBE final regulation issued in December 2012, where banks should disclose the detailed calculation of their CAR elements in their financial statements (disclosure notes). Furthermore, a corporate governance regulation was issued in July 2011, providing the structure through which the objectives of the company were set, and the means of attaining those objectives and monitoring performance were determined.

6.4 Iran

6.4.1 Regulator

Bank Markazi Jomhouri Islami Iran is the central bank of the Islamic Republic of Iran (Bahmani-Oskooee & Shiva, 1998). The Central Bank of Iran (CBI) was established in 1960 (1339 solar year). As stated in the Monetary and Banking Act of Iran (MBAI), the CBI is responsible for the design and implementation of monetary and credit policies with due regard to the general economic policy of the country. Four major objectives of CBI as stated in the MBAI are as follows:

- Maintaining the value of a national currency.
- Maintaining the equilibrium in the balance of payments.
- Facilitating trade-related transactions.
- Improving the growth potential of the country.

To achieve the objectives as stated in the MBAI, CBI is endowed with the responsibility of fulfilling the following functions:

- Issuance of notes and coins
- Supervision of banks and credit institutions
- Formulation and regulation of foreign exchange policies and transactions
- Regulation of gold transactions
- Formulation and regulation on transactions and inflow/outflow of Domestic currency

6.4.1.1 Objectives of Central Banks

As a banker to the government, the CBI is mandated to keep the government accounts and grant loans and credits to state enterprises and agencies. The CBI also covers such functions as lending facilities to banks, purchase and sale of government participation papers as well as other legal banking operations.

After the Islamic Revolution of Iran laws and regulations pertaining to money and banking institutions and monetary policy design and implementation were amended to reflect the priorities and principles as set out in the Constitution of the Islamic Republic of Iran (Zahedi & Azadi, 2018). At present, the CBI is responsible for the design and conduct of monetary policy within the context of the government's five-year development plan and annual budget. In line with the articles of the constitution, the monetary and credit policies are formulated and implemented in consistent with the MBAI as amended, the Usury-Free Banking Act of 1983, the Banks Nationalization Act of 1979 and the Law for the Administration of Banks, of 1979.

According to the Law of Riba from the Iran Central Bank, the Objectives of the Banking System are as follows:

- The establishment of a monetary and credit system based on rightness and justice (as delineated by Islamic jurisprudence) for the purpose of regulating the sound circulation of money and credit to enhance the health and growth of the country's economy (Anwar, 1992).
- Availing itself of monetary and credit mechanisms to engage in activities conducive to the attainment of the economic goals, policies and plans of the Government of the Islamic Republic.
- Creation of necessary facilities for the extension of cooperation and Gharz-al-hasaneh among the general public through the attraction and absorption of surplus funds, reserves, savings and deposits and the mobilization thereof in the provision of conditions and opportunities for gainful employment and investments, as stipulated in Clauses (2) and (9), Article (43) of the Constitution.
- Maintenance of the currency value and equilibrium in the balance of payments and facilitating the commercial exchanges.
- Facilitating payments and receipts, exchanges, transactions and other services to be performed by the banks, as determined by the Law.

6.4.1.2 Duties of Banking Systems

The duties of the Banking System as written in Article 2 are as follows:

- Issuance of notes and coins as legal tender, in conformity with the Law and regulations.
- Regulating, controlling and guidance of the circulation of money and credit, in accordance with the Law and regulations (Khan & Mirakhor, 1990).
- Performance of all banking operations in foreign exchange and local currency, and undertaking or guaranteeing the foreign exchange payments of the Government, according to the Law and regulations.
- Supervised transactions in gold and foreign exchange and the inflow or outflow of Iranian currency and foreign exchange, and the formulation of regulations governing thereof, in accordance with the Law.
- Performance of operations relating to valuable papers and documents according to the Law and regulations.

- Carrying out the monetary and credit policies, in accordance with the Law and regulations.
- Banking operations related to those parts of the approved economic plans are to be conducted through the monetary and credit system.
- Opening various Gharz-al-hasaneh (current and savings) accounts and accepting term investment deposits and issuance of relevant certificates, as required by the Law and regulations.
- Granting of loans and credits free of interest charges in accordance with the Law and regulations.
- Granting of loans and credits and provision of other banking services to the legally established cooperatives, for the realization of the provisions of Clause (2), Article (43) of the Constitution.
- Conducting transactions in gold and silver and holding and management of foreign exchange and gold reserves, with due observance of the relevant Laws and regulations.
- Holding the Rial balances of international monetary and financial institutions or similar organizations and/or their affiliates, according to the Law and regulations.
- Entering into payment arrangements in order to effect monetary, trade and transit agreements concluded between the Government and other countries in accordance with the Law and regulations.
- Accepting and holding in trust gold, silver, valuables securities and official documents for real or legal persons and leasing of safe deposit boxes.
- Issuing, confirming and accepting Rial or foreign exchange guarantees for customers. (16) Performance of the services of attorney or guardian, in accordance with the Law and regulations.

6.4.2 Islamic Banking Laws

Article 3(a) of the Monetary and Banking Act states that the Government is the sole authority having the right of issuing notes and coins and this right is hereby vested exclusively in Bank Markazi Iran Subject to the provisions of this Act. Another clause of Article 3(b) specifies that the maximum value to which coins may be used in the discharge of debts, as well as the manner in which notes and coins are called in and the conditions under which they are withdrawn from circulation shall be determined by Bank Markazi Iran, subject to confirmation by the Currency and Credit Council and the approval of the Minister of Finance, and shall be communicated to the public through the official Gazette, at least one of the mass-circulation daily papers of the Capital, and the radio and television networks. Iran has developed five additional principles in the profit-loss category, namely,

1 civil partnership
2 legal partnership
3 direct investment
4 mozaarah
5 mosaqat

While civil partnership, legal partnership and direct investment deal with the legal aspects of the formation of new ventures, both mozaarah and mosaqat are methods in agricultural financing (Nomani, 2003). Similarly, Pakistan has additional principles such as equity participation, participation terms certificate, modaraba certificate and rent sharing. All these principles, however, can be clustered together within the principles of mudaraba and musharaka. The Bai Salam of IBBL of Bangladesh is a principle in which the bank will make an advance purchase and the customer will deliver the goods at a later date.

This principle is similar to the principle of forward delivery transaction which is used by Islamic banks in Iran. The principle of bai al-dayn used at BIMB of Malaysia is for the transactions involving the sale and purchase of trade documents such as bills of exchange and bankers' acceptance. This principle is similar to the principle of debt purchasing in Iran. The principle of bai-mua'zzal or cost-plus sales under deferred payment is similar to the principle of bai bithaman ajil of BIMB and the principle of instalment sales which is used by Islamic banks in Iran. This facility is nevertheless incorporated within the principle of murabaha.

In Pakistan, the principles of mark-up and buy-back arrangement also operate in line with the principle of murabaha. Iran also has an additional principle within this category called "joalah." This concept refers to the undertaking of one party ja'el or employer (either bank or customer) to pay a specified amount of money or wage to another party in return for rendering a specified service in accordance with the terms of the contract (Zarrokh, 2010). This principle, therefore, is similar to the principles of commission and service charges in other countries.

6.4.2.1 The Law for Usury (Interest) - Free Banking (21 March 1984)

Guarantee of deposits and disbursement of prizes to the qard hasan customer is allowed by the religious council of Iran. In order to attract and mobilize deposits, the banks may, through promotional methods, give the following rewards to the depositors:

1 Non-fixed bonuses in cash or in kind to gharz-al-hasaneh deposits.
2 Exempting the depositors from, or granting discounts thereto, in payment of commissions and/or fees.
3 According priority to depositors in the use of banking facilities

6.4.2.2 The Banking and Monetary Act 1339 (27 May 1960)

The Act was to enable the establishment of the Currency and Credit Council. The duties of the Currency and Credit Council shall be as follows:

1 To co-operate with the Governor of the Bank Markazi Iran in the general management of the Bank as provided under Chapter Three of this Act.
2 To direct and supervise all the banking affairs, as provided under Chapter Four of this Act; In the discharge of this duty, the Council may demand any

information from, or issue any directive to other banks, whether their articles of association have been enacted by the Parliament or drawn up under the provisions of the Banking Act. All banks shall implement the Council's directives. No information, however, may be demanded by the Council in respect of the relation of any particular customer with a bank or of his account, except in cases specified in this Act.

3 To act as an advisory body to the Government in matters of public finance, monetary policy and national economy. The Government shall consult the Council on matters of, and decisions relating to, monetary and fiscal policies.
4 To co-operate with the High Economic Council in order to ensure coordination in decision-making on matters of monetary and fiscal policies.
5 To investigate and carry out research on problems related to its work as well as on subjects referred to it by the Government.
6 To perform such other duties and to exercise such other powers as have been specified for it by this Act.

6.4.2.3 The Law for the Issuance of Participation Papers (2 March 1376)

In order to increase public participation in the implementation of profitable government development projects included in government annual budget laws, as well as profitable productive, construction and services projects, the government, public corporations, municipalities, non-public institutions and entities, public utility corporations, and their affiliates, joint-stock and public joint-stock companies and producers' cooperatives are allowed, in line with this law, to finance part of the financial resources required for the implementation of the projects including financial resources required for the procurement of raw materials required by the productive units through the issuance and public offerings of participation papers. Participation papers are registered or bearer securities with a specified nominal value issued for a specific period of time and assigned to investors for the implementation of the projects mentioned in Article 1. Holders of these papers are entitled to their shares of profit, proportionate to the nominal value and the period of their participation. These papers are transacted directly or through the stock exchange.

Below are several laws governing the practice of the banking industry.

- Guidelines for credit institutions
- Gudidelines to establish a foreign bank in Iran
- Gudelines to establish, supervise and dissolve foreign bank branches in Iran
- Gudelines to overseas foreign banks participation in banks capital
- Gudelines for base capital of banks and credit institutions
- Gudelines to monitor banking operations in the free-trade industrial zones

6.5 Kuwait

The primary law is Law No. 30 of 2003 on the addition of a special section on Islamic banks to Chapter III of Law No. 32 of 1968 concerning currency, the Central Bank of Kuwait and the Organisation of Banking Business (Al-Muharrami, 2008). The Laws permit the establishment of full-fledged Islamic financial institutions and prohibit conventional banks from opening Islamic banking windows. A Shari'ah board is required for each Islamic financial institution. Disputes over Shari'ah issues shall be referred to the Ministry of Awqaf. The Central Bank has also issued rules and regulations concerning the opening of branches of Islamic banks within and outside Kuwait, direct investment, instructions regarding financial investment policy, and regulations governing the purchase of their own shares by Islamic financial institutions.

6.5.1 The Regulator

The Islamic banking and capital market system is regulated by the Central Bank of Kuwait (CBK). The first Islamic financial institution was Kuwait Finance House, which was established in 1977 (Çakmak, 2023). Other licensed Islamic financial institutions are Boubyan Bank and Al- Safa Bank. Amendment was made to CBK Law No 32 1968 to establish the following:

1. Prudent regulation and supervisory policies and framework which conform to international best practices and standards
2. Procedures for Islamic finance in general and Islamic banking in particular

On top of that, the CBK law also sets the requirement for the Shariah governance tributary. Approval of products and services is undertaken by the individual bank's supervisory board (SBB). However, any disagreement or conflict of opinion among members of the SBB concerning a Shariah rule may be referred to the Fatwa Board in the Ministry of Awqaf and Islamic Affairs which is the final authority on the matter

In line with the Central Bank's responsibility to ensure the sound progress of the banking business in accordance with the rules of Law No. 32 of the year 1968 concerning Currency, CBK and the Organization of Banking Business, and its amendments, and as the competent authority to consider local banks applications for opening branches in various areas of Kuwait and abroad, the CBK has established the following rules and regulations that local Islamic bank have to comply with when they want to open banking branches within and outside the State of Kuwait (Arslan et al., 2019).

6.5.1.1 Regulations to Open a Bank in Kuwait

The bank desiring to open a local bank branch shall submit a written application to the CBK in this respect, together with an economic feasibility study for opening the branch, showing the extent to which the area in which the bank is applying to open a branch, is in need of banking services (Hussain et al., 2007). Such a study shall cover at least the following points:

- The objectives of opening the subject branch.
- The extent to which the area, in which the branch is to be opened, is in need of banking services and the considerations on which basis the branch site has been selected.

- Expectations regarding the increase of the banking services in that area and the period after which the business activity of the branch will be considered profitable.
- The extent to which the area needs the services of more than one branch, if any other bank desires to open a branch there.
- The impact of the activities of the requested branch on the activities of other branches of the bank, as well as on its head office.
- The financial assumptions of the requested branch are based on the new branch's anticipated business size, development, construction cost, operating cost and results over an appropriate time period of not less than three years.
- Any other matters the bank has taken into consideration when making the decision to open the branch.

6.5.1.2 Shariah Supervisory Board

Based on the rules of articles (93 and 97) of Law No. 32 of the year 1968 and its amendments, and given the prime importance of the supervisory role of the Shariah Supervisory Board in Islamic Banks, particularly in relation to preserving the reputation of those banks and the confidence of their customers, as well as with regard to compliance of Islamic banks with their articles of association through observing the rules and principles of Islamic shariah in their business and operations, the Central Bank Board of Directors resolved, at its meeting of 15/6/2003, to set up the following rules and conditions that regulate the appointment and responsibilities of the Shariah Supervisory Boards in Islamic banks (Tashkandi, 2023).

These rules and conditions are intended to ensure the efficiency required for the members of the Shariah Supervisory Board in order to discharge their functions in the appropriate manner. The instructions establish the framework for performing those duties while providing independence for the members of the Shariah Supervisory Board in expressing their opinions and defining the role of the external auditor in relation to the bank's compliance with the rules and principles of the Islamic Shariah.

The Central Bank of Kuwait provides the Summary of the Rules of Law No. 30 of the year 2003 on adding a special section on Islamic banks to the Third Chapter of Law No. 32 of the year 1968 Concerning Currency, CBK and the Organization of the Banking Business (Al-Shammari, 2013).

6.6 Lebanon

After a long history, Lebanon managed to establish a Central Bank. Since the Convention of May 29, 1937, concerning the concession to issue the Lebanese pound was coming to an end in 1963, Lebanon decided to create a central bank (Finger & Hesse, 2009). The Code of Money and Credit was promulgated by Legislative Decree NÂ° 13513 of August 1, 1963. It established the Banque du Liban as a moral person of public law, bestowed with the privilege of issuing the national currency. The Code repealed restrictions on exchange operations and capital movements, and the rate of the Lebanese pound was set in pure gold, as provided in Article 2. On the other hand, Article 229 provides for the following:

Until a new rate for the Lebanese pound is set in agreement with the IMF, a real exchange rate that is as close as possible to the market rate – called the "provisional legal rate" of the Lebanese pound – was to be adopted against the dollar whose rate was set at 0.888671 g of pure gold (Dibeh, 2009). The gold coverage of banknotes issued by the Central Bank was to be recorded at the "provisional legal rate." The differences between the value of the coverage in gold and in foreign currencies at the provisional legal rate, on the one hand, and the real rate of these gold and foreign-currency assets, on the other hand, were to be governed by the provisions of Decree No. 15105/K of 27 May 1949, until the transfer of these elements to the Central Bank.

6.6.1 The Regulator

The main authority is the Banking Control Commission – Banque du Liban, and the primary law is Law No. 575 of 2004 on Establishing Islamic Banks in Lebanon and the Central Bank's Basic Decisions No. 8828 and 8829 both dated 26 August 2004 (Bizri, 2014). Law No. 575 provides the definition of Islamic banks and sets out rules for the licensing of domestic Islamic banks, the registration of foreign Islamic banks, applicable laws, permissible activities, exemptions from existing laws, minimum disclosure frameworks, as well as Shari'ah compliance supervision. In addition, with regards to Islamic banking products, there has been issued Circular 94 8/6/2004 on Practice of Islamic Banking in Lebanon, Circular No. 95 8/26/2004 on Conditions of Islamic Banking Establishment in Lebanon, Circular 96 10/20/2004 on Operations of Murabahah and Circular 97 1/19/2005 Operations of Musharakah or Participation Operations executed by Islamic Banks. In addition, a number of CBL Decisions were issued to support the implementation of Islamic banking, specifically, CBL Basic Decision No. 8870 of 20/10/2004 regulates Murabahah operations, Article 69 determines the coverage of the Lebanese pound in the following manner:

- In order to ensure the coverage of the Lebanese pound, the assets of the Central Bank must include funds in gold and foreign currencies that are, at least, equal to 30% of the total value of the issued currency and sight deposits, on condition that the ratio of coverage in gold and foreign currencies is not inferior to 50% of the value of the issued currency. However, the assets of the Central Bank in Lebanese currency are not taken into account in computing both ratios.
- Periodical financial statements are to be prepared in accordance with Decision No. 883 of 27 March 1973, taken by the Minister of Finance in agreement with the Banque du Liban, on the basis of the US dollar valued at 42.2222 Troy ounces of pure gold, and on the basis of the average closing rates of foreign currencies at the corresponding dates.
- The procedures used for adopting a new gold rate against the US dollar and for adopting foreign-currency market rates against the Lebanese pound were confirmed by Legislative Decree No. 6104 of 5 October 1973.

- In accordance with Legislative Decree No. 6105 of 5 October 1973, the Government was empowered to set, in consultation with the Banque du Liban and the IMF, a new provisional legal rate for the Lebanese pound for a six-month period from the promulgation date of the said law. (The Government did not use this prerogative).

6.6.2 Islamic Banking Laws

Lebanon has gazetted a law called Law No. 575 outlining legal requirements for establishing Islamic banks in Lebanon. The law consists of 10 Articles explaining, in general, the nature of Islamic banking businesses, the real assets rights and the customer's right to informed investment.

Article 1 of the law states that Islamic banks are the business entities whose Articles of Association require them to comply, in their operations, with the provisions of Islamic Law (Sharia), particularly with the prohibition to pay or receive interest. Unless otherwise specified in this Law, Islamic banks shall be governed by all legal and regulatory provisions in force in Lebanon, particularly those related directly or indirectly to banks, including the Code of Land Trade, the Code of Money and Credit and the Banking Secrecy Law. Interestingly, the Article mentions the particular prohibition confining giving and receiving interest.

Islamic banks are entitled to offer and provide all banking, commercial, financial and investment services and operations, including the establishment of corporations and the participation in established projects or projects under establishment (Khan & Bhatti, 2008). Unless agreed with clients to link their deposits to the bank's annual results or to the results of its operations according to a procedure set for this purpose by the BDL Central Council, cash deposits received by Islamic banks are governed by the provisions of Article 307, Par. 1 and 2, of the Code of Land Trade, and by the provisions of Section 2 of Law 28/67 of 9 May 1967. Deposits received in accordance with Article 3, Par. 2, of this Law, and in accordance with Article 307, Par. 1 and 2, of the Code of Land Trade, should have a minimum term of six months. However, Islamic banks are entitled to open checking accounts in the names of their clients for recording cash deposit and withdrawal operations, securities purchase and sale operations and other Islamic banks' operations.

Based on Article 3 Law No 575 above, Islamic banks are not allowed to accept deposits for a period of time less than six months, which is a large burden in attracting customer deposits.

6.7 Oman

6.7.1 The Regulator

The establishment of the Central Bank of Oman (CBO) in the beginning of 1974 was a natural outcome of the steady evolution of the monetary system in the Sultanate of Oman coupled with the prospects of the vast economic developments the country was about to witness (Basiddiq & Rana, 2024). Until 1970, when His Majesty, Sultan Qaboos bin Said took over, there was no national authority responsible for the supervision of the incipient banking system. The number of banks was small and the banking activities were limited in scale.

The two monetary authorities that preceded the establishment of the CBO , namely the Muscat Currency Authority in 1970 and the Oman Currency Board in 1972, were not vested with full banking status, but they had well-prepared the ground for the emergence of the CBO (Magd & McCoy, 2014). However, the major event heralding the eminent creation of the CBO was the launching of the Banking Law in 1974 (which was amended vide Royal Decree No. 114/2000).

6.7.1.1 Regulator's Objectives

The objectives of the bank are as follows:

- Establish & implement sound monetary & credit policies.
- Foster a stable & sound banking financial system using dynamic regulatory & supervisory roles.
- Promote stabilization of the value of the local currency.
- Act as banker & financial advisor to the Government.
- Promote, regulate & supervise payment & settlement systems.
- Manage & safeguard the country's foreign exchange reserves.
- Collect, analyze & publish monetary/economic statistical data.

6.7.2 Islamic Banking Law

Oman Central Bank has issued the Islamic Banking Regulatory Framework (IBRF) which consists of ten Titles with different provisions in each title elaborating the operation of Islamic banking in detail. This was issued pursuant to the Islamic Banking Royal Decree 69/2012.

The status of Islamic banking transactions with regard to taxation and land law issues is illustrated in above mentioned Article 124 and following Article (Mubeen et al., 2014) 125 of the Islamic Banking Law. Art. 124 of the amended Banking Law mandates the transactional base of Islamic banking in Oman, stating that, without prejudice to the restrictions set by the Board of Governors, the banks licensed to practice Islamic banking shall in the context of such practice, conduct all transactions – without limitation – in due consistency with the provisions of Islamic Shari'ah Law, and in particular the following:

1 Accepting deposits and management of joint or specific investment accounts with or without fees and profits.
2 Financing and investing in Mudaraba, Musharaka, Murabaha, ljara, Salam, Istisna or Qard Hassan and other Shari'ah formulas.
3 Issuing instruments guaranteed by assets and projects, and the investment therein.

4 Dealing in real estate and movable property through buying, selling, investment, letting and leasing, in exception to the restrictions set forth in the Banking Law and the relevant laws and Royal Decrees.

The Art.124 authorization is supplemented by Art.125, which states that "Banks licensed to practice Islamic banking shall be exempted from fees levied on transactions related to the appropriation of real estate and movable property, or the letting or leasing thereof; conducted for the purpose of conducting Islamic banking in accordance with the provisions of this Law".

The first paragraph of Art.126 reads that a bank licensed to practice Islamic banking shall have a Shari'ah supervisory committee; and the regulations shall determine its composition, competence and work procedure, along with the requisite qualifications of its members, whose appointment and remuneration shall be decided by a resolution of the General Assembly of the Bank. The form, prerogatives and functioning of this Shari'ah supervisory boards are governed by the Islamic Banking Framework.

Additionally, according to Art.126, para 2 the Board of Governors shall establish a High Authority for Shari'ah Supervision, and the establishing decision shall specify its composition, competence and the work procedure therein, along with the requisite qualifications of its members and their remuneration. The authority's goal is to ensure the compatibility of Islamic banking and takaful products with Shari'ah principles.

The Board of Governors shall set the regulations, circulars and instructions relating to Islamic banking whether concerning licensing, regulation, management and Shari'ah supervision, capital, credit, investment and exposure limits, accounting, reporting, disclosure, risk management or otherwise (Aqeeq, 2014). The Central Bank shall issue the license to conduct Islamic banking activities through Islamic banks or Islamic windows of conventional banks. As noted above, the CBO issued the Islamic Banking Framework on 18 December 2012 pursuant to Circular IB.

By virtue of Royal Decree 69/2012, published on 6 December 2012, amending the Banking Law of 2000, the CBO issued the Islamic IBRF, regulating the following areas:

- Licensing requirements,
- General Obligations and Governance,
- Accounting Standards and Auditor Reports,
- Supervision and Control,
- Capital Adequacy,
- Credit Risks,
- Market Risks,
- Operational Risks,
- Liquidity Risks, and
- Miscellaneous

6.8 Palestine

6.8.1 The Regulator

The Palestine Monetary Authority (PMA) is an independent public institution responsible for the formulation and implementation of monetary and banking policies to maintain price stability and low inflation, foster financial stability and safeguard the banking sector and promote sustainable growth of the national economy (Alshaer et al., 2021). PMA works to achieve these goals through:

- Development and execution of monetary policy designed to ensure low inflation and achieve price stability.
- Effective and transparent regulation and supervision of banks, specialized lending institutions and money changers operating in Palestine.
- Overseeing the development, implementation and operation of modern, efficient payment systems.

PMA operates by an act of the Palestine Legislative Council's PMA Law Number (2) of 1997, which stipulated the proper authority and autonomy of the PMA and the Banking Law Number 9 of 2010 (Iqtait, 2021). An authority called the Palestinian Monetary Authority shall hereby be established with the status of an independent legal entity and full legal capacity to engage in all the activities and actions necessary to achieve the objectives for which it was established, including the acquisition of the real estate and movable property necessary to carry out its work, in accordance with the provisions of the law.

The objectives of the Monetary Authority are ensuring the soundness of banking operations, maintaining monetary stability and encouraging economic growth in Palestine, in accordance with the general policy of the National Authority (Powers, 2024). To achieve these objectives, it shall have the following duties and powers, in accordance with the provisions of the law:

1 Issue the national currency and coins in due course, in accordance with the terms and conditions to be determined by the law on issuing currency and securing the required cash reserve.
2 Regulate banking activities, issue and cancel bank licenses, control and supervise banks, and impose penalties on them.
3 Prepare, regulate and publish the balance of payments.
4 Provide liquidity to banks within the legally established limits.
5 Develop, regulate and implement monetary and credit policies and policies on dealing in foreign currencies, in accordance with the provisions of the Currency and Credit Law.
6 Maintain and manage the National Authority's reserve of gold and foreign currencies.
7 Provide financial and economic advice to the National Authority, carry out economic and monetary analyses on a regular basis and publish the results.

8 Serve as a financial agent to the National Authority and Palestinian public institutions inside and outside Palestine.
9 Develop and implement regulations, decrees and instructions to ensure the maintenance of an effective, secure and sound banking system.
10 Regulate the quantity, quality and cost of credit in order to meet the requirements of economic growth and monetary stability, in accordance with the provisions of the Currency and Credit Law.
11 Serve as the bank for licensed banks, specialized lending institutions and finance companies and supervise them to ensure the soundness of their financial position and the protection of depositors' rights.
12 Regulate activities of money-changers, finance companies and development and investment funds; issue them the relevant licenses and control and supervise them.
13 Carry out any other tasks assigned to it by any relevant legislation, law, or agreement, provided that they do not conflict with the provisions of this law.

6.8.2 Islamic Banking Laws

In Palestine, the whole Islamic banking system is regulated by the PMA, while the primary pieces of legislation are the Banking Law No. 2 of 2002 and the PMA Law No. 2 of 1997 (Migdad, 2017). To date, there are three full-fledged Islamic financial institutions: the Arab Islamic Bank, the Palestine Islamic Bank and the Al-Aqsa Islamic Bank.

The banking practice is regulated under the No. (2) of 2002. Islamic bank is defined as any publicly owned company whose objective is to compile the savings of corporeal and non-corporeal persons for use in funding and investment for the benefit of the person or group in accordance with the rulings of Islamic law.

6.9 Qatar

On 2 December 1971, Dubai became a part of the UAE. Therefore, it was decided to relinquish the Qatar Dubai Currency Board in accordance with Amiri Decree No. 6 of May 1973 (Zahlan, 2016). On 13 May 1973, Law No. 7 of 1973 was issued, establishing the Qatar Monetary Agency (QMA) to assume the duties of a central bank. Further Amiri Decree – No. 24 of 1973 – was issued authorizing the redemption of QDR and the issuance of a new currency known as the Qatari Riyal (QR), with the same par value against gold as the QDR.

QMA was responsible for maintaining the stability of the QR exchange rate against, and its free convertibility into, other currencies (Alsamara et al., 2019). In 1975, as per Decree No. 60 of 1975, the QR was pegged to the Special Drawing Rights (SDR) at a rate of 0.21 SDR per QR with a fluctuation margin of ± 2.25% (QR 4.7619 ± 0. 2.25% per an SDR unit). Over the period 1973–1993, QMA adopted the United States dollar (USD) as an intervention currency to fix the daily value of the QR. The QR exchange rate against the USD was to be determined on the basis of the latter's exchange rate against SDR as determined by the IMF. The

QR exchange rate against other currencies was to be determined on the basis of the USD exchange rate against other currencies in the international financial markets. The tolerable flux margin of the QR against the SDR was increased to ± 7.25% in early 1976 due to the appreciation of the USD against the SDR in late 1975. Hence, the QR exchange rate used to fluctuate against currencies other than the USD, in line with the fluctuations of the latter against those currencies.

During the second half of the1970s, QMA revalued the QR against the USD many times in order to stabilize its value against other major currencies of Qatar's trading partners and to alleviate pressures of imported inflation (Mirzaei & Moore, 2016). Over the period of March 1976 to June 1980, the QR was revalued 12 times against the USD. On the whole, the QR was revalued against the USD by 8.5%, compensating for 13.4% depreciation in the value of the latter against the SDR. When the USD started its upward trend (in July 1980) vis-à-vis other major currencies, QMA had maintained a de facto exchange rate of QR 3.64 per USD unaltered. The immediate impact of this link was the appreciation of the QR with the appreciation of the USD against major currencies of Qatar's trading partners, particularly the European countries.

Established in August 1993, the Qatar Central Bank (QCB) has inherited the QMA monetary strategy of targeting the exchange rate. QCB kept adopting the policy of a fixed exchange rate against the USD at the same rate of QR 3.64 per USD.

6.9.1 The Regulator

The main regulator of the Islamic banking and capital market is the QCB, while the regulator for the players at the Qatar Financial Centre (QFC) is the QFC Regulatory Authority (QFCRA). The primary law is the Amiri Decree No. 15 of 1993, which was issued to establish the central bank as the monetary authority, and mandates it to formulate monetary banking and credit policies to achieve certain financial and economic objectives. All commercial banks, including Islamic financial institutions, are governed under this Decree (Sharar & Khulaifi, 2016). To date, there are two full-fledged Islamic financial institutions in Qatar: Qatar Islamic Bank and Qatar International Islamic Bank; both were established in 1991. Apart from these two, commercial banks, e.g. Commercial Bank, Doha Bank and Qatar National Bank, offer Islamic windows. The QFCRA places great importance on cooperation with other regulators and international organisations. This is achieved through bilateral and multilateral forums.

6.9.1.1 International Commitment

The QFCRA is a member of the International Association of Insurance Supervisors (IAIS) and is a signatory to the IAIS Multilateral Memorandum of Understanding(MMoU). The QFCRA has continuously worked to strengthen its efforts with the IAIS. This has been achieved through the QFCRA's membership in the IAIS macro-prudential Policy Surveillance Working Group and the Signatories Working Group. The QFCRA is also an Associate Member of the International

Organisation of International Securities Commissions ("IOSCO") and contributes to IOSCO's work through participation in the Africa/Middle-East Regional Committee (Antoniades et al., 2022). The QFCRA is also a full signatory to the IOSCO MMoU Concerning Consultation and Cooperation and the Exchange of Information. Multilateral Agreements contracted are as follows:

- IAIS Multilateral Memorandum of Understanding
- IOSCO Multilateral Memorandum of Understanding

The Qatar Financial Centre (QFC) is the financial and business centre established by the Government of Qatar in 2005 to attract international banking, insurance business and other financial services in order to grow and develop the financial services sector in Qatar and the region. The QFC offers domestic and international firms the opportunity to establish a broad range of banking, asset management and insurance businesses under a legal and regulatory regime that meets international best practice.

The QFCRA is the independent regulator of the QFC, established to authorise and regulate firms and individuals conducting financial services in or from the QFC. It was established by Law No.7 of 2005 of the State of Qatar (the "QFC Law") and has built a principles-based regulatory regime aligned with widely used common law, embracing transparency, predictability and accountability.

The QFCRA's objectives include the promotion and maintenance of efficiency, transparency, integrity and confidence in the QFC, as well as the maintenance of financial stability and reduction of systemic risk. It also strives to develop financial awareness and protection for customers and investors.

6.9.2 Banking Laws and Regulations

The Government of Qatar recently issued the new Central Bank Law, Law No. 13 of 2012 (the "QCB Law"), which came into force in January 2013. This new piece of legislation is a comprehensive overhaul of the legislation that existed previously, Law No. 33 of 2006 (the "Old QCB Law"). QCB is the primary regulator of financial institutions operating in Qatar, and the QCB Law has now expanded its supervisory powers to cover the insurance sector and the Qatar Financial Centre (Al-Khatib & Al-Abdulla, 2001). The QCB Law addresses many issues for the first time, including Islamic banking, mergers and acquisitions of financial institutions, credit rating agencies, insurance, treating customers fairly and resolution of failing banks.

The definition of the term "Financial Services" under the QCB Law is wider than that of the Old QCB Law. In addition to the scope of activities under the Old QCB Law, it also covers insurance (conventional and Islamic), reinsurance business and IF.

The significant difference between the Old QCB Law and the QCB Law from a cross-border perspective is that Article 205 of the QCB Law imposes a specific penalty for providing financial services without being duly licensed by the QCB.

Article 205 provides that any entity providing financial services without a license from the QCB is liable to be imprisoned for not more than three years and fined for not more than QAR 5,000,000 (Hussain et al., 2007). Under the Old QCB Law, a specific penalty did not exist for providing financial services in Qatar without a license. Any breach under the Old Law would result in a fine of QAR 5,000 per day.

Therefore, it is clear that the QCB has taken note of cross-border activities. If the QCB views such activities as being provided in or from the State of Qatar, it now has the power to impose the penalties under Article 205 of the QCB Law. The QCB Law came into force only in January 2013, and thus far, we are not aware of any instances wherein the penalties under Article 205 of the New QCB Law have been imposed.

6.9.2.1 Shariah Principles

The QCB Law seeks to define the scope of Islamic banking as including (1) non-interest-based banking with the aim of generating profit; (2) developing innovative Shari'a-compliant savings and investment products; (3) providing financial services with the aim of benefiting society at large and based on the concept of mutual benefit; and (4) deepening the financial institutions understanding of Shari'a-compliant banking and contributing to the existing body of knowledge in this area.

The QCB Law clarifies that an Islamic financial institution may acquire title to assets and lease both land and movable assets in order to achieve its objectives. Each Islamic financial institution is required to have a Sharia Supervisory Authority (consisting of at least three members) appointed by the general body of shareholders of that financial institution. These members can be from various fields, including experts in law, Sharia law, financial services and business (Yeung & Al-Barashdi, 2022). The Shariaa Supervisory Authority is required to supervise all aspects of the business of the financial institution to ensure that all such activities are conducted in accordance with the Sharia Principles. It is also required to issue opinions on whether the products, services and agreements provided and entered into by the financial institution are consistent with Sharia principles.

As part of Qatar's National Vision for 2030, the State of Qatar has committed itself to creating a dynamic and sustainable economy by increasing economic diversification through the re-investment of its significant energy wealth into various industries. Through its robust legal and regulatory framework, the QFCRA provides means into one of the fastest-growing economies in the world.

The governance framework of the Board of Directors of the QFCRA is made up of elements set out in the relevant legislation and in the Governance Resolution adopted by the Board in September 2006 and reviewed and amended by the Board in March 2012 and in March 2015.

The relevant legislation is:

- Article 8 of the Qatar Financial Centre Law (Law No. 7 of 2005) ("the QFC Law"); and
- the Financial Services Regulations ("the FSR") made by the Minister under Article 9 of the QFC Law.

6.9.2.2 Dispute Resolution

The QCB Law establishes a Disputes Resolution Committee. The Disputes Resolution Committee has the power to review complaints filed against the decisions of the QCB, review fines and punishments issued by the QCB, review the decision of the QCB to appoint an administrator to take control of a failing bank of financial institution and amicably settle disputes between financial institutions (Dahdal, 2018). The Disputes Resolution Committee will consist of three judges of the Court of Appeal and two individuals nominated by the QCB's Board. Decisions of the Disputes Resolution Committee can be appealed to the Court of Appeal. CDRS decisions are binding on QFC-authorised firms if they are accepted by the customer. When the Independent Adjudicator makes a decision, the customer has 14 days to decide if they accept the decision. If the customer does not make a decision within 14 days, then it is assumed that they have rejected the decision. A rejected decision has no effect. When a decision is accepted by the customer, it is binding and cannot be appealed by either party.

6.10 Saudi Arabia

Saudi Arabia is an Islamic monarchy ruled by King Abdullah bin Abdulaziz Al-Saud. The uniqueness of the system is founded in the modified civil code system based on Islamic Shariah (Van Eijk, 2010). This system has its origins in the Holy Koran, the teachings of the Prophet Mohammed (Sunna) and the texts of the Islamic religious scholars commenting on and explaining such teachings. The concept of "law" in Saudi Arabia is synonymous with shariah, which is the basic and primary law of the country, as defined in Article 7 of the Basic Law of 1992 (the Basic Law) which reads: "The powers of ruling the Kingdom of Saudi Arabia originate from the Book of God and the Sunna of his Apostle, both of which reign supreme over this and all other laws of the State." Article 23 of the Basic Law further provides that "The state shall protect the Islamic Creed and shall cater to the application of shariah. The state shall enjoin good and forbid evil, and shall undertake the duties of the call to Islam."

6.10.1 The Regulator

6.10.1.1 The Saudi Arabian Monetary Agency

SAMA or Saudi Arabian Monetary Authority was established in 1952. The Agency shall be a corporation with continuing succession (Young, 1960). The Agency is authorized to take such action as may be necessary and appropriate to give effect to this Charter, including, but without thereby limiting its authority, the power to make contracts, to acquire and hold and pledge assets, and to incur such liabilities as are necessary and appropriate to the conduct of its operations. Its main role was to act as the central bank for the government. Its duties included, but were not limited to issuing the national currency (Saudi Riyal), supervising commercial banks, managing the country's foreign exchange reserves, conducting the monetary policy and promoting the growth and soundness of the financial system. SAMA derives its power from the SAMA Charter.

6.10.1.1.1 OBJECTIVES

The objectives of the Saudi Arabian Monetary Agency shall be:

1 To issue and strengthen the Saudi currency and to stabilize its internal and external value;
2 To deal with the banking affairs of the Government;
3 To regulate commercial banks and exchange dealers.

Article 2 of the Charter states that the Saudi Arabian Monetary Agency shall not pay nor receive interest, but it shall only charge certain fees on services rendered to the public and to the Government, in order to cover the Agency's expenditures. Such fees shall be charged in accordance with a regulation passed by the Board of Directors and approved by the Minister of Finance. The Agency shall not have capital and shall, therefore, repay to the Government its entire capital.

6.10.1.1.2 SHARIAH COMPLIANCE

Constitutionally, Shariah supervision was not identified among its responsibilities. The mandate to monitor and supervise falls under individual banks (Alnori & Alqahtani, 2019). The Gold Sovereign was not put into circulation in the same year it was minted (1370H) compliant with advice by Young, an Economic & Financial Advisor Heading an American delegation tasked with assisting in the development of the Kingdom's financial system (Banafe & Macleod, 2017). This American delegation was asked to focus mainly on monetary matters besides economic matters in general. Mr. Young advised that the two and a half million pieces of Gold Sovereigns were to be kept in custody until the establishment of a competent institution capable of managing the financial affairs of the country. So, these Gold Sovereigns were held until the foundation of the SAMA in 1372H (1952G).

SAMA is responsible for ensuring the soundness of the banking system and its effectiveness in carrying out its functions for the service of the country in general and users of its services and stakeholders (Al-Jasser & Banafe, 1999). In pursuance of this responsibility, SAMA has established the Banking Control Department for this purpose. The Banking Control Department is concerned with supervising and controlling the banking sector in the Kingdom of Saudi Arabia under Royal Decree No. M/5 dated 22/2/1386H and Royal Decree No. 23 dated 23/5/1377H, particularly Articles 1 (C) and 3 (D) of the Banking Control Law which entrust SAMA with controlling banks and money exchangers and issuing instructions thereof.

6.10.1.1.3 REGULATIONS FOR OPENING A BANK IN SAUDI ARABIA

Opening a bank in Saudi must fulfil the following criteria:

- It shall be a Saudi joint.
- The paid-up capital shall not be less than SAR 2.5 million and all subscriptions towards share capital shall be payable in cash.
- The founders and members of the board of directors shall be persons of good reputation.

The memorandum and articles of association shall be approved by the Minister of Finance and National Economy. In the case of a Foreign Bank wishing to set up a branch or branches in the Kingdom, the grant of a license shall be subject to such conditions as the Council of Ministers may stipulate upon the recommendation of the Minister of Finance and National Economy. The license shall in all cases be issued by the Minister of Finance and National Economy after the approval of the Council of Ministers.

The deposit liabilities of a bank shall not exceed 15 times its reserves and paid-up or invested capital. If the deposit liabilities exceed this limit, the bank must within one month of the date of submission of the statement referred to in paragraph 1 of Article 15 either increase its capital and reserves to the prescribed limit or deposit 50% of the excess with SAMA.

Every bank shall maintain with SAMA at all times a statutory deposit of a sum not less than 15% of its deposit liabilities. SAMA may, if it deems it to be in the public interest, vary the aforesaid percentage provided that it shall not be less than 10% nor more than 17.5%. SAMA may, however, exceed these two limits with the approval of the Minister of Finance and National Economy (Ronall, 1967). In addition to the statutory deposit provided, every bank shall maintain a liquidity reserve of not less than 15% of its deposit liabilities. Such reserve shall be in cash, gold or assets, which can be converted into cash within a period not exceeding 30 days. SAMA may, if deemed necessary, increase the aforesaid percentage up to 20%.

Every bank shall, before declaring distribution of any profits, transfer a sum equal to not less than 25% of its net profits, to the statutory reserve, until the amount of that reserve equals as a minimum of the paid-up capital.

The 1970s witnessed the establishment of a number of Islamic banks in the Middle East, such as Dubai Islamic Bank (1975), Faisal Islamic Bank of Sudan (1977) and the Bahrain Islamic Bank (1979). However, many people were still uncertain about the legitimacy of Islamic banking, a situation that remained through the 1980s. In 1978, Al Rajhi Investment Co. in Saudi Arabia was set up as the first purely Islamic financial institution providing foreign exchange and money transfer services to a large segment of foreigners and locals in the country. Based on the regulations issued by the SAMA, it seems that Shariah compliance is not regulated. Market and Bank regulators in the Kingdom do not regulate or statutorily impose any Shariah-related duties. Each institution is free to recognize its Shariah criteria, Shariah board and acceptable established practices. These practices have widened the maximum innovation, peer review and development of Shariah-compliant products.

6.10.2 Dispute Resolution in Banking and Finance

The disputes in banking and finance are regulated under the Insurance Dispute Committee under SAMA (the Saudi Arabian Monetary Agency), the SAMA Committee for the Resolution of Banking Disputes and the Committee for the Resolution of Securities Disputes (Nor, 2012). The Committee for the Settlement of Banking Disputes (the "Committee"), operating for three decades under the aegis of the Saudi Arabian Monetary Agency, has issued a new landscape in muamalat financial systems. With its jurisdiction over banking disputes and the application of its own particular principles to disputes (which include giving effect to the transaction agreed between the parties and the upholding of recognised, international banking practices), it allows banking claims to be enforced in a way which would not be possible were the cases to be heard in the general court system of the Kingdom. In 2011, the Royal Decree was issued that the Committee's judgments should be considered final and prevent any court from hearing disputes relating to them.

Arbitrations seated in Saudi Arabia are governed by the Law of Arbitration, adopted in 1983 by Royal Decree No. M/46, and by the more detailed Implementing Regulations, adopted in 1985 by Royal Decree No. M/7/2021.55 (Muneeza & Mustapha, 2019). The Implementing Regulations provide details on procedural issues not covered by the principles outlined in the Law of Arbitration. The Law of Arbitration provides a framework for commercial arbitration that applies equally to domestic and international disputes. Under the Law, the "body which has original jurisdiction over the dispute" – typically, the Board of Grievances at present – supervises all stages of the arbitration proceedings and monitors compliance with the Law of Arbitration and its Implementing Regulations. Notwithstanding its limitations, arbitration has been a familiar form of dispute resolution in Saudi Arabia for centuries.

Indeed Arbitration has historically served as a method of resolving certain tribal disputes in Saudi. The practice has evolved as a valid alternative dispute resolution mechanism concerning oil concessions (Meyer-Reumann, 1995). The arbitration was applied in a landmark case of Saudi Arabia v. Arabian American Oil Co (Aramco). In this, the tribunal held that "as Saudi law" contained "no particular rules which define mining concessions in general and petroleum concessions in particular," Saudi law had to be "interpreted or supplemented by general principles of law, by the custom and practice in the oil business and by notions of pure jurisprudence."

This case has resulted in the issuance of the Council of Ministers Resolution No. 58, dated 03/02/1383 AH (25 June 1963); see also Article 8 of the Implementing Regulations, in which the Council of Ministers Resolution barred government agencies from entering into arbitration without the approval of the President of the Council of Ministers and required that contracts with Saudi government agencies be subject to Saudi law. In furtherance of the above, another Royal Decree M/8, dated 22/3/1394 AH (28 September 1979), was issued, precludes investment disputes involving oil and acts of sovereignty from being adjudicated under the terms of the ICSID Convention.

6.10.3 Arbitration

In January 2012, the Shura Council approved a new draft regulation to govern arbitration in Saudi Arabia, which is based in part on the Council Committee's review of contemporary legal systems and the opinions of legal experts. An arbitration Agreement is defined as an agreement between two or more parties to refer to arbitration all or certain disputes that have arisen or which may arise between them in respect of a defined legal relationship, whether contractual or non-contractual. An arbitration agreement may be in the form of an arbitration clause in a contract or in the form of a separate arbitration agreement (Alamer, 2020). If both parties to the arbitration agree to subject the relationship between them to the provisions of any document (model contract, international convention, etc.), then the provisions of such document, including those related to arbitration, shall apply, provided this is not in conflict with the provisions of Sharia.

Compared to the Arbitration Laws of other countries, Saudi imposes a Shariah Law degree in order to be appointed as an arbitrator. An arbitrator should also comply with other conditions such as legal capacity and a man of good conduct and reputation. If the arbitration tribunal is composed of more than one arbitrator, it is sufficient for the chairman to have a Shariah Law degree. An action to nullify an arbitration award shall not be admitted except in the following cases:

1 If no arbitration agreement exists, or if such agreement is void, voidable or terminated due to expiry of its term;
2 If either party, at the time of concluding the arbitration agreement, lacks legal capacity, pursuant to the law governing his capacity;
3 If either arbitration party fails to present his defence due to lack of proper notification of the appointment of an arbitrator or of the arbitration proceedings or for any other reason beyond his control;
4 If the arbitration award excludes the application of any rules which the parties to arbitration agree to apply to the subject matter of the dispute;
5 If the composition of the arbitration tribunal or the appointment of the arbitrators is carried out in a manner violating this Law or the agreement of the parties;
6 If the arbitration award rules on matters not included in the arbitration agreement. Nevertheless, if parts of the award relating to matters subject to arbitration can be separated from those not subject thereto, then nullification shall apply only to parts not subject to arbitration.
7 If the arbitration tribunal fails to observe conditions required for the award in a manner affecting its substance, or if the award is based on void arbitration proceedings that affect it.

6.10.3.1 New York Convention on Enforcement of Arbitral Award

Saudi became a member of the New York Convention on Enforcement of Arbitral Award in 1994. During the 1950s, Saudi Arabian courts refused to enforce many non-Saudi Arabian arbitral awards, finding them degrading and disrespectful to Saudi Arabia's Islamic legal system. This resulted in the issuance of a Decree in 1963 that banned Saudi Arabian governmental agencies from using arbitration in resolving international commercial disputes (Roy, 1994).

A contract containing Riba will not be enforced by a Saudi Arabian court. Any practice of gharar such as insurance is prohibited. Therefore, the contract is ill-defined for being speculative. Realising the importance of insurance companies in the country, the Saudi Arabian government requires private insurance companies to invest their profit in land development or business enterprises with the Kingdom of Saudi Arabia. In 1986, the government began to recognize the use of cooperative insurance, wherein each policyholder is a partner in the company and all company funds are invested. This type of insurance is not seen as immoral because there is no Riba. In the case of any losses, there will be coverage for the policyholders. If not, the policyholder receives a return on his investment. This is designed to avoid any uncertainty of return. The cooperative is strictly a non-profit organization designed to meet the need for insurance without compromising Islamic principles. Saudi Arabia now permits parties to utilize insurance as an investment tool, with the condition thatinsurance companies invest all insurance profits within the borders of Saudi Arabia.

6.11 Syria

The primary law is Legislative Decree No. 35 of 2005 on Establishment of Islamic Banks, which, among others, requires an Islamic financial institution to have a minimum capital of SP1.5 billion. It also imposes compliance with AAOIFI accounting standards, the International Financial Services Board capital adequacy standards, as well as the Shari'ah. The whole Islamic banking system is regulated by Banque Centrale Syria (Wilson, 1992). The first Islamic financial institution is reported to be Qatar-Syria Islamic Bank, established in 2007. The Syrian Insurance Supervision Committee also licensed three takaful operators, the Syrian Qatari Insurance Co., Nour Insurance Co. and Al Aqila Insurance Co. in 2007.

Recently, the European Council added 17 ministers and the Governor of the Central Bank of Syria to the list of those targeted by EU restrictive measures against the Syrian regime for being responsible for the violent repression against the civilian population in Syria, benefiting from or supporting the regime, and/or being associated with such persons. This decision follows the Council conclusions of 17 October 2016 and the European Council conclusions of 20–21 October 2016. This decision brings to 235 the total number of persons subject to a travel ban and an asset freeze for the violent repression against the civilian population in Syria.

In addition, 69 entities are targeted by an asset freeze. More broadly, sanctions currently in place against Syria include an oil embargo, restrictions on certain investments, a freeze of the assets of the Syrian central bank within the EU, export restrictions on equipment and technology that might be used for internal repression as well as on equipment and technology for the monitoring or interception of internet or telephone communications. These measures were last extended on 27 May 2016 and are in place until 1 June 2017.

6.12 Turkey

Turkey has established the Association as a professional public institution that has been established as per the previous Banking Law no. 4389 in 2002. According to the Banking, Law No.5411 put into practice in November 2005, "Special Finance Houses" has been transformed into "Participation Banks," gaining the "bank" statute (Gürsel, 2019). The current statute of the Association as well has been regulated with article 81 of Banking Law No.5411 and publicly announced via the Official Circular on 28 February 2006. According to the banking law in question, to be a member of the association is a must for participating banks. The Association has currently five members, namely, Albaraka Türk, Bank Asya, Kuveyt Türk, Türkiye Finans and Ziraat Katilim. The head office of the association is located in Istanbul without any branches yet. "The Statute of Turkish Participation Banks' Association" was resolved by the Council of Ministers on 03/02/2006 upon the official letter dated 23/01/2006 and no. 110 of the Offices of Minister of State and Deputy Prime Minister based upon the proposal of BRSA (Banking Regulation and Supervision Agency), as per article 81 of the Banking Law no. 5411.

6.12.1 The Regulator

As the country's monetary authority, the Central Bank of the Republic of Turkey (CBRT) is responsible for implementing monetary and exchange rate policies and printing banknotes in Turkey (Bakir & Öniş, 2010). The CBRT was established as a joint stock company on 11 June 1930 (Demiralp & Demiralp, 2019). The primary objective of the Bank is to achieve price stability. The CBRT is also responsible for taking measures to sustain the stability of the financial system in Turkey. Particularly, the Bank strives to contain the macro-financial risks stemming from global imbalances. Accordingly, maintaining financial stability is defined as the supporting objective of the Bank.

The privilege of printing banknotes in Turkey was initially vested with the Grand National Assembly of Turkey. The Assembly has transferred the privilege of printing and issuing banknotes exclusively and indefinitely to the CBRT.

One of the main responsibilities of the CBRT is to determine the exchange rate regime jointly with the government. The CBRT is responsible for and authorized to design and implement the exchange rate policy in line with the agreed exchange rate regime. Since 2001, the floating exchange rate regime has been implemented in Turkey.

The CBRT is in charge of managing the gold and FX reserves of Turkey. Reasons for holding international reserves are to establish and maintain confidence in monetary and exchange rate policies, provide FX liquidity for the Treasury's domestic and foreign debt services, reduce the economy's susceptibility to endogenous and exogenous shocks and boost the confidence of international markets in the Turkish economy.

The CBRT is also responsible for establishing and securing the uninterrupted functioning of payment, security transfer and settlement systems to enable safe and swift transfer of money and securities (Şahin, 2012).

During the Ottoman Empire, treasury operations, money and credit transactions and trade in gold and foreign currencies were executed in varying degrees by the Treasury, the Mint, jewellers, moneylenders, foundations and guilds. Within this structure, which continued till the second half of the 19th century, the Ottoman Empire minted gold coins. The Ottoman Empire exercised the right of seigniorage for the first time in 1839. Hand-made cash banknotes called "Kaime-i nakdiye-i mutebere," served more like interest-bearing debt instruments or Treasury bills. During the Crimean War in 1854, the Empire resorted to foreign borrowing for the first time in its history.

Thus, the need arose for an institution that would act as an intermediary between the Ottoman Government and European countries in foreign debt repayment. For this purpose, in 1856, the London-based "Ottoman Bank (Bank-Osmanî)" was founded with British capital. The powers of the Bank were limited to extending small loans, making advance payments to the government and discounting certain Treasury bills of Turkey Central Bank. Later, the Ottoman Bank dissolved itself and was renamed "Imperial Ottoman Bank (Bank-ı Osmanî-I Şahane or Osmanlı Bankası)." In 1863, it was restructured to become a state bank as a British–French partnership. The Bank was granted the privilege of issuing banknotes for 30 years. Moreover, the Bank would act as the treasurer of the state and collect the state revenues, perform the payments of the Treasury and discount Treasury bills as well as make the interest and principal payments regarding domestic and foreign debts.

However, the foreign capital ownership of Osmanli Bankası led to heated debates over the years. Reactions intensified in the Second Constitutional Monarchy when the idea of establishing a national central bank was formed. Attempts to establish a domestic capital-based central bank culminated in the establishment of the "Ottoman National Credit Bank (Osmanlı İtibar-ı Milli Bankası)" on 11 March 1917. However, due to the defeat of the Ottoman Empire in the First World War, the Bank failed to achieve its purpose of becoming a national bank that would replace the Osmanlı Bankası. Following the end of the First World War, a tendency to establish central banks to issue currency and manage monetary policies emerged throughout the world. In the same period, Turkey also witnessed efforts towards establishing a central bank to reinforce the political independence achieved by the War of Independence (1919–1922) with economic independence (Herrera & Ozbay, 2005). The issue was first raised at the İzmir Economic Congress in 1923, with a special emphasis on the establishment of "a national state bank" that would function as "the bank of issue." In this framework, Turkey exchanged views with central banks of other countries on establishing the Turkish Central Bank.

In 1928, Vissering, Head of the Board at De Nederlandsche Bank (Central Bank of the Netherlands), was invited to Turkey. In his report, Vissering underlined the need for establishing an independent central bank. Likewise, Count Volpi, Italy's former Minister of Finance, maintained that establishing a central bank was essential to ensure the stability of the Turkish currency. Draft bill on the establishment of a central bank was prepared with the contribution of Prof. Leon Morf from the University of Lausanne. Finally, the Grand National Assembly of Turkey passed "Law No. 1715 on the Central Bank of the Republic of Turkey" on 11 June 1930 (Berument & Malatyali, 2000).

The Central Bank started to operate on 3 October 1931. As a manifestation of its independence and a peculiarity that distinguishes the Central Bank from other public institutions, the Bank was established as a joint stock company and was structured as a General Directorate. Bank shares were divided into class A, B, C and D shares (Şahin, 2012). Class A shares belong solely to the Treasury. Law No. 1715 stipulated that to reinforce the Bank's independence, the amount of these shares would not exceed 15% of the capital. Class B shares were allocated to national banks operating in Turkey, while class C shares were allocated to banks other than national banks and privileged companies; and class D shares were allocated to Turkish commercial institutions and legal and real persons of Turkish nationality. During the establishment process, a publicity campaign was carried out to encourage the public, especially civil servants, to purchase shares of the Bank.

The main body of the Bank was the General Assembly of Shareholders. Other bodies of the Bank were The Board of Directors, the Bank's highest executive body; the Commission of Auditors authorized to audit the operations and accounts of the Bank; the Discount and Credit Committee, entrusted with regulating discount and interest rates besides credit affairs; the Administrative Committee to regulate the appointment of employees as well as credit and discount procedures; the General Directorate and Branches. According to Law No. 1715, the primary objective of the Central Bank was to bolster the economic development of the country. The Bank was vested with the exclusive privilege of issuing banknotes in Turkey. Efforts towards improving the Central Bank's efficiency accelerated in the 1960s, and a new era began in the history of the Central Bank with the enactment of a new founding law. With this law, the Central Bank, albeit partially, gained a structure that adopted innovations both in the economic field and the field of central banking. "Law No. 1211 on the Central Bank of the Republic of Turkey," which was enacted on 14 January 1970, introduced significant changes in the legal status and organizational structure as well as the powers and duties of the Bank.

While the legal status of the Bank was retained as a joint stock company, it was decided that the Treasury's shares would not be less than 51% of the Bank's capital. One of the significant changes made in the organizational structure was the formation of the chair of the Governor to replace the title "General Manager" to achieve equivalence in foreign representation and protocol. Accordingly, Mr. Naim Talu became the first person to assume the title of "Governor." Moreover, a new decision-making body called the "Executive Committee," consisting of the Governor and Deputy Governors was established. The highest decision-making body of the Bank, the eight-member Board of Directors was transformed into the six-member "Board."

Besides, the General Assembly of Shareholders was renamed as the "General Assembly"; the Commission of Auditors as the "Auditing Committee"; the term General Directorate was replaced by "Head Office." Law No. 1211 enhanced the Bank's powers and duties. First, the Bank's control over direct and indirect monetary policy instruments has been increased. In addition, it was stipulated that the Government, while taking measures regarding money and credits, shall ask the opinion of the Bank. With the amendment to the Law, the Bank was authorized to conduct open market operations to regulate money supply and liquidity. Moreover, the Bank was permitted to conduct rediscount operations and extend medium-term loans to support investments and economic development.

The amount of advances available to the Treasury was raised to 15% of the budget appropriation for the year in question. The 1980s were marked by developments that could be defined as milestones for both the Turkish economy and the Central Bank. The decision of 24 January 1980 introduced a structural transformation for the Turkish economy. Price controls were abandoned to allow prices to be determined under the market mechanism and foreign trade has been liberalized. With the launch of the financial liberalization process, owing to which, important steps were taken towards establishing an infrastructure that would enable the Central Bank to conduct monetary and exchange rate policies in tune with the free market economy. Concurrently, it was decided that interest rates for deposits and credits shall be determined under market conditions. Moreover, the Turkish lira was devalued and the fixed exchange rate regime was abandoned. Another important development of the 1980s was the amendment to the Law on the CBRT in 1983 to authorize the Bank to manage the national gold and foreign exchange reserves effectively.

Furthermore, an article was added to the Law stipulating that the Bank shall carry out its primary functions parallel with the key requirements of the economy as well as to ensure price stability. In 1987, the Central Bank began to conduct open market operations. In that period, the Bank pioneered to establishment of money and foreign exchange markets in the modern sense. For this purpose, the Money Markets and Fund Management Department was established on 22 October 1987. In 1989, Decree No. 32 allowed economic units to conduct operations in foreign currency and, the Turkish lira was recognized as a convertible currency and a more flexible exchange rate regime was thus adopted.

6.12.2 Central Bank Law

The Central Bank Law was gazetted on 25 April 2001. The most notable of these amendments was the inclusion of an article stipulating that the primary objective of the Bank shall be to achieve and maintain price stability (Gormez, 2008). In this context, the Law stipulates that the Bank shall determine, at its discretion, the monetary policy that it will implement and the monetary policy tools that it is going to use. Hence, the Bank has gained instrument independence. The Law also states that the Bank shall support growth and employment policies of the Government, provided that they are not in conflict with the objective of price stability. Another important amendment made to the Law was the prohibition of granting advances and extending credits to the Treasury or to other public institutions. Purchasing

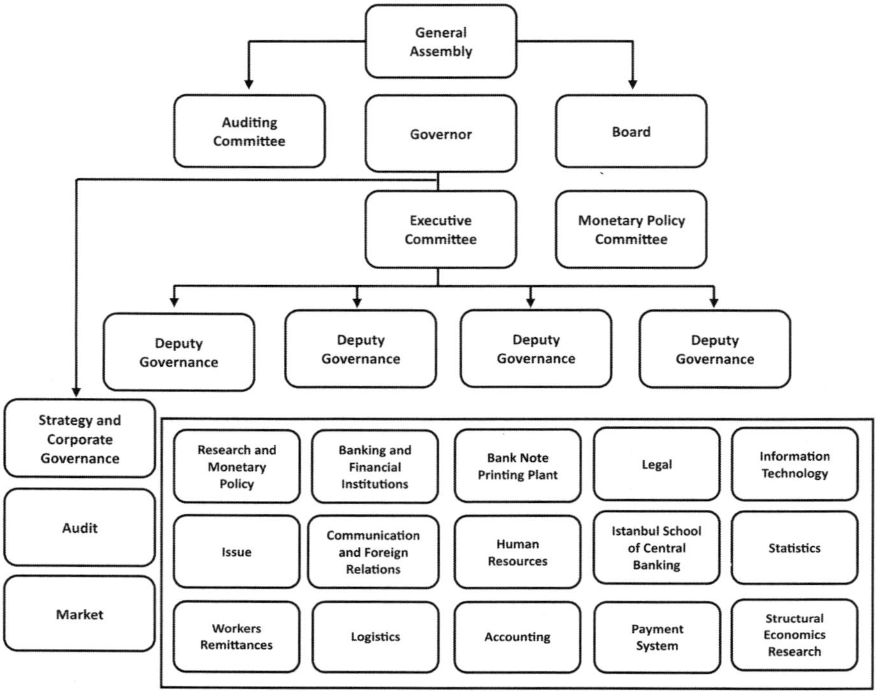

Figure 6.1 Financial regulators in Turkey.

debt instruments in the primary market issued by the Treasury and public establishments and institutions was also prohibited. Thus, the practice of using Central Bank resources in financing public sector deficits was abandoned. This amendment is important as it brings an end to the monetization of public deficits, one of the primary causes of inflation. Central Bank independence brought about the concept of "accountability." The practice of this principle has been outlined in Article 42 of the Bank's Law. In this respect, every year in April and October, the Governor submits a report to the Council of Ministers on the operations of the Bank and the monetary policy pursued. The Bank bi-annually provides the Planning and Budget Commission of the Grand National Assembly of Turkey with information regarding its operations. Figure 6.1 outlines the financial regulatory bodies in Turkey.

6.12.3 Bank in Turkey

Banks in Tukey shall fulfil the following requirements for licensing and operations validity (Basar et al., 2017);

- It should be established as a joint stock company.
- Its shares should be issued against cash and name.
- The founders should meet the requirements indicated in the Banking Law d. Its members of the board of directors shall bear the qualifications set out in the

corporate governance provisions in Banking Law and shall have the professional experience required for carrying out the planned activities.
- Its envisaged fields of activity shall be in harmony with the planned financial, managerial and organizational structure.
- Its paid-up capital, consisting of cash and free of all kinds of fictitious transactions, should not be less than TRY 30.000.000.
- Its articles of association shall not be in conflict with the provisions of the Banking Law. There should be a transparent and open partnership structure and organizational chart that will not constitute an obstacle to the efficient supervision of the institution.
- There should not be any element that hampers its consolidated supervision.
- The work plans for the envisioned fields of activity, the projections regarding the financial structure of the institution including capital adequacy, the budgetary plan for the first three years and an activity program including internal control, risk management and internal audit system showing the structural organization must be submitted.
- Any bank established abroad that desires to operate in Turkey by opening a branch within the framework of the principles and procedures set by the BRSA should meet the following conditions:
    - Its primary activities must not have been prohibited in the country where they are headquartered.
    - The supervisory authority in the country, wherein the headquarters of the bank is located should not have negative views regarding its operation in Turkey.
    - The paid-in capital reserved for Turkey should not be less than the amount indicated in Article 7 of the Banking Law No 5411.
    - The members of the board of managers should have adequate professional experience to be able to satisfy the requirements laid down in the corporate governance provisions and to perform the planned activities.
    - It must submit an activity program indicating work plans for the fields of activity covered by the permission, the budgetary plan for the first three years as well as its structural organization.
    - The group including the bank must have a transparent partnership structure.
    - An application for operating permission cannot be granted for the activities prohibited due to the violation of the local legislation in the country where such institutions are headquartered.

6.12.4 Banking Legal Framework in Turkey

The regulator gained their power from Law No. 1211 Date of Acceptance: 14 January 1970. The Central Bank mandate is stipulated in the provisions of the Act (Yanīkkaya & Pabuçcu, 2017). The fundamental duties and powers of the Bank as written in Art 4 shall be as follows:

- to carry out open market operations,
- to take necessary measures to protect the domestic and international value of the Turkish Lira and to establish the exchange rate regime to determine the parity of the Turkish Lira against gold and foreign currencies jointly with the

Government, to execute spot and forward purchase and sale of foreign exchange and banknotes, foreign exchange swaps and other derivatives transactions to determine the value of the Turkish Lira against foreign currencies,
- determine the procedures and principles of reserve requirements and liquidity requirements by taking into consideration the liabilities of banks and other financial institutions to be deemed appropriate by the Bank,
- to conduct rediscount and advance operations,

  e to manage the gold and foreign exchange reserves of the country,
  f (As amended by Law No. 6493 of June 20, 2013) to regulate the volume and circulation of the Turkish Lira, to establish payment, securities transfer and settlement systems, to ensure the uninterrupted operation and oversight of the systems established and to be established and to make necessary regulations, to determine the methods and instruments including electronic environment that shall be used for payments,

- to take precautions for enhancing the stability of the financial system and to take regulatory measures for money and foreign exchange markets,
- to monitor the financial markets,
- to determine the terms and types of deposits in banks and the terms of participation funds in special finance houses.

Law No 6493 is the law to regulate the procedures and principles regarding payment and securities settlement systems, payment services, payment institutions and electronic money institutions. The law to regulate market conduct is Law No. 5411, officially Gazetted on 1 November 2005. The objective of this law is to regulate the principles and procedures of ensuring confidence and stability in financial markets, the efficient functioning of the credit system and the protection of the rights and interests of depositors. The deposit banks, participation banks, development and investment banks, the branches in Turkey of such institutions established abroad, financial holding companies, Banks Association of Turkey, Participation Banks Association of Turkey, BRSA, Savings Deposit Insurance Fund and their activities shall be subject to provisions of this law. The provisions of this law shall also apply to banks that have been established as per their special laws, on the condition to preserve the provisions of their special laws. The general provisions shall apply to cases for which provisions are not set out in this law.

6.12.5 ***Banking Regulatory Framework***

In Turkey, there are five major regulators for financial institutions:

- *Central Bank of Turkey*

  The CBRT is responsible for securing the objectives of price stability and financial system stability. It is also responsible for the operation, regulation and oversight of payment systems in Turkey. The details of the CBRT's role are described in Section 6.12.1.
- *The Banking Regulation and Supervision Agency (BRSA)*

The BRSA, which was established under the Banks Law (Law No 4389 enacted in 1999; annulled by the Banking Law, Law No 5411 enacted in 2005), issues licences and regulates and supervises all major financial institutions, including banks, financial holding companies, leasing, factoring and consumer finance companies. The BRSA also ensures that independent external auditing firms and rating agencies offering services to financial institutions act following the principles and methods set by the Agency.

The BRSA aims to safeguard the rights and benefits of depositors and to create the proper environment in which banks can operate with market discipline in a healthy, efficient and globally competitive manner, thus contributing to the achievement of long-run economic growth and stability of the country.

6.12.5.1 The Capital Markets Board of Turkey

The Capital Markets Board of Turkey (CMB) is the regulatory and supervisory authority in charge of securities markets in Turkey. Authorised by the Capital Markets Law (Law No 2499 enacted in 1981), the CMB is responsible for regulating and supervising securities markets and institutions, determining the operational principles of capital markets, and protecting the rights and interests of investors.

- In this framework, the CMB regulates and supervises corporations offering securities to the public, securities market intermediaries, securities investment funds (mutual funds), investment companies (including real estate investment companies and venture capital investment companies), independent external auditing firms offering services to capital market institutions, stock exchanges and secondary markets in general, precious metal exchanges and derivatives exchanges and other related institutions operating on the capital markets, such as rating agencies, clearing and depository institutions.
- The regulation of capital market instruments, including futures and options, also falls under the regulatory scope of the CMB. In this framework, it also assumes responsibility for introducing and developing new instruments.
- Based on the overall objectives of ensuring the fair and orderly functioning of the markets and protecting the rights of investors, the major goal of the CMB is to take necessary measures to foster the development of capital markets, hence contributing to the efficient allocation of financial resources in the country while ensuring investor protection.
- The General Directorate of Insurance under the Turkish Treasury is responsible for the regulation and supervision of the insurance business, both life insurance and casualty insurance, under the Insurance Law (Law No 7397 on "Supervision of Insurance Companies", enacted in 1959).
- The Turkish Competition Authority (TCA), which was established by the Protection of Competition Law (Law No 4054 enacted in 1994) and commenced operation in 1997, is responsible for preventing proposed mergers and acquisitions that would create a dominant position in the market and adversely affect competition. Under the Regulation on Mergers and Acquisitions Requiring

Authorisation by the TCA (Communiqué No 1997/1 enacted in 1997), authorisation by the TCA is obligatory when the total market share of the undertakings, including banks, carrying out the merger or acquisition exceeds 25% of the relevant product market throughout the country or a part thereof, or when their total turnover exceeds TRY 25 million. The TCA may authorise a proposed merger or acquisition on condition that other measures deemed appropriate are taken and certain obligations are met.

- The legal framework for the payment and settlement systems Although there are no laws dealing specifically with payment and settlement systems, several laws have a bearing on payment-related activities and institutions:

The CBRT Law (Law No 1211 enacted in 1970; last amended by Law No 4651 enacted in 2001) regulates the operations of the CBRT, whose major functions and responsibilities derive from this Law – inter alia the exclusive right to issue banknotes in Turkey (Cetin, 2011). The Law authorises the CBRT to establish payment, securities transfer and settlement systems, to enact regulations to ensure the uninterrupted operation and oversight of existing or future systems and to determine the methods and instruments, including the electronic environment for payments.

The Banking Law (Law No 5411 enacted in 2005; amended by Law No 5472 enacted in 2006) regulates the business activities of all banks, namely deposit banks, participation banks and development and investment banks, in Turkey and Turkish branches of such institutions established abroad. The Law lays down rules and procedures governing the incorporation, management, transactions, transfer, merger, liquidation and supervision of banks, acceptance of deposits and participation funds to protect the rights and interests of depositors (Bakir, C., & Öniş, 2010). Its purpose is also to ensure the efficient functioning of the credit system by giving due consideration to confidence and stability in financial markets. All activities of financial holding companies, the Banks Association of Turkey (BAT), the Participation Banks Association of Turkey (TKBB), the BRSA and the Savings Deposits Insurance Fund (SDIF) are also governed by the provisions of this Law.

Article 79 of the Banking Law makes it obligatory for all deposit, development and investment banks subject to this Law to become a member of the BAT, and for all participating banks to become members of the TKBB, within one month of receipt of their operating licence.

The Capital Markets Law (Law No 2499 enacted in 1981; amended by Law No 4487 enacted in 1999) regulates and controls the secure, transparent and stable functioning of the capital markets and protects the rights and interests of investors to ensure efficient and widespread participation by the public in the development of the economy through investment of savings in the securities market. Capital market instruments and their public offering and sale, exchange markets and other markets organised according to the Law, capital market activities, capital market institutions and the Capital Markets Board are subject to the provisions of the Law. The Law focuses on shares offered to the public and does not apply to the issuance of shares not offered to the public by joint stock corporations that are not publicly held.

The Cheque Law (Law No 3167 on "Arrangement of Payments by Cheque and Protection of Cheque Bearers," enacted in 1985; amended by Law No 4814 enacted in 2003) governs all aspects of payment by cheque and provides the framework for issuance, acceptance and payment of cheques. Article 6 of the Law governs the establishment and functioning of the interbank clearing houses (ICHs). Under the Central Bank Law, the CBRT is responsible for their supervision.

The Bank Cards and Credit Cards Law (Law No 5464 enacted in 2006) aims to ensure the efficient functioning of card-based payment systems by establishing rules and procedures governing the issuance, usage and clearing of bank and credit cards. The Law also governs the issuance and revocation of operating licences, the obligations of card issuers and the penalties and obligations of card bearers.

The Public Finance and Debt Management Law (Law No 4749 enacted in 2002; last amended by Law No 5335 enacted in 2005) establishes rules and procedures for domestic and foreign public debt management. Article 6, paragraph 4, of the Law, authorises the Undersecretariat of the Treasury and the CBRT to regulate the issuance of government bonds and treasury bills in the book entry form. All securities transfers under this Law are registered with the Electronic Securities Transfer and Settlement System (TIC-ESTS), which is the securities settlement system of the CBRT. The final settlement of the related payments is provided by the national Real-time Gross Settlement System (TIC-RTGS).

The Turkish Commercial Code (Law No 6762 enacted in 1956) defines in Articles 557–815 the form and usage of negotiable instruments, namely promissory notes, bills and cheques, which are used widely for commercial payments.

The rules and procedures governing participation in TIC-RTGS and TIC-ESTS are governed by a special agreement between the CBRT and participating banks. The banks sign a participation contract with the CBRT agreeing to comply with the provisions of the TIC-RTGS & ESTS Operational Rules. These rules define the relationship between the CBRT and the participating banks and between the participants concerning the services provided by TIC-RTGS and TIC-ESTS. In particular, they specify the roles of the various parties involved in the system. The operational procedures, rights, obligations and a clause stipulating arbitration as a means of dispute resolution are also governed by these rules.

References

Abdel-Baki, M. A. (2012). The impact of Basel III on emerging economies. *Global Economy Journal*, *12*(2), 1850256.

Alamer, S. A. (2020). Interest and banks in Saudi Arabia: A dilemma. *Arab Law Quarterly*, *36*(1-2), 49–85.

Ali, S. A., Al-Absy, M. S. M., Al Astal, A. Y. M., & Gharaibeh, A. M. O. (2024). Impact of financial technology on future of banking in Bahrain: Opportunities and challenges. Digital technology and changing roles. In *Managerial and Financial Accounting: Theoretical Knowledge and Practical Application*, *36*, 141–153.

Al-Jasser, M., & Banafe, A. (1999). Monetary policy instruments and procedures in Saudi Arabia. *BIS Policy Papers*, *5*, 203–217.

Al-Khatib, F., & Al-Abdulla, M. (2001). The state of Qatar: A financial and legal overview. *Middle East Policy*, *8*(3), 110–126.

Al-Mashat, R., & Billmeier, A. (2008). The monetary transmission mechanism in Egypt. *Review of Middle East Economics and Finance*, *4*(3), 32–82.

Al-Muharrami, S. (2008). Testing the contestability in Kuwait banking industry. *Studies in Economics and Finance*, *25*(4), 253–266.

Alnori, F., & Alqahtani, F. (2019). Capital structure and speed of adjustment in non-financial firms: Does sharia compliance matter? Evidence from Saudi Arabia. *Emerging Markets Review*, *39*, 50–67.

Alsamara, M., Mrabet, Z., Jarallah, S., & Barkat, K. (2019). The switching impact of financial stability and economic growth in Qatar: Evidence from an oil-rich country. *The Quarterly Review of Economics and Finance*, *73*, 205–216.

Alshaer, H., Md. Said, M. H., & Rajamanickam, R. (2021). The role of the Palestine monetary authority in combating money laundering. *Journal of Money Laundering Control*, *24*(4), 762–774.

Al-Shammari, B. (2013). An investigation of voluntary disclosure by Kuwaiti Shariah-compliant companies. *Journal of Economic and Administrative Sciences*, *29*(1), 21–41.

Antoniades, A., Al-Jassim, R., & Gharatkar, K. (2022). The blockade against Qatar: A blessing in disguise?. In *Reflecting on the GCC crisis* (pp. 107–125). Routledge.

Anwar, M. (1992). Islamic banking in Iran and Pakistan: A comparative study. *The Pakistan Development Review*, *31*(4), 1089–1097.

Aqeeq, M. A. (2014). Islamic Banking in Oman–Today & The Way Forward. Available at SSRN 2577369.

Arslan, I., Bozgeyik, Y., & Al-Azaki, Z. (2019). The role of bank credit in Kuwait economic growth. *International Journal of Academic Research in Business and Social Sciences*, *9*(2).

Awad, I. L. (2008). Towards measurement of political pressure on central banks: The case of the central bank of Egypt. *Prague Economic Papers*, *17*(3), 254–275.

Bahmani-Oskooee, M., & Shiva, R. (1998). A method of detecting whether a central bank engages in the black market for foreign exchange: Evidence from Iran. *Economics Letters*, *60*(1), 97–103.

Bakir, C., & Öniş, Z. (2010). The regulatory state and Turkish banking reforms in the age of post-Washington Consensus. *Development and Change*, *41*(1), 77–106.

Banafe, A., & Macleod, R. (2017). *The Saudi Arabian Monetary Agency, 1952-2016: Central Bank of Oil*. Springer.

Basar, A., Kabak, Ö, & Ilker Topcu, Y. (2017). A decision support methodology for locating bank branches: A case study in Turkey. *International Journal of Information Technology & Decision Making*, *16*(01), 59–86.

Basiddiq, H., & Rana, S. (2024). Banking System in Oman. In *Handbook of banking and finance in the MENA Region* (pp. 171–192).

Berument, H., & Malatyali, K. (2000). The implicit reaction function of the Central Bank of the Republic of Turkey. *Applied Economics Letters*, *7*(7), 425–430.

Bizri, M. R. (2014). A study of Islamic banks in the non-GCC MENA region: Evidence from Lebanon. *International Journal of Bank Marketing*, *32*(2), 130–149.

Çakmak, C. (2023). Central Bank of Kuwait. In The Palgrave Encyclopedia of Islamic finance and economics (pp. 1–2). Springer International Publishing.

Cetin, N. (2011). Enforcement of prudential standards in Turkish banking law. *Journal of Money Laundering Control*, *14*(3), 254–265.

Dahdal, A. (2018). Finance and fairness: Enhancing the customer dispute resolution scheme (CDRS) in the Qatar financial centre (QFC). *Law and Financial Markets Review*, *12*(3), 133–140.

Demiralp, S., & Demiralp, S. (2019). Erosion of central bank independence in Turkey. *Turkish Studies*, *20*(1), 49–68.

Dibeh, G. (2009). The political economy of central banking in the MENA region with special reference to Lebanon. *In Monetary policy and Central Banking in the Middle East and North Africa* (pp. 21–38). Routledge.

El-Faham, K. M. D. (2020). Financial risk and financial performance of banks in Egypt, Doctoral dissertation, Cardiff Metropolitan University.

Fayed, M. E. (2013). Comparative performance study of conventional and Islamic banking in Egypt. *Journal of Applied Finance and Banking*, *3*(2), 1.

Finger, M. H., & Hesse, M. H. (2009). *Lebanon-determinants of commercial bank deposits in a regional financial center*. International Monetary Fund.

Fouad, J., Said, M., Sherif, W., & Zaki, C. (2022). *Public banks and development in Egypt: Overview, issues and the way forward*. Economic Research Forum (ERF).

Galal Abdullah Mouawad, S. (2009). The development of Islamic finance: Egypt As a case study. *Journal of Money Laundering Control*, *12*(1), 74–87.

Ghanem, A. (2017). The impact of Basel II on the banking strategies in the Middle East and North African (MENA) region. *Journal of Banking Regulation*, *18*, 48–60.

Gormez, Y. (2008). Banking in Turkey: history and evolution. http://dx.doi.org/10.2139/ssrn.4165562 [Accessed 18 June 2023].

Gürsel, Y. (2019). Handbook for Foreigners to Conduct a Banking Business Under Turkish Legal System: Establishment of a Subsidiary. Available at SSRN 3440045.

Handy, H. (2001). Monetary policy and financial sector reform in Egypt: The record and the challenges ahead (Vol. 51). Egyptian Center for Economic Studies.

Herrera, A. M., & Ozbay, P. (2005). *A dynamic model of central bank intervention: Evidence from Turkey*. Unpublished manuscript, Michigan State University.

Hussain, M. M., Mohamed, E. K., Islam, M. M., & Rahman, M. (2007). Banking policies and regulations: Comparative study of Kuwait, UAE and Qatar. *International Journal of Financial Services Management*, *2*(3), 214–234.

Iqtait, A. (2021). The Palestinian authority political economy: The architecture of fiscal control. In *Political economy of Palestine: Critical, interdisciplinary, and decolonial perspectives* (pp. 249–270).Springer International Publishing.

Khan, M., & Bhatti, I. M. (2008). Islamic banking and finance: On its way to globalization. *Managerial Finance*, *34*(10), 708–725.

Khan, M. S., & Mirakhor, A. (1990). Islamic banking: Experiences in the Islamic Republic of Iran and in Pakistan. *Economic Development and Cultural Change*, *38*(2), 353–375.

Magd, H. A., & McCoy, M. P. (2014). Islamic finance development in the sultanate of Oman: Barriers and recommendations. *Procedia Economics and Finance*, *15*, 1619–1631.

Meyer-Reumann, R. (1995). The banking system in Saudi Arabia. *Arab Law Quarterly*, *10*(3), 207–237.

Migdad, A. M. (2017). CSR practices of Palestinian Islamic banks: Contribution to socio-economic development. *ISRA International Journal of Islamic Finance*, *9*(2), 133–147.

Mirzaei, A., & Moore, T. (2016). Banking performance and industry growth in an oil-rich economy: Evidence from Qatar. *The Quarterly Review of Economics and Finance*, *60*, 58–69.

Mubeen, S. A., Kulkarni, N. A., & Al Hussaini, Y. K. (2014). The future of Islamic banking in Sultanate of Oman. *International Journal of Economics and Finance*, *6*(5), 203–209.

Muneeza, A., & Mustapha, Z. (2019). Settlement of Islamic finance disputes in the Kingdom of Saudi Arabia. In *Dispute Resolution in Islamic finance* (pp. 120–135). Routledge.

Nomani, F. (2003). The problem of interest and Islamic banking in a comparative perspective: The case of Egypt, Iran and Pakistan. *Review of Middle East Economics and Finance*, *1*(1), 36–69.

Nor, M. Z. M. (2012, May). Cross-country analysis of Islamic finance dispute resolution: Malaysia and Saudi Arabia. *In 2012 International Conference on Innovation Management and Technology Research* (pp. 290–295). IEEE.

Powers, C. (2024). Banking in/on the occupation. Resisting Domination in Palestine: Mechanisms and Techniques of Control, Coloniality and Settler Colonialism, 108

Ramady, M. A. (2015). Effective regulatory regimes: A comparative analysis of GCC financial regulators. *Journal of Financial Regulation and Compliance*, *23*(1), 2–17.

Ronall, J. O. (1967). Banking Regulations in Saudi Arabia. *The Middle East Journal*, *21*(3), 399–402.
Roy, K. T. (1994). New York Convention and Saudi Arabia: Can a country use the public policy defense to refuse enforcement of non-domestic arbitral awards. *The. Fordham Int'l LJ*, *18*, 920.
Şahin, S. B. (2012). Central bank independence in Turkey: A neo-Gramscian analysis. *Cooperation and Conflict*, *47*(1), 106–123.
Sarea, A. M. (2019). Establishment of the first waqf bank in Bahrain: Exploratory study. *The Journal of Developing Areas*, *53*(2), 221–228.
Sharar, Z. A. A., & Khulaifi, M. A. (2016). The courts in Qatar Financial Centre and Dubai International Financial Centre: A comparative analysis. *Hong Kong LJ*, *46*, 529.
Tashkandi, A. A. (2023). Shariah supervision and corporate governance effects on Islamic banks' performance: Evidence from the GCC countries. *Journal of Business and Socio-Economic Development*, *3*(3), 253–264.
Van Eijk, E. (2010). Sharia and national law in Saudi Arabia. Sharia incorporated: A comparative overview of the legal systems of twelve Muslim countries in past and present, 139–180.
Wilson, R. (1992). Islamic banking and finance. *The Middle East and North Africa*, *321*, 180–84.
Yanīkkaya, H., & Pabuçcu, Y. U. (2017). Causes and solutions for the stagnation of Islamic banking in Turkey. *ISRA International Journal of Islamic Finance*, *9*(1), 43–61.
Yeung, H., & Al-Barashdi, S. (2022). Institutional development of the Dubai International Financial Centre and the Qatar Financial Centre: *The insolvency law framework*. In *Bank insolvency law in developing economies* (pp. 127–148). Routledge.
Zahedi, R., & Azadi, P. (2018). Central banking in Iran. *Stanford Iran*, *2040*, 1–37.
Zahlan, R. S. (2016). *The creation of Qatar*. Routledge.
Zarrokh, E. (2010). Iranian Islamic banking. *European Journal of Law and Economics*, *29*(2), 177–193.

7 Islamic Financial Regulations in Africa

7.1 Introduction

Most of the African nations have well-established laws to foster the Islamic finance (IF) industry. However, the banking system of a few African countries still uses conventional banking laws. The regulatory authorities in these African countries have started modifying their regulatory frameworks to provide Islamic financial services to customers. This chapter aims to analyse the key regulators, laws and their relevance to Islamic financial institutions. We focus on exploring diverging laws and their impact on developing IF in these developing economies.

7.2 Algeria

Modern banking in Algeria began in the mid19th century. The French authorities established many commercial and merchant banks to facilitate the financing of commerce (Hacini & Dahou, 2018). Those banks operated only in major cities; they did not serve rural areas. Some projects acquired credit through issuing, deposit and merchant banks; people's banks funded small firms; and the Caisse de Crédit Agricole Mutuel served the financial needs of the agricultural sector. After 1871, many small discount houses were set up for agricultural purposes in Algeria under the umbrella of the Banque de Al'geri. The bank granted seasonal credit to the colonists and became a true agricultural bank and gradually bogged down due to the farming crisis in 1880. After the Second World War, Societes Indigenes de Prevoyance (SIP) and Societes Agricole de Prevoyance (SAP) were established to provide for the Muslim population in Algeria. Algeria has about 29 banking institutions, 7 of them are government-owned public banks, and more than 20 foreign banks, from the Gulf countries in particular, others are French and one is British. Islamic banking services are considered as financing real estate purchases (lands and houses), cars and consumables (furniture and equipment) as well as financing small investment projects with limited amounts of money (Toumache & Rouaski, 2016).

7.2.1 The Regulator

The Central Bank of Algeria was created by the number 62-144 bill passed by the Constituent Assembly on 13 December 1962, establishing and laying down the

DOI: 10.4324/9781003471035-7

statutes of the Central Bank. Adjustments were made in the 1970s and early 1980s Reform of the financial system, both in its management and in its powers, nevertheless became imperative (Boucekkine et al., 2021). Law No. 86-12 of 19 August 1986 concerning the trademark banking system started the overhaul of the Algerian banking system. Thus, the Central Bank covers prerogatives in defining and implementing the monetary and credit policy, at the same time that were reviewed its relationship with the Treasury.

These developments, however, have proven ill-adapted to the new socio-economic context marked by deep reforms. Law No. 90-10 of 14 April 1990 as amended and supplemented on money and credit would completely redefine the configuration of the Algerian banking system (Bakhouche, 2007). The law thus confers considerable autonomy, both organic and functional to the Central Bank, now called the Bank of Algeria. A few other notable financial regulations were Ordinance No. 03-11 of 26 August 2003 on money and credit repeals Act No. 90-10 of 14 April 1990 and Ordinance No. 10-04 of 26 August 2010 amending and supplementing Ordinance No. 03-11 of 26 August 2003 on money and credit.

Management, administration and supervision of the Bank are performed respectively by the Governor, the Board of Directors, chaired by the Governor and two censors. The Board is composed of three Vice-Governors and three top officials appointed for their competence in economic and financial matters. The Governor is appointed by presidential decree. The three Vice-Governors are appointed under the same conditions. The other members of the Board are appointed by executive decree. The censors are appointed by presidential decree, on the proposal of the Minister of Finance. The Board of Directors enjoys the classical prerogatives of an organ of this kind.

7.2.2 Islamic Banking Laws

Article 2 of the Algerian Constitution states that Islam is the religion of the state. However, the Algerian market share of Islamic banks is still marginal which does not exceed 1% (Benamraoui, 2008). The Ordinance No. 03-11 is used as a primary law to address the currency and credit regulations in Algeria. It also covers the regulations about local currency, banking operation, required documentations for the approval of banks and exchange controls. The Ordinance is complemented by 80 regulations, which are further complemented and implemented by notes and instructions. All banking and financial activities in Algeria are regulated by the Central Bank, i.e. Banque d'Algerie. Before commencing IF operations, the financial institutions must obtain approval from the Central Bank pursuant to Regulation No. 93-01 of 3 January 1993 as modified by Regulation No. 2000-02 of 2 April 2000. In 1991, the first Islamic bank opened its doors in Algeria, the Banque al Baraka Algeria and operations are limited in the scope and scale. However, with the passage of time, the number of borrowers and savers of the banks has steadily increased. In 2017, Algeria's Prime Minister, Ahmed Ouyahia, announced that Islamic banking and financial services are to be approved in two public government banks before the end of this year and will be approved in four other banks in 2018. This announcement came during Ouyahia's response to the deputies of the People's National Assembly (the first chamber of the parliament) during the ratification session of the government's plan of action.

The quantitative study conducted by Elhachemi in 2018 found that the obstacles to developing Islamic finance in Algeria are divided into three types: social, banking (technical) and legal. The results indicate that more than 70% of respondents agree that legislation is the most critical factor in determining the success or failure of the establishment of Islamic banking in Algeria. Also, the lack of legislation laws about Islamic banking in Algeria forms the primary barrier to the establishment of the Islamic banking system in Algeria. Additionally, about 80% of respondents agree with the non-existence of Islamic technical and development plans as well as the mechanisms for promoting the Islamic banking and finance system. There is a shortage of qualified individuals in the field of Islamic banking, and finance is determined as the primary factor that hinders the development of Islamic banking in Algeria. The failure of the media to play their roles in educating the public on the Islamic banking system and finance in Algeria is also recognised as one of the factors contributing to the delay in implementing Islamic banking in Algeria.

7.2.3 Basel Adoption

The regulation concerning the implementation of Basel II is issued by the monetary authority (Conseil de la monnaie et du crédit) on 16 February 2015. It includes Basel II recommendations (Basel II framework 2006) and some from Basel III. As the regulation issued indicates (N° 14-01 on 16 February 2014), the final rule entered into force on the First of October 2014 (Habiba, 2023). The regulation N° 14-01 is published in the *Journal Officiel* of the People's Democratic Republic of Algeria. The evaluation for credit, operational and market risks is based on the standardised method.

7.3 Djibouti

The primary law is Law No. 92/AN/05 which relates to the opening, activities and control of credit institutions (Triki et al., 2017). The law provides that the banking system is regulated by the Central Bank of Djibouti, and the same law which is applicable to the conventional banks applies equally to Islamic financial institutions. However, certain specific operations relating to the credit and exchange activities carried out by Islamic financial institutions will be regulated by the Central Bank (McCabe, 2019). The first Islamic financial institution was Bank Albaraka Djibouti (reported to be in liquidation, in 1998).

The expansion of Islamic finance in Djibouti has also prompted regular commercial banks to explore the sharia-compliant segment as a further avenue for growth (Abdallah Ali et al., 2022). In 2011, the Central Bank reinforced IF regulation and allowed for this possibility.

However, the regulation states that any commercial bank wishing to open an Islamic finance branch will have to keep its assets, operations and management separate from regular banking operations. The minimum capital for these Islamic banking branches within commercial banks was set at DJF300m ($1.7m).

The first regular bank set to take advantage of the law was Banque pour le Commerce et l'Industrie – Mer Rouge (BCI-MR), one of the country's oldest banks, which has announced plans to open an Islamic banking branch. According to the BCD, the necessary documentation has already been submitted and the new IF branch is expected to begin operations in 2016. A law regulating Islamic insurance was published in 2014, for example. For the time being, a market study commissioned by an insurance pool is underway, with results due by the end of 2016 on the feasibility of marketing sharia-compliant products. If successful, this may add new challenges in terms of banking regulation.

7.4 Gambia

The Central Bank of The Gambia (CBG) was established in 1971 when it took over the assets of the then Gambia Currency Board which was formed on 01 October 1964 (Sillah, 2005).

7.4.1 The Regulator

In 2005, the Central Bank law was reviewed to make it conform to current best practices in modern central banking law (Jawo et al., 2023). This current trend focuses on protecting the Central Bank from the Government's short-term influence on monetary policy whilst at the same time ensuring that the Central Bank is more accountable for its conduct of monetary policy and the use of its resources.

7.4.2 Islamic Banking Laws

In Gambia, the Banking Act 2009 has replaced the Financial Institutions Act 2003 (FIA) allowing for the operations of Islamic banks. It defines "Islamic banking business" as the practice of all banking activities in conformity with Sharia and mainly avoiding all types of interest deals (usury or riba) as a lender or a borrower (Sonko, 2020). The Act also sets the requirement of approvals from the Central Bank before any Islamic financing instruments can be offered. The whole Islamic banking and insurance system is regulated by the Central Bank of Gambia. Mudarabah has been included as part of contractual obligations as profit and sharing contracts in section 2.

Section 2 further explains that "usury," when used within Islamic banking includes any payment of interest at any rate, whether the rate is excessive or not.

The Act further states that, No local financial institution shall carry on an Islamic banking business in or outside The Gambia and no foreign financial institution shall carry on an Islamic banking business in The Gambia without a licence issued by the Bank authorising the institution to do so (Ibrahim et al., 2017).

The Bank shall not issue a licence to a financial institution proposing to carry on an Islamic banking business unless the proposed memorandum and articles of association of the financial institution contain conditions stating that the financial institution shall:

1 as a joint venture with the depositors, bear any losses resulting from any cause for which it is legally liable, including any case where authority is exceeded or insufficient care or caution is exercised by members of its Board of Directors, or its managers or employees, provided that if the loss arose through no fault of the financial institution, the liability shall be jointly shared by the depositors and the financial institution;
2 at the end of each financial year, announce by public notice, the general percentage of profit to be allocated to the general funds participating in joint investment; and
3 not carry on any business which involves usury or which is not approved by Sharia.

A financial institution carrying on Islamic banking business is entitled to:

1 the remaining percentage of profits after the deduction of the amount allocated to investors; and
2 participate in the profits of joint investments in proportion to the amount of its funds or the funds which it is authorised to risk in joint investment.

Section 14(1) restricts financial institutions carrying on Islamic banking business from dealing with banks involved in conventional banking engage in banking operations that involve usury. Anyone who contravenes the provision commits an offence and is liable on conviction to a fine of not less than one hundred thousand dalasis, which is equivalent to 1,908.40 USD.

Section 15(1) of FIA states operations on fees/ujrah contract that allows Islamic banking to carry businesses such commissions or fees as may be necessary. The funds received as commissions and fees shall constitute the banks' income and shall not be divided among the depositors.

The first Islamic financial institution was the Arab Gambia Islamic Bank, which was established in 1997. The banking law is governed under the Banking Act 2009. Gambia has one Islamic bank and a Takaful company. The Arab Gambian Islamic Bank is one of West Africa's most dynamic Islamic financial institutions (Mbawuni & Nimako, 2017). The bank was incorporated under the Companies Act 1955 and has been licensed by the Central Bank of Gambia to operate based on Islamic principles under the Financial Institutions Act 1992, as amended in 1993.

The Gambian government has issued the Financial Reporting Act 2013 to adopt the international standards of accounting and auditing following IFRS under the International Accounting Standard Board.

7.5 Kenya

7.5.1 The Regulators

The banking sector of Kenya comprises 42 licensed commercial banks, 1 mortgage finance institution and 8 authorised non-operating holding companies which account for assets of US$26.08 billion (Beck et al., 2010). Some of the legislations implemented in the banking system of Kenya are:

- The Constitution of Kenya 2010

  This is the supreme law of Kenya whose supremacy is declared in Article 2(1). This law unites all individuals and state organs together at the government level (Akech, 2010). The supremacy of this law grants authority to Parliament and legislative assemblies in the county to enact laws for the development of society including commerce. A liberal interpretation of Article 170 of this law allows the application of Islamic law to financial matters.
- The Banking Act

  This is the main Act that covers the regulation of both conventional and Islamic banks in Kenya (Waweru & Kalani, 2008). This Act differentiates conventional and Islamic banking business based on the definition of a bank which is defined in Section 2(1) of the Banking Act.
- The Central Bank Act

  The Parliament of Kenya enacted this act to establish a central bank to supervise banking operations. The main purpose of this Act was to formulate and implement monetary policy to ensure general stability and enhance liquidity, solvency and stable functioning of the banking industry. Section 4A of this Act specifies the functioning of the bank. It also includes formulation, implementation of foreign exchange, licensing and supervision of authorised dealers.
- The Judicature Act

  This law is hierarchical and represents the undocumented transfer of African customs and traditions from generation to generation (Buluma & Mursat, 2008). This law contradicts the existence of the Islamic banking system as Islamic banking laws are well documented as Quran and Hadith are the source of Islamic banking laws.
- The Kenya Deposit Insurance Act 2012

  This Act was assented to by the President to transform The Deposit Protection Fund Board (DPFB), which was initially designed for the protection of customers' deposits in a financial institution. The Kenya Deposit Insurance Corporation (KDIC) emerged due to this Act to take over the authority, duties, assets and liabilities of DPFB.
- The Prudential Guidelines and Risk Management Guidelines (2013)

  This legislation was introduced by CBK to establish a new set of Prudential Guidelines and Risk Management Guidelines to implement in all licensed banks, financial institutions and mortgage financing companies. The matters of corporate governance, liquidity management, licensing requirements, board composition, capital adequacy requirements, stress testing, foreign exchange exposure limits, prohibited business, consumer protection and enforcement of banking laws and regulations are covered under the umbrella of this legislation.

The banking industry of Kenya is regulated by various authorities some of these regulators are.

7.5.1.1 The Central Bank

According to Section 4 of The Central Bank of Kenya Act, Cap 491 Laws of Kenya, the Central Bank is the main institution to supervise, formulate monetary policy and maintain price stability of the banking industry in Kenya. The core functions of the Central Bank include oversight of monetary policy, financial markets, national payments system, banking services and currency services. The chairperson and directors of the Central Bank are appointed for four years by the President with the approval of parliament. Any financial institution aiming to conduct banking or mortgage business in Kenya must obtain approval from the Central Bank.

7.5.1.2 Deposit Protection Fund Board

A DPFB was established after the banking crises of 1985/86 to protect depositors from losing their deposits in cases where a bank fails to provide confidence to depositors about the payment of insured deposits.

7.5.1.3 The Financial Reporting Centre

The financial reporting centre commended its operation in 2009 under the Act of Crime and Anti-Money Laundering (AML) to combat financial crime and money laundering.

7.5.1.4 Sharia Boards

Islamic banks in Kenya have established a Shariah Advisory Board to ensure the Shariah compliance of Islamic banking products and services. This board audits and makes sure that all Islamic financial institutions operate and implement Islamic principles.

7.5.2 Islamic Banking in Kenya

Islamic banking in Kenya holds 2% market share where Muslims account for 40.3 million of its overall population of 46.60 million. The historical development of Islamic banking shows that the emergence of Islamic banking in Kenya was an outcome of financial innovation (Kinyanjui, 2013). The Central Bank of Kenya's (CBK) research shows that plenty of unaccounted money was in circulation in the economy. Most of this money was owned by Muslims who were reluctant to save their money in conventional banks as they do not comply with Shariah rules and regulations. This fascinating discovery led to the foundation of

Table 7.1 Events of Islamic banking development in Kenya.

Date	*Event*
21 December 2005	Launching of La-Riba account
29 May 2007	The first Community Bank (FCB) was issued an Islamic banking license
2008	Gulf African Bank (GAB) was issued an Islamic banking license
2014	Dubai Islamic Bank (DIB) was issued an Islamic banking license

Source: Authors, 2024.

establishing Islamic banking in Kenya to cater to their needs and target a potential source of economic support. Table 7.1 represents the chronological development of Islamic banking in Kenya.

On 21 December 2005, Barclays Bank launched an interest-free account with the name La-Riba and became the first bank to offer Islamic banking products in Kenya. Later on, CBK granted a full Islamic banking license to the First Community Bank (FCB) on 29 May 2007 and officially started to function as a full-fledged Islamic bank in 2008. Subsequently, another two banks the Gulf African Bank (GAB) and DIB Bank were granted Islamic banking license in 2008 and 2014.

The Islamic banking windows owned by conventional banks such as Barclays Bank, National Bank, Chase Bank and Standard Chartered Bank also provide Shariah-compliant financing services to individuals and corporations. National bank's "Amanh," Barclays' "La-Riba" and Standard Chartered's "Sadiq" windows offer Shariah-compliant financing for vehicles, mortgages and assets purchasing. There is a Takaful Islamic insurance company known as Crescent Takaful Sacco Society Limited (CTS), it performs front office operations and provides deposit-taking and lending facilities. Moreover, there is a Shariah-compliant mutual fund and two cooperatives yet, a giant step in the development of Islamic banking in Kenya was becoming a member of the Islamic Financial Services Board (IFSB). A key development was the amendment of the Finance Act of 2010 Section 45 of the CBK Act which allowed CBK to recognise the payment of return rather than interest on government securities to open up a spectrum of Sharia-compliant investment in the country.

7.5.2.1 Current Overview of Islamic Banking

Islamic banking has established a solid foundation in Kenya and steadily growing its strength by attracting investors, especially from the Middle East. Some of the tipping points for the steady growth of the Islamic banking sector are an increase in the number of middle-class people, escalating education level and awareness of Islamic banking and amendments in the Kenyan banking regulatory framework. Due to the rising demand for Islamic banking products, CBK has issued Islamic banking windows licenses to several conventional banks (Adano, 2010). Recently, the country's best conventional bank, Kenya Commercial Bank (KCB), was issued a license to operate an Islamic banking window.

7.5.2.2 Islamic Banking Challenges in Kenya

Islamic banking in Kenya faces various challenges which may diminish its growth and pose serious threats to the survival of the country's banking industry. These challenges occur due to the internal weaknesses in the Islamic financial institutions or lack of support from the external environment such as regulatory and government policies. One of the major challenges is the skills and expertise of the members of the Shariah advisory board. The regulations of Kenya's Banking Act and Central Banking Act are also a major challenge in the progression of Islamic banking as The Banking Act contains reporting requirements. Similarly, the Central Bank Act does not define, acknowledge and cover the presence of Islamic banks (Abubakar & Aduda, 2013). Although the government has attempted to amend these regulations to provide room for Islamic banks, yet it is seen as an uncertainty in the long-term existence of Islamic banking in some industries. In addition to these challenges a few other challenges can be:

- Lack of Open Market Operations
- Lender of Last Resort and the Inter-Bank borrowing
- Lack of coordination between the Central Bank and Sharia Boards
- Absence of Sharia-Compliant Legal Framework
- Lack of Effective Prudential Regulation
- Dependence of Conventional Banking Benchmarks
- Inadequate Consumer Protection
- Lack of Trained Personnel in Islamic Banking

Even though Islamic financial products in Kenya have high demand, financial institutions offering Shariah-compliant services are insufficient to cater to the needs of customers. The ultimate reason for the shortage of Islamic banking institutions is the lack of support from the legal and regulatory system governing the banking industry of Kenya. Therefore, it is essential to analyse the existing legislation and regulatory structure to determine a policy framework for Islamic banks.

7.6 Libya

The Central Bank of Libya (CBL) started its operations on 1 April 1956, replacing the formerly established "Libyan Currency Committee," which was established in 1951 (Baryun, 2023). The functions of the committee were limited to backing the issued local currency with Sterling assets, thus having no role in controlling money supply, credit or commercial bank supervision (Blowers & McLeod, 1952).

The CBL derives its functions and powers from the Bank law no. 1 2005. The primary mission of the CBL is to achieve and maintain price stability within the context of monetary policy in the interest of balanced and sustainable economic growth in Libya. It also plays a pivotal role in ensuring financial stability.

The Board of Directors is in charge of the management of the CBL; consisting of the Governor as chairman, Deputy Governor as Vice-Chairman and seven other members, who represent other financial and economic interests. The Governor is the chief executive officer responsible for the implementation of the CBL's policies and the management of its affairs; he also represents the CBL in all its associations with other parties.

7.6.1 Islamic Banking Laws

In June 2009, Islamic Banking Law number 9 was introduced to allow commercial banks to open Islamic windows or full-fledged Islamic subsidiaries. CBL Regulations also address the issue of the Shari'ah Board and the basis of accounting to be used by Libyan commercial banks (Albashir et al., 2018). It has adopted for that purpose the Accounting & Auditing Islamic Financial Institutions (AAOIFI) and the IFSB standards. Law No. 46 of 2012 was issued by the Libyan Transitional Council to amend Act No. 1 of 2012. The new Act added provisions concerning practices and operation of Islamic banking. This is further supported by the Law No. 1 of 2013 which prohibits the practice of riba. Unlike the Gulf and several Asian countries, the political and regulatory authorities in North Africa seem to be reluctant to support the development of IF. Political backing is a strong motivation for success. For example, the Central Bank of Egypt has not so far issued any regulation for the Islamic banking industry. Consequently, there have been no initiatives for the creation of Islamic banks. As claimed by Bashir al-Amin, the absence of investment law for Islamic asset management resulted in a dearth of Islamic private equity. Indeed, the absence of a secondary market for short-term investors to sell Islamic products has deterred institutional investors (El-Brassi et al., 2018).

7.7 Mauritius

Initially, under the Bank of Mauritius Act 1966, the purposes of the Bank of Mauritius were to "safeguard the internal and external value of the currency of Mauritius and its internal convertibility' and to 'direct its policy towards achieving monetary conditions conducive to strengthening the economic activity and prosperity of Mauritius." The Bank of Mauritius Act 1966 was revoked with the enactment of a new Bank of Mauritius Act in 2004 (Zafar, 2011). The main purpose of the Bank of Mauritius in that Act evolved to that of maintaining price stability and promoting the orderly and balanced economic development of Mauritius. The Bank of Mauritius is also responsible for the formulation and execution of monetary policy consistent with stable price conditions. There is a Monetary Policy Committee which formulates and determines the monetary policy. This committee sits under the chairmanship of the Governor.

7.7.1 The Regulator

The Bank of Mauritius was established in September 1967 as the central bank of Mauritius. It was modelled on the Bank of England and was, in effect, set up with the assistance of senior officers of the Bank of England.

The then Premier and Minister of Finance, Sir Seewoosagur Ramgoolam, in his inaugural speech, stated that "the time, however, had come for a rather more modern and sophisticated organisation to look after the broader technical aspects of our monetary affairs. It was logical, therefore, that a central bank should be established."

The Bank of Mauritius was conceived along the line of the well-known Radcliffe Report as "a separate organisation with a life of its own, capable of generating advice, views and proposals that are something more than a mere implementation of its superior's instructions."

7.7.1.1 Objectives of the Bank

Besides its primary objective, the Bank of Mauritius also has the responsibility to ensure the stability and soundness of the financial system of Mauritius as well as the regulation of credit and currency in the best interests of the economic development of Mauritius.

Section 6 of the Mauritius Bank Act 2004 states that the power of the Central Bank is to promote the development of the money market of Mauritius, including the Islamic money market through the issue of such Shariah-compliant instruments as may be determined by the Bank. The Bank of Mauritius is independent. Its Board is not subject to the direction or control of any other person or authority. The Bank of Mauritius has to be transparent in the conduct of its operations and also has to act equitably and uniformly and under sound administrative principles.

7.7.2 Islamic Banking Laws

The Banking Act 2004 ('2004 Act') was enacted in October 2004 and came into operation by proclamation on 10 November 2004. The 2004 Act is the enabling legislation for all financial institutions falling under the purview of the Bank of Mauritius. The 2004 Act repealed the former Banking Act 1971 and Foreign Exchange Dealers Act 1995, consolidated the laws relating to the business of banking and other financial institutions and provided for a single licensing regime for banks. The Banking Act 2004 strengthened the regulatory framework to lay the whole banking system on more solid foundations and to better face systemic risks and monetary shocks (Juwaheer et al., 2012). The Banking Act 2004 was amended in December 2007 and 2010 to enable the Bank to issue specialised licences to banks wishing to carry on any or all of the following: banking business, Islamic banking business, private banking business and investment banking business. The basic objectives of the Banking Act 2004 are to maintain a sound banking system in Mauritius and to protect the interests of depositors.

It incorporates the following general principles of prudential regulation and supervision of licensed financial institutions:

- Licensing of banks and other financial institutions
- Issuance of guidelines to financial institutions
- Capital structure and requirements
- Quality of management, including responsibilities of directors and other officers
- Liquidity management
- Electronic banking
- Limitation on operations
- On-site Examinations and off-site Surveillance
- Role of External Auditors
- Compliance with Anti-Money Laundering/Combatting the Financing of Terrorism (AML/CFT)
- Conservatorship, voluntary and compulsory liquidation of financial institutions

Banks, like any other private enterprise, are profit maximisers. In the pursuit of profit-making, they may become less cautious and take greater risks in their lending operations. This may endanger the safety of depositors' money. From the supervisory angle, one of the most important concerns of the Bank of Mauritius is therefore to ensure the maintenance of a sound banking system to sustain stability in the financial system. As part of its functions, the Bank of Mauritius is thus required to ensure:

1 the promotion of adequate and reasonable services to the public,
2 a high standard of conduct and management throughout the banking system, and
3 the pursuit of such policies as may be in the national interest

The words Islamic are mentioned 18 times in the Act. The act applies to conventional as well as Islamic banking.

Section 2 of the Banking Act 2004 defines a "bank" as a company incorporated under the Companies Act, or a branch of a company incorporated abroad, which is licensed under section 7(5) of the Act to carry on any of the following – (a) banking business; (b) Islamic banking business; (c) private banking business. "Islamic banking business" means any financial business, the aims and operations of which are, in addition to the conventional good governance and risk management rules, in consonance with the ethos and value system of Islam (Beebeejaun, 2023).

"Islamic deposit" means a sum of money or monies worth received by or paid to any person, under which the receipt and repayment shall be, by the terms of an agreement, made on any basis including custody or profit sharing.

7.8 Nigeria

7.8.1 The Regulator

The Central Bank of Nigeria Act (CBN) 2007 was enacted by the National Assembly of the Federal Republic of Nigeria (Daud et al., 2011). This new

Act came into enforcement on 25 May 2007 and repealed the Central Bank of Nigeria Act 1991. By the power given under section 33(b), the Central Bank issued Guidelines on Shariah governance for non-interest financial institutions in Nigeria.

This guideline has given a clear mandate for the establishment of Shariah Advisory Council. Consequently, all non-interest banks and other financial institutions under the purview of the CBN, designated as Non-Interest Financial Institutions (NIFIs) in Nigeria, are required to establish a Shariah advisory body as part of their governance structure, to be known as "Shariah Advisory Committee" (SAC). In recognition of the foregoing, the CBN has developed the following guidelines for the appointment, duties and responsibilities of the Shariah advisory committees of NIFIs. Hence, to effectively play its role, the SAC shall operate as an independent body, with the principles of competence, confidentiality and consistency properly enshrined in its operations. It is expected that an independent SAC will command public confidence, thereby promoting the growth and development of the industry (Aliyu, 2012). The objectives of the guidelines are as follows:

1 To set rules, regulations and procedures in the establishment of a Shariah Advisory Committee of a NIFI;
2 To define the role, scope of duties and responsibilities of the Committee and its members;
3 To outline the functions relating to Shariah review and audit processes; and
4 To define the relationship and working arrangement between the Committee and CBN Shariah Council (CSC).

An offence or breach of the rules or guidelines issued by CBN is punishable on conviction by imprisonment not exceeding three years or with a fine not exceeding N10,000,000 (est. USD56,000) or both for every false information. Also, the offender could be imprisoned or given a fine not exceeding N2,000,000 (est. USD12,000) for failure to comply with the requirements of the Bank. Both Acts establish a Financial Ombudsman Scheme for handling disputes in conventional as well as Islamic finance.

7.8.1.1 Objectives of Regulator

The mandate of CBN is derived from the 1958 Act of Parliament, as amended in 1991, 1993, 1997, 1998, 1999 and 2007 (Dafe, 2019). The CBN Act of 2007 of the Federal Republic of Nigeria charges the Bank with the overall control and administration of the monetary and financial sector policies of the Federal Government. The objects of the CBN are as follows:

- to ensure monetary and price stability;
- to issue legal tender currency in Nigeria;
- to maintain external reserves to safeguard the international value of the legal tender currency;

- to promote a sound financial system in Nigeria; and
- to act as a Banker and provide economic and financial advice to the Federal Government.

Consequently, the Bank is charged with the responsibility of administering the Banks and Other Financial Institutions (BOFI) Act (1991) as amended, with the sole aim of ensuring high standards of banking practice and financial stability through its surveillance activities, as well as the promotion of an efficient payment system.

The CBN has also taken responsibility for nurturing the money and capital markets. In furtherance of this, the CBN introduced treasury bills in 1960, a treasury certificate in 1968, and facilitated the establishment of the Lagos Stock Exchange in 1961 and the capital issue committee now known as the Securities & Exchange Committee in the early 1970s.

In addition to its core functions, CBN has, over the years, performed some major developmental functions, focussed on all the key sectors of the Nigerian economy (financial, agricultural and industrial sectors). Overall, these mandates are carried out by the Bank through its various departments.

In January 2012, the CBN granted Jaiz Bank approval in principle to operate as a regional interest-free bank in northern Nigeria. As a result of that, Jaiz bank became the first and the only full-fledged Islamic banking in Nigeria. Islamic banking is based on the principles of profit and loss sharing.

The development of the Guideline for the supervision and regulation of Islamic banking as a category of the Non-Interest Financial system is based on the provision of Section 33 (1) (b) of the Central Bank of Nigeria Act, (CBN), 2007, which empowers the Central Bank of Nigeria to "issue guideline to any person and any institution under its supervision" to carry out banking services. Section 23(1) re-asserted this legal framework with a provision in that "profit and loss sharing banks" are less affected by the provision that binds other banks. The Section exonerates "profit and loss sharing banks" from the requirement of mandatory display of information on the interest rates of their lending and deposits and the rendering of such information to the Central Bank of Nigeria.

The Guideline of Non-Interest Financial Institution, 2011, Non-Interest Financial Institution is defined as a bank or other financial institution under the purview of CBN, which transacts banking business, engages in trading, investment and commercial activities as well as the provision of financial products and services in accordance with the Shariah principles and rules of Islamic commercial jurisprudence.

7.9 South Africa

7.9.1 The Regulator

South African banking sector comprised 17 registered banks, 3 mutual banks, 15 local branches of foreign banks, 2 cooperative banks and 36 foreign banks with approved local representative offices. The regulatory framework of the South African banking industry is illustrated in Figure 7.1.

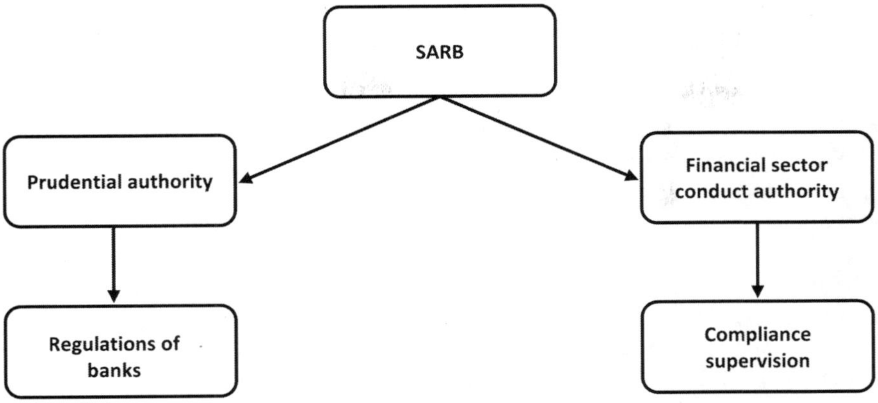

Figure 7.1 South Africa's financial regulators.

7.9.1.1 South African Reserve Bank

The South African Reserve Bank (SARB) is the central regulatory authority responsible for the banks' regulation, supervision and promotion the soundness of the domestic banking system through effective and efficient implementation of regulatory and supervisory standards (Maredza & Ikhide, 2013). SARB not only issues banking license to financial institutions but also monitors their activities under the Banks Act 94 of 1990 and the Mutual Banks Act 124 of 1993. SARB actively participates in several international forums such as G20 and the Financial Stability Board (FSB) to achieve the country's commitment to developing an effective and efficient global financial regulatory architecture. The key functions of SARB are;

- Formulation and implementation of monetary policy
- Issuance of banknotes and coin
- Supervision of banking sector
- Ensuring the effective function of the national payment system
- Official management of gold and foreign exchange reserves
- Liaison as a banker to the government
- Administration of remaining exchange controls
- Lenders last resort in exceptional circumstances

7.9.1.2 Twin-peaks Regulatory System

For many years, the National Treasury has proposed a twin-peak regulatory system to further ensure the safety of the financial sector (Godwin et al., 2017). Under the umbrella of the twin-peaks model of the Financial Sector Regulation Act of 2017, SARB will be responsible for oversight safety and soundness of banks, insurers and financial conglomerates and macro-financial stability. This act provides new financial services regulatory architecture through the establishment of two financial sector regulators known as the Prudential Authority and the

Financial Sector Conduct Authority. The Prudential Authority became effective in April 2018. This authority will act as a jurist and shall be administrated by SARB through four departments;

- Financial conglomerates' supervision department
- Banking insurance and financial market infrastructure supervision department
- Risk support department
- Policy and statistics and industry support department

This Act introduced three major changes in the regulation of the financial sector. Firstly, it gave SARB an obvious mandate to maintain and improve its financial stability. Secondly, it created a prudential regulator known as the Prudential Authority under the umbrella of SARB, which took responsibility for the regulation of banks, insurers, cooperative financial institutions, financial conglomerates and certain market infrastructures. Lastly, a new market conduct regulator evolved which is responsible for supervising conduct perspective across the financial services industry. It is anticipated that a devoted market conduct regulator will result in enhanced supervisory attention in the wider areas of market conduct and consumer protection which will establish the financial system's business model like the UK model.

The banking sector of South Africa is well-regulated to ensure proper oversight over its operations (Malan & Pretorius, 2001). The vastly used legislation to regulate the banking industry includes;

- The Banks Act of 1989
- Reserve Bank Amendment Act, 2010
- Currency and Banking Act, 1920
- The National Payment System Act
- The Financial Intelligence Centre Act (FICA);
- The Financial Intermediary and Advisory Services Act (FAIS)
- The National Credit Act
- The Consumer Protection Act
- The Home Loan and Mortgage Disclosure Act
- The Competition Act

In addition, to the above legislation banks are required to comply with the King Code on Corporate Governance and Basel III. Further, there are various regulations and protocols available to achieve quick and effective dispute resolution between customers and banks in a fair, unbiased and confidential manner.

7.9.1.3 Basel III

After the shocks of the global financial crisis of 2008, international standard-setting bodies formulated various strategies to reveal the fundamental weaknesses in the financial system. Among these strategies, the most common was the implementation of new amendments (Basel III) in the existing regulatory framework where banks need to hold more capital of higher quality and allocate sufficient liquid assets to cover funds outflow (Adesina & Mwamba, 2016). Basel III was implemented in January 2013 and South Africa being a member of G20 had to follow the rest of the world to ensure the stability of the financial system. South African banks were well prepared to implement a new regulatory framework as after the shocks of financial crises banks reassessed their strategies and formulated new pricing to comply with more capital and liquidity requirements.

7.9.2 Islamic Banking

South Africa is the first Sub-Saharan African country to establish itself as a potential hub for Islamic banking in the region. Al Baraka part of the Saudi Arabian-based Dallah Al-Baraka group was the first Islamic bank granted a license by the SARB in 1989; however, it was liquidated in 1998 due to corporate governance issues. Later on, WesBank launched an Islamic banking window and offered Motor vehicle and asset financing services. Another bank, First National Bank, followed and started to provide deposits and transactional banking services. The bank ABSA also launched Islamic banking windows in November 2005. Other IF institutions in South Africa are Oasis Crescent Equity Fund, Symmetry Islamic Fund, Element Islamic Equity Fund and Stanlib Shari'ah Equity Fund. The first Sovereign Sukuk was issued in 2014 with a $500 million issuance priced at a profit rate of 3.90% (Gelbard et al., 2014).

The Islamic banking system is operational in South Africa through two channels; fully fledge Islamic banks and conventional banks offering Islamic banking services through Islamic banking windows. Even though the Islamic banking system in South Africa began four decades ago, yet it is still in its infancy in terms of market penetration and its length of operations. Currently, there are five full-fledged Islamic banks and two conventional banks (ABSA and FNB) providing Islamic financial services to a relatively small Muslim population of 1.5%. Besides offering basic account facilities, ABSA and FNB offer a variety of services to their customers such as:

ABSA	*FNB*
• Islamic Target Save	• Islamic Personal & Business Accounts
• Islamic Value Bundle	• Islamic Investments
• Vehicle & Asset Finance	• Islamic Wealth Management
• Islamic Term Deposit	• Islamic Property Finance
• Islamic Wills, & Profit Share	• Islamic Vehicle & Asset Finance

IF is moving slowly but steadily in South Africa due to several reasons such as the comparatively small Muslim population and the political instability. However, due to the fairness of the Islamic banking business model, the regulatory authorities are showing great interest in amending their regulatory framework (Kholvadia, 2017). The government of South Africa modified its taxation law in

2010 to attract inward foreign investment through capital looking for a Shariah-compliant destination and to incorporate the legal definitions for mushyraka and mudaraba in the draft Taxation Laws Amendment Bill. The main focus of the Bill was not the creation of comprehensive legal frameworks regulating IF. Instead, the Bill focused on the introduction of the definition of musharaka and murabaha for taxation. The question remains unresolved whether the elimination of riba and the utilisation of Islamic contracts are sufficient for financial institutions to be regarded as compliant with Shariah law.

7.10 Sudan

In the past, some of the functions of the Central Bank of Sudan were divided between the Ministry of Finance, the Sudan Currency Board and the National Bank of Egypt (Branson & Braga de Macedo, 1987). The Ministry of Finance maintained part of the official foreign exchange balances through two accounts, one in USD and the other in Sterling Pounds which were managed respectively by the National Bank of Egypt and Barclays Bank D.C.O. The Currency Board is used to perform the task of issuing and managing the currency and keeping a part of the country's balance in foreign currency as a cover to the national currency.

Also, the branch of the National Bank of Egypt used to manage the banking activities of the government besides discharging its main role as the bank of commercial banks (bank of the banks) at the same time. During this period, the Egyptian and British currencies were prevailing until the first national currency was issued.

After the independence of Sudan, there was a need for having a Central Bank to replace the existing bodies. Its functions included regulating the process of the issuance of the national currency, formulating and directing monetary and finance policies to serve various economic sectors, building up a strong, efficient and effective banking system that meet the development needs of the country, maintain government accounts, acting as an adviser for the government on financial affairs and providing foreign currency for the development projects adopted by the government at that time. To achieve the abovementioned objectives, a committee of three experts from the US Federal Reserve was formed in December 1956. The experts were:

- 1/Oliver Weale, Vice chairman of the Federal Reserve Bank of Francisco as president.
- 2/Alan R. Holmes from the Federal Reserve Bank of New York as a member.
- 3/Andrew F. Primer from the Federal Reserve Bank of New York as a member.

The Committee was requested to conduct a study on the possibilities of establishing a Central Bank in Sudan. It submitted its report in March 1957. This was followed by the issuance of the Bank of Sudan Act of 1959. The Bank opened for business on 22 February 1960.

The Act stipulated that the Bank of Sudan shall have an independent corporate personality, legal personality and perpetual succession and a common seal and may litigate in its name as a plaintiff defendant. Following the establishment of the Bank of Sudan, Sudanese officials were appointed to replace Egyptian officials, while the junior staff that used to work for the National Bank of Egypt were retained (Kireyev, 2001). The first Governor of the Bank of Sudan was His Excellency/Mamoun Behairy. Due to the special nature and purpose of the Central Bank, several qualified employees with university degrees were recruited, in addition to those who were seconded from the Ministry of Finance.

7.10.1 The Regulator

The Regulator is the Central Bank of Sudan under the Bank of Sudan Act, 2002. The Bank shall have the following objects, to:

- issue currency of the types thereof, organise, control and supervise the same;
- issue monetary and financing policies and manage the same, in consultation with the Minister, in such way, as may achieve the national objectives of the national economy;
- organise banking business, monitor and supervise, strive to promote and develop and raise the efficiency thereof, in such way, as may achieve the balanced economic and social development;
- strive to achieve economic stability and the stability of the par value of the Sudanese Dinar;
- act, in its capacity as the Government banker, as an advisor and agent thereof, in the monetary and financial affairs;
- abide, in the discharge of the duties, achievement of the objects, the exercise of the powers and supervision thereby, of the banking system, by the ordinances of the Islamic Sharia.

7.10.2 Banking Laws

AAOIFI standards are fully adopted, which among others require a separate *Shari'ah* board for each Islamic financial institution. The Central Bank has its own High *Shari'ah* Supervisory Board. The first Islamic financial institution was the Faisal Islamic Bank of Sudan, which was established in 1977. Apart from that, the insurance market is regulated by the Insurance Commission of Sudan. The governing law is the Insurance Act 2003. Essentially, Sudan is the pioneer of modern takaful operations. The Act contains a comprehensive regulatory framework to be complied with by takaful operators. The first takaful operator was the Islamic Insurance Co. of Sudan, which was established in 1979.

7.11 Tanzania

7.11.1 The Regulators

The Tanzanian banking sector is a mixture of different institutions which provide a variety of products to an increasing population. As of June 2017, there are 3 development financial institutions, 11 community banks, 5 microfinance banks,

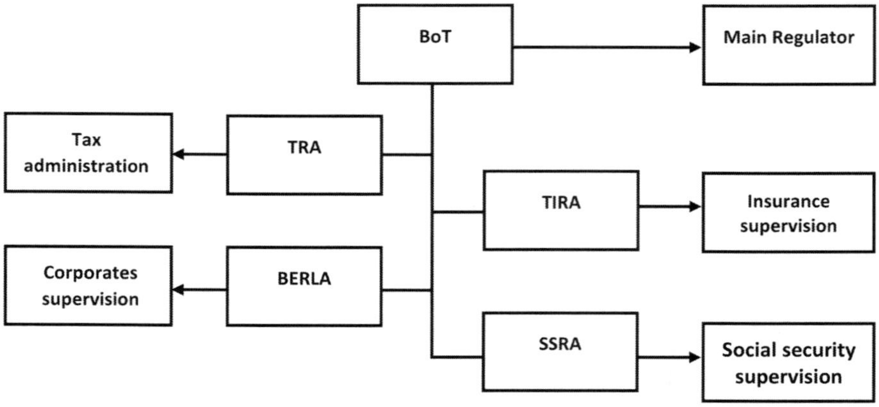

Figure 7.2 Financial regulators in Tanzania.

2 credit reference bureaus, 3 financial leasing companies and 40 licensed commercial banks including multinational banking giants such as Standard Charted, Citibank and Barclays which operate and compete with deep-rooted local banks. Tanzanian banks are regulated through various banking and subsidiary legislations. Figure 7.2 outlines the main regulators in the banking industry.

The Central Bank known as the Bank of Tanzania (BoT) is the main regulatory body. The BoT has the ultimate authority to supervise and regulate the operations of banks and other financial institutions which means no other institution is legally permitted to participate or be involved with the BoT Lwiza & Nwankwo, 2002). The method used by the BoT for regulation and supervision of banks is known as inspection. It applies to all commercial banks and is done by two methods, on-site inspection and off-site inspection.

- In an on-site inspection, banks are inspected based on their capital adequacy, asset quality, management quality, earning capability and liquidity which are collectively known as CAMEL. Each bank is required to maintain core capital and total capital. A core capital of not less than TZS15,000,000,000 is required to start a banking business and each bank must maintain not less than 10% of its total risk-weighted assets and off-balance sheet exposure.
- In the off-site inspection method, the BoT analyses a detailed assessment of the soundness of financial institutions. In this regard, various statistical returns are analysed and warning reports are produced and financial institutions are required to submit weekly, monthly, quarterly, semi-annually and yearly financial reports.

7.11.2 Financial Regulation

Tanzania's financial and subsidiary legislation gives a mandate to the BoT to regulate and supervise the financial institutions (Simpasa, 2011). The supervisory and regulatory mandate of the BoT is guided through the Acts of Parliament of the

United Republic of Tanzania which are enacted to cater for the financial operation. The key regulations of the financial institutions are as follows:

7.11.2.1 The BoT Act, No. 4 of 2006

This was the first legislation for banking regulation enacted after several amendments and repeals to fit into the economic needs. This legislation began as the BoT Act of 1965 which established BoT. This Act provided a mandate to the BoT to formulate monetary policy and act as a supervisor of banks and financial institutions (Kessy et al., 2017). This Act not only established BoT but also gave authority to the Central Bank to issue currency and perform supervisory and regulatory functions.

7.11.2.2 The BFIA Act, No. 5 of 2006

The Act No. 12 of 1991 for the Banking and Financial Institutions initiated the establishment of this regulation. Later on, it was replaced by the Banking and Financial Institutions Act No.5 of 2006 which is applied currently for harmonising the operations of all financial institutions in Tanzania. The sole purpose of this legislation is to foster sound banking operations, regulate credit operations and provide for other matters incidental to or connected with these purposes.

7.11.2.3 The Foreign Exchange Act, No. 1 of 1992

This legislation was passed by the parliament to establish a sound playing field, efficient administration and management to handle the acts related to gold, foreign currency, securities, payments, debts, import, export, transfer or settle property.

7.11.2.4 The Law of Contract Act RE 2002

The Companies Act No. 12 of 2002 and the Law of Contract Act, Cap 345 of 2002 was established for the regulation of contractual agreements, formalisation, enforcement and other related issues.

7.11.2.5 The Anti-Money Laundering Act 2006

This legislation seeks to improve efforts to prevent and prohibit money laundering under the Anti-Money Laundering (Amendment) Act 2012.

7.11.2.6 Insurance Act 2009

This Act was designed for the regulation of insurance institutions. The regulations of 2009 and the Micro-Insurance regulations of 2013 established the Tanzanian Insurance Regulatory Authority.

The principal regulatory bodies for financial services in Tanzania are as follows:

7.11.2.6.1 THE BANK OF TANZANIA (BoT)

The Bank of Tanzania was established under the legislation of the Bank of Tanzania Act 2006. BoT is the central bank and the main monetary, financial and regulatory authority in the country. The primary objectives of the BoT are to formulate, define and implement monetary policy with the economic objective of maintaining domestic price stability favourable for a balanced and sustainable growth of the national economy. Under the subsidiary functions, it is responsible for the process of money supply, the quantity of money in circulation and credit injected into the economy. Additionally, the BoT's subsidiary role is to maintain realistic interest rates above the inflation rate. BoT not only acts as a government bank but also functions as an advisor to the government for protecting the country's international reserves.

7.11.2.6.2 TANZANIA INSURANCE REGULATORY AUTHORITY (TIRA)

The TIRA was established under the Insurance Act of No.10 of 2009. The main purpose of this authority is to coordinate policy and supervision in the insurance industry. TIRA specifies a code of conduct for the members and monitors insurers, brokers and agents.

7.11.2.6.3 SOCIAL SECURITY REGULATORY AUTHORITY (SSRA)

SSRA was established in 2010 under the Social Security (Regulatory Authority) Act, Cap 135 (Revised Edition 2015). It is responsible to regulate and supervise provisions of social security services in Tanzania. SSRA issues guidelines for government, managers, custodians and public and social security schemes developers to ensure efficient and effective operation of the social security sector.

7.11.2.6.4 BUSINESS REGISTRATIONS AND LICENSING AGENCY (BRELA)

Government Executive Agencies Act no. 30 of 1997 established the BERLA. Its primary objective is to register corporates and monitor legal compliance issues such as business names, IP rights, patents, industrial designs, trade and service marks and the issuance of industrial licenses. The Ministry of Industry and Trade directly regulates the operations of BERLA.

7.11.2.6.5 TANZANIA REVENUE AUTHORITY (TRA)

It was established in 1996 under the Act of Parliament No.11 of 1995. TRA is regulated by the Tax Administration Act 2015, Cap 332 which provides a mandate to perform its statutory functions of tax administration in the country.

7.11.2.6.6 LICENSING CONDITIONS TO ESTABLISH A BANK

The BoT has established certain criteria for individuals or companies wishing to establish a bank or a financial institution in Tanzania. The licensing applications must be submitted to the Directorate in a prescribed format and must satisfy capital, liquidity and governance requirements.

7.11.3 Islamic Banking

Islamic banking in Tanzania has gained remarkable popularity since Muslims account for 35.2% of its overall population of 60.33 million. The first attempt to introduce Islamic banking in Tanzania was made by Kenya Commercial Bank in 2008. Afterwards, two more banks the National Bank of Commerce and Stanbic successfully launched Islamic banking in 2010 (Tegambwage & Kasoga, 2022). The development of Islamic banking continued to grow as locally established banks such as People's Bank of Zanzibar and Amana Bank started to offer Islamic banking services. Amana Bank is the only full-fledged Islamic bank which offers exclusive Islamic banking services. The total assets of Amana Bank were TZS204 billion at the end of year 2017.

The central bank (Bank of Tanzania) has allowed conventional banks to offer Shariah banking through Islamic banking windows. Currently, there is only one full-fledged Islamic and four conventional banks with Islamic banking windows which are functional in Tanzania. However, these banks do not provide comprehensive Islamic financial services such as investment and financing.

Due to the increase in demand and the potential of growth for Islamic banking, significant progress can be seen which indicates the shift in approach to diversify operations by adopting technology. People's Bank of Zanzibar (PBZ) deployed conventional banks' Islamic Windows universal software to host the PBZ conventional and Islamic banking facilities. Bahrain's Al-Baraka Banking Group announced last year that; it is planning to establish in Tanzania. The Bank of Tanzania became a member of IFSB in 2016 which will assist in shaping Islamic banking services and education to be on par with international standards. Tanzanian Insurance Regulatory Authority is in the legislation process to allow conventional insurance companies to establish a separate structure to offer Takaful products.

Despite multiple efforts to develop Islamic banking, several facts hinder and pose a serious threat to the growth of the Islamic banking industry in Tanzania. Some of these facts are as follows:

- Tax and Accounting

  One of the major challenges to establishing and developing Islamic banking services is the lack of Shariah framework in the governance of most African countries. Practically, Islamic banking products are unable to comply with local statutory legislation as it is against the fundaments of Islamic banking which is a threat to product development (Mzee, 2016). Currently, accounting regulations do not require much changes; however, it will be ideal for Islamic banks if Tanzanian financial regulatory authorities design mandatory guidelines for all financial institutions that offer Islamic financial products. Additionally, the current tax framework of the Tanzanian Revenue Authority (TRA) does not support IF growth. TRA has taken no initiative to draft a tax amendment bill to establish a playing field for Islamic banks within existing tax legislation. It signifies the need for reforms, especially regarding the potential double-tax effect of murabaha transactions. The ideal tax amendments would be similar to were made in South Africa.

- Takaful and re-takaful

  Tanzania does not have a specific regulation for takaful which means that it is regulated by the laws governing conventional insurance. Islamic banks have been approved to be established in Tanzania and it is necessary to pay attention to key Islamic banking products. One of the key challenges to developing takaful is the lack of support from regulators to promote and enhance public awareness of takaful.

- Regulatory issues

  The Central Bank of Tanzania is working with the World Bank to develop a supervisory, regulatory and legal framework for Islamic banking. The World Bank's task force is consistently working with local financial authorities to develop reporting requirements and the establishment of an independent Shariah supervisory function. The existing regulations indicate the need for an internal auditing structure, which includes an independent assessment of control procedures addressing the specific profile of IF. Islamic banks require an integrated and holistic approach to risk management which should include designing a liquidity management framework.

- Cross-border financing

  The foreign exchange market in Tanzania is vastly affected due to the restrictions imposed by the government on the transfer of Tanzanian sheilings out of the country. Moreover, the central bank has imposed restrictions on access to credit facilities by non-residents.

References

Abdallah Ali, M., Mughal, M., & Chhorn, D. (2022). Microfinance and poverty reduction: Evidence from Djibouti. *Review of Development Economics*, *26*(2), 1189–1224.

Abubakar, A. S., & Aduda, J. (2013). *Islamic Banking and investment financing: A case of Islamic Banking in Kenya. School of Business, University of Nairobi.*

Adano, S. K. (2010). Factors influencing development of Islamic banking in Kenya (Doctoral dissertation, University of Nairobi, Kenya).

Adesina, K. S., & Mwamba, J. M. (2016). Do Basel III higher common equity capital requirements matter for bank risk-taking behaviour? Lessons from South Africa. *African Development Review*, *28*(3), 319–331.

Akech, M. (2010). Institutional reform in the new constitution of Kenya. International Center for Transnational Justice. https://www.ictj.org/sites/default/files/ICTJ-Kenya-Institutional-Reform-2010-English.pdf [accessed 22 March 2024].

Albashir, W. A., Zainuddin, Y., & Panigrahi, S. K. (2018). The acceptance of Islamic banking products in Libya: A theory of planned behavior approach. *International Journal of Economics and Financial Issues*, *8*(3), 105.

Aliyu, S. U. R. (2012). Islamic banking and finance in Nigeria: issues, challenges and opportunities. *Munich Personal RePEc Archive.* https://mpra.ub.uni-muenchen.de/42573/ [accessed 13 March 2024].

Bakhouche, A. (2007). Does the financial sector promote economic growth? A case of Algeria. *Savings and Development*, *31*(1), 23–44.

Baryun, N. A. (2023). The development of the Banking system in Libya. In *The economic development of Libya* (pp. 183–194). Routledge.

Beck, T., Cull, R., Fuchs, M. J., Getenga, J., Gatere, P. K., Randa, J., & Trandafir, M. (2010). Banking sector stability, efficiency, and outreach in Kenya. *World Bank Policy Research Working Paper*, (5442).

Beebeejaun, A. (2023). Compliance with anti-money laundering procedures by banks—the case of Mauritius. *Journal of Banking Regulation*, *24*(4), 474–483.

Benamraoui, A. (2008). Islamic banking: The case of Algeria. *International Journal of Islamic and Middle Eastern Finance and Management*, *1*(2), 113–131.

Blowers, G. A., & McLeod, A. N. (1952). Currency unification in Libya. *Staff Papers-International Monetary Fund*, *2*(3), 439–467.

Boucekkine, R., Laksaci, M., & Touati-Tliba, M. (2021). Long-run stability of money demand and monetary policy: The case of Algeria. *The Journal of Economic Asymmetries*, *24*, e00217.

Branson, W. H., & Braga de Macedo, J. (1987). Smuggler's blues at the Central Bank: lessons from Sudan. *National Bureau of Economic Research, Working Paper No: 2220*. https://www.nber.org/system/files/working_papers/w2220/w2220.pdf [accessed 27 March 2024].

Buluma, B., & Mursat, M. (2008). The case for the supremacy of customary law in Kenya. *Law Journal*, *4*(1), 61–68.

Dafe, F. (2019). Fuelled power: Oil, financiers and central bank policy in Nigeria. *New Political Economy*, *24*(5), 641–658.

Daud, M., Yussof, I. M., & Abideen, A. (2011). The establishment and operation of Islamic banks in Nigeria: Perception study on the role of the Central bank of Nigeria. *Australian Journal of Business and Management Research Vol*, *1*(2).

El-Brassi, M. A. A., Bello, N., & Alhabshi, S. M. (2018). Conversion of conventional banks to islamic banks in Libya: Issues and challenges. *International Journal of Accounting*, *3*(15), 25–39.

Gelbard, E., Hussain, M., Maino, R., Mu, Y., & Yehoue, E. B. (2014). Islamic Finance in Sub-Saharan Africa: Status and Prospects. *IMF working paper, WP/14/149*.

Godwin, A., Howse, T., & Ramsey, I. (2017). Twin peaks: South Africa's financial sector regulatory framework. *South African Law Journal*, *134*(3), 665–702.

Habiba, M. (2023). Basel Agreements in the efficiency of Algerian banks financial evolution and interdisciplinary research. *Financial Markets, Institutions and Risks*, *7*(1), 123–135.

Hacini, I., & Dahou, K. (2018). The evolution of the Algerian banking system. *Management Dynamics in the Knowledge Economy*, *6*(1/19), 145–166.

Ibrahim, M. F., Mohammad, N. M. F. N., & Manjang, M. (2017). Comparative analysis of regulation and supervision of Islamic banking between Malaysia and Gambia. *Labuan e-Journal of Muamalat and Society (LJMS)*, *11*.

Jawo, A., Jebou, M., & Bayo, L. F. (2023). The relationship between inflation, exchange rate, money supply and economic growth in the Gambia. *Technium Soc. Sci. J*, *40*, 213.

Juwaheer, T. D., Pudaruth, S., & Ramdin, P. (2012). Factors influencing the adoption of internet banking: A case study of commercial banks in Mauritius. *World Journal of Science, Technology and Sustainable Development*, *9*(3), 204–234.

Kessy, P. J., Nyella, J., & Stephen, A. C. (2017). Monetary policy in Tanzania. Africa: Policies For Prosperity Series, 241.

Kholvadia, F. (2017). Islamic banking in South Africa–form over substance? *Meditari Accountancy Research*, *25*(1), 65–81.

Kinyanjui, S. N. (2013). Challenges facing the development of Islamic banking. Lessons from the Kenyan experience. *European Journal of Business and Management*, *5*(22), 94–102.

Kireyev, A. (2001). Financial reforms in Sudan: streamlining bank intermediation. *IMF Working Paper No. 01/53.*

Lwiza, D. R., & Nwankwo, S. (2002). Market-driven transformation of the banking sector in Tanzania. *International Journal of Bank Marketing, 20*(1), 38–49.

Malan, F. R., & Pretorius, J. T. (2001). The reserve bank, banks, and clearing houses in South African Law: Part 1. *South African Mercantile Law Journal, 13*, 35.

Maredza, A., & Ikhide, S. (2013). Measuring the impact of the global financial crisis on efficiency and productivity of the banking system in South Africa. *Mediterranean Journal of Social Sciences, 4*(6), 553–568.

Mbawuni, J., & Nimako, S. G. (2017). Determinants of Islamic banking adoption in Ghana. *International Journal of Islamic and Middle Eastern Finance and Management, 10*(2), 264–288.

McCabe, R. (2019). Policing the seas: Building constabulary maritime governance in the Horn of Africa–the case of Djibouti and Kenya. *African Security, 12*(3-4), 330–355.

Mzee, M. M. (2016). The legislative challenges of Islamic banking in Tanzania. *JL Pol'y & Globalization, 45*, 131.

Sillah, B. (2005). Banking and economic growth: Case of the Gambia (Doctoral dissertation).

Simpasa, A. M. (2011). Competitive conditions in the Tanzanian commercial banking industry. *African Development Review, 23*(1), 88–98.

Sonko, M. (2020). Customers' perceptions on Islamic Banking: A case study in the Gambia. *Journal of Islamic Finance, 9*(1), 13–23.

Tegambwage, A. G., & Kasoga, P. S. (2022). Antecedents of customer loyalty in Islamic banking: Evidence from Tanzania. *Journal of Islamic Accounting and Business Research, 13*(4), 701–713.

Toumache, R., & Rouaski, K. (2016). Prospective analysis of the Algerian economic growth by 2025: Structural analysis. *Journal of Applied Business Research, 32*(3), 791.

Triki, T., Kouki, I., Dhaou, M. B., & Calice, P. (2017). Bank regulation and efficiency: What works for Africa? *Research in International Business and Finance, 39*, 183–205.

Waweru, N., & Kalani, V. M. (2008). Commercial banking crises in Kenya: Causes and remedies. *Global Journal of Finance and Banking Issues, 3*(3).

Zafar, A. (2011). Mauritius: An economic success story. Yes Africa can: Success stories from a dynamic continent. *In Growth and emerging prospects of international Islamic banking* (pp. 61–77). IGI Global.

8 Regulatory Framework of Financial Institutions in Europe

8.1 Introduction

This chapter exhibits the financial regulations of major European countries and analyses whether these regulations support Islamic financial provisions. The authors present an overview of the financial frameworks of a few European countries namely Austria, Germany, Luxemburg, the Netherlands and the United Kingdom. This chapter expects to diversify the conceptual context of this book as the authors highlight the regulatory provisions of the countries dominated by conventional banking. The overview of these regulatory provisions will allow readers to critically analyse the structural modifications in financial laws for Islamic financial services.

8.2 Austria

8.2.1 The Regulators

The Oesterreichische National Bank (OeNB) is the central bank of the Republic of Austria (Dub, 1902). It operates under the legal framework provided by the Treaty on the Functioning of the European Union (TFEU), the Protocol on the Statute of the European System of Central Banks (ESCB) and the European Central Bank (ECB) and, most importantly, the Federal Act on the Oesterreichische Nationalbank (Nationalbank Act), as amended. The OeNB is a stock corporation. Its nominal capital of EUR12 million is held in its entirety by the central government.

The OeNB is a fully independent national central bank. In pursuing the objectives and tasks of the ESCB, neither the OeNB nor a member of its bodies may seek or take instructions from EU institutions, the governments of Member States or any other body. Likewise, the bodies and institutions of the EU and the governments of Member States must respect this principle and must not seek to influence the members of ECB bodies or national central bank representatives in the performance of their tasks within the ESCB (Oesterreichische Nationalbank, 2014).

8.2.2 Banking Law

Art 1 of the Federal Banking Act "credit institution" refers to an institution authorised to carry out banking transactions based on Article 4 or Article 103 no. 5 of this

DOI: 10.4324/9781003471035-8

federal act, or on the basis of special provisions under Austrian federal law. Banking transactions include the following activities if carried out for commercial purposes (Arpa et al., 2001):

1 The acceptance of funds from other parties for administration or as deposits (deposit business);
2 The provision of non-cash payment transactions, clearing services and current-account services for other parties (current account business);
3 The conclusion of money-lending agreements and the extension of monetary loans (lending business);
4 The purchase of cheques and bills of exchange, and in particular the discounting of bills of exchange (discounting business);
5 The safekeeping and administration of securities for other parties (custody business);
6 The issuance and administration of payment instruments such as credit cards and traveler's cheques;
7 Trading for one's own account or on behalf of others in:

   a Foreign means of payment (foreign exchange and foreign currency business);
   b Money-market instruments;
   c Financial futures contracts, including equivalent instruments settled in cash as well as call and put options on the instruments listed in lit. a and d to f, including equivalent instruments settled in cash (futures and options business);
   d Interest-rate futures contracts, forward rate agreements (FRAs), interest-rate and currency swaps as well as equity swaps;
   e Transferable securities including securities business;
      Derivative instruments based on lit. b to e;
      unless these instruments are traded for private assets;
      7a. trading in financial instruments under Article 1 para. 1 no. 6 lit. e to g and j Securities Supervision Act 2007 (Wertpapieraufsichtsgesetz 2007 – WAG 2007, Federal Law Gazette I No. 60/2007), for the credit institution's account or on behalf of others, except in the case of trading conducted by persons under Article 2 para. 1 nos. 11 and 13 Securities Supervision Act 2007;

8 The assumption of suretyships, guarantees and other forms of liability for other parties where the obligation assumed is monetary (guarantee business);
9 The issuance of mortgage bonds, municipal bonds and covered bank bonds as well as the investment of proceeds from such instruments following the applicable legal provisions (securities underwriting business);
10 The issuance of other fixed-income securities to invest the proceeds in other banking transactions (miscellaneous securities underwriting business);
11 Participation in underwriting third-party issues of one or more of the instruments listed under no. 7 lit. b to f as well as related services (third-party securities underwriting business);

12 The acceptance of building savings deposits and the extension of building loans following the Building Society Act (Bausparkassengesetz – BSpG) (building savings and loan business);

Art 1(4) allows the Federal Minister of Finance to issue regulations that amend or expand the list of activities under paras. 1 and 2 where necessary, due to sufficiently defined obligations arising from the Republic of Austria's accession to the European Union. In cases where the list of activities under para. 2 is amended or expanded, the Federal Minister of Finance must issue such regulations in consultation with the Federal Minister of Economics and Labour (Fritzer, 2006).

8.2.3 Islamic Banking in Austria

BAWAG PSK envisioned piloting an Islamic current account in the future which will grant it the status of the first Austrian bank to do so. It plans to offer current accounts which will be consistent with the principles of Islamic law or sharia (Lamrani, 2019). This prohibits banks from dealing with businesses that are considered sinful, such as alcohol, weapons, gambling and pornography. Usury is also prohibited under sharia law so in principle banks cannot charge fees or interest for money lending.

Banks generally get around this by lending out their capital through lending practices where interest and fees are not explicitly stated. So, if a client wants to buy a house, the bank buys it for them and sells it back to them at a profit but allows them to pay in instalments.

There will be three different BAWAG sharia accounts available which will not pay or charge any interest, but customers will pay fixed charges of €4.90, €11.90 or €34.90 a month for the account, debit card and overdraft facility. There are now an estimated 600,000 Muslims living in Austria, making up 7% of the population.

8.3 Germany

8.3.1 The Regulators

Being a state of the European Union (EU), the German regulatory framework follows EU directives and regulations. The most significant directives among these directives are Capital Requirement Directive (CRD IV, Directive 2013/2013/36/EU) and prudential requirement directive for credit and investment institutions (EU 575/2013) (Detzer & Herr, 2015). The EU is an active participant in multiple international regulatory convergences on the Basel Committee of Banking Supervision at the Bank of International Settlements (BIS). The established rules of Basel II are incorporated into EU directives and exclusively practiced by the member countries in their national laws. EU conducted a comprehensive scanning of the current regulatory framework (Basel III) and implemented it effective on 1 January 2014. The regulatory structure of the German banking industry is outlined in Figure 8.1.

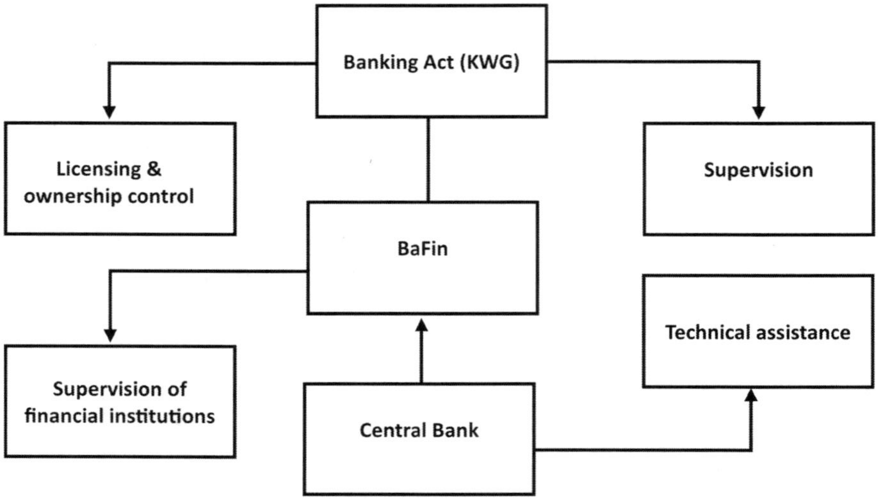

Figure 8.1 The financial regulatory structure of Germany.

The main item of German banking regulation is the Banking Act (*Kreditwesengesetz, KWG*). It covers, in particular, licensing requirements, ownership control and supervision. For capital adequacy, the main law is now the Capital Requirements Regulation (Westrup, 2007). The banks in Germany are categorised as follows:

- State-owned banks

  There are more than 400 state-owned banks that are directly or indirectly owned by the state. These banks are known as public banks (Lander) and offer services to large populations outside city areas where private commercial banks do not exist. Additionally, there is a group of small banks such as Kreditanstalt für Wiederaufbau (KfW) owned by the federal government; it grants promotional loans and offers subsidised financing for friendly environmental investments.
- Universal banks, commercial and retail banks

  This covers 1,900 fragmented credit institutions which provide mortgages for building societies.
- Universal banks

  There are three large universal banks in Germany which include Deutsche Bank, Commerzbank and HypoVereinsbank
- Investment banks

  Recently, investment banks have been growing in numbers as there is flexibility to establish an investment bank in Germany. In most cases, the initial capital is provided by the bankers who previously had worked for a leading institution.
- Private banks

  These banks focus on asset management. Usually, they are closely held and some of these banks have been in operation for many years.

- Co-operative banks

  This is the third most significant group of banks after public and large universal banks. There are 976 co-operative banks which accounted for 16% of market share in 2016.

The principal legislations to regulate the German Banking industry are (Krahnen & Schmidt, 2004):

- The Bank Act (*Kreditwesengesetz, KWG), for licensing, ownership and supervision*
- Capital Requirements Regulation for capital adequacy
- Special acts Pfandbrief Act (*Pfandbriefgesetz - PfandBG*)
- The Safe Custody Act (*Depotgesetz - DepotG*)
- The Building and Loan Associations Act (*Bausparkassengesetz - BauSparkG*)
- The Savings Banks Acts (*Sparkassengesetz - SpkG*) of the Federal States

Germany is a member of the Single Supervisory Mechanism (SSM) recognised by all EU member states whose currency is the Euro. The main purpose was to centralise the prudential supervision of banks especially the ECB whose headquarter is in Frankfurt. SSM applies to 126 institutions and banking groups in the EU including 28 German institutions.

8.3.2 Federal Financial Services Authority

Federal Financial Services Authority (*Bundesanstalt für Finanzdienstleistungsaufsicht, BaFin*) is the main authority to authorise and supervise the banking operation in Germany. The headquarters of BaFin is in Bonn and Frankfurt. The Bonn office supervises banking and insurance, and the Frankfurt office supervises securities markets (Fischer & Pfeil, 2003). There are over 2,500 employees. The budget for BaFin is allocated from the mandatory contribution from banks, insurance and securities companies.

- Any party that wishes to conduct banking business in Germany requires written authorisation from BaFin, which follows sections 32 and 33 of the Banking Act to authorise a banking business license. Some of the conditions to obtain a banking license in Germany are:

  - Must show an initial capital of at least 730,000 euros to establish an investment bank and at least 5 million euros to establish a deposit-taking credit institution.
  - Must satisfy a minimum criterion of regulatory structure such as at least two senior managers (executive directors). Both persons must be "fit" and "proper." Being a "fit" refers to satisfying minimum theoretical and practical knowledge and experience for the job and properly refers to meeting the character requirements which shall be verified through the central Commercial Register (*Gewerbezentralregister*).

- The significant participating interests of individuals must be declared.
- Must propose a three-year viable business plan, clearly indicating the proposed business, organisational structure, proposed internal control systems and implication of required organisational measures to carry out safe business.

8.3.2.1 Central Bank

BaFin and Central Bank (Deutsche Bundesbank) share banking supervision under section 6(1) of the Banking Act. BaFin conducts administrative activities to supervise the financial institutions, whereas Deutsche Bundesbank provides technical assistance to BaFin such as analyses of reports and returns, assessment of capital and risk management process in conjunction with ongoing process for authorisation. Deutsche Bundesbank and BaFin have issued a guideline for the supervision of credit and financial services institutions to follow especially on execution and quality assurance.

8.3.3 Islamic Banking in Germany

Germany is the third most populous country after France and the United Kingdom with a population of just over 83 million and is a key member of the European Union and most of its economic, political and defence organisations. Although Germany is home to the majority of Muslims in Europe with more than 4.3 million, it was not until 2015 that steps were taken to implement Islamic Banking in Germany.

The Turkish Kuveyt Turk (KT) Bank was the first bank to function as a full-fledged Islamic bank in Germany which offers full Islamic banking services to its customers. Initially, KT bank inaugurated its first two branches in Berlin and Mannheim and aimed to establish its business in other German and EU cities (Volk & Pudelko, 2010). The bank expected massive success and penetration in the German market as there is a potential market for Islamic banking services in the form of different religious groups.

Even though the first decade witnessed significant development of Islamic finance in Germany, it was not until the summer of 2015 that Islamic finance officially kicked off. A close scanning of the last two decades shows that various efforts were made in the past by Islamic finance to establish in Germany, for example, the sukuk issuance in 2004 and the first unlicensed subsidiary of Kuwait Finance House (Kuveyt Turk Katilim Bankasi) started Islamic bank branch in Mannheim. Later on, the limited brokering license of deposit business for the collection of deposits was granted in 2010 to operate outside the European Economic Area (Gümüşay et al., 2024). Nonetheless, the range of products was limited as it did not have a full banking license. KT Bank formally applied for a full banking license in 2012 without a blueprint; however, after two and half years, German Banking Supervisory Authority (Bundesanstalt für Finanzdienstleistungsaufsicht, BaFin) granted KT Bank a full Islamic banking license. The reasons for such a long delay in obtaining a banking license remain unclear. Perhaps the absence of a rulebook to establish an Islamic bank in Germany and the assessment required about the nature of products and services of Islamic banks by the regulatory authorities.

An Islamic bank looking to establish itself in Germany must comply with the following set of rules set by the German banking regulation authority.

- Compliance with Shariah rules particularly for the handling of corporate governance issues.
- Incorporation of two tire systems; administration ('Vorstand') and supervision/advice ('Aufsichtsrat').
- The wording of the objects clause to include the Islamic orientation of the financial institution.
- Integration of Shariah standards (e.g. AAOIFI) in its business policy (code of conduct, board regulation).

Currently, various conventional German banks such as Deutsche Bank, UniCredit (Hypovereinsbank), Bahrain's Al-Baraka and Commerz Bank, or insurance companies such as Allianz offer Islamic banking services. Indeed, Islamic finance is penetrating in German financial market through investment funds like Meridio Funds, which was established under the law of Luxembourg and through the foundation of consultation firms such as Amanah Institute for Islamic Banking and Finance (Khan & Porzio, 2010). There are heaps of trends and opportunities and trends for the growth of Islamic banking in Germany such as, it is the fifth largest economy in the world in PPP terms, has the highest savings rate in the EU, second largest stock market capitalisation after the United Kingdom and also has the second in terms of financial assets and the highest Muslim population in Europe.

8.4 Luxemburg

The development of Islamic finance in Luxembourg commenced in the late 1970s with the establishment of the Islamic Banking System as the Holdings Limited Luxembourg, the first Islamic Financial Institution in Europe. Luxembourg is an International Hub for Financial Services and Investment Funds. In 1978, Luxembourg hosted the first Islamic finance institution to be established in a Western country (the Islamic Banking System). Leading European financial centre for investment funds (second worldwide, after the United States) (Belouafi & Belabes, 2010)

1. Leading Financial centre for wealth management in the Eurozone.
2. Currently having 149 banks. The number of investment funds (31 August 2010) was 3,614.

Assets under management for Luxembourg investment funds was 2,068.990 billion EUR. In 1983, the first *shariah*-compliant insurance company in Europe was established in Luxembourg. In 2002, Luxembourg Stock Exchange was the first European stock exchange to list a *sukuk*. The Commission de Surveillance du Secteur Financier (CSSF) is the Luxembourg Supervisory Authority of the Financial Sector (Shiblaq & Stainier, 2014);

1 The CSSF is a Public Institution with a legal personality and financial autonomy.
2 It is placed under the political authority of the Minister of Finance.
3 It is responsible for the supervision of the entire Luxembourg Financial Sector, except for the Insurance Sector.
4 Currently, the CSSF employs 348 persons.

Today, there are as many *sukuks* listed on the Luxembourg Stock Exchange as on the London Stock Exchange. Luxembourg is currently the fourth largest Financial Centre for Islamic Investment Funds Worldwide and the largest in the European Union;

1 Over 40 Luxembourg Islamic Investment Fund Units with Assets.
2 Under management of almost 500 million USD many of the Luxembourg Islamic Investment Funds are UCITS the Luxembourg legal and regulatory framework is well adapted for the set up of Islamic financial instruments, Islamic transactions and Islamic investment funds the Luxembourg government has repeatedly stressed that Islamic finance is one of the important fields it wants to develop given its policy for the financial sector.
3 The CSSF has regular contacts with certain Supervisory Authorities of Jurisdictions that are active in the context of Islamic Finance.
4 The CSSF has signed Memoranda of Understanding (MoUs) with certain Supervisory Authorities of Jurisdictions that are active in the context of Islamic Finance.

In January 2010, a new tax circular was issued on the tax treatment of a wide range of Islamic products including Murabahah, Musyarakah, Mudharabah, Istisna', Ijarah, Ijarah wa Iqtina or Ijarah Muntahia Bitamlik, as well as Sukuk. The circular confirmed the classification of sukuk as debt for Luxembourg tax purposes (Cattelan, 2013). Consequently, the yield payments under the sukuk are treated under domestic tax law as deductible interest expenses at the level of the paying entity. It means that no Luxembourg withholding tax is levied on yield payments. In February 2010, the Luxembourg Stock Exchange and Bursa Malaysia signed a MOU to cooperate in the field of funds and sukuk listings, especially in dual listings.

8.4.1 The Regulator

The Banque centrale du Luxembourg is part of the ESCB. The statute of the ESCB and the ECB is annexed to the Treaty on the European Union and to the TFEU. In particular, they aim at implementing the monetary policy of the Eurosystem in

Luxembourg. The terms and conditions also organise the cash operations as well as the conditions for participation in payment systems.

8.5 Britain

In the United Kingdom, the banking, insurance and capital market is regulated by the Financial Services Authority, UK (FSA) guided by the primary legislation, Financial Services and Markets Act 2000, which applies to all sectors (Ferran, 2014). As the FSA adopts a position that they are a financial, not a religious, regulator, all banks including Islamic financial institutions are governed by the same Act. One of the most important issues for Islamic financial institutions arises from the legal definition of a deposit in the United Kingdom as being "a sum of money paid on terms under which it will be repaid either on demand or in circumstances agreed by the parties." In other words, money placed on a deposit must be capital-certain (Treanor, 2007). Hence, there is a conflict between the UK laws, which require capital certainty and the Shari'ah laws, which require the customer to accept risk losses to have the possibility of a return. Islamic financial institutions resolve this problem by offering full repayment of the investment but informing the customer how much should be repayable to comply with the risk-sharing formulation. This allows customers to choose not to accept full repayment as their religious convictions dictate otherwise.

8.5.1 The Regulator

The Bank of England was founded as a private bank in 1694 to act as a banker to the Government. Today, it is the United Kingdom's central bank (Francis, 1862). The timeline presents a snapshot of some of the highlights of the long history of the Bank of England. Following the financial crisis of 2007–2008, the government decided to bring in massive regulatory reform. Reforms to the financial system came through the Financial Services Act (2012) which announced:

1 An independent Financial Policy Committee (FPC) was created at the Bank of England.
2 A new prudential regulator, the Prudential Regulation Authority, was created as a subsidiary of the Bank.
3 New Bank of England's responsibilities for supervising financial market infrastructure providers.

8.5.2 Islamic Banking

The first full-fledged Islamic financial institution was the Islamic Bank of Britain, established in 2004, although some High Street banks are known to have offered limited Islamic banking products much earlier (Andreades, 2013). The second full-fledged Islamic financial institution was the European Islamic Investment Bank, established in 2006.

The establishment of enabling fiscal and regulatory framework in the United Kingdom for Islamic finance since 2003 has resulted in several advantages:

1 The removal of double taxation on Islamic mortgages and the extension of tax relief on Islamic mortgages to companies, as well as individuals.
2 Reform of arrangements for issues of debt so that returns and income payments can be treated as interest. This makes London a more attractive location for issuing and trading sukuk.
3 Initiatives by the Financial Services Authority to ensure that regulatory treatment of Islamic finance is consistent with its statutory objectives and principles.
4 The establishment of Shariah compliant financial institutions that are authorised as deposit-taking entities.

European Islamic Investment Bank plc (EIIB) was incorporated on 11 January 2005 and received authorisation from FSA on 8 March 2006 (Riaz et al., 2017). The shares were introduced to the market on the London Stock Exchange (AIM market) on 17 May 2006. The bank deals solely with wholesale and claims to be the first Islamic investment bank in the United Kingdom.

Bank of London and The Middle East plc (BLME) bank was incorporated in August 2006 and received FSA authorisation in July 2007. The company that was originally launched did not trade publicly; rather newly issued shares are offered as private placements (to a select group of financiers). It is a bank aimed at the wholesale market. Their business strategy is "the building of an institution which will provide business solutions, using innovative Sharia'a compliant investment and financing products, whilst addressing all the financial needs of our customers." The bank issued shares on several occasions: in August 2006, February 2007, November 2007, December 2007 and (not covered by this chapter) February 2008. Some shares were issued to employees as incentive schemes (November and December 2007).

On 25 June 2014, the government confirmed that £200 million of Sukuk, maturing on 22 July 2019, have been sold to investors based in the United Kingdom and the major hubs for Islamic finance around the world. The United Kingdom's first sovereign Sukuk received very strong demand, with orders totalling around £2.3 billion, and allocations have been made to a wide range of investors including sovereign wealth funds, central banks and domestic and international financial institutions. Investors from the major centres for Islamic finance in the Middle East, Asia and Britain were all represented in the final allocation. The profit rate on the Sukuk has been set at 2.036% in line with the yield on gilts of similar maturity. The Chancellor of the Exchequer George Osborne said:

> Today's issuance of Britain's first sovereign Sukuk delivers on the government's commitment to becoming the western hub of Islamic finance and is part of our long-term economic plan to make Britain the undisputed centre of the global financial system.

> We have seen very strong demand for the Sukuk, resulting in a price that delivers good value for money for the taxpayer. I hope that the success of this government issuance will encourage further private sector issuances of Sukuk in the UK.

By issuing sovereign Sukuk, the government has demonstrated that it is possible to create a successful British base for Islamic finance. Britain's sovereign Sukuk uses the Al-Ijara structure, the most common structure for sovereign Sukuk, with rental payments on the property providing income for investors. The Sukuk is underpinned by three central government properties. Today's issue will settle on 2 July 2014 and will be listed on the London Stock Exchange.

Section 157 of the Finance Act 2008 authorised Her Majesty Treasury "by regulations [to] make provision for raising money through alternative finance arrangements" and to specify a purpose under the National Loans Act 1968, thereby bringing sukuk into the box of monetary instruments available to HMT and its executive arm the Debt Management Office DMO. More generally, this work culminated in the Finance Act 2009 where Schedule 61 implemented the architecture that would ensure that from a tax perspective, sukuk could be issued by the UK companies using real estate as the underlying asset.

One of the critical components of any sukuk is the availability of genuine assets that are sufficiently Shari'a compliant to underpin the issuance. HMT identified UK government-owned and occupied buildings in London (referred to as the Properties) that could be used to support the Issue (Bellalah & Masood, 2013). The Properties were valued (to ensure that they would match or exceed the value being raised by the Issue), assessed by the financial advisor's Shari'a Supervisory Board (to ensure that no objectionable activities were being carried out) and subjected to a title due diligence exercise (to ensure there were no encumbrances or other adverse interests).

Unlike many sukuk in the Middle East, the underlying qualifying interests in the Properties were transferred by way of a registered and perfected lease granted by the Department of Communities and Local Government ("DCLG") to HMT UK Sovereign Sukuk PLC (the wholly owned special purpose company established by HMT for the purposes of the Issue (the Trustee) to hold the Properties as a trustee on behalf of the sukuk holders. A long leasehold interest of 99 years was granted by DCLG in favour of the Trustee in exchange for the payment of a premium. The structure had to balance the requirements of the Finance Act 2009 (which required that where the underlying assets consisted of land the "qualifying interest" should be a freehold interest or a leasehold interest longer than 21 years) against the concerns of government (to ensure a continuing reversionary interest in the Property) and the concerns of Shari'a scholars (that a genuine right in property should underpin the Issue). To generate the necessary cashflows that would allow the Trustee to make periodic distributions to holders of the Certificates, the Trustee would grant a short sublease (five years) in favour of DCLG with periodic payments of rental due thereunder. The arrangements under which DCLG allows various governmental departments to occupy the Properties would be able to continue after the structure had been implemented as the Shari'a scholars satisfied themselves that there were no pre-existing lease concerns to be addressed.

8.5.3 Islamic Banking Laws

Finance Act 2009 is the statute that sets out the conditions to secure the relief available under alternative finance investment bonds when dealing with real estate (see in particular Schedule 61). The statutory requirements are prescriptive and the arrangements not only have to be entered into following their terms at the outset but complied with at maturity of the sukuk (or earlier if an event triggering an early termination occurs), when they have to be exited as prescribed by statute to obtain relief (Akbar et al., 2012). The provisions are designed as anti-avoidance measures to ensure sukuk are not used to circumvent taxes that would otherwise apply to the underlying arrangements. A new Statutory Instrument (Government Alternative Finance Regulations 2014 (S.I. 2014/1327) (the "SI") under the Finance Act 2008, which came into effect on 16 June 2014, confirmed the necessary powers and aligned the sukuk fundraising with certain requirements of the National Loans Act 1968.

References

Akbar, S., Zulfiqar Ali Shah, S., & Kalmadi, S. (2012). An investigation of user perceptions of Islamic banking practices in the United Kingdom. *International Journal of Islamic and Middle Eastern Finance and Management*, *5*(4), 353–370.

Andreades, A. M. (2013). *History of the Bank of England*. Routledge.

Arpa, M., Giulini, I., Ittner, A., & Pauer, F. (2001). The influence of macroeconomic developments on Austrian banks: Implications for banking supervision. *BIS Papers*, *1*, 91–116.

Bellalah, M., & Masood, O. (2013). *Islamic banking and finance*. Cambridge Scholars Publishing.

Belouafi, A., & Belabes, A. (2010). Islamic finance in Europe: The regulatory challenge. *Islamic Economic Studies*, *17*(2).

Cattelan, V. (2013). *Islamic finance in Europe: Towards a plural financial system*. Edward Elgar Publishing.

Detzer, D., & Herr, H. (2015). Financial regulation in Germany. In *Financial Regulation in the European Union* (pp. 41–77). Routledge.

Dub, M. (1902). The National Debt of Austria-Hungary. *The North American Review*, *174*(545), 565–576.

Ferran, E. (2014). The break-up of the Financial services authority in the UK. In *Institutional Structure of Financial Regulation* (pp. 109–136). Routledge.

Fischer, K. H., & Pfeil, C. (2003). Regulation and competition in German banking: An assessment (No. 2003/19). CFS Working Paper.

Francis, J. (1862). History of the Bank of England, Its Times and Traditions, from 1694 to 1844. *Office of the Banker's Magazine*.

Fritzer, F. (2006). The financial system and the institutional environment as determinants of economic performance: Austria In comparison. *Monetary Policy & the Economy*, (1), 136–158.

Gümüşay, A. A., Meyer, R. E., & Höllerer, M. A. (2024). Committed actors, institutional complexity, and pathways to compromise: The emergence of Islamic Banking in Germany. *Journal of Management Studies*. https://doi.org/10.1111/joms.13061

Khan, M. F., & Porzio, M. (2010). *Islamic banking and finance in the European Union: A challenge*. Edward Elgar Publishing.

Krahnen, J. P., & Schmidt, R. H. (2004). *The German financial system*. OUP Oxford.
Lamrani, H. (2019). Status of Islamic Finance in Austria in 2019. Islamic Finance in Europe: A cross analysis of 10 European Countries, 79–108.
Oesterreichische Nationalbank. (2014). Facts on Austria and its banks. https://facts-on-austria.oenb [Accessed 13 July 2023].
Riaz, U., Burton, B., & Monk, L. (2017). Perceptions on the accessibility of Islamic banking in the UK—Challenges, opportunities and divergence in opinion. *In Accounting Forum*, *41*(4), 353–374.
Shiblaq, B., & Stainier, F. (2014). Luxembourg: The European Hub for cross-border finance, sukuk issuances, and Islamic funds. In Pacaud, J. M. The Islamic Finance Handbook, pp. 287–302, Earnst & Young.
Treanor, W. M. (2007). The UK FSA: Nobody does it better. *Fordham Journal of Corporate & Financial Law*, *12*, 259.
Volk, S., & Pudelko, M. (2010). Challenges and opportunities for Islamic retail banking in the European context: Lessons to be learnt from a British–German comparison. *Journal of Financial Services Marketing*, *15*, 191–202.
Westrup, J. (2007). The politics of financial regulatory reform in Britain and Germany. *West European Politics*, *30*(5), 1096–1119.

9 Financial Regulations in Australia and North America

9.1 Introduction

This chapter aims to highlight the key financial regulations in Australia and North America which are relatively under-studied areas in terms of regulations about Islamic financial services. Therefore, an overview of financial regulations will allow authors reinforcing the claims that Western jurisdictions' (Australia, Europe and North America) regulatory frameworks support the provisions of Islamic financial services to meet the escalating demand.

9.2 Australia

In 1911, legislation established the Commonwealth Bank of Australia (1968). In 1959, this original body corporate was preserved as the Reserve Bank of Australia (RBA) in legislation, specifically to carry on the central banking functions; at that same time, the commercial and savings banking functions were transferred into a new institution, which carried on the old name of Commonwealth Bank of Australia (Attard, 1992).

With the Federation of the Australian States into the Commonwealth of Australia, the Australian Parliament assumed power to make laws for banking and currency. In 1911, the first Commonwealth Bank Act gave the Bank only the ordinary functions of commercial and savings banking; the Bank did not specifically have a central banking remit and it was not responsible for the note issue. Management of the Bank was vested in the Governor (Fitz-Gibbon & Gizycki, 2001). The Bank opened for business in mid-1912. At that time, the note issue was administered by the Australian Department of the Treasury, which had taken it over from the private trading banks and the Queensland Government.

In 1920, responsibility for the note issue was transferred from the Treasury to a Notes Board (consisting of four members, appointed by the Government). The Governor of the Bank was ex officio a member of the Notes Board. The administration of the note issue was undertaken by the Bank, though the Bank and the Notes Board were formally independent of each other.

In 1924, the Commonwealth Bank Act was amended and the Bank was given control over the note issue. Management was then vested in a board of eight

DOI: 10.4324/9781003471035-9

directors, including an official, the Governor and the Secretary to the Treasury (Merrett, 1989). From this time until 1945 (when there were major changes to the legislation), the Bank gradually evolved its central banking activities, initially in response to the pressures of the Depression in the early 1930s and later by formal, albeit temporary, expansion of its powers under wartime regulations. These included exchange control and a wide range of controls over the banking system (including the authority to determine advance policy and interest rates and to require private banks to lodge funds with it in special accounts).

The new Commonwealth Bank Act and the Banking Act, both of 1945, formalised the Bank's powers to the administration of monetary and banking policy, and exchange control (Bopp, 1952). Under the 1945 legislation, there ceased to be a board, which was replaced by an advisory council of six, comprising entirely officials from the Bank and the Treasury; the legislation specified that the Governor was responsible for managing the Bank.

However, legislation in 1951 established a new board (at that time of ten members), including the Governor, Deputy Governor and the Secretary to the Treasury, and maintained the responsibility of the Governor for managing the Bank. With minor variations in the number of members, this has been the structure of the Bank's Board since that time.

As indicated above, the Reserve Bank Act 1959 preserved the original corporate body, under the new name of the RBA, to carry on the central banking functions of the Commonwealth Bank, which had evolved; other legislation separated the commercial banking and savings banking activities into the newly created Commonwealth Banking Corporation. The Reserve Bank Act 1959 took effect on 14 January 1960.

9.2.1 The Regulator

The RBA is Australia's central bank and derives its functions and powers from the Reserve Bank Act 1959. It must contribute to the stability of the currency, full employment and the economic prosperity and welfare of the Australian people (Kim, 2008). It does this by setting the cash rate to meet an agreed medium-term inflation target, working to maintain a strong financial system and efficient payments system and issuing the nation's banknotes (Gollan, 1968).

The RBA provides certain banking services as required to the Australian Government and its agencies, and several overseas central banks and official institutions. Additionally, it manages Australia's gold and foreign exchange reserves.

Australia has begun to see a steady stream of property deals using Islamic financing as the attraction of low-risk tenants and a weak Australian dollar offset concerns about the lack of a welcoming tax environment for such transactions (Mirza & Halabi, 2003). National Australia Bank Ltd (NAB.AX), one of the most active banks in the sector, this month helped fund an A$160 million ($114 million) Brisbane property purchase, its third Islamic financing transaction.

TH Properties Sdn Bhd, a unit of Malaysia's pilgrims fund Tabung Haji, completed a A$220 million Sydney development in November helped by A$96 million in financing from Maybank Islamic Bank. It also has more Australian projects in the pipeline with a gross development value of A$800 million, including an A$500 million residential project in Sydney. Prompted by investor demand, NAB has designed a funding tool using an agency-based contract known as wakala, the first such dedicated funding platform in the country (Rammal & Zurbruegg, 2007). That and other structures have now been developed that can suit commercial investment deals as well as development financing.

9.3 Canada

Islamic finance is moving slowly in Canada despite being home to a consistently increasing Muslim population. To date, there is no formal Islamic bank in the country even though Canada has an established legal framework, a potential customer base and a warming appetite for Islamic banking services (Hamid et al., 2017). The centralised banking system provides stability to the financial system of the country and that is why during global financial crises, Canadian banks managed to weather the storm without any external assistance. The banking system in Canada functions within the legislation of the Bank Act (Graham, 2014). The office of the Superintendent of Financial Institutions is the main regulatory authority to ensure the smooth functioning of the banks.

9.3.1 The Regulator

Banking in Canada falls under federal jurisdiction which indicates that the parliament of Canada is the ultimate authority to oversight banking, incorporation of banks and the issuance of paper money. Canada's banking system is more centralised as compared to the United States, as it allows it to focus on macroprudential regulation and stability of the financial system (Dixon, 1935). All aspects of banking are governed under the Bank Act which is the federal statute for the banking industry. The purpose of centralised banking system is to foster a resilient and efficient banking sector comprising competitive financial institutions, protect the depositors' and consumers' interests and maintain public confidence and system stability.

9.3.2 Regulatory Structure

The main regulatory authority (central bank) is the Bank of Canada, which implements monetary policy aimed maintaining the inflation rate below 2% and designing a policy of non-intervention in a flexible foreign exchange. Canada strongly supports the Financial Stability Board and became a leading jurisdiction in the adoption of the Basel III international regulatory framework (Jones, 1985). The regulatory structure of the Canadian banking industry is represented in Figure 9.1.

Multiple regulators supervise banking operations in Canada, with the Office of the Superintendent of Financial Institutions (OSFI) sustaining financial regulation, and the Financial Consumer Agency of Canada (FCAC) assuring

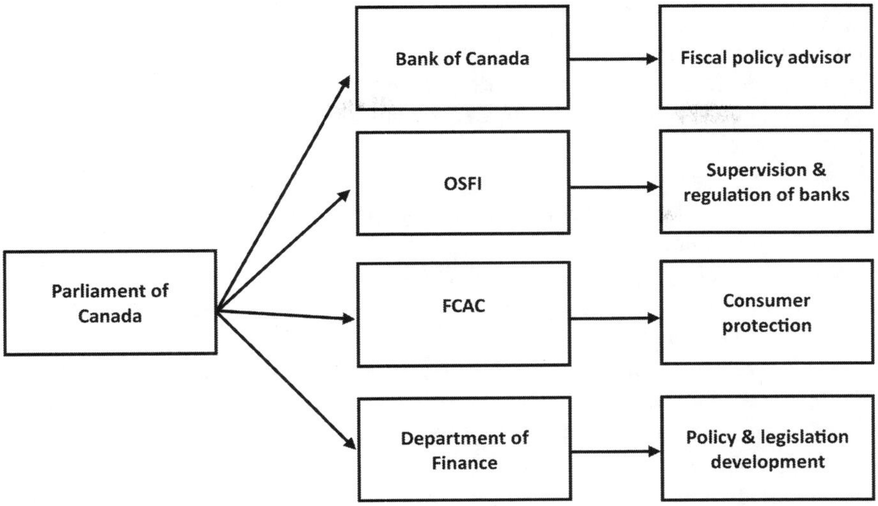

Figure 9.1 Overview of Canada's financial regulatory landscape.

consumer protection. OSFI holds a supervisory and regulatory role within the established legislative framework. FCAC is responsible for monitoring the compliance of federally regulated financial institutions to take measures of consumer protection and help to keep consumers informed (Black, 2004). The Compliance and Enforcement Branch of FCAC has the authority to assess and evaluate the possible issues and force compliance. The Department of Finance supports the government in the development and implementation of financial sector policy and legislation. The Bank of Canada is owned by the federal government and acts as a fiscal agent to the federal government in keeping inflation below par and promoting an efficient banking system. The additional financial regulatory authorities are as follows:

- The Bank of Canada
- The Office of the Superintendent of Financial Institutions (OSFI)
- The Financial Consumer Agency of Canada (FCAC)
- Department of Finance
- Canadian Payment Association (PC)
- The Financial Transactions and Reports Analysis centre of Canada (FINTRAC)
- The Canada Deposit Insurance Corporation (CDIC)
- The Ombudsman for Banking Services and Investment (OBSI)
- The Canadian Bankers Association
- Office of the Privacy Commissioner of Canada
- FISC (members of OSFI, The Bank of Canada, FCSC, Department of Finance, CDIC and FCAC)

9.3.2.1 Office of the Superintendent

The federal government of Canada enacted the Office of the Superintendent for Financial Institutions Act, which established OSFI as the primary regulatory authority to oversight the banks in Canada (Lessambo, 2023). Additionally, it is responsible to:

- Safeguard the Bank Act and make sure that banks function within the published supervisory framework.
- Assess the safety and soundness of banks.
- Provides feedback and intervenes if required.
- Protect depositors against loss.
- Analyses the material risk to banks on a collective basis which involves an examination of all material entities of the banks (including subsidiaries, branches and joint ventures within domestic and international settings).

The OSFI requires all federally regulated banks to disclose monthly, quarterly and annually facts such as balance sheets, deposits and interbank exposures not later than the last day of each month. Moreover, the Bank requires OSFI to determine compliance and regulations of each bank and assess financial conditions on an annual basis. During 2015–2016, the higher domestic household indebtedness, declining interest rates, sustained low oil prices and ongoing financial uncertainty were examined as a systematic vulnerability (Anand & Green, 2012). OSFI designed a task force team to address the impact of these challenges so that it advises federally regulated banks to implement strategic priorities in the areas of corporate and commercial lending, retail lending, outsourcing, cyber-risk and risk management.

The banks in Canada are regulated by the OSFI which introduced revised capital adequacy requirements in 2011. The domestic and foreign operating banks and foreign bank subsidiaries in Canada are regulated under the Bank Act, and Foreign bank subsidiaries are controlled by eligible foreign institutions. There are 88 banks, 44 trust companies, 18 loan companies, 68 life insurance companies, 12 fraternal benefit societies, 152 property and casualty insurance companies which are federally regulated and 23 foreign banks representative offices. Some of the regulations to oversight the performance of these financial institutions are as follows

- The Bank Act
- Office of the Superintendent OSFI Act
- Trust and Loan Companies Act
- Cooperative Credit Associations Act
- Insurance Companies Act
- Pension Benefits Standards Act, 1985
- Pooled Registered Pension Plans Act

9.3.3 Islamic Banking in Canada

Canada is a second home for Muslims from all over the world either as an immigrant or a refugee. Currently, there are 1.3 million Muslims and the anticipations are that the population is likely to increase to 3 million by 2030. Islamic finance growth is slow and limited like other Western countries due to significant differences between the regulatory framework and Islamic business model. In Canada, a few private institutes provide Islamic financial services as most of the major Canadian banks due to a lack of regulatory framework have avoided tapping into this unique market (Khan & Bhatti, 2008).

The first attempt to introduce Islamic finance in Canada was made in 2004 by the Royal Bank of Canada by offering Shariah-compliant financing product which was withdrawn after some time due to lack of demand. However, the market trend has significantly changed in recent times as Islamic finance geared up in Canada in the form of financing and Sukuk issuance. The key drivers of Islamic finance's development are the growing Muslim population, increased awareness of Islamic finance and Canada's demand to attract foreign investment to contribute to infrastructure development. Global Banking and Finance Review (2013) shows that five private and public listed institutions provide Islamic financial services. These financial institutions are as follows:

- UM Group
- Islamic Co-Operative Housing Corporation Ltd.
- Qurtaba Housing Coop & Al-Ittihad Investments Inc.
- The Bank of Tokyo-Mitsubishi UFJ, Ltd.
- Islamic Bank of Britain

Additionally, there are a few Shariah-compliant mortgage providers such as:

- Ijara Community Development Corp
- Zero Mortage
- IHC Canada (formerly ISNA Housing Co-Operative)
- Ansar & Islamic Co-Operative Housing Corporation
- An-Nur Cooperative Corporation

There are ample growth opportunities for Islamic finance in Canada as it is driven by deposit products (a safe and Shariah compliant place for consumers to bank) and home financing. According to the national housing agency, Canada Mortgage and Housing Corporation (CHHC), there are no legal obstacles to Islamic mortgage products, therefore, offering Shariah-compliant housing mortgages will help Canadian banks embark on new business opportunities even on a pilot basis. Canada may become a leading financial hub for Islamic finance in North America by designing a grand regulatory framework and promoting Islamic finance to enhance awareness. The recent G20 group agenda emphasises member countries to explore infrastructure investment through Sukuk which can be a giant step towards wide acceptance of Islamic finance. Sukuk issuance will allow Canadian corporates to explore new Islamic capital markets and diversify their sources of funding.

9.4 The United States of America

The first Islamic financial institution was the LARIBA Finance House, which is based in Pasadena and was established in 1987. Other players include Guidance Financial Group in Revlon, Devon Bank in Chicago, University Bank in Ann Arbor and HSBC Bank in New York. HSBC, University Bank in Ann Arbor, Michigan and Devon Bank of Chicago all now offer Islamic banking products in the United States (Chiu et al., 2005). There are also several non-bank mortgage and finance companies offering these services. Freddie Mac and Fannie Mae have purchased shari'a-compliant mortgages from a number of these providers, supplying crucial liquidity that has enabled these Islamic financial institutions to originate additional mortgages (Goud & Hassan, 2012).

The main regulators of the banking system are the Federal Reserve State Reserve Banks, and the Office of the Comptroller of the Currency (OCC). The OCC charters, regulates and supervises all national banks and federal savings associations as well as federal branches and agencies of foreign banks (Warburg, 1914). The OCC is an independent bureau of the US Department of the Treasury. To ensure that national banks and federal savings associations operate safely and soundly, provide fair access to financial services, treat customers fairly and comply with applicable laws and regulations.

The main piece of legislation is the National Bank Act. Although the Federal Reserve has not licensed any Islamic financial institution, Islamic finance has been present in the United States through Islamic financial institutions licensed by State Reserve Banks, hence limiting more nationwide presence (Poole, 1976). The OCC, in response to queries from the United Bank of Kuwait, permitted ijarah-based financing products in 1997 and murabahah-based financing products in 1999.

9.4.1 Compliance

In April 2014, the US Congress approved the Dodd-Frank Act (Nwogugu, 2015). The final rules generally would prohibit banking entities from:

1 engaging in short-term proprietary trading of securities, derivatives, commodity futures and options on these instruments for their account;
2 owning, sponsoring or having certain relationships with hedge funds or private equity funds, referred to as "covered funds."

As required by section 619 of the Dodd-Frank Act, the final rules, adopted under the Bank Holding Company Act, provide exemptions for certain activities, including market making, underwriting, hedging, trading in certain government obligations and organising and offering a hedge fund or private equity fund, among others. Like the Dodd-Frank Act, the final rules limit these exemptions if they involve a material conflict of interest; a material exposure to high-risk assets or trading strategies; or a threat to the safety and soundness of the banking entity or to US financial stability (Acharya & Richardson, 2012). The United States has introduced FATCA for Foreign Financial institutions that promotes cross-border tax compliance by implementing an international standard for the automatic exchange of information related to US taxpayers. FATCA regulations require tax authorities to obtain detailed account information for US taxpayers on an annual basis (De Simone et al., 2020).

FATCA is intended to increase transparency for the Internal Revenue Service (IRS) for US persons who may be investing and earning income through non-US institutions. While the primary goal is to gain information about US persons, FATCA imposes tax withholding where the applicable documentation and reporting requirements are not met. Introduced by the Internal Revenue Service (IRS), FATCA prevents US taxpayers who hold financial assets in non-US financial institutions and offshore accounts from avoiding taxation on their income and assets. Foreign financial institutions agree to report on these US account holders or face a 30% withholding on all US income.

A global reporting standard known as CRS is set for the automatic exchange of information (AEoI) set forth by the Organisation for Economic Cooperation and Development (OECD). More than 96 countries share information on residents' assets and incomes in conformation with reporting standards. CRS is wider reaching than FATCA and requires a unified, cross-team effort to ensure readiness and compliance.

References

Acharya, V. V., & Richardson, M. (2012). Implications of the Dodd-Frank act. *Annual Review of Financial Economics*, *4*(1), 1–38.

Anand, A., & Green, A. (2012). Regulating financial institutions: The value of opacity. *McGill Law Journal*, *57*(3), 399–427.

Attard, B. (1992). The Bank of England and the origins of the Niemeyer mission, 1921–1930. *Australian Economic History Review*, *32*(1), 66–83.

Black, J. (2004). The development of risk based regulation in financial services: Canada, the UK and Australia. *ESRC Centre for the Analysis of Risk and Regulation Research Report*.

Bopp, K. R. (1952). Commonwealth Bank-The Growth of a Central Bank: The development of The Commonwealth Bank of Australia, 1924–1945. By LF Giblin. Melbourne: University Press. 1951. Pp. xi, 363. 30s. *The Journal of Economic History*, *12*(3), 303–303.

Chiu, S., Newberger, R., & Paulson, A. (2005). Islamic finance in the United States. *Society*, *42*(6), 64–68.

De Simone, L., Lester, R., & Markle, K. (2020). Transparency and tax evasion: Evidence from the foreign account tax compliance act (FATCA). *Journal of Accounting Research*, *58*(1), 105–153.

Dixon, F. J. (1935). Monetary and Banking Legislation in Canada, 1934. *The American Economic Review*, *25*(1), 73–84.

Fitz-Gibbon, B., & Gizycki, M. (2001). The 1970s| RDP 2001-07: A History of Last-Resort Lending and Other Support for Troubled Financial Institutions in Australia. *Reserve Bank of Australia Research Discussion Papers*, (December).

Gollan, R. (1968). *The Commonwealth Bank of Australia; origins and early history. Australian National University Press.*

Goud, B., & Hassan, M. K. (2012). Islamic finance in the United States. In Hassan, M. K., & Mahlknecht, M. *Islamic capital markets: Products and strategies*, pp. 279–306. John Wiley & Sons, Ltd. https://doi.org/10.1002/9781119206040.ch14

Graham, J. (2014). Canada: A promising emerging market for Islamic finance. In Thiagaraja, S., Morgan, A., Tebbutt, A., & Chan, G. *The Islamic Finance Handbook*, pp. 77–92. John Wiley & Sons Pte. Ltd.

Hamid, B. A., Ali, M., Azmi, W., & Anwer, Z. (2017). Introducing Islamic Finance in Unchartered Economies: The Case of Canada. In Rizvi, S., A., & Irum Saba, I. *Developments in Islamic Finance*, pp. 121–145. Palgrave Macmillan.

Jones, C. (1985). The fiscal motive for monetary and banking legislation in Argentina, Austrália and Canada before 1914. In *Argentina, Australia and Canada: Studies in comparative development 1870–1965* (pp. 123–138). Palgrave Macmillan UK.

Khan, M., & Bhatti, I. M. (2008). Islamic banking and finance: On its way to globalization. *Managerial Finance*, *34*(10), 708–725.

Kim, S. J. (2008). The reaction of the Australian financial markets to the interest rate news from the Reserve Bank of Australia and the US Fed. *Research in International Business and Finance*, *22(*3), 378–395.

Lessambo, F. I. (2023). *Bank of Canada*. In *Fintech regulation and supervision challenges within the Banking industry: A comparative study within the G-20* (pp. 175–185). Springer Nature Switzerland.

Merrett, D. T. (1989). Australian banking practice and the crisis of 1893. *Australian Economic History Review*, *29*(1), 60–85.

Mirza, A. M., & Halabi, A. K. (2003). Islamic banking in Australia: Challenges and opportunities. *Journal of Muslim Minority Affairs*, *23*(2), 347–359.

Nwogugu, M. I. (2015). Failure of the Dodd-frank act. *Journal of Financial Crime*, *22*(4), 520–572.

Poole, W. (1976). A proposal for reforming bank reserve requirements in the United States. *Journal of Money, Credit and Banking*, *8*(2), 137–147.

Rammal, H. G., & Zurbruegg, R. (2007). Awareness of Islamic banking products among Muslims: The case of Australia. *Journal of Financial Services Marketing*, *12*, 65–74.

Warburg, P. M. (1914). A United Reserve Bank of the United States. *Proceedings of the Academy of Political Science in the City of New York*, *4*(4), 75–115.

Index

Note: *Italicized* and **bold** pages refer to figures and tables respectively.

9781032748108